Rock Slope Engineering
Civil Applications
Fifth Edition

T0144181

Other Books by Duncan C. Wyllie

- Foundations on Rock
- Rock Fall Engineering
- Landslides—Investigation and Mitigation (co-author)

Rock Slope Engineering
Civil Applications
Fifth Edition

Duncan C. Wyllie
Wyllie & Norrish Rock Engineers
Vancouver, Canada

Based on the third edition by Evert Hoek and John Bray

CRC Press
Taylor & Francis Group
Boca Raton London New York

CRC Press is an imprint of the
Taylor & Francis Group, an **informa** business

About the Cover: The cover is an image of Todgha Gorge, in the Atlas Mountains of Morocco.

CRC Press
Taylor & Francis Group
6000 Broken Sound Parkway NW, Suite 300
Boca Raton, FL 33487-2742

© 2018 by Taylor & Francis Group, LLC
CRC Press is an imprint of Taylor & Francis Group, an Informa business

No claim to original U.S. Government works

Printed on acid-free paper

International Standard Book Number-13: 978-1-4987-8627-0 (Paperback)

This book contains information obtained from authentic and highly regarded sources. Reasonable efforts have been made to publish reliable data and information, but the author and publisher cannot assume responsibility for the validity of all materials or the consequences of their use. The authors and publishers have attempted to trace the copyright holders of all material reproduced in this publication and apologize to copyright holders if permission to publish in this form has not been obtained. If any copyright material has not been acknowledged, please write and let us know so we may rectify in any future reprint.

Except as permitted under U.S. Copyright Law, no part of this book may be reprinted, reproduced, transmitted, or utilized in any form by any electronic, mechanical, or other means, now known or hereafter invented, including photocopying, microfilming, and recording, or in any information storage or retrieval system, without written permission from the publishers.

For permission to photocopy or use material electronically from this work, please access www.copyright.com (http://www.copyright.com/) or contact the Copyright Clearance Center, Inc. (CCC), 222 Rosewood Drive, Danvers, MA 01923, 978-750-8400. CCC is a not-for-profit organization that provides licenses and registration for a variety of users. For organizations that have been granted a photocopy license by the CCC, a separate system of payment has been arranged.

Trademark Notice: Product or corporate names may be trademarks or registered trademarks, and are used only for identification and explanation without intent to infringe.

Library of Congress Cataloging–in–Publication Data

Names: Wyllie, Duncan C., author.
Title: Rock slope engineering : civil applications / Duncan Wyllie.
Description: Fifth edition. | Boca Raton : Taylor & Francis, CRC Press, 2017.
Identifiers: LCCN 2017011060| ISBN 9781498786270 (pbk.) | ISBN 9781498786287
(e-book)
Subjects: LCSH: Rock slopes.
Classification: LCC TA706 .W98 2017 | DDC 624.1/5132--dc23
LC record available at https://lccn.loc.gov/2017011060

Visit the Taylor & Francis Web site at
http://www.taylorandfrancis.com

and the CRC Press Web site at
http://www.crcpress.com

Printed and bound in the United States of America by Sheridan

This book is dedicated to

Dr. Evert Hoek and Dr. John Bray

For their pioneering work in the field of rock slope engineering

Contents

List of tables

List of figures

Introduction to the fifth edition

This fifth edition of *Rock Slope Engineering* has been prepared 42 years after the first edition authored by Evert Hoek and John Bray in 1974, and 12 years after the fourth edition in 2004. This 42-year period is an expression of the evolution of this field of engineering – the first three editions, published by the Institute of Mining and Metallurgy in London, were prepared for the design of open-pit mines that were at that time becoming several hundred metres deep as the result of the development of drills, shovels and trucks that made it economically feasible to mine low-grade ore deposits. Open-pit mines are now reaching depths of 1000 metres and new procedures are being used to design slopes for these pits for which the rock mass strength is more important than the geologic structure. These procedures are documented in publications such as the *Guidelines for Open Pit Slope Design* (2009).

In civil engineering, rock slopes usually range in height up to several hundred metres, and for these conditions the geological structure is usually the most important design parameter. Because of the divergence in the scale of, and design procedures for civil and mining rock slopes, it has been decided that this fifth edition of *Rock Slope Engineering* will only address civil applications. The structure of the book and technical approach originally developed by Evert Hoek and John Bray in the first three editions have been retained in this edition.

In preparing this new edition, numerous minor edits and corrections have been made, and for information on many of these, I am grateful for messages that I have received from readers around the world; their diligence is greatly appreciated. It is interesting that virtually every page of this edition has changed in some way from the 2004 edition.

In addition to the edits, the following significant additions have been made:

- *Weathered rock* – a significant omission of the four earlier editions was the subject of weathered rock. These geologic conditions exist in the tropical regions of the world, and design procedures for slopes in weathered rock are somewhat different from those in fresh rock. Therefore, a new chapter (Chapter 3) has been added to address the occurrence and characteristics of weathered rock. Also, the properties of weathered rock are addressed in other chapters, such as shear strength, ground water and stability analysis.
- *Seismic design* – a new chapter (Chapter 11) has been added on seismic design of rock slopes, which is an expansion of the material presented in the chapter on planar failure in the fourth edition. This more detailed information has been prepared because of the important influence of seismic ground motions on rock slope stability, and because much of the literature on this subject addresses soil slopes that is not directly applicable to rock slopes.
- *Numerical analysis* – the chapter (Chapter 12) on numerical analysis of rock slopes has been significantly updated to address civil engineering applications, whereas the chapter in the fourth edition focused on open-pit applications. This chapter has been

prepared by Loren Lorig of the Itasca Corporation in Minneapolis and Douglas Stead of Simon Fraser University in Burnaby, Canada.

- *Dip slopes* – an important class of rock slopes are those where the bedding or other persistent structure is oriented parallel to the face such that instability requires break-out of a surface between the structure and the face. This material, based on work by Dr. Brendan Fisher, is covered in Chapter 7.
- *Developments in rock slope engineering* – in the 12 years since the fourth edition was prepared, the most significant developments in this field have been in remote sensing such as LiDAR and the preparation of digital terrain models. These techniques have allowed geological mapping and movement monitoring to be carried out remotely and to a high degree of accuracy; these new techniques are addressed in this edition. However, it has been found that the basic methods of stability analysis – planar, wedge, circular and toppling that were first developed by Evert Hoek and John Bray – are still applicable today. The change that has occurred over the years is the development of computer programs for the analysis of these slopes.

For the fourth edition, my co-author was Chris Mah. Chris and I have now worked together, and have shared an office, for nearly 20 years and I appreciate his contribution to the material covered in this book. The reason that I am the sole author of the fifth edition is that I have developed an obsession, perhaps unfortunately, with the study of the minutiae of rock mechanics and writing about it. However, it is only possible to write books such as this if one has this obsession.

In the preparation of this new edition, I want to thank many persons who have contributed to its development. First, the chapter on weathered rock involved several visits to tropical areas where I observed the special techniques that are used for the excavation of cuts in weathered rock. I am most grateful for the information provided by Mr. Shaik Wahed in Kuala Lumpur, Malaysia, and by Dr. Roberto Coutinho of the Federal University of Pernambuco in Recife and by Dr. Tarcisio Celestino and Mr. Jaoa Pimenta and their colleagues at Themag Engenharia in Sao Paulo, Brazil.

The chapter on the effects of seismic ground motions on slope stability has benefitted greatly from my discussions over many years with Dr. Randall Jibson and his colleagues at the U.S. Geological Survey in Golden, Colorado, as well as Dr. Ed Kavazanjian at Arizona State University in Tempe, Arizona. I also appreciate reviews by Dr. Uppal Atukorala and Dr. Bob Pyke.

The new chapter on numerical analysis has been prepared by Loren Lorig of Itasca and Dr. Doug Stead of Simon Fraser University; I am most grateful for the time and effort that they have put into the preparation of this document.

Other invaluable assistance has been provided by Ms. Sonia Skermer who has prepared many of the drawings in this book, and previous books. Her accuracy and attention to detail has ensured that the illustrations are seamless in all editions. Ms. Calla Jamieson and Ms. Glenda Gurtina have been of great assistance in assembling the final document. Finally, this book could not have been prepared without the assistance of Ms. Cheng-wen Tina Chen who worked both on technical matters such as slope stability analysis and shear strength of weathered rock, and organizational matters such as the references and preparing and checking the manuscript.

Finally, I must thank my family for putting up with yet another book writing struggle.

Duncan C. Wyllie
Vancouver, Canada
2016

Foreword to fifth edition

It is a delight to see that the 5th edition of the classic book *Rock Slope Engineering* is now available. The first edition was prepared in 1974 by Professor Evert Hoek and Dr. John Bray, then of Imperial College in London. However, since the 1970s, rock engineering in general and the specific subject of rock slope engineering have advanced in leaps and bounds, so that there has been a continuing increase in the theoretical and practical aspects of the subject. This has necessitated updating of the book through editions 2 to 4, resulting in it being the key publication on rock slopes for both civil and mining applications. In this context, another book on rock slopes, *Guidelines for Open Pit Slope Design*, was published in 2009 concentrating on just mining applications. As a result, Dr. Wyllie made the sensible decision to restrict this 5th edition of *Rock Slope Engineering* to rock slopes associated with civil engineering projects.

The structure of the book is particularly helpful in following the logic of rock slope engineering: that is, the principles of slope design, geology, site investigation, rock strength, groundwater, modes of slope failure, seismic analysis, numerical analysis, slope excavation and stability, plus monitoring and case examples. In addition to providing up-to-date information, Dr. Wyllie has also included new subjects, and in particular numerical modelling/simulation. The spectacular advances in this area, together with the associated graphics, have led to the current situation where almost all large-scale rock engineering projects use some form of numerical modelling throughout the design and construction process. Thus, a particularly helpful feature of this book is that the different computer codes and their outputs are so well explained.

Taylor & Francis has a wide range of books in rock engineering subjects available and this new volume on civil rock slope engineering with its extended contents fits well in their portfolio and will ensure that the reader has maximum access to the information, especially through the inclusion of illustrative worked examples. Moreover, special care has been taken with the balance of the text and figure contents, plus the presentation of the mathematical equations, so as to maximise readability. The book is comprehensive, well written, has a sound geological and engineering basis resulting from Dr. Wyllie's almost 50 years' practical experience, and has successfully incorporated modern knowledge into the many explanations throughout.

Because the book is comprehensive and the content so clearly organised and presented, it is highly recommended as a primary handbook source for civil rock slope practitioners, whether clients, consulting engineers or contractors. Also, it will serve as an excellent book for supporting civil engineering students.

Emeritus Professor John A. Hudson, Imperial College London, UK
President of the International Society for Rock Mechanics 2007–2011

Foreword to fourth edition

My work on rock slope engineering started more than 30 years ago while I was professor of rock mechanics at the Imperial College of Science and Technology in London, England. A four-year research project on this topic was sponsored by 23 mining companies around the world, and was motivated by the need to develop design methods for the rock slopes in large open-pit mines which were becoming increasingly important in the exploitation of low-grade mineral deposits. *Rock Slope Engineering*, co-authored with my colleague Dr. John Bray, was first published in 1974 and then revised in 1977 and again in 1981.

While rock slope engineering remains an important subject, my own interests have shifted towards tunnelling and underground excavations. Consequently, when Duncan Wyllie suggested that he and Christopher Mah were willing to prepare a new book on rock slopes I felt that this would be an excellent move. They have had a long involvement in practical rock slope engineering, mainly for civil engineering construction projects, and are familiar with recent developments in methods of analysis and stabilization. The evolution of their text, from *Rock Slope Engineering* to a manual on rock slope design for the U.S. Federal Highway Administration, is described in their introduction and need not be repeated here.

The text which follows is a comprehensive reference work on all aspects of rock slope engineering and, while it embodies all the original concepts of my work, it expands on these and introduces a significant amount of new material, for both mining and civil engineering and several new case studies. As such, I believe that it will be an important source of fundamental and practical information for both students and designers for many years to come.

I commend the authors for their efforts in producing this volume and I look forward to having a copy on my own bookshelf.

Evert Hoek
Vancouver, 2003

Author

Duncan C. Wyllie is the president of Wyllie & Norrish Rock Engineers based in Vancouver, Canada. He has been working exclusively in the field of applied rock engineering for almost 50 years, initially in underground and open-pit mining and then in civil engineering. He has worked on slope, landslide, tunnel, blasting and foundation projects throughout North America, as well as in countries such as India, Sri Lanka, New Zealand, Ethiopia, Turkey and Peru. The major portion of this work is for highways, railways and power projects, and involves both new construction and the evaluation and remediation of existing slopes and tunnels. Dr. Wyllie is also the author of *Rock Fall Engineering* (2014) and *Foundations on Rock* (1999), both published by CRC Press. He has prepared manuals on rock slope engineering for the Federal Highway Administration, and taught about 80 courses throughout the United States for the National Highway Institute in Washington, DC.

Nomenclature

A	Area of plane (m²)
a	Material constant for rock mass strength; ground acceleration (m/s²)
a_H	Peak horizontal ground acceleration, PHGA (g)
B	Burden distance for blast holes (m)
b	Distance of tension crack behind crest of face (m); joint spacing (m)
C_d	Dispersion coefficient
c	Cohesion (kPa, psi); damping coefficient
D	Disturbance factor for rock mass strength; depth (m)
d	Diameter (mm); displacement (in., cm)
E_m	Deformation modulus for rock mass (GPa)
e	Joint aperture (mm)
F	Shape factor
F_{PGA}	Site factor for zero period of acceleration spectrum
F_v	Site factor for long period range of acceleration spectrum
FS	Factor of safety
G	Shear modulus (GPa)
GSI	Geological strength index
g	Acceleration due to gravity (m/s²)
H	Height of slope, face (m)
h	Water level head above datum (m)
I_a	Arias intensity (m/s)
i	Roughness angle of asperities (degrees)
JRC	Joint roughness coefficient
K	Bulk modulus (GPa); hydraulic conductivity (cm/s); corrosion service life constant (μm), rate of slope movement constant
K	Seismic coefficient; attenuation constant for blast vibrations; stiffness
K_y	Yield acceleration (g)
K_{max}	Maximum acceleration (g)
L	Length of scan line, face, borehole (m)
l	Persistence of discontinuity (m); unit vector
M	Earthquake magnitude
M_L	Earthquake local magnitude or Richter scale
M_w	Earthquake moment magnitude
m	Mass (lb, kg); unit vector
m_b	Material constant for rock mass strength
m_i	Material constant for intact rock
N	Standard penetration resistance (blow/ft); number of joints, readings
n	Unit vector

P	Probability
PGA	Peak ground acceleration (g)
PGD	Peak ground displacement (cm, in.)
PGV	Peak ground velocity (cm/s)
p	Pressure (kPa, psi); probability
Q	External load (kN, kips)
R	Resultant vector; hole radius (mm); return period (years); rupture distance (km)
R_H	Hypocentre distance (km, miles)
r	Radius of piezometer (mm)
S	Shape factor; spacing distance for blast holes (m)
S_1	Spectral acceleration at period 1 s
SD	Standard deviation
s	Spacing of discontinuity (m, ft); material constant for rock mass strength
T	Rock bolt tension force (kN, kips)
t	Time (s, year)
t_0	Duration (s)
U	Uplift force on slip plane due to water pressure (kN, kips)
V	Thrust force in tension crack due to water pressure (kN, kips); velocity of seismic waves (m/s), slope movement
v_s	Shear wave velocity (m/s, ft/s)
W	Weight of sliding block (kN, kips); explosive weight per delay (kg, lb)
X	Loss of element thickness, corrosion (μm)
z	Depth of tension crack (m, ft)
z_w	Depth of water in tension crack (m, ft)
α	Dip direction of plane (degrees); seismic slope height reduction factor
β	Attenuation constant for blast vibrations; factor in calculating seismic height reduction factor
δ	Displacement (mm)
ε	Coefficient of thermal expansion ($^\circ C^{-1}$)
ϕ	Friction angle (degrees)
φ	Load and resistance factors (LRFD design)
γ	Unit weight (kN/m^3, lb/ft^3)
μ	Poisson's ratio
σ	Normal stress (kPa, psi)
σ_{cm}	Compressive strength of rock mass (kPa, psi)
σ_{ci}	Compressive strength of intact rock (kPa, psi)
σ_1'	Major principle effective stress
σ_3'	Minor principle effective stress
τ	Shear stress (kPa, psi)
ψ	Dip of plane (degrees)
υ	Viscosity of water (m^2/s)

Note

The recommendations and procedures contained herein are intended as a general guide, and prior to their use in connection with any design, report, specification or construction procedure, they should be reviewed with regard to the full circumstances of such use. Accordingly, although every care has been taken in the preparation of this book, no liability for negligence or otherwise can be accepted by the author or the publisher.

Chapter 1

Principles of rock slope design

1.1 INTRODUCTION

A variety of engineering activities requires excavation of rock cuts. In civil engineering, projects include transportation systems such as highways and railways, dams for power production and water supply, and industrial and urban development. The design of rock cuts, and their excavation and stability are closely related to site geological conditions that may vary from strong, blocky rock containing well-defined discontinuities, to weak, highly weathered rock in which the remnant discontinuities, if any, have little influence on stability. Between these two extremes, a wide spectrum of conditions exists that must be quantified and incorporated appropriately into the design.

Figures 1.1 and 1.2 show contrasting examples of rock slopes with respect to rock strength and structural geology.

In Figure 1.1, the rock is a very strong, blocky granite containing a set of high-persistence, planar joints dipping from right to left at about 30°. Movement of blocks of rock on these smooth, planar joints had opened tension cracks on a steeply dipping orthogonal joint set. Previous instability in this area had occurred with no warning because only a few millimetres of movement was required for the shear strength on the smooth joints to be reduced from peak to residual. Stabilization of the slope involved the construction of a reinforced concrete buttress with the lower part secured with tensioned rock anchors to stable rock below the sliding plane, and the upper part acting as a cantilever to support the sliding blocks.

Figure 1.2 shows an excavated face at an angle of 56° in highly weathered schist with a portion of the face reinforced with soil nails (steel dowels fully embedded in cement grout) on a 2-m square pattern and covered with a 100-mm (4 in.)-thick layer of reinforced shotcrete. The bolt lengths alternate between 5 and 6 m (16–20 ft) to avoid creating a potential plane of weakness at the end of the bolts that would occur if they were all the same length.

Other stabilization measures at this site include erosion control of this low-strength rock by installing lined ditches on each bench, and growing grass on the exposed rock faces.

In addition to man-made excavations, the stability of natural rock slopes in mountainous terrain may also be of concern. One of the factors that may influence the stability of natural rock slopes is the regional tectonic setting. For example, factors of safety of natural slopes may be only slightly greater than unity where rapid uplift of the land mass, and corresponding down cutting of the water courses, is occurring. Also, earthquake ground motions may loosen surficial rock and cause slope displacement. Such conditions exist in seismically active areas such as the Pacific Rim, the Himalayas and Central Asia.

The required stability conditions of rock slopes will vary depending on the type of project and the consequence of failure. For example, for cuts above a highway carrying high-traffic volumes, it will be important that both the overall slope be stable, and that few, if any, rock falls reach the traffic lanes. This will often require careful blasting during construction, and

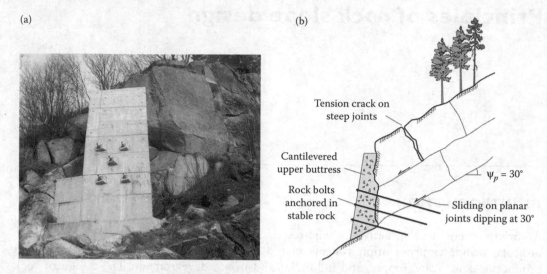

Figure 1.1 (**See colour insert**.) Rock slope in very strong granite in which upper blocks of rock can slide from right to left on planar joints dipping at about 30° (near Agassiz, British Columbia, Canada). (a) Concrete buttress and tie-back anchors; (b) slope geometry showing dimensions of sliding blocks, bolt lengths and buttress configuration (Gygax Engineering).

the installation of stabilization measures such as rock anchors. Because the useful life of such stabilization measures may be only 10–30 years, depending on the climate and rate of rock degradation, periodic maintenance may be required for long-term safety.

In the design of cut slopes, it is usually not possible to adjust the orientation of the slope to suit the geological conditions encountered in the excavation. For example, in the design of a highway, the alignment is primarily governed by factors such as available right-of-way, gradient and vertical and horizontal curvature. Therefore, the slope design must accommodate the particular geological conditions that are encountered along the highway, which may require differing slope designs around curves, or on opposite sides of through cuts. Circumstances where geological conditions may dictate modifications to the highway design

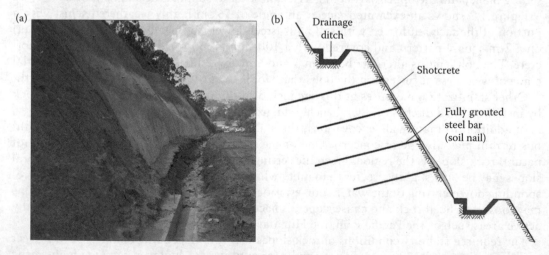

Figure 1.2 Excavated slope at 56° in highly weathered rock for highway construction: (a) image showing reinforcement of slope with soil nails and shotcrete (Sao Paulo, Brazil); (b) layout of soil nails in slope face.

include the need for relocation where the alignment intersects a major landslide that could be activated by construction.

The common design requirement for rock cuts is to determine the maximum safe cut face angle compatible with the planned maximum height. The design process is a trade-off between stability and economics. That is, steep cuts are usually less expensive to construct than flat cuts because the volume of excavated rock is minimized, less acquisition of right-of-way is required and smaller cut face areas are created. However, with steep slopes, it may be necessary to install extensive stabilization measures such as rock bolts and shotcrete in order to minimize both the risk of overall slope instability and rock falls during the operational life of the project.

1.1.1 Scope of book

The design of rock cuts involves the collection of geotechnical data, the use of appropriate design methods and the implementation of excavation methods and stabilization/protection measures suitable for the particular site conditions. In order to address all these issues, the book is divided into three distinct sections that cover *design data*, *design methods* and *excavation/support procedures*. Details of the main topics covered in each section are as follows:

1. Design data
   a. *Geological data*, of which structural geology in strong rock is usually the most important. This information includes the orientation of discontinuities, and their characteristics such as length, spacing, roughness and infilling; Chapter 2 discusses interpretation of these data. Chapter 3 discusses weathered rock, its occurrence, classification and differing weathering products related to lithology. Chapter 4 describes methods of data collection for both fresh and weathered rock.
   b. *Rock strength*, with the most important parameter being the shear strength. Slope conditions may include discontinuity surfaces in strong rock, or rock masses with either very closely spaced discontinuities or in weathered rock where individual discontinuities do not control stability (Chapter 5).
   c. *Ground water* conditions comprise ground water level within the slope, water pressures that are generated on potential sliding surfaces, and procedures to drain the slope, if necessary (Chapter 6).

2. Design methods
   a. Design methods for rock slopes fall into two groups – *limit equilibrium analysis* (LEA) and *numerical analysis*. LEA calculates the factor of safety of the slope based on the ratio of the resisting and driving forces. Different LEA procedures are used for plane, wedge, circular and toppling failures, with the type of failure being defined by the structural geology of the slope (Chapters 7 to 10). The adaption of LEA analysis methods for seismic design of rock slopes, including Newmark displacement analysis, is discussed in Chapter 11. Numerical analysis examines the stresses and strains developed in the slope, and stability is assessed by comparing the stresses in the slope with the rock strength (Chapter 12).

3. Excavation and stabilization
   a. *Blasting* issues relevant to slope stability include production blasting, controlled blasting on final faces, and in urban areas the control of damage from ground vibrations, flyrock and noise (Chapter 13).
   b. *Stabilization* methods include rock reinforcement with rock anchors and dowels, rock removal involving scaling and trim blasting, and rock fall protection measures comprising ditches, fences and sheds (Chapter 14).

    c. *Monitoring* of slope movement in order to identify acceleration that can be a precursor to failure of the slope. Monitoring is often an important part of slope management in open pit mines, as well as large, slow-moving landslides above important infrastructure. Surface and sub-surface monitoring methods are discussed, as well as interpretation of the data (Chapter 15).

    d. *Civil applications* – Chapter 16 describes slope designs for six civil projects, including stabilization methods. The examples illustrate the design procedures discussed in the earlier chapters for geological conditions that include both strong, blocky rock and weak, closely jointed rock.

Also included in the book are a series of example problems demonstrating both data analysis and design methods.

1.1.2 Socioeconomic consequences of slope failure

Failures of rock slopes, both man-made and natural, include rock falls, overall slope instability and landslides, as well as debris flows and shallow landslips in weathered rock. The consequence of such failures can include direct costs such as removing the failed rock and stabilizing the slope, and possibly a wide variety of indirect costs. Examples of indirect costs include damage to vehicles and injury to passengers on highways and railways, traffic delays, business disruptions, loss of tax revenue due to decreased land values, and flooding and disruption to water supplies where rivers are blocked by slides.

The cost of slope failures is greatest in urbanized areas with high-population densities where even small slides may destroy houses and block transportation routes (Transportation Research Board, 1996). In contrast, slides in rural areas may have few indirect costs, except perhaps the costs due to the loss of agricultural land. An example of a landslide that resulted in severe economic costs is the 1983 Thistle Slide in Utah that resulted in losses of about $200 million when the landslide dammed the Spanish Fork River severing railways and highways, and flooding the town of Thistle (University of Utah, 1984). An example of a landslide that resulted in both loss of life and economic costs is the Vajont slide in Italy in 1963. The slide inundated a reservoir, sending a wave over the crest of the dam that destroyed five villages and took about 2000 lives (Kiersch, 1963; Hendron and Patton, 1985).

Japan, which has both highly developed infrastructure and steep mountainous terrain, experiences high costs of rock falls and landslides. In addition, frequent triggering events occur such as high rainfall, freeze–thaw cycles and ground shaking due to earthquakes. Documentation of major landslides between 1938 and 1981 recorded total losses of 4834 lives and 188,681 homes (Japan Ministry of Construction, 1983). Similar conditions occur in South Korea where on average 60 deaths and damage of up to $1 billion are recorded annually (Lee, 2012).

In South America, where bedrock can be deeply weathered and intense rainfall events occur, similar damage and loss of life occur. For example, in Medellin, Columbia, 74% of the deaths resulting from natural hazards are due to slope instability, and one event in 1974 resulted in at least 770 fatalities (Carvajal, Restrepo and Azevedo, 2012).

1.2 PRINCIPLES OF ROCK SLOPE ENGINEERING

This section describes issues that need to be considered in rock slope design for civil projects. The principal requirement for projects such as highways and railways is that a high-degree

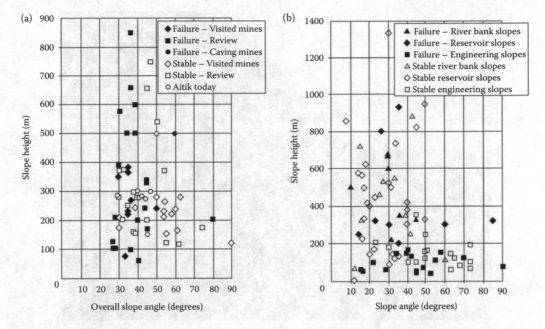

Figure 1.3 Relationship between slope height and slope angle for open pits, and natural and engineered slopes. (a) Pit slopes and caving mines (Sjöborg, 1999). (b) Natural and engineered slopes in China (data from Chen, 1995a,b).

of reliability is required because slope failure, or even rock falls, can rarely be tolerated. Furthermore, the design life of these slopes is usually decades, and can exceed 100 years, and the owner/operator will generally want to minimize maintenance costs and disruptions to operations during this period. In order to meet these requirements, it is common practice to use controlled blasting on final walls to minimize damage to rock behind the face, and to excavate ditches at the base of the cut that are wide enough to contain rock falls. The extra construction cost of these measures is usually justified by the resulting reduced long-term operating costs.

As a frame of reference for rock slope design, Figure 1.3 shows the results of surveys of the slope height and angle and stability conditions for natural, engineered slopes, as well as open pit mine slopes. The plots show that many slopes are unstable at flatter angles and lower heights than the maximum stable values because weak rock or adverse structure can result in instability of even low slopes at shallow slope angles. Importantly, the plots also show that no relationship exists between stable slope height and face angle, and that every slope must be evaluated on its own merits.

1.2.1 Structural geology control of stability

The design of rock cuts for projects such as highways and railways is usually concerned with details of the structural geology, that is, the orientation and characteristics (such as continuous length [or persistence], roughness and infilling materials) of the joints, bedding and faults that occur behind the rock face. For example, Figure 1.4 shows a cut slope in shale containing smooth bedding planes that are continuous over the full height of the cut and dip at an angle of about 50° towards the highway. Since the friction angle of these discontinuities is about 20–25°, any attempt to excavate this cut at a steeper angle than the

Figure 1.4 (**See colour insert**.) Cut face coincident with continuous, low-friction bedding planes in shale (Trans Canada Highway near Lake Louise, Alberta). (Image by A. J. Morris.)

dip of the beds would result in blocks of rock sliding from the face on the beds; the steepest unsupported cut that can be made is equal to the dip of the beds. However, as the alignment of the road changes so that the strike of the beds is at right angles to the cut face (right side of photograph), it is not possible for sliding to occur on the beds, and a steeper face can be excavated.

For most rock cuts on civil projects where the slope heights are less than about 100 m (300 ft), the stresses in the rock are much less than the rock strength so that fracturing of intact rock is unlikely. Therefore, slope design is primarily concerned with the stability of blocks of rock formed by the discontinuities. Intact rock strength, which is used indirectly in slope design, relates to the shear strength rock masses, roughness of discontinuities, as well as excavation methods and costs.

Figure 1.5 shows a range of geological conditions and their influence on stability, and illustrates the types of information that are important to design.

Slopes (a) and (b) show typical conditions for sedimentary rock, such as sandstone and limestone containing continuous beds on which sliding can occur if the dip of the beds is steeper than the friction angle of the discontinuity surface. In (a), the beds 'daylight' on the steep cut face and blocks may slide on the bedding, while in (b), the face is coincident with the bedding and the face is stable.

For slope (c), the overall face is also stable because the main discontinuity set dips into the face. However, some risk of instability exists for surficial blocks of rock formed by the conjugate joint set that dips out of the face, particularly if blasting during construction has damaged the rock behind the face.

For slope (d), the main joint set also dips into the face but at a steep angle to form a series of thin slabs that can fail by toppling where the centre of gravity of the block lies outside the base.

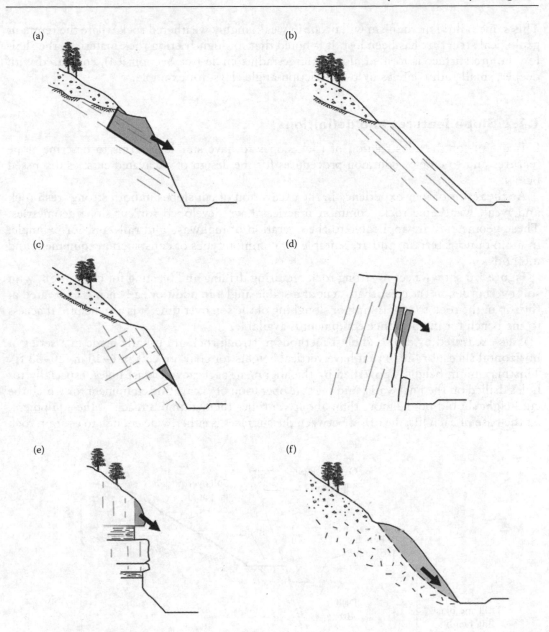

Figure 1.5 Influence of geological conditions on stability of rock cuts: (a) potentially unstable – discontinuities 'daylight' in face; (b) stable slope – face excavated parallel to discontinuities; (c) stable slope – main discontinuities dip into face; (d) toppling failure of thin beds dipping steeply into face; (e) in horizontally bedded sedimentary formations, weathering of shale undercuts strong sandstone beds to form overhangs; (f) potentially shallow circular failure in closely fractured, weak rock or weathered rock.

Slope (e) shows a typical horizontally bedded sandstone–shale sequence in which the shale weathers considerably faster than the sandstone to form a series of overhangs that can fail suddenly where vertical stress relief joints form in the sandstone.

Slope (f) is cut in weak rock containing closely spaced but low-persistence joints that do not form a continuous sliding surface. A steep slope cut in this weak rock mass may fail along a shallow circular surface, partially along joints and partially through intact rock.

This same failure mechanism will occur in weak, highly weathered rock where the remnant geological structure has been lost. It is noted that for high-friction rock materials, the shallow failure surface is formed along a large-radius circle (see Section 9.4), compared with deeper, small-radius circles in lower-friction-angle clays, for example.

1.2.2 Slope features and definitions

It is useful for the construction of rock slopes to have standard terms to describe slope features and to employ common procedures for the design of excavated cuts, as discussed below.

As the result of long experience in the excavation of cut slopes in both strong fresh rock and weak weathered rock, common practices have developed for cut slope geometries. These geometries are applicable to the operation of highways and railways, for example, in mountainous terrain, and are suitable for common types of construction equipment and methods.

Figure 1.6 shows a cut in strong rock requiring drilling and blasting for excavation, with soil overburden at the crest that is cut at a stable angle. In addition, a bench is excavated at the top of the rock to contain minor sloughing of the soil over time; it is also helpful if access to the bench for maintenance equipment is available.

Cuts excavated by drill and blast methods in strong rock are usually made in a series of horizontal slices (or 'lifts') that have vertical heights for civil projects of 6–10 m (20–33 ft). This maximum height is governed by the alignment accuracy of blast holes, especially for holes drilled on the final walls, and the safe operation of excavating equipment for which the cut height should not be more than about 1.5 times the maximum reach of the equipment. At the base of each lift, the offset between the cut faces is either wide enough to contain rock

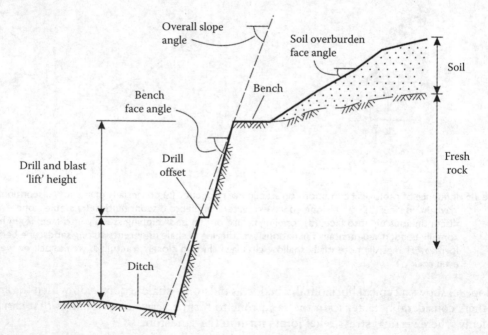

Figure 1.6 Examples of rock cut geometry – two bench, drill and blast excavation with shallow layer of soil along the crest, and a bench at the top of the rock cut to contain soil sloughing.

falls and provide access for excavating equipment (minimum of 5 m, 16 ft), or is as narrow as possible to avoid creating 'ski jumps' for rock falls that could fall on to the road or railway, for example. The minimum offset of about 0.75 m (2 ft) is required for clearance of the head of the drill when drilling blast holes at the angle of the final face.

A common face angle for rock cuts is 76° (1/4H:1V). It is found that cuts at this angle in strong, massive rock are usually stable, and the face can accommodate some misalignment of final line blast holes that can be corrected on the next lift. In contrast, excavation of a vertical face requires very precise drill hole alignment to avoid overhanging faces (holes inclined into face), or loss of excavation width (holes inclined out of face).

Where necessary, a ditch to contain rock falls is excavated at the base of the slope, with the width and depth of the ditch determined by the height and face angle of the rock face (see Section 14.6.2). It is preferable to have a ditch at the base of the slope where it can be readily cleaned, rather than a bench on the slope where equipment access may be difficult.

Terminology: In North America, the breaks in the slope at each lift are defined by the bench height and bench width, and the interbench face angle. Equivalent terms in Australia are bench height, berm width and batter angle.

1.3 STANDARD TERMS FOR LANDSLIDES

The International Association of Engineering Geology has prepared definitions of landslide features and dimensions as shown in Figures 1.7 and 1.8 (IAEG Commission on Landslides, 1990; Transportation Research Board, 1996). Although the diagrams depicting the landslides show soil-type slides with circular sliding surfaces, many of these landslide features are applicable to both rock slides and slope failures in weak and weathered rock. The value of the definitions shown in Figures 1.7 and 1.8 is to encourage the use of consistent terminology that can be clearly understood by others in the profession when investigating and reporting on rock slopes and landslides.

1.4 ROCK SLOPE DESIGN METHODS

This section summarizes four different procedures for designing rock slopes, and shows the basic data that are required for analyzing slope stability.

1.4.1 Summary of design methods

A basic feature of all slope design methods is that shear takes place along either a discrete sliding surface, or within a zone, behind the face. If the shear force (displacing force) is greater than the shear strength of the rock (resisting force) on this surface, then the slope will be unstable. Instability could take the form of displacement that may or may not be tolerable, or the slope may collapse either suddenly or progressively. The definition of instability will depend on the application. For example, a landslide in a rural area may undergo several metres of displacement without being a hazard provided that movement is not accelerating, while a slope supporting a bridge abutment would have little tolerance for movement. Also, a single rock fall from a slope above a highway may be of little concern if the ditch at the base of the slope is adequate to contain the fall, but a failure of a significant portion of the slope that reaches the travelled surface could have serious consequences.

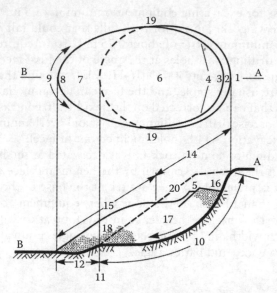

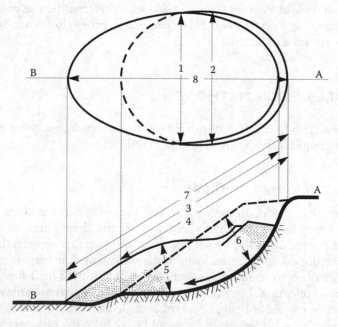

Figure 1.7 Definitions of landslide features: upper – plan of typical landslide in which dashed line indicates trace of rupture surface on original ground surface; lower – section in which hatching indicates undisturbed ground and stippling shows extent of displaced material. Numbers refer to dimensions defined in Table 1.1 (IAEG Commission on Landslides, 1990).

Figure 1.8 Definitions of landslide dimensions: upper – plan of typical landslide in which dashed line indicates trace of rupture surface on original ground surface; lower – section in which hatching indicates undisturbed ground, stippling shows extent of displaced material, and broken line is original ground surface. Numbers refer to dimensions defined in Table 1.2 (IAEG Commission on Landslides, 1990).

Table 1.1 Definition of landslide features

No.	Name	Definition
1	Crown	Displaced material adjacent to highest parts of main scarp
2	Main scarp	Steep surface on undisturbed ground at upper edge of landslide caused by movement of displaced matter (13, stippled area) away from undisturbed ground; it is the visible part of rupture surface (10)
3	Top	Highest point of contact between displaced material (13) and main scarp (2)
4	Head	Upper parts of landslide along contact between displaced material and main scarp (2)
5	Minor scarp	Steep surface on displaced material of landslide produced by differential movements within displaced material
6	Main body	Part of displaced material of landslide that overlies surface of rupture between main scarp (2) and top of surface of rupture
7	Foot	Portion of landslide that has moved beyond toe of surface of rupture (11) and overlies original ground surface (20)
8	Tip	Point on toe (9) farthest from top (3) of landslide
9	Toe	Lower, usually curved margin of displaced material of a landslide, most distant from main scarp (2)
10	Surface of rupture	Surface that forms (or that has formed) lower boundary of displaced material (13) below original ground surface (20); mechanical idealization of surface of rupture is called sliding surface in stability analysis
11	Toe of surface of rupture	Intersection (usually buried) between lower part of surface of rupture (10) of a landslide and original ground surface (20)
12	Surface of separation	Part of original ground surface (20) now overlain by foot (7) of landslide
13	Displaced material	Material displaced from its original position on slope by movement of landslide; forms both depleted mass (17) and accumulation (18); it is stippled in Figure 1.7
14	Zone of depletion	Area of landslide within which displaced material (13) lies below original ground surface (20)
15	Zone of accumulation	Area of landslide within which displaced material lies above original ground surface (20)
16	Depletion	Volume bounded by main scarp (2), depleted mass (17) and original ground surface (20)
17	Depleted mass	Volume of displaced material that overlies surface of rupture (10) but underlies original ground surface (20)
18	Accumulation	Volume of displaced material (13) that lies above original ground surface (20)
19	Flank	Undisplaced material adjacent to sides of surface of rupture; compass directions are preferable in describing flanks, but if left and right are used, they refer to flanks as viewed from crown
20	Original ground surface	Surface of slope that existed before landslide took place

Based upon these concepts of slope stability, the stability of a slope can be expressed in one or more of the following terms:

1. *Factor of safety (FS)* – stability quantified by the factor of safety that is the ratio of the summed resisting forces to driving forces, termed *limit equilibrium* of the slope, which is stable if $FS > 1$.
2. *Allowable strain* – failure defined by onset of strains great enough to prevent safe operation of the slope. Acceptable strain values depend on the application such as foundations for bridges or buildings that have a low tolerance for movement (Wyllie, 1999).

Table 1.2 Definitions of landslide dimensions

No.	Name	Definition
1	Width of displaced mass, W_d	Maximum breadth of displaced mass perpendicular to length, L_d
2	Width of surface rupture, W_r	Maximum width between flanks of landslide perpendicular to length, L_r
3	Length of displaced mass, L_d	Minimum distance from tip to toe
4	Length of surface of rupture, L_r	Minimum distance from top of surface of rupture crown
5	Depth of displaced mass, D_d	Maximum depth of surface of rupture below original ground surface measured perpendicular to plane containing W_d and L_d
6	Depth of surface of rupture, D_r	Maximum depth of surface of rupture below original ground surface measured perpendicular to plane containing W_r and L_r
7	Total length, L	Minimum distance from tip of landslide to crown
8	Length of centre line, L_{cl}	Distance from crown to tip of landslide through points on original ground surface equidistant from lateral margins of surface of rupture and displaced material

3. *Probability of failure* – stability quantified by probability distribution of difference between resisting and displacing forces (margin of safety), which are each expressed as probability distributions.
4. *Load and resistance factor design (LRFD)* – stability defined by the factored resistance being greater than or equal to the sum of the factored loads.

At this time (2016), the factor of safety is the most common method of slope design for civil engineering, and it is widely applied to all types of geological conditions, for both rock and soil. Furthermore, factor of safety values that are generally accepted for slopes excavated for different purposes promotes the preparation of reasonably consistent designs. The ranges of minimum total factors of safety as proposed by Terzaghi and Peck (1967) and the Canadian Geotechnical Society (1992) are given in Table 1.3.

In Table 1.3, the upper values of the factors of safety apply to usual loads and service conditions, while the lower values apply to maximum loads and the most unfavourable expected geological conditions. For temporary slopes with little consequence in the event of instability, the factor of safety generally used is in the range of 1.2–1.4.

Although probabilistic design for rock slopes was first developed in the 1970s (Harr, 1977; Canada DEMR, 1978), it is not widely used (as of 2016). A possible reason for this lack of acceptance is a term such as '5% probability of failure' is not well understood, and experience is limited on acceptable probabilities to use in design (see Section 1.4.5). A more acceptable term for owners of facilities incorporating slopes would be

Table 1.3 Values of minimum total safety factors

Failure type	Category	Safety factor
Shearing	Earthworks	1.3–1.5
	Earth retaining structure, excavations	1.5–2.0
	Foundations	2.0–3.0

'95% reliability', but in reality, no one wants to hear that it is possible that their slope may fail.

The calculation of strain in slopes is the most recent advance in slope design. The technique has resulted from the development of numerical analysis methods, and particularly those that can incorporate discontinuities (Starfield and Cundall, 1988). It is most widely used in the design of open pits with depths several hundred metres where movement is tolerated, and the stresses in the slope are high enough to generate strains in the rock mass that may initiate slope instability (see Chapter 12). In contrast, in slopes for civil projects with heights up to about 100 m (330 ft), stresses generated in the slope are usually considerably less than the rock strength.

The LRFD method has been developed for structural design, and is now being extended to geotechnical designs such as foundations and retaining structures where structural engineers are using LRFD in their work. Further details of this design method are discussed in Section 1.4.6.

The factor of safety, probability of failure or allowable strain method that is used in design should be appropriate for each site. The design process requires a considerable amount of judgement because of the variety of geological and construction factors that must be considered. Conditions that would require the use of factors of safety at the high end of the ranges quoted in Table 1.3 include:

- A limited drilling programme that does not adequately sample conditions at the site, or drill core in which extensive mechanical breakage or core loss has occurred.
- Absence of rock outcrops so that mapping of geological structure is not possible, and no history of local stability conditions is available.
- Inability to obtain undisturbed samples for strength testing, or difficulty in extrapolating laboratory test results to *in situ* conditions.
- Absence of information on ground water conditions, and/or significant seasonal fluctuations in ground water levels.
- Uncertainty in failure mechanisms of the slope and the reliability of the analysis method. For example, plane-type failures can be analyzed with considerable confidence, while the detailed mechanism of toppling failures is less well understood.
- Concern regarding the quality of construction, including materials, inspection and weather conditions.
- The consequence of instability, with higher factors of safety being used for dams and major transportation routes, and lower values for temporary structures or industrial roads for logging and mining operations.

This book does not cover the use of rock mass rating systems (Haines and Terbrugge, 1991; Duran and Douglas, 1999) for slope design. At this time (2016), it is considered that the frequent influence of discrete discontinuities on stability can, and should be, incorporated directly into stability analyses.

A vital aspect of all rock slope design is the quality of the blasting used in excavation. Designs usually assume that the rock mass comprises intact blocks, the shape and size of which are defined by naturally occurring discontinuities. Furthermore, the properties of these discontinuities should be predictable from observations of surface outcrops and drill core. However, if excessively heavy blasting is used that results in damage to the rock behind the face, stability could be dependent on the condition of the fractured rock. Since the properties of the fractured rock are unpredictable, stability conditions will also be unpredictable. Blasting and the control of blast damage are discussed in Chapter 13.

1.4.2 Limit equilibrium analysis (deterministic)

The stability of rock slopes for the geological conditions shown in Figure 1.5a and f depends on the shear strength generated along the sliding surface. For all shear-type failures, the rock can be assumed to be a Mohr–Coulomb material in which the shear strength is expressed in terms of the cohesion c and friction angle ϕ. For a sliding surface on which an effective normal stress σ' is acting, the shear strength τ developed on this surface is given by

$$\tau = c + \sigma' \cdot \tan\phi \tag{1.1}$$

Equation 1.1 is expressed as a straight line on a normal stress – shear stress plot (Figure 1.9a), in which the cohesion is defined by the intercept on the shear stress axis, and the friction angle is defined by the slope of the line. The effective normal stress is the difference between the normal component of the vertical stress due to the weight of the rock lying above the sliding plane, and the uplift due to any water pressure acting on this surface.

Figure 1.9b shows a slope containing a continuous joint dipping out of the face and forming a sliding block. Calculation of the factor of safety for the block shown in Figure 1.9b involves resolution of the force acting on the sliding surface into components acting perpendicular and parallel to this surface. That is, if the dip of the sliding surface is ψ_p, its area is A and the weight of the block lying above the sliding surface is W, then the normal and shear stresses on the sliding plane are

$$\text{Normal stress, } \sigma = \frac{W \cdot \cos\psi_p}{A}; \quad \text{and shear stress, } \tau_S = \frac{W \cdot \sin\psi_p}{A} \tag{1.2}$$

and Equation 1.1 can be expressed as

$$\tau = c + \frac{W \cdot \cos\psi_p \cdot \tan\phi}{A} \tag{1.3}$$

or

$$\tau_s \cdot A = W \cdot \sin\psi_p, \quad \text{and} \quad \tau \cdot A = c \cdot A + W \cdot \cos\psi_p \cdot \tan\phi \tag{1.4}$$

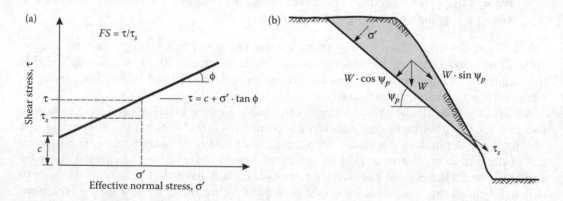

Figure 1.9 LEA method of calculating factor of safety of sliding block: (a) Mohr diagram showing shear strength defined by cohesion c and friction angle ϕ; (b) resolution of force W due to weight of block into components parallel and perpendicular to sliding plane (dip ψ_p).

In Equation 1.4, the term $(W \cdot \sin \psi_p)$ defines the resultant force acting down the sliding plane and is termed the 'driving force' $(\tau_s \cdot A)$, while the term $(c \cdot A + W \cos \psi_p \cdot \tan \phi)$ defines the total shear strength forces acting up the plane that resist sliding and are termed the 'resisting forces' $(\tau \cdot A)$. The stability of the block in Figure 1.9b can be quantified by the ratio of the resisting and driving forces, which is termed the factor of safety, *FS*. Therefore, the expression for the factor of safety is

$$FS = \frac{\text{Resisting force}}{\text{Driving force}} \tag{1.5}$$

$$FS = \frac{c \cdot A + W \cdot \cos \psi_p \cdot \tan \phi}{W \cdot \sin \psi_p} \tag{1.6}$$

The displacing shear stress τ_s and the resisting shear stress τ defined by Equation 1.4 are plotted in Figure 1.9a, which shows that the resisting stress exceeds the displacing stress $(\tau > \tau_s)$, so the factor of safety is greater than 1.0 and the slope is stable.

If the sliding surface is clean and contains no infilling, then the cohesion is likely to be zero and Equation 1.6 reduces to

$$FS \cdot \frac{\sin \psi_p}{\cos \psi_p} = \tan \phi \tag{1.7}$$

or

$$FS = 1 \quad \text{when } \psi_p = \phi \tag{1.8}$$

Equations 1.7 and 1.8 show that for a dry, clean surface with no support installed, the block of rock will slide when the dip angle of the sliding surface equals the friction angle of this surface, and that stability is independent of the size of the sliding block. That is, the block is at a condition of 'limiting equilibrium' when the driving forces are exactly equal to the resisting forces and the factor of safety is equal to 1.0. Therefore, the method of slope stability analysis described in this section is termed *limit equilibrium analysis* (LEA).

LEA can be applied to a wide range of conditions and can incorporate forces such as water pressures acting on the sliding surface, as well as external reinforcing forces supplied by tensioned rock anchors. Figure 1.10a shows a slope containing a sliding surface with area A and dip ψ_p, and a vertical tension crack. The slope is partially saturated such that the tension crack is half filled with water, and the water table exits where the sliding surface daylights on the slope face. The water pressures that are generated in the tension crack and on the sliding surface can be approximated by triangular force diagrams where the maximum pressure p at the base of the tension crack and the upper end of the sliding surface is given by

$$p = \gamma_w \cdot h_w \tag{1.9}$$

where γ_w is the unit weight of water and h_w is the vertical height of water in the tension crack. Based on this assumption, the water forces acting in the tension crack V and on the sliding plane U are as follows (see Figure 1.10a):

$$V = \frac{1}{2} \gamma_w \cdot h_w^2 \quad \text{and} \quad U = \frac{1}{2} \gamma_w \cdot h_w \cdot A \tag{1.10}$$

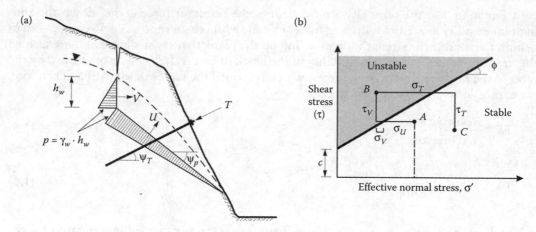

Figure 1.10 Effect of ground water and bolt forces on factor of safety of rock slopes: (a) ground water and bolting forces acting on sliding surface; (b) Mohr diagram of stresses acting on sliding surface showing stable and unstable stability conditions.

and the factor of safety of the slope is calculated by modifying Equation 1.6 as follows:

$$FS = \frac{c \cdot A + (W \cdot \cos \psi_p - U - V \cdot \sin \psi_p) \tan \phi}{W \cdot \sin \psi_p + V \cdot \cos \psi_p} \qquad (1.11)$$

Similarly, an equation can be developed for a reinforced slope in which a tensioned rock anchor has been installed with the anchor below the sliding plane. If the tension in the anchor is T and it is installed at an angle ψ_T below the horizontal, then the normal and shear forces acting on the sliding plane due to the anchor tension are, respectively,

$$N_T = T \cdot \sin(\psi_T + \psi_p); \quad \text{and } S_T = T \cdot \cos(\psi_T + \psi_p) \qquad (1.12)$$

and the equation defining the factor of safety of the anchored, partially saturated slope is

$$FS = \frac{c \cdot A + (W \cdot \cos \psi_p - U - V \cdot \sin \psi_p + T \cdot \sin(\psi_T + \psi_p)) \tan \phi}{W \cdot \sin \psi_p + V \cdot \cos \psi_p - T \cdot \cos(\psi_T + \psi_p)} \qquad (1.13)$$

Figure 1.10b shows on a Mohr diagram the magnitude of the normal and shear stresses on the sliding surface developed by the water and bolting forces, and their influence on the factor of safety. For a drained slope where the shear resistance on the sliding plane due to the cohesion and friction exceeds the driving force, the slope is stable (point A). Introduction of water into the slope develops destabilizing forces that decrease the normal stress and increase the shear stress, and may cause the resultant of the forces to be above the limiting strength line, indicating instability (point B). In contrast, stabilizing forces (bolting and drainage) increase the normal stress and decrease the shear stress, and cause the resultant to be below the limiting strength line, indicating stability (point C).

The diagram in Figure 1.10b can also be used to show that the optimum plunge angle for the bolts ($\psi_{T(opt)}$), that is, the plunge that produces the greatest factor of safety for a given rock anchor force, is

$$\psi_{T(opt)} = (\phi - \psi_p), \quad \text{or} \quad \phi = (\psi_p + \psi_{T(opt)}) \qquad (1.14)$$

Application of Equation 1.14 may show that the anchor should be installed above the horizontal, that is, ψ_T is negative. However, in practice, it is usually preferable to install anchors below the horizontal because this facilitates drilling and grouting, and provides a more reliable installation.

These examples of LEA to calculate the stability of rock slopes show that this is a versatile method that can be applied to a wide range of conditions. One limitation of the limit equilibrium method is that all the forces are assumed to act through the centre of gravity of the block, and that no moments are generated.

The analysis described in this section is applicable to a block sliding on a plane. However, under certain geometric conditions, the block may topple rather than slide, in which case a different form of LEA must be used. Figure 1.11 shows the conditions that differentiate stable, sliding and toppling blocks in relation to the width Δx and height y of the block, the dip ψ_p of the plane on which it lies and the friction angle ϕ of this surface. In this figure, the assumed friction angle is 35°, so the block will only slide when the dip of the base plane is greater than 35°. Sliding blocks are analyzed either as plane or wedge failures (see Chapters 7 and 8, respectively), while the analysis of toppling failures is discussed in Chapter 10.

Figure 1.11b shows that toppling occurs for tall, thin blocks resting on an inclined plane, at a dip angle less than the friction angle of this surface, such that the line of the centre of gravity of the block lies outside the base. As shown by the relatively small area for toppling instability on this chart, the conditions under which toppling occurs are limited, and this is a less common type of failure compared to sliding failures.

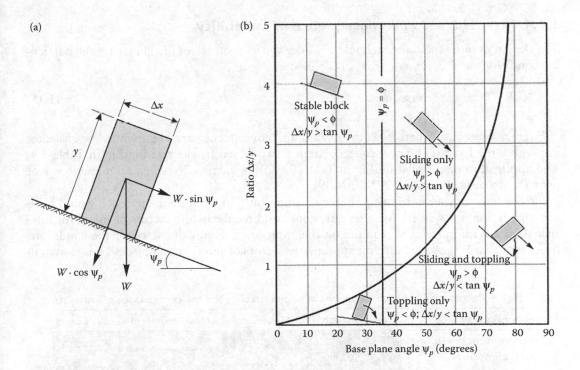

Figure 1.11 Identification of sliding and toppling blocks: (a) geometry of block on inclined plane; (b) conditions for sliding and toppling of block on an inclined plane.

1.4.3 Sensitivity analysis

The factor of safety analysis described above involves the selection of a single value for each of the parameters that define the driving and resisting forces in the slope. In reality, each parameter has a range of values, as well as uncertainty regarding the true value. A method of examining the effect of this variability and uncertainty on the factor of safety is to carry out sensitivity analyses using upper and lower bound values for those parameters considered critical to design to determine if conditions exist where the factor of safety may be too low. However, to carry out sensitivity analyses for more than three parameters is cumbersome, and it is difficult to examine the relationship between each of the parameters. Consequently, the usual design procedure involves a combination of analysis and judgement in assessing the influence on stability of variability in the design parameters, and then selecting an appropriate factor of safety.

An example of sensitivity analysis is shown in Section 5.4 in Chapter 5 that describes the stability analysis of a quarry slope in which sensitivity analyses were carried out for both the friction angle (range 15–25°) and the water pressure – fully drained to fully saturated (Figure 5.20). This plot shows that water pressures have more influence on stability than the friction angle. That is, a fully drained, vertical slope is stable for a friction angle as low as 15°, while a fully saturated slope is unstable at a face angle of 60° even if the friction angle is 25°.

The value of sensitivity analysis is to assess which parameters have the greatest influence on stability. This information can then be used in planning investigation programmes to collect data that will define this parameter(s) more precisely. Alternatively, if uncertainty exists in the value of an important design parameter, this can be accounted for in design by using an appropriate factor of safety.

1.4.4 Risk: Hazard and consequence of instability

The design of most rock slopes should consider the level of risk of instability for the particular conditions at the site, where risk is defined as

$$\text{Risk} = (\text{hazard}) \cdot (\text{consequence}) \tag{1.15}$$

High-risk sites are where both the hazard and consequence are high, while low-risk sites are where the hazard and consequence are low, as shown by the relationship in Table 1.4. The application of the relationship in Table 1.4 can be qualitative when no numeric values for risk can be established (CSIRO, 2009).

On the basis of the relationship shown in Equation 1.15, the design approach would be to use higher factors of safety for high-risk slopes and relatively low factors of safety for low-risk slopes, with judgement required on the appropriate factor of safety for the conditions (refer to Table 1.3). A guideline on appropriate factors of safety for slope design is shown in

Table 1.4 Relationship between the hazard of slope instability and the consequence of instability

Hazard/consequence	Very high hazard	High hazard	Moderate hazard	Low hazard
Very high consequence				
High consequence				
Moderate consequence				
Low consequence				

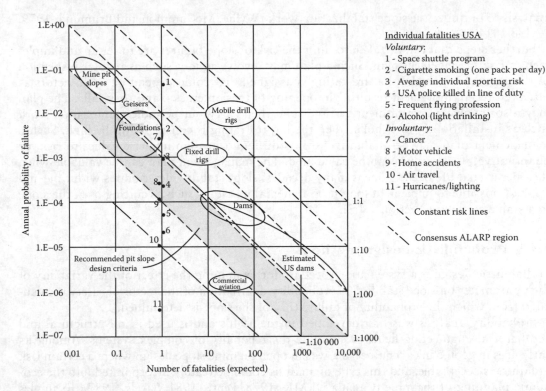

Figure 1.12 Risks for selected engineering projects, with comparison with individual risk (adapted from Whitman, 1984; Steffan et al., 2015).

Figure 1.12 that shows a well-established relationship for the annual probability of failure and the number of fatalities that may occur in the event of a slope failure. A discussion on probabilistic design and Figure 1.12 is presented in Section 1.4.5.

Examples of the types of hazards and consequences that may occur for rock slopes are as follows. High-hazard slopes generally occur where the rock contains persistent discontinuities that dip out of the face, where precipitation levels are high, and earthquake ground motions are common. In contrast, low-hazard slopes usually occur in strong, massive rock in desert climates. It may be possible to quantify the hazard using probabilistic design to calculate an approximate annual probability of failure as discussed in Section 1.4.5. Alternatively, some highway and railway systems keep records of past rock fall frequency that can be used to quantify the hazard such as the annual probability of instability on each rock slope in a certain geographic area.

The consequence of instability relates to the facilities located at the base of the slope. That is, permanently occupied housing and high-traffic transportation routes would be high-consequence conditions, whereas logging roads are likely to be low-consequence conditions. The consequence can be quantified by determining the cost of instability such as damage and injuries, as well as indirect costs of slope stabilization and loss of business due to traffic delays.

Once the annual probability of failure and the cost (consequence) of instability have been established, Equation 1.15 can be used to calculate the risk, which may be expressed as the expected cost of an event. Because of the uncertainty in both the probability of failure and the cost, the risk will also be a probability distribution, which may then be used in decision analysis to optimize the slope stabilization work (Wyllie, McCammon and Brumund, 1979; Wyllie 2014a,b).

Further steps that can be taken to limit the risk of slope failures are to set up and implement a risk management plan. Such a plan may involve actions such as regular inspection of the slope(s) in which a hazard rating is assigned that could depend on such factors as observed recent movement resulting in opening of tension cracks, and rock falls. The plan may also involve a commitment to fund, and then carry out periodic stabilization work such as installation of rock bolts when the hazard rating is considered to be high. Such a management plan may be qualitative, if it is difficult to assign numeric values to risk, or quantitative, if risk values can be calculated. The management plan can be valuable when the operational life of the slope is many decades, slope stability deteriorates with time and it is not possible to implement long-term remedial measures such as cutting back the slope to a stable angle (see Section 1.4.3).

1.4.5 Probabilistic design methods

Probabilistic design is a systematic procedure for examining the effect of the variability of each parameter on slope stability. A probability distribution of the factor of safety is calculated from which the probability of failure (PF) of the slope is determined.

Probability analysis was first developed in the 1940s and is used in the structural and aeronautical engineering fields to examine the reliability of complex systems. Among its early uses in geotechnical engineering was in open pit mine slope design where a certain risk of failure is acceptable, and this type of analysis could be readily incorporated into the economic planning of the mine (Canada DEMR, 1978; Pentz, 1981; Savely, 1987). Examples of its use in civil engineering are in the planning of slope stabilization programmes for transportation systems (Wyllie, McCammon and Brumund, 1979; McGuffey, Athanasiou-Grivas, Lori et al., 1980) and landslide hazards (Fell, 1994; Cruden, 1997) and in the design of storage facilities for hazardous waste (Roberds, 1984, 1986).

Sometimes, probabilistic design is used with reluctance when the amount of design data is limited and may not be representative of the population. In these circumstances, it is possible to use subjective assessment techniques that provide reasonably reliable probability values from small samples (Roberds, 1990). The basis of these techniques is the assessment and analysis of available data, by an expert or group of experts in the field, in order to arrive at a consensus on the probability distributions that represent the opinions of these individuals. The degree of defensibility of the results tends to increase with the time and cost that is expended on the analysis. For example, the assessment techniques range from, most simply, informal expert opinion, to more reliable and defensible techniques such as Delphi panels (Rohrbaugh, 1979). A Delphi panel comprises a group of experts who are each provided with the same set of data and are required to produce a written assessment of these data. These documents are then provided anonymously to each of the other assessors who are encouraged to adjust their assessments in the light of their peer's results. After several iterations of this process, it should be possible to arrive at a consensus that maintains anonymity and independence of thought.

The use of probability analysis in design requires that generally accepted ranges of probability of failure exist for different types of structure, as exist for factors of safety. To assist in selecting appropriate probability of failure values, Figure 1.12 gives a relationship between levels of annual probability of failure for a variety of engineering projects, and the consequence of failure in terms of lives lost (F-N diagram). For civil geotechnical projects, the acceptable annual probability of failure is 10^{-2} for foundations with little consequence of failure, while for dams where failure could result in the loss of several hundred lives, the annual probability of failure should not exceed about 10^{-4}–10^{-5}. It is expected that the target probability of failure for civil slopes will be less than that of pit mine slopes, and equal to that of foundations where the consequences of failure are high.

As also shown in Figure 1.12, these probabilities for engineering projects can be referenced to individual fatalities for both voluntary and involuntary activities. Despite the wide range of values shown in Figure 1.12, this approach provides a useful benchmark for the ongoing development of probabilistic design (Salmon and Hartford, 1995). Figure 1.12 includes two lines defining the 'ALARP (as low as reasonably practicable) zone', the upper limit of which is a constant risk of 1:1000. For a risk to be ALARP, it must be possible to demonstrate that the cost required to reduce the risk further would be grossly disproportionate to the benefit gained.

1. *Probability distribution functions*: In probability analysis, for each parameter such as the friction angle and cohesion for which the design value is uncertain, a range of values is assigned with the range defined by a probability density function. The types of distribution functions that are appropriate for geotechnical data include normal, beta, lognormal and triangular distributions. The most common type of function is the normal distribution in which the mean value is the most frequently occurring value (Figure 1.13a). The density of the normal distribution is defined by

$$f(x) = \frac{1}{SD \cdot \sqrt{2\pi}} e^{-1/2((x-\bar{x})/(SD))^2} \tag{1.16}$$

where $\bar{x}$ is the mean value given by

$$\bar{x} = \frac{\sum_{x=1}^{n} x}{n} \tag{1.17}$$

and SD is the standard deviation given by

$$SD = \left[\frac{\sum_{x=1}^{n} (x - \bar{x})^2}{n} \right]^{1/2} \tag{1.18}$$

As shown in Figure 1.13a, the scatter in the data, as represented by the width of the curve, is measured by the standard deviation. Important properties of the normal distribution are that the total area under the curve is equal to 1.0. That is, the probability is unity that all values of the parameter fall within the bounds of the curve. Also, 68% of the values will lie within a range of one standard deviation either side of the mean, and 95% will lie within two standard deviations either side of the mean.

Conversely, it is possible to determine the value of a parameter defined by a normal distribution by stating the probability of its occurrence. This is shown graphically in Figure 1.13b where $\Phi(z)$ is the distribution function with mean 0 and standard deviation 1.0. For example, a value which has a probability of being greater than 50% of all

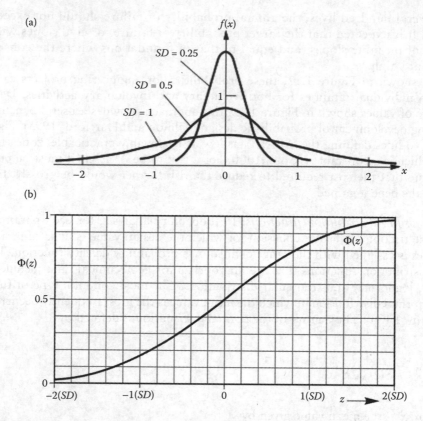

Figure 1.13 Properties of the normal distribution: (a) density of the normal distribution with mean $\bar{x} = 0$ and standard deviations (SD) of 0.25, 0.5 and 1.0; (b) distribution function $\Phi(z)$ of the normal distribution with mean 0 and standard deviation 1.0 (Kreyszig, 1976).

values is equal to the mean, and a value which has a probability of being greater than 16% of all values is equal to the mean plus one standard deviation.

The normal distribution extends to infinity in both directions, but this is often not a realistic expression of geotechnical data for which the likely upper and lower bounds of a parameter can be defined. For these conditions, it is appropriate to use the beta distribution which has finite maximum and minimum points, and can be uniform, skewed to the left or right, U-shaped or J-shaped (Harr, 1977). However, where little information is available on the distribution of the data, a simple triangular distribution can be used which is defined by three values: the most likely, and the minimum and maximum values. Examples of probability distributions are shown in the worked example in Section 7.6.

2. *Probability of failure*: The probability of failure is calculated in a similar manner to that of the factor of safety in that the relative magnitudes of the displacing and resisting forces in the slope are examined (see Section 1.4.2). Two common methods of calculating the probability of failure, and the corresponding coefficient of reliability, are the margin of safety method, and the Monte Carlo method as discussed below.

The *margin of safety* is the difference between the resisting and displacing forces, with the slope being unstable if the margin of safety is negative. If the resisting and displacing

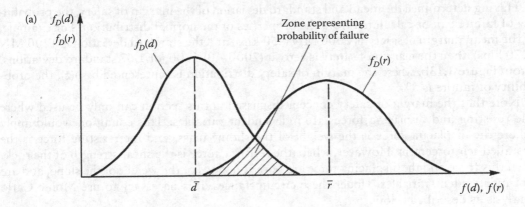

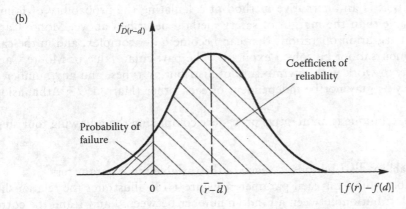

Figure 1.14 Calculation of probability of failure using normal distributions: (a) probability density functions of the resisting force f_R and the displacing force f_D in a slope; (b) probability density function of difference between resisting and displacing force distribution $[f(r) - f(d)]$ (margin of safety).

forces are mathematically defined probability distributions – $f_D(r)$ and $f_D(d)$, respectively, in Figure 1.14a – then it is possible to calculate a third probability distribution for the margin of safety. As shown in Figure 1.14a, a probability of failure will occur if the lower limit of the resisting force distribution $f_D(r)$ is less than the upper limit of the displacing force distribution $f_D(d)$. This is shown as the shaded area in Figure 1.14a, with the probability of failure being proportional to the area of the shaded zone. The method of calculating the area of the shaded zone is to calculate the probability density function of the margin of safety: the area of the negative portion of this function is the probability of failure (Figure 1.14b). If the resisting and displacing forces are defined by normal distributions, the margin of safety is also a normal distribution, the mean and standard deviation of which are calculated as follows (Canada DEMR, 1978):

$$\text{Mean, margin of safety} = \overline{f_r} - \overline{f_d} \tag{1.19}$$

$$\text{Standard deviation, margin of safety} = \left(SD_r^2 + SD_d^2\right)^{1/2} \tag{1.20}$$

where $\overline{f_r}$ and $\overline{f_d}$ are the mean values, and SD_r and SD_d are the standard deviations of the distributions of the resisting and displacing forces, respectively. Note that the definition of the deterministic factor of safety is given by $(\overline{f_r}/\overline{f_d})$.

Having determined the mean and standard deviation of the margin of safety, the probability of failure can be calculated from the properties of the normal distribution. For example, if the mean margin of safety is 2000 MN (450 kips) and the standard deviation is 1200 MN (270 kips), then the margin of safety is zero at (2000 – 0)/1200, or 1.67 standard deviations. From Figure 1.13b, where the margin of safety distribution is represented by $\phi(z)$, the probability of failure is 5%.

Note that the margin of safety concept discussed in this section can only be used where the resisting and displacing forces are independent variables. This condition would apply where the displacing force is the weight of the sliding mass, and the resisting force is the installed reinforcement. However, where the resisting force is the shear strength of the rock, then this force and the displacing force are both functions of the weight of the slope, and are not independent variables. Under these circumstances, it is necessary to use Monte Carlo analysis as described below.

Monte Carlo analysis is an alternative method of calculating the probability of failure which is more versatile than the margin of safety method described above. Monte Carlo analysis avoids the integration operations that can become quite complex, and in the case of the beta distribution cannot be solved explicitly. The particular value of Monte Carlo analysis is the ability to work with any mixture of distribution types, and any number of variables, which may or may not be independent of each other (Harr, 1977; Athanasiou-Grivas, 1979, 1980).

The Monte Carlo technique is an iterative procedure comprising the following four steps (Figure 1.15):

1. Estimate probability distributions for each of the variable input parameters.
2. Generate random values for each parameter; Figure 1.13b illustrates the relationship for a normal distribution between a random number between 0 and 1 and the corresponding value of the parameter.
3. Calculate values for the displacing and resisting forces and determine if the resisting force is greater than the displacing force.
4. Repeat the process N times (N > 100) and then determine the probability of failure PF from the ratio

$$PF = \frac{N - M}{N} \tag{1.21}$$

where M is the number of times the resisting force exceeded the displacing force (i.e. the factor of safety is greater than 1.0).

An example of the use of Monte Carlo analysis to calculate the coefficient of reliability of a slope against sliding is given in Section 7.6 in Chapter 7. This example shows the relationship between deterministic and probabilistic analyses. The factor of safety is calculated from the mean or most likely values of the input variables, while the probabilistic analysis calculates the distribution of the factor of safety when selected input variables are expressed as probability density functions. For the unsupported slope, the deterministic factor of safety has a value of 1.4, while the probabilistic analysis shows that the factor of safety can range from a minimum value of 0.69 to a maximum value of 2.52. The proportion of this distribution with a value less than 1.0 is 3.4%, which represents the probability of failure of the slope. This probability can be expressed as an annual value by assuming that the slope has a design life of 50 years, so that the annual probability of failure is ($p = 0.034/50 = 0.0007$, or 7E – 4), which lies within the range of probabilities for 'foundations' and 'fixed drill rigs' in Figure 1.12.

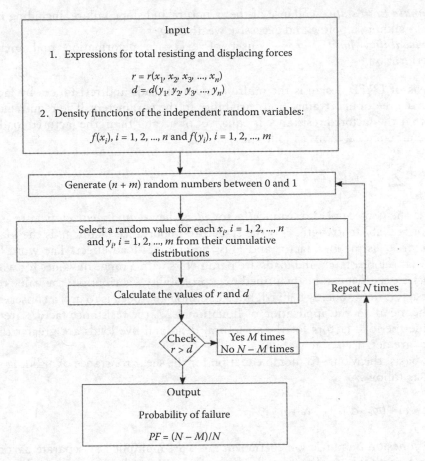

Figure 1.15 Flow chart for Monte Carlo simulation to calculate probability of failure of a slope (Athanasiou-Grivas, 1980).

1.4.6 Load and resistance factor design

The LRFD method is based on the use of probability theory to develop a rational design basis for structural design that accounts for variability in both loads and resistance. The objective is to produce a uniform margin of safety for steel and concrete structures such as bridges, and geotechnical structures such as foundations, under different loading conditions. The LRFD method has been developed in structural engineering and is becoming widely used in the design of major structures such as bridges (Canadian Standards Association, 1988; European Committee for Standardization, 1995; TRB, 1999; FHWA, 2011; AASHTO, 2012).

The LRFD method has been extended to geotechnical engineering, primarily for foundation design, in order to achieve consistency with the structural design of bridges and buildings (Fenton, Naghibi, Dundas et al., 2015).

Early LRFD work in geotechnical engineering was carried out by Meyerhof (1984) who used the term *limit states design*, and defined the two limit states as follows. First, the structure and its components must have, during the intended service life, an adequate margin of safety against collapse under the maximum loads that might reasonably occur. Second, the structure and its components must serve the designed functions without excessive deformations and deterioration. These two service levels are the ultimate and serviceability limit states, respectively, and are defined as follows:

- *Ultimate limit state* – collapse of the structure and slope failure, including instability due to sliding, toppling and excessive weathering
- *Serviceability limit state* – onset of excessive deformation and unacceptable deterioration

The basis of LRFD design is the multiplication of loads and resistances by factors that reflect the degree of uncertainty and variability in the parameters. The requirement of the design is that the factored resistance is equal to, or greater than, the factored loads. This is stated in mathematical terms as follows:

$$\varphi_g \cdot R \geq \sum (I_i \cdot \eta_i \cdot \alpha_i \cdot F_i) \tag{1.22}$$

where φ_g is the geotechnical resistance factor, R is the nominal geotechnical resistance, or the rock or soil shear strength, I_i is the structure importance factor, η_i is the load combination factor, α_i is the load factor and F_i is the nominal load effect. The word 'nominal' indicates that the resistance and loads are parameters with a range of values for which some uncertainty exists that can be quantified by a probability distribution; the value of R to be used in design is not explicitly defined, but a common approach is to use 'a conservative estimate of the mean'. In the application of Equation 1.22, the resistance factors are less than unity, while the load factors for the structural dead and live loads are greater than unity where they are detrimental to stability.

On this basis, the Mohr–Coulomb equation for the shear resistance of a sliding surface is expressed as follows:

$$\tau = (\varphi_c \cdot c) + (\varphi_\sigma \cdot \sigma') \cdot (\varphi_\phi \cdot \tan\phi) \tag{1.23}$$

where the cohesion c and friction coefficient $\tan\phi$ are multiplied by separate partial factors φ_c and φ_Φ, respectively, with values less than unity. For rock masses, the cohesion is usually more difficult to define than the friction angle, so φ_c would have a smaller value than φ_ϕ. The effective normal stress σ' on the sliding surface (weight of the sliding mass less any uplift water force) is related to the mass of the slope that is beneficial to shear resistance so the partial load factor φ_σ is greater than unity with the value depending on how precisely the dimensions of the slide can be defined.

At present (2016), LRFD methods for geotechnical design are mainly used for foundations as a component of structural design. The use of LRFD for slope design is limited by the difficulty in calibrating the design parameters because of their wide, and uncertain, range of values compared with the strengths of steel and concrete, and structural dead and live loads, for example. Also, service and ultimate limit states are difficult to apply to rock slopes where it is rarely possible to define an acceptable amount of movement.

As an alternate to LRFD design, it is possible for slope design to be carried by probabilistic analysis using explicit probability distributions for each design parameter and Monte Carlo simulation to calculate the probability of failure (or coefficient of reliability). The calculated probability of failure can be compared with the generally accepted risks for engineering projects shown in Figure 1.12.

An example of probabilistic design for a planar failure is shown in Section 7.6.

Chapter 2

Structural geology and data interpretation

2.1 OBJECTIVES OF GEOLOGICAL INVESTIGATION

The stability of rock slopes is often significantly influenced by the structural geology of the rock in which the slope is excavated. Structural geology refers to naturally occurring breaks in the rock such as bedding planes, joints and faults, which are generally termed *discontinuities*. The properties of discontinuities relevant to stability include orientation, persistence, spacing, roughness and infilling. The significance of discontinuities is that they are planes of weakness in the much stronger, intact rock so failure tends to occur preferentially along these surfaces. The discontinuities may directly influence stability such as the slopes shown in Figure 2.1. In Figure 2.1a, the face is formed by the bedding planes that are continuous over the full height of the cut – this condition is termed a *plane* failure and is discussed in detail in Chapter 7. Alternatively, slope failure may occur on two discontinuities that intersect behind the face (Figure 2.1b) – this condition is termed a *wedge* failure and is discussed in detail in Chapter 8.

Alternatively, discontinuities may only indirectly influence stability where their length is much shorter than the slope dimensions, such that no single discontinuity controls stability. However, the properties of the discontinuities will affect the strength of the rock mass in which the slope is excavated.

Almost all rock slope stability studies should address the structural geology of the site, and such studies involve two steps, as follows. First, determine the properties of the discontinuities, which involves mapping outcrops and existing cuts, if any, and examining diamond drill core, as appropriate for the site conditions. Second, determine the influence of the discontinuities on stability, which involves studying the relationship between the orientations of the discontinuity and the face. The objective of this study, which is termed *kinematic analysis*, is to identify possible modes of slope instability.

The overall purpose of geological mapping is to define a set or sets of discontinuities, or a single feature such as a fault, which will control stability on a particular slope. For example, the bedding may dip out of the face and form a plane failure, or a pair of joint sets may intersect to form a series of wedges. It is common that the discontinuities will occur in three orthogonal sets (mutually at right angles), with possibly one additional set. It is suggested that four sets are the maximum that can be incorporated into a slope design, and that any additional sets that may appear to be present are more likely to represent scatter in the orientation of the sets. Discontinuities that occur infrequently in the rock mass are not likely to have a significant influence on the stability of the overall slope, and only cause local instability, and so can be discounted in design. However, it is important to identify a single feature such as a through-going, adversely orientated fault that may be a controlling feature for stability.

(a)

(b)

Figure 2.1 (**See colour insert.**) Rock faces formed by persistent discontinuities: (a) plane failure formed by bedding planes in shale parallel to face with continuous lengths over the full height of the slope on Route 19 near Robbinsville, North Carolina; (b) wedge failure formed by two intersecting planes in sedimentary formation dipping out of the face on Route 60 near Phoenix, Arizona. (Image by C. T. Chen.)

In certain geological conditions, the discontinuities may be randomly orientated. For example, basalt that has cooled rapidly while still flowing may have a 'crackled' structure in which the joints are not persistent and have no preferred orientation. Also, some volcanic rocks exhibit sets that only extend over a few metres on the face, and then different sets occur in adjacent areas.

In the design of rock slopes on civil projects, it is usually advisable that the full length of each slope be designed at a uniform slope angle; it is not practical to excavate a slope with varying slope angles because this will complicate surveying and the layout of blast holes. This will require that in applying geological data to slope design the dominant geological structure, such as bedding or orthogonal joint sets, be used for the design. An exception to this guideline may be where a significant change in rock type occurs within the cut, and it may be appropriate to prepare a separate design for each. However, even in these circumstances, it may be more economical in terms of construction costs to cut the entire slope at the flatter of the two slope designs, or to install support in the less stable material.

Another issue in planning a geological mapping programme is to decide how many discontinuities should be mapped in order to define the design sets. It is usually possible, by inspection of a natural face or existing cut, to ascertain whether the structure occurs in sets or is randomly oriented. For locations where rock is well exposed and the structure is uniform, as few as 20 measurements should provide information on the orientation of the sets, with a further 50–100 measurements required to define typical properties such as persistence, spacing and infilling. Conditions in which a greater number of discontinuities should be mapped include faulted or folded structure, or contacts between different rock types. In these cases, several hundred features may need to be mapped in order to define the properties of each unit. A detailed procedure for determining the number of joints to be mapped is described by Stauffer (1966).

Structural geology is specifically addressed in two chapters in the book. Chapter 2 describes the properties of discontinuities and how they are used in kinematics analysis, and Chapter 4 discusses methods of collecting structural geological data, including mapping and drilling.

2.2 MECHANISM OF JOINT FORMATION

All the rock observed in outcrops and excavations has undergone a long history of modification over a time of hundreds of millions, or even billions of years. The sequence of modification is that a sedimentary rock, for example, typically undergoes deposition at the surface, gradual burial to depths of up to several kilometres with imposition of heat and pressure, and then uplifts to the surface (Figure 2.2). Throughout this sequence, the rock may also be subject to deformation comprising folding and faulting. These processes usually result in the stresses in the rock exceeding its strength a number of times, causing the rock to fracture and form joints and faults. In sedimentary rock, bedding planes that are coincident with breaks in the continuity of sedimentation will also form.

Figure 2.2a shows how the stresses in rock increase with burial, assuming that no water, thermal or tectonic pressures are acting in the rock (Davis and Reynolds, 1996). The vertical stress, which is the major principal stress σ_1, is equal to the weight of the overlying rock and is given by

$$\sigma_1 = \gamma_r \cdot H \tag{2.1}$$

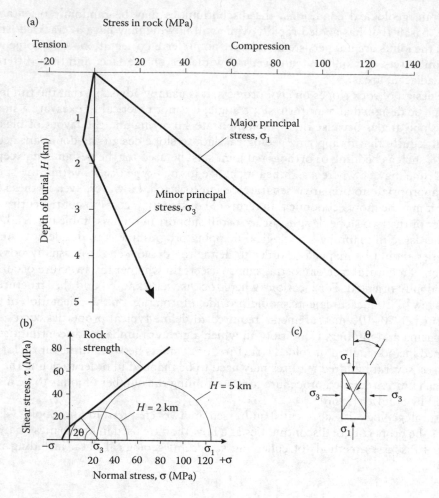

Figure 2.2 Development of jointing due to burial and uplift of rock: (a) stress changes in rock during burial; (b) Mohr diagram showing conditions for rock fracture; (c) inclination of joints with respect to stress direction (adapted from Davis and Reynolds, 1996).

where γ_r is the unit weight of the rock and H is the depth of burial. The horizontal stress, which is the minor principal stress σ_3, also increases with the depth of burial due to the effect of Poisson's ratio μ, and any temperature increase that occurs. In ideal conditions, σ_3 is related to σ_1 as follows:

$$\sigma_3 = \left(\frac{\mu}{1-\mu} \cdot \sigma_1 \right) + \left(\frac{E}{1-\mu} \cdot \varepsilon \cdot \Delta T \right) \qquad (2.2)$$

where E is the modulus of deformation of the rock, ε is the coefficient of thermal expansion and ΔT is the temperature rise. The first component of Equation 2.2 shows the value of the horizontal stress due to gravitational loading; if Poisson's ratio is 0.25, for example, then $(\sigma_3 = 0.33 \cdot \sigma_1)$. If the rock were not free to expand, and a temperature rise of 100°C occurs in a rock with a modulus value of 50 GPa (7.2E6 psi) and an ε value of 15E-6 per °C, then a thermal stress of 100 GPa (14.4E6 psi) will be generated. In reality, the values of the principal stresses will be modified by tectonic action resulting in deformation such as folding and faulting.

In Figure 2.2a, the value of σ_1 is defined by Equation 2.1, and the value of σ_3 varies with depth as follows. The value of σ_3 is tensile at depths less than 1.5 km where the sediments have not been consolidated into rock, and below this depth, σ_3 increases as defined by Equation 2.2, assuming no temperature change.

The formation of joints in rock during the burial–uplift process shown in Figure 2.2a will depend on the rock strength in comparison to the applied stresses. A method of identifying conditions that will cause the rock to fracture is to use a Mohr diagram (Figure 2.2b). In Figure 2.2b, the rock strength is shown as a straight line under compressive stress, and curved under tensile stress because microfractures within the rock act as stress concentrators that diminish the strength when a tensile stress is applied. On the Mohr diagram, circles represent the σ_1 and σ_3 stresses at different depths, and where the circle intersects the strength line, failure will occur. The stress conditions show that fracture will occur at a depth of 2 km ($\sigma_1 = 52$ MPa [7500 psi]; $\sigma_3 = 0$) because no horizontal confining stress is acting. However, at a depth of 5 km where the rock is highly confined ($\sigma_1 = 130$ MPa [18,900 psi]; $\sigma_3 = 25$ MPa [3620 psi]), the rock will not fracture because the strength exceeds the stress. For conditions where σ_3 is low or tensile (negative), failure occurs more readily compared to conditions at greater depth where both σ_1 and σ_3 are compressive (positive).

The Mohr diagram also shows the orientation of the fracture with respect to the stress direction (Figure 2.2c). Since principal stresses are oriented mutually at right angles, sets of joints tend to form in orthogonal directions.

2.3 EFFECTS OF DISCONTINUITIES ON SLOPE STABILITY

While the orientation of discontinuities is the prime geological factor influencing stability, and is the subject of this chapter, other properties such as persistence and spacing are significant in design. For example, Figure 2.3 shows three slopes excavated in a rock mass containing two joint sets: set J1 dips at 45°, and out of the face, and set J2 dips at 60° into the face. The stability of these cuts differs as follows. In Figure 2.3a, set J1, which is widely spaced and has a persistence greater than the slope height, forms a potentially unstable plane failure over the full height of the cut. In Figure 2.3b, both sets J1 and J2 have low persistence and are closely spaced so that, while small blocks ravel from the face, no overall slope failure occurs. In Figure 2.3c, set J2 is persistent and closely spaced, and forms a series of thin slabs dipping into the face that create a toppling failure.

The significance of Figure 2.3 is that, while an analysis of the orientation of joint sets J1 and J2 would show identical conditions on a stereonet, other characteristics of these discontinuities must also be considered in design. These characteristics, which are discussed further in Chapter 4, should be described in detail as part of the geological data collection programme for rock slope design.

2.4 ORIENTATION OF DISCONTINUITIES

The first step in the investigation of discontinuities in a slope is to analyse their orientation and identify sets of discontinuities, or single discontinuities that could form potentially unstable blocks of rock. Information on discontinuity orientation may be obtained from

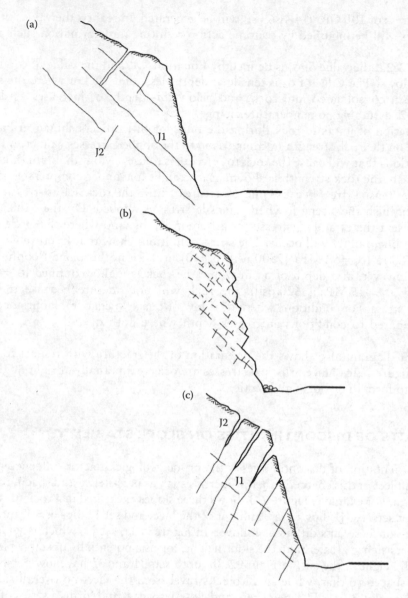

Figure 2.3 Effects of joint properties on slope stability: (a) persistent J1 joints dipping out of face form
potentially unstable sliding blocks; (b) closely spaced, low-persistence joints cause ravelling of
small blocks; (c) persistent J2 joints dipping into face form potential toppling slabs.

such sources as surface and underground mapping, diamond drill core and geophysics, and
it is necessary to combine these data using a procedure that is readily amenable to analysis.
This analysis is facilitated by the use of a simple and unambiguous method of expressing the
orientation of a discontinuity. The recommended terminology for orientation is the *dip* and
dip direction which are defined as follows, and shown schematically in Figure 2.4a and b.

1. *Dip* is the maximum inclination of a discontinuity to the horizontal (angle ψ).
2. *Dip direction* or *dip azimuth* is the direction of the horizontal trace of the line of dip,
   measured clockwise from north (angle α).

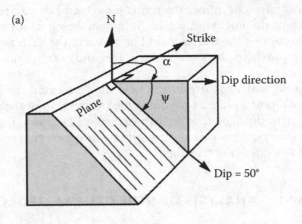

(a)

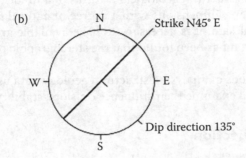

(b)

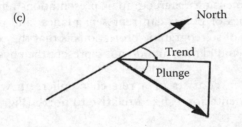

(c)

Figure 2.4 Terminology defining discontinuity orientation: (a) isometric view of plane (dip and dip direction); (b) plan view of plane; (c) isometric view of line (plunge and trend).

As will be demonstrated in Section 2.5, the dip/dip direction system facilitates field mapping, plotting stereonets, and analysis of discontinuity orientation data.

Strike, which is an alternative means of defining the orientation of a plane, is the trace of the intersection of an inclined plane with a horizontal reference plane.

The strike is at right angles to the dip direction, and the relationship between the strike and the dip direction is illustrated in Figure 2.4b, where the plane has a strike of N45°E and a dip of 50°SE. In terms of dip and dip direction, the orientation of the plane is 50/135°, which is considered to be a simpler nomenclature that also facilitates the use of stereographic analysis. By always writing the dip as two digits and the dip direction as three digits, for example, 076 for 76°, no confusion can arise as to which set of figures refers to which measurement. Strike and dip measurements can be readily converted into dip and dip direction measurements, if this mapping system is preferred.

In defining the orientation of a line, the terms *plunge* and *trend* are used (Figure 2.4c). The plunge is the dip of the line, with a positive plunge being below the horizontal and a negative plunge being above the horizontal. The trend is the direction of the horizontal projection of the line measured clockwise from north, and corresponds to the dip direction of a plane.

When mapping geological structure in the field, it is necessary to distinguish between the true and apparent dip of a plane. True dip is the steepest dip of the plane, and is always steeper than the apparent dip. The true dip can be found as follows. If a pebble or a stream of water is run down the plane, it will always fall in a direction that corresponds to the dip direction; the dip of this line is the true dip.

2.5 STEREOGRAPHIC ANALYSIS OF STRUCTURAL GEOLOGY

The previous sections describe structural geological features that influence rock slope stability. This data often occur in three dimensions with a degree of natural scatter, and in order to be able to use the data in design, it is necessary to have available an analysis technique that can address these matters. It has been found that the stereographic projection is an ideal tool for this application.

This section describes methods of analysing structural geology data using the stereonet to identify discontinuity sets, and examine their influence on slope stability.

2.5.1 Stereographic projection

The stereographic projection allows the three-dimensional orientation data to be represented and analysed in two dimensions. Stereographic presentations remove one dimension from consideration so that lines or points can represent planes, and points can represent lines. An important limitation of stereographic projections is that they consider only angular relationships between lines and planes, and do not represent the physical position or size of the feature.

The stereographic projection consists of a reference sphere in which its equatorial plane is horizontal, and its orientation is fixed relative to north (Figure 2.5). Planes and

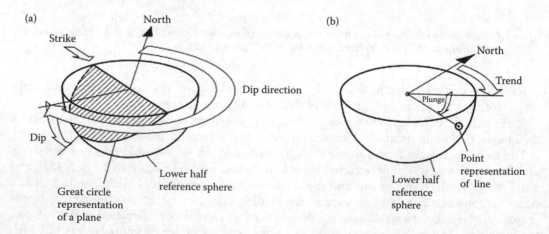

Figure 2.5 Stereographic representation of plane and line on lower hemisphere of reference sphere: (a) plane projected as great circle; (b) isometric view of line (plunge and trend).

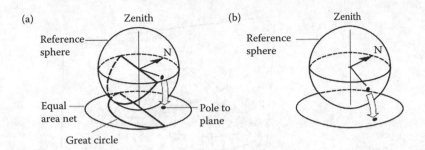

Figure 2.6 Equal area projections of plane and line: (a) plane projected as great circle and corresponding pole; (b) line projected as pole.

lines with a specific plunge and trend are positioned in an imaginary sense so that the axis of the feature passes through the centre of the reference sphere. The intersection of the feature with the lower half of the reference sphere defines a unique line on the surface of the reference hemisphere. For a plane, this intersection with the reference sphere is a circular arc called a great circle, while for a line, the intersection with the reference sphere is a point. In order to develop a stereographic projection of a plane or line, the intersection with the reference sphere is rotated down to a horizontal surface at the base of the sphere (Figure 2.6). The rotated lines and points are unique locations on the stereonet that represent the dip (plunge) and dip direction (trend) of the feature. In slope stability analysis using stereonets, planes are used to represent both discontinuities and slope faces.

An alternative means of representing the orientation of a plane is the pole to the plane (Figure 2.6a). The pole is the point on the surface of the reference sphere that is pierced by a radial line in a direction normal to the plane. The value of the pole projection is that a single point can represent the complete orientation of a plane. As described in Section 2.5.2, the use of poles facilitates the analysis of a large number of planes compared with the use of great circles.

As an aid to interpreting the information shown on stereonets, it can be seen from Figures 2.5 and 2.6 that planes and lines with shallow dips have great circles and points that plot near the circumference of the stereonet, and those with steep dips plot near the centre. In contrast, the pole of a shallow dipping plane plots close to the centre of the circle, and the pole of a steep plane plots close to the perimeter.

The two types of stereographic projections used in structural geology are the polar and equatorial projections as shown in Figure 2.7. The polar net can only be used to plot poles, while the equatorial net can be used to plot both planes and poles as described below. In the case of the equatorial projection, the most common type of stereonet projection is the equal area or Lambert (Schmidt) net. On this net, any point on the surface of the reference sphere is projected as an equal area on the stereonet. This property of the net is used in the contouring of pole plots to find concentrations of poles that represent preferred orientations, or sets of discontinuities. The other type of equatorial projection is the equal angle or Wulff net; both the Wulff and Lambert nets can be used to examine angular relationships, but only the Lambert net can be used to develop contours of pole concentrations.

Polar and equatorial stereonets are included in Appendix I in a size that is convenient for plotting and analysing structural data. For hand plotting of structural data, the usual procedure is to place tracing paper on the nets and then draw poles and planes on the tracing

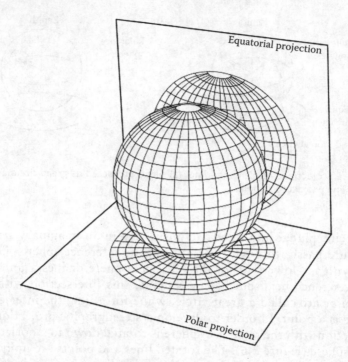

Figure 2.7 Polar and equatorial projections of a sphere.

paper as shown in Figure 2.8. Since plotting of great circles requires that the tracing paper be rotated on the net, as described below, a thumb tack is placed at the centre point so that the curves can be plotted without distortion.*

Further details on stereographic projections are described by Phillips (1971) who discusses the theoretical background to this technique, and Leyshon and Lisle (1996) who demonstrate applications of this technique to geological mapping. Goodman and Shi (1985) demonstrate stereographic techniques for identifying wedges of rock that can slide from the face, or are 'removable'; this technique is termed *key block theory*.

2.5.2 Pole plots and contour plots

The pole to a plane, as shown in Figure 2.6, allows a point to represent the orientation of the plane. Pole plots, in which each plane is represented by a single point, are the most convenient means of examining the orientation of a large number of discontinuities. The plot provides an immediate visual depiction of concentrations of poles representing the orientations of sets of discontinuities, and the analysis is facilitated by the use of different symbols for different types of discontinuities.

Poles can be plotted by hand on a polar net as shown in Figure 2.9. On the net, the dip direction scale (0–360°) around the periphery has the zero mark at the bottom of the vertical axis and the 180° mark is at the top of the net. This is a convenience for plotting such that poles can be plotted directly without the need for rotating the tracing paper; it can

---

* For occasions in remote hotel rooms when the computer is out of power and not available for preparing stereonets, it has been found that a television screen provides an adequate substitute. If the set is tuned to 'snow', the static electricity holds the papers to the screen and allows the tracing sheet to be rotated on the stereonet.

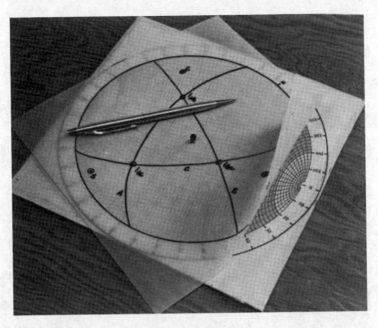

Figure 2.8 Geological data plotted and analysed on a piece of tracing paper that is located over the centre of the stereonet with a pin allowing the paper to be rotated.

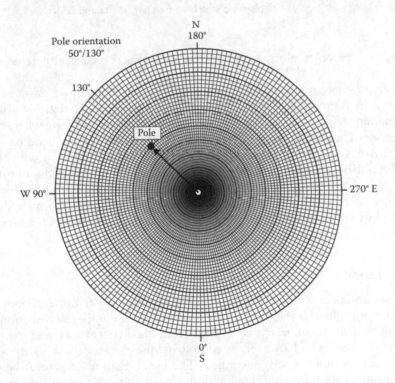

Figure 2.9 Plotting poles on a polar net. Plot pole of plane orientated at 50/130 – locate dip direction of 130° clockwise around the circumference of a circle starting at the lower end of the vertical axis. At 130° radial line, count 50° out from the centre of the net, and plot a point at the intersection between 130° radial line and 50° circle.

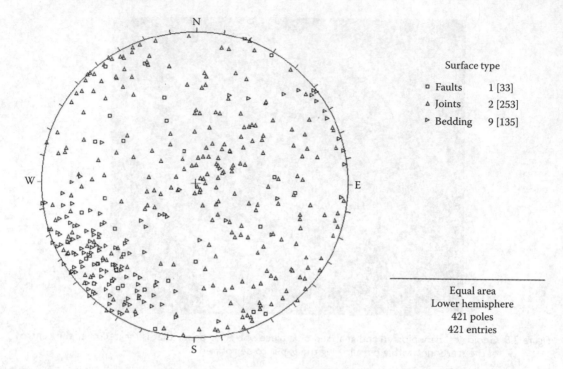

Surface type		
□ Faults	1	[33]
▵ Joints	2	[253]
▷ Bedding	9	[135]

Equal area
Lower hemisphere
421 poles
421 entries

Figure 2.10 Example of pole plot of 421 planes comprising bedding, joints and faults.

be demonstrated that poles plotted on both the polar and equatorial nets are in identical positions.

Pole plots are commonly generated by stereographic computer programs, an example of which is shown in Figure 2.10. This is a lower hemisphere, equal angle projection of 421 original poles mapped over an area of about a square kilometre, at a site where the rock is a bedded limestone. The rock contains discontinuity sets comprising the bedding and two sets of joints, together with a number of faults that are generally coincident with the bedding. In Figure 2.10, a different symbol is used for each of the three types of discontinuity. While considerable scatter occurs in the pole orientations, careful examination of this plot shows some clustering particularly in the southwest quadrant. In order to identify discontinuity sets on poles plots with considerable scatter, it is necessary to prepare contours of the pole density, as described in the next section.

2.5.3 Pole density

All natural discontinuities have a certain amount of variation in their orientations that results in scatter of the pole plots. If the plot contains poles from a number of discontinuity sets, it can be difficult to distinguish between the poles from the different sets, and to find the most likely orientation of each set. However, by contouring the plot, the most highly concentrated areas of poles can be more readily identified. The usual method of generating contours is to use the contouring package contained with most stereographic projection computer programs. Contouring can also be carried out by hand using a counting net such as the Kalsbeek net that consists of mutually overlapping hexagons, each with an area of 1/100 of the total area of the full stereonet (Leyshon and Lisle, 1996); a Kalsbeek counting net is shown in

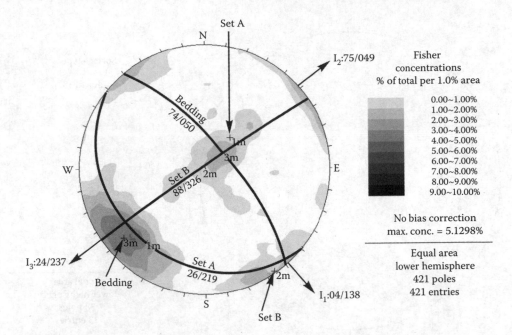

Figure 2.11 Contoured plot of data shown in Figure 2.10, with great circles corresponding to mean orientation of bedding and two orthogonal joint sets, and lines of intersection (I_n) between planes.

Appendix I. Contouring is carried out by overlaying the counting net on the pole plot and counting the number of poles in each triangle. For example, if eight poles out of a total of 421 poles occur in one triangle, then the concentration in that triangle is 2%. Once the percent concentration in each triangle has been determined, contours can be drawn.

Figure 2.11 shows a contour plot of the poles plotted in Figure 2.10. The contour plot shows that the orientation of the bedding has relatively little scatter – the maximum concentration is 5%–6%, and that the mean orientation of the bedding has a dip of 74° and a dip direction of 050°. In contrast, the joint orientations show more scatter, and on the pole plot it is difficult to identify discontinuity sets. However, on the contoured plot, it is possible to clearly distinguish two sets of orthogonal joints. Set A has a shallow dip of about 26°, and a dip direction of about 219° that is in a direction at 180° to the bedding. Set B has a near vertical dip and a dip direction of 326°, approximately at right angles to set A. The poles for set B occur in two concentrations on opposite sides of the contour plot because some dip steeply to the north-west and others steeply to the south-east.

In Figure 2.11, the different pole concentrations are shown by symbols for each 1% contour interval. The percentage concentration refers to the number of poles in each 1% area of the surface of the lower hemisphere.

A further use of the stereographic projection computer program in analysing structural data is to prepare plots of data selected from the total data collected. For example, joints with lengths that are only a small fraction of the slope dimensions are unlikely to have a significant influence on stability.

Faults usually have greater persistence and lower friction angle than joints. Therefore, it would facilitate design to prepare a stereographic plot showing only faults (Figure 2.12). This plot shows that 33 discontinuities are faults, and that their orientations are similar to those of the bedding. Selections can also be made, for example, of discontinuities that have a certain type of infilling, or are slickensided, or show evidence of seepage, provided that the

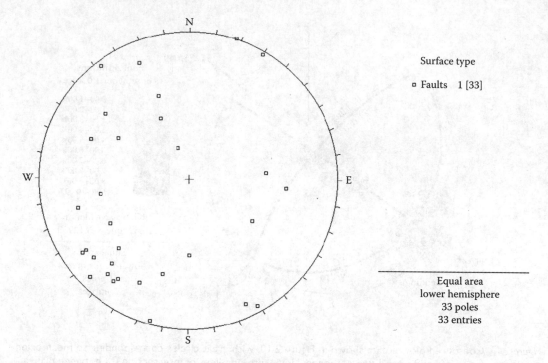

Surface type

□ Faults   1 [33]

Equal area
lower hemisphere
33 poles
33 entries

Figure 2.12 Pole plot of faults selected from data plotted in Figure 2.10.

mapping identifies this level of detail for each surface. Appendix II contains field mapping sheets for recording details of discontinuity properties by the use of codes that are input directly into the stereographic analysis computer program.

The assignment of poles into discontinuity sets is usually achieved by a combination of contouring, visual examination of the stereonet, and knowledge of geological conditions at the site, which will frequently show trends in orientation of the sets. It is also possible to identify discontinuity sets by rigorous and less subjective analysis of clusters in orientation data. A technique presented by Mahtab and Yegulalp (1982) identifies clusters from random distributions of orientations using Poisson's distribution. However, in applying such techniques, a result that identifies more than about four concentrations should be carefully examined before being used in design.

2.5.4 Great circles

Once the orientation of the discontinuity sets, as well as important single discontinuities such as faults, has been identified on the pole plots, the next step in the analysis is to determine if these discontinuities form potentially unstable blocks in the slope face. This analysis is carried out by plotting great circles of each of the discontinuity set orientations, as well as the orientation of the face. In this way, the orientation of all the surfaces that have an influence on stability is represented on a single diagram. Figure 2.11 shows the great circles of the three discontinuity sets identified by contouring the pole plot in Figure 2.10. It is usually only possible to have a maximum of five or six great circles on a plot, because with a greater number, it is difficult to identify all the intersection points of the circles.

While computer-generated great circles are convenient, hand plotting is of value in developing an understanding of stereographic projections. Figure 2.13 illustrates the procedure

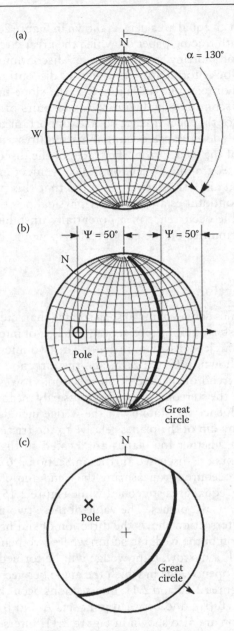

Figure 2.13 Construction of great circle and pole representing a plane with orientation 50° (dip)/130° (dip direction) on an equal area net: (a) with the tracing paper located over the stereonet by means of the centre pin, trace the circumference of the net and mark the north point. Measure off the dip direction of 130° clockwise from north and mark this position on the circumference of the net; (b) rotate the net about the centre pin until the dip direction mark lies on the W-E axis of the net, that is, the net is rotated through 40° counter clockwise. Measure 50° from the outer circle of the net and trace the great circle that corresponds to a plane dipping at this angle. The position of the pole, which has a dip of (90–50°), is found by measuring 50° from the centre of the net as shown, or alternatively 40° from the outside of the net. The pole lies on the projection of the dip direction line which, at this stage of the construction, is coincident with the W-E axis of the net; (c) the tracing is now rotated back to its original position so that the north mark on the tracing coincides with the north mark of the net. The final appearance of the great circle and the pole representing a plane dipping at 50° in a dip direction of 130° is as illustrated.

for drawing great circles on an equal area net. As shown in Figure 2.8, the procedure involves overlaying the stereonet with tracing paper on which the great circles are plotted.

The primary purpose of plotting great circles of discontinuity sets in a slope is to determine the shape of blocks formed by intersecting discontinuities, and the direction in which they may slide. For example, in Figure 2.1, the slope failures only occurred for conditions where single discontinuities (Figure 2.1a), or pairs of intersecting discontinuities (Figure 2.1b), dip out of the face. It is, of course, important to identify such potential failures before movement and collapse occurs. This requires an ability to visualise the three-dimensional shape of the wedge from the traces of the discontinuities on the face of the original slope. The stereographic projection is a convenient means of carrying out the required three-dimensional analysis, keeping in mind that this procedure examines only the orientation of the discontinuities and not their physical position or dimensions. If the stereonet shows the possible occurrence of a potentially unstable block, examination of the location of the discontinuities on the geological map would help to determine if they intersect the slope.

2.5.5 Lines of intersection

The intersection of two planes defines a line in space that is characterised by a trend (0–360°) and plunge (0–90°). In the stereographic projection, this line of intersection is defined as the point where the two great circles cross (Figure 2.14). The two intersecting planes may form a wedge-shaped block as shown in Figure 2.1b, and the direction in which this block may slide is determined by the trend of the line of intersection. However, the existence of two intersecting great circles on the stereonet does not necessarily mean that a wedge failure will occur. The factors that influence the stability of the wedge include the direction of sliding relative to the slope face, the dip of the planes relative to the friction angle, external forces such as ground water, and whether the planes are located so that they actually intersect behind the face. These factors are discussed further in Section 2.6.3.

Figure 2.14 shows the procedure for measuring the trend and plunge of the line of intersection of two planes on the equal area stereonet, while Figure 2.15 shows the procedure for measuring the angle between two planes. The value of these two measurements is that the orientation of the line of intersection shows the direction of sliding, and the angle between the planes gives an indication of the wedging action where two planes intersect. If the angle between the planes is small, a narrow, tight wedge will be formed with a higher factor of safety compared to a wide, open wedge in which the angle between the planes is large.

For the data shown in Figures 2.10 and 2.11, intersections occur between the bedding and joint set A (I_1) and joint set B (I_2), and between joint sets A and B (I_3). The orientations of the three lines of intersection are also shown in Figure 2.11. Intersection line I_3 has a trend of 237° and a plunge of 24°, and joint sets A and B could together form a wedge failure that would slide in the direction of the trend. Intersection line I_1 is almost horizontal (dip 4°) so the wedge formed by the bedding and joint set A is unlikely to slide, while intersection line I_2 is steep (dip 75°) near vertical and would form a thin wedge in the face.

2.6 IDENTIFICATION OF MODES OF SLOPE INSTABILITY

Different types of slope failure are associated with different geological structures and it is important that the slope designer be able to recognise potential stability problems during the early stages of a project. Some of the structural patterns that should be identified when examining pole plots are outlined in the following pages.

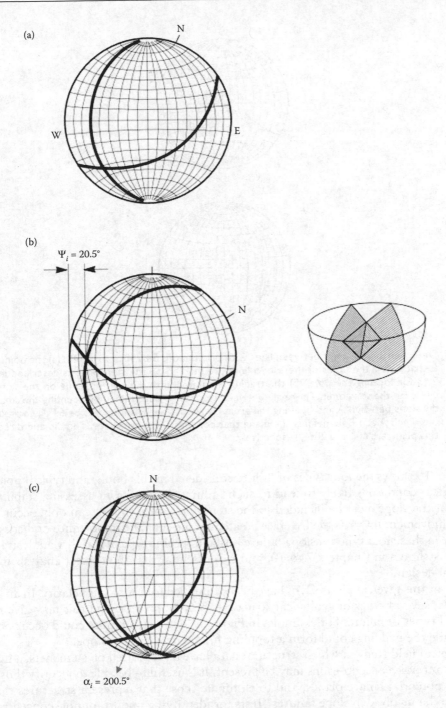

Figure 2.14 Determination of orientation (plunge and trend) of line intersection between two planes with orientations 50/130 and 30/250: (a) the first of these planes has already been drawn in Figure 2.13. The great circle defining the second plane is obtained by marking the 250° dip direction on the circumference of the net, rotating the tracing until the mark lies on the W-E axis and tracing the great circle corresponding to a dip of 30°; (b) the tracing is rotated until the intersection of the two great circles lies along the W-E axis of the stereonet, and the plunge of the line of intersection is measured as 20.5°; (c) the tracing is now rotated until the north mark coincides the north point on the stereonet and the trend of the lines of intersection is found to be 200.5°.

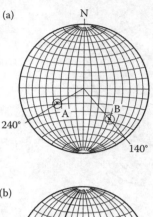

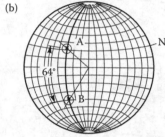

Figure 2.15 Determination of angle between lines with orientations 54/240 and 40/140: (a) the points A and B that define the poles of these two lines are marked on the stereonet as described in Figure 2.13 for locating the pole; (b) the tracing is rotated until the two poles lie on the same great circle on the stereonet. The angle between the lines is determined by counting the small circle divisions between A and B, along the great circle; this angle is found to be 64°. The great circle on which A and B lie defines the plane that contains these two lines. The dip and direction of this plane are 60° and 200°, respectively.

Figure 2.16 shows the four types of failure considered in this book, and typical pole plots of geological conditions likely to lead to such failures. Note that in assessing stability, the cut face of the slope must be included in the stereo plot since sliding can only occur as the result of movement towards the free face created by the cut. The importance of distinguishing between these four types of slope failure is that a specific type of stability analysis is used for each as shown in Chapters 7 to 10, and it is essential that the correct analysis method be used in design.

The diagrams given in Figure 2.16 have been simplified for the sake of clarity. In an actual rock slope, several types of geological structures may be present, and this may give rise to additional types of failure. For example, in Figure 2.11, a plane failure could occur on joint set A, while the bedding could form a toppling failure on the same slope.

In a typical field study in which structural data have been plotted on stereonets, a number of significant pole concentrations may be present. It is useful to be able to identify those that represent potential failure planes, and to eliminate those that represent structures that are unlikely to be involved in slope failures. Tests for identifying important pole concentrations have been developed by Markland (1972) and Hocking (1976). These tests establish the possibility of a wedge failure in which sliding takes place along the line of intersection of two planar discontinuities as illustrated in Figure 2.16b. Plane failure shown in Figure 2.16a is also covered by this test since it is a special case of wedge failure. For a wedge failure, contact is maintained on both planes and sliding occurs along the line of intersection between the two planes. For either plane or wedge failure to take place, it is fundamental that the dip

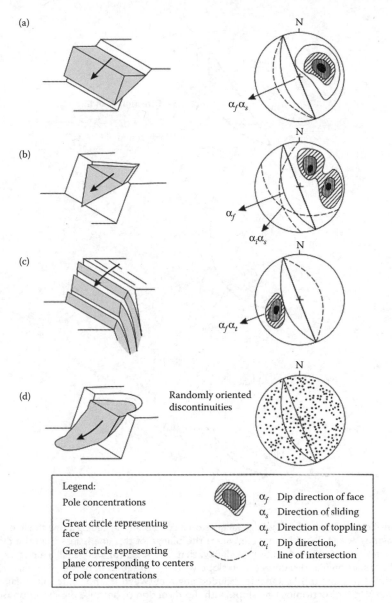

Figure 2.16 Main types of block failures in slopes, and structural geology conditions likely to cause these failures: (a) plane failure in rock containing persistent joints dipping out of the slope face, and striking parallel to the face; (b) wedge failure on two intersecting discontinuities; (c) toppling failure in strong rock containing discontinuities dipping steeply into the face; and (d) circular failure in rock fill, very weak rock or closely fractured rock with randomly oriented discontinuities.

of the sliding plane in the case of plane failure, or the plunge of the line of intersection in the case of wedge failure, be less than the dip of the slope face (i.e. $\psi_i < \psi_f$) (Figure 2.17a). That is, the sliding surface 'daylights' in the slope face.

The test can also differentiate between sliding of a wedge on two planes along the line of intersection, or along only one of the planes such that a plane failure occurs. If the dip directions of the two planes lie outside the included angle between α_i (trend of intersection line) and α_f (dip direction of face), the wedge will slide on both planes (Figure 2.17b). If the

(a)

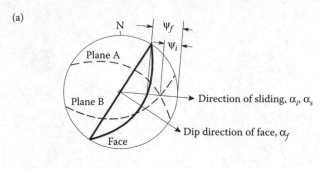

(b)

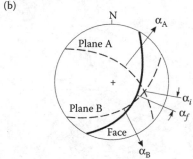

(c)

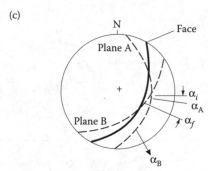

Figure 2.17 Identification of plane and wedge failures on a stereonet: (a) sliding along the line of intersection of planes A and B (α_i) is possible where the plunge of this line is less than the dip of the slope face, measured in the direction of sliding, that is $\psi_i < \psi_f$; (b) wedge failure occurs along the line of intersection (dip direction α_i) on slope with dip direction α_f because dip directions of planes A and B (α_A and α_B) lie outside included angle between α_i and α_f; (c) plane failure occurs on plane A (dip direction α_A) on slope with dip direction α_f because dip direction of planes A lies inside included angle between α_i and α_f.

dip direction of one plane (A) lies within the included angle between α_i and α_f, the wedge will slide on only that plane (Figure 2.17c).

2.6.1 Kinematic analysis

Once the type of block failure has been identified on the stereonet, the same diagram can also be used to examine the direction in which a block will slide and give an indication of stability conditions. This procedure is known as kinematics analysis. An application of kinematic analysis is the rock face shown in Figure 2.1b where two joint planes form a wedge which has slid out of the face and towards the photographer. If the slope face had been less steep than the line of intersection between the two planes, or had a strike at 90° to

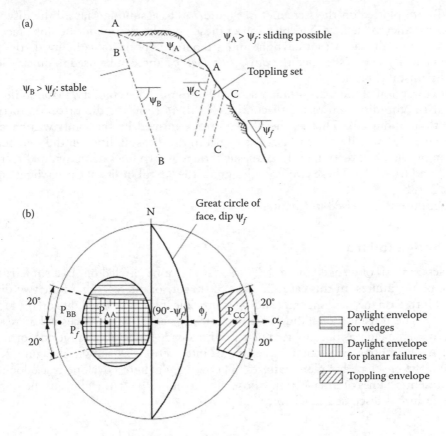

Figure 2.18 Kinematic analysis of blocks of rock in slope: (a) discontinuity sets in slope; and (b) daylight envelopes on equal area stereonet.

the actual strike, then although the two planes form a wedge, it would not have been able to slide from the face. This relationship between the direction that the block of rock will slide and the orientation of the face is readily apparent on the stereonet. However, while analysis of the stereonet gives a good indication of stability conditions, it does not account for external forces such as water pressures or reinforcement comprising tensioned rock bolts, which can have a significant effect on stability. The usual design procedure is to use kinematics analysis to identify potentially unstable blocks, followed by detailed stability analysis of these blocks using the procedures described in Chapters 7 to 10.

An example of kinematics analysis is shown in Figure 2.18, where a rock slope contains three sets of discontinuities. The potential for these discontinuities to result in slope failures depends on their dip and dip direction relative to the face; stability conditions can be studied on the stereonet as described below.

2.6.2 Planar failure

In Figure 2.18a, a potentially unstable planar block is formed by plane AA, which dips at a flatter angle than the face ($\psi_A < \psi_f$) and is said to 'daylight' on the face. However, sliding is not possible on plane BB which dips steeper than the face ($\psi_B > \psi_f$) and does not daylight. Similarly, discontinuity set CC dips into the face and sliding cannot occur on these planes, although toppling is possible. The poles of the slope face and the discontinuity sets (symbol P) are plotted on the stereonet in Figure 2.18b, assuming that all the discontinuities strike parallel to the face. The position of these poles in relation to the slope face shows that the poles of all planes that daylight and are potentially unstable lie inside the pole of the slope face. This area is termed the *daylight envelope* and can be used to quickly identify potentially unstable blocks.

The dip direction of the discontinuity sets will also influence stability. Plane sliding is not possible if the dip direction of the discontinuity differs from the dip direction of the face by more than about 20°. That is, the block of rock formed by the joints will have intact rock at one end that will have sufficient strength to resist instability. On the stereonet, this restriction on the dip direction of the planes is shown by two lines defining dip directions of $(\alpha_f + 20°)$ and $(\alpha_f - 20°)$. These two lines designate the lateral limits of the daylight envelope in Figure 2.18b.

Planar failure is discussed in Chapter 7.

2.6.3 Wedge failure

Kinematics analysis of wedge failures (Figure 2.16b) can be carried out in a similar manner to that of plane failures. In this case, the pole of the line of intersection of the two discontinuities is plotted on the stereonet and sliding is possible if the pole daylights on the face, that is, $(\psi_I < \psi_f)$. The direction of sliding of kinematically permissible wedges is less restrictive than that of plane failures because two planes with a wide range of orientations form release surfaces. A daylighting envelope for the line of intersection, as shown in Figure 2.18b, is wider than the envelope for plane failures. The wedge daylight envelope is the locus of all poles representing lines of intersection whose dip directions lie in the plane of the slope face.

Wedge failure is discussed in Chapter 8.

2.6.4 Toppling failure

For a toppling failure to occur, the dip direction of the discontinuities dipping into the face must be within about 20° of the dip direction of the face so that a series of slabs are formed parallel to the face. Also, the dip of the planes must be steep enough for interlayer slip to occur. If the faces of the layers have a friction angle ϕ_j, then slip will only occur if the direction of the applied compressive stress is at angle greater than ϕ_j with the normal to the layers. The direction of the major principal stress is parallel to the slope face (dip angle ψ_f), so interlayer slip and toppling failure will occur on planes with dip ψ_p when the following conditions are met (Goodman and Bray, 1976):

$$(90° - \psi_f) + \phi_j < \psi_p \tag{2.3}$$

These conditions on the dip and dip direction of planes that can develop toppling failures are defined in Figure 2.18b. The envelope defining the orientation of these planes lies at the opposite side of the stereonet from the sliding envelopes.

Toppling failure is discussed in Chapter 10.

2.6.5 Friction cone

Having determined from the daylight envelopes whether a block in the slope is kinematically permissible, it is also possible to examine the stability conditions on the same stereonet. This analysis is carried out assuming that the shear strength of the sliding surface comprises only

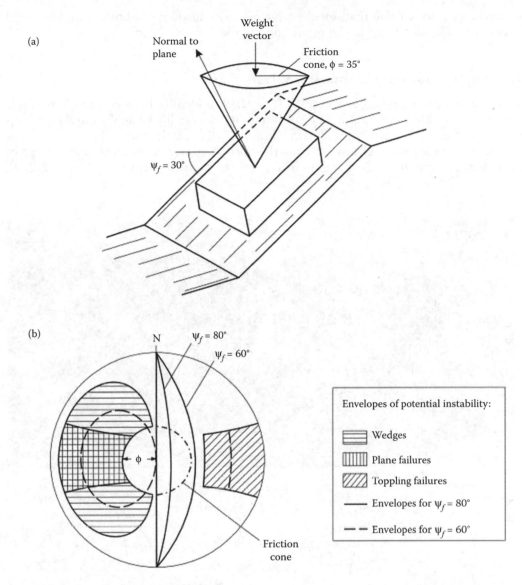

Figure 2.19 Combined kinematics and simple stability analysis using friction cone concept: (a) friction cone in relation to block at rest on an inclined plane (i.e. $\phi > \psi_p$); and (b) stereographic projection of friction cone superimposed on 'daylighting' envelopes.

friction and the cohesion is zero. Consider a block at rest on an inclined plane with a friction angle of ϕ between the block and the plane (Figure 2.19a). For an at-rest condition, the force vector normal to the plane must lie within the friction cone. When the only force acting on the block is gravity, the pole to the plane is in the same direction as the normal force, so the block will be stable when the pole lies within the friction circle.

The envelopes in Figure 2.19b show the possible positions of poles that may form unstable blocks. Envelopes have been drawn for slope face angles of 60° and 80° which show that the risk of instability increases as the slope becomes steeper as indicated by the larger envelopes for the steeper slope. Also, the envelopes of potential instability become larger as the friction angle diminishes showing that instability is more likely on low-friction-angle planes.

The envelopes also indicate that, for the simple gravity loading condition, instability will only occur in a limited range of geometric conditions.

2.6.6 Applications of kinematic analysis

The techniques demonstrated in Figures 2.16 to 2.19 to identify both potentially unstable blocks of rock on the slope and the type of instability can readily be applied to the preliminary stages of slope design. This is illustrated in the two examples that follow.

Highway: A proposed highway on a north-south alignment passes through a ridge of rock in which a through-cut is required to keep the highway on grade (Figure 2.20a).

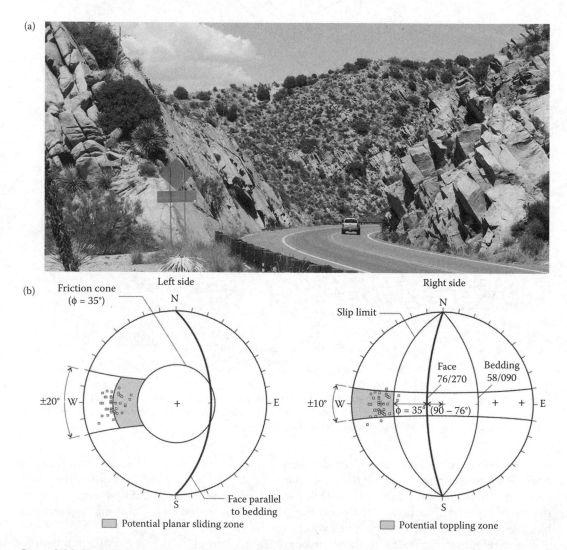

Figure 2.20 (**See colour insert.**) Relationship between structural geology and stability conditions on slope faces in through-cut: (a) photograph of through-cut showing two failure mechanisms – plane sliding on left side (west) and toppling on right side (east) on Route 60 near Globe, AZ (image by C. T. Chen); (b) stereographic plots showing kinematics analysis of left and right cut slopes.

Diamond drilling and mapping shows that the geological conditions in the ridge are consistent so that the same structure will be exposed in each face. The predominant geological structure is the bedding that strikes north-south, parallel to the highway alignment and dips to the east at angles of between 70° and 80° (i.e. dip/dip direction of 70–80°/090°).

The stereonets in Figure 2.20b show poles representing the dip and dip direction of the bedding, and great circles representing the orientations of the left and right cut faces. Also plotted on the stereonets is a friction cone representing a friction angle of 35° on the bedding. These stereonets show that on the left (west) face, the bedding dips towards the excavation at a steeper angle than the friction angle so sliding can occur on the bedding. The cut face has been made along the bedding to create a stable face.

On the right (east) face, the bedding dips steeply into the face and it is possible that the slabs formed by these joints will fail by toppling. According to Equation 2.3, toppling is possible if $[(90° - \psi_f) + \phi_j < \psi_p]$. If the face is cut at 76° (0.25V:1H) and the friction angle is 35°, then the left side of this equation equals 49°, which is less than the dip of the bedding (70–80°). The potential for toppling is indicated by the poles to the bedding planes lying inside the toppling envelope.

This preliminary analysis shows that the right (east) cut slope has potential toppling stability problems and that a more detailed investigation of structural geology conditions would be required before finalising the design. The first step in this investigation would be to examine the spacing of the bedding planes and determine if the centre of gravity of the slabs will lie outside the base, in which case toppling is likely (see Figure 1.11). Note that it is rarely possible to change the alignment sufficiently to overcome a stability problem, so it may be necessary to either reduce the slope angle on the right side, or stabilise the 76° face.

Quarry slopes: During the feasibility study for a proposed quarry, an estimate of safe slope angles is required for the preliminary pit layout. The only structural information that may be available at this stage is that obtained from the mapping of surface outcrops, and any oriented diamond drill cores that may be available. Although this information is limited, it does provide a basis for preliminary slope design.

A contour plan of the proposed quarry is presented in Figure 2.21 and contoured stereo plots of available structural data are superimposed on this plan. Two district structural regions, denoted by A and B, have been identified and the boundary between these regions has been marked on the plan. For the sake of simplicity, major faults have not been shown. However, it is essential that any information on faults should be included on large-scale plans of this sort and that the potential stability problems associated with these faults should be evaluated.

Overlaid on the stereonets are great circles representing the orientations of the east and west quarry faces, assuming an overall slope angle of 45°. Also shown on the stereonets is a friction cone of 30°, which is assumed to be the average friction angles of the discontinuity surfaces.

For the western and southern portions of the quarry, the stereonets show that the slopes are likely to be stable at the proposed slope of 45° because joint set B_2, which is oriented parallel to the slope, dips steeper than 45° and does not daylight in the face. This suggests that if the rock is strong and free of major faults, these slopes could probably be steepened or, alternately, this portion of the quarry wall could be used for the haul road with steep faces above and below the haul road.

However, the northeastern portion of the quarry contains a number of potential stability problems. The northern face is likely to suffer from plane sliding on discontinuity set A_1, since this set will daylight in the face at an angle steeper than the friction angle. Wedge

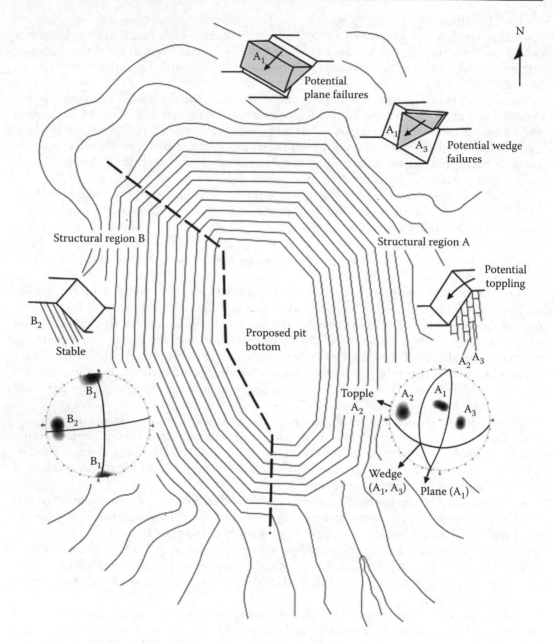

Figure 2.21 Presentation of structural geology on stereonets, and preliminary evaluation of slope stability of proposed quarry.

failures on the intersection of sets A_1 and A_3 are possible in the northeastern corner of the pit, and toppling failure on set A_2 may occur in the eastern slope.

The potential instability in the northeastern part of the proposed quarry would suggest that the slope angle would need to be less than 45° to reduce the possibility of planar and wedge instability.

It is interesting to note that three types of structurally controlled slope failure can occur in the same structural region, depending upon the orientation of the slope face.

2.7 GEOLOGIC STRUCTURE IN WEATHERED ROCK

Deep weathering of rock is common in tropical climates where a combination of consistently high temperatures (greater than 18°C), heavy rainfall (greater than 600 mm [24 in.] per year) and stable geology progressively alters the rock over time. The main process that causes weathering are chemical reactions that weaken the intact rock and eventually change the rock to a variety of clays and sandy soils. Other weathering processes are physical degradation that breaks the rock into smaller fragments with an increased surface area that enhances the chemical attacks, and biological processes such as root growth and production of humic acid from rotting vegetation (see Chapter 3). The depth of weathering can be tens to hundreds of metres.

The weathering process starts at the joint and bedding planes by reducing the rock strength on these surfaces, and progressively alters more of the intact rock until all the rock is converted to soil. This is an *in situ* process in which the original discontinuities are retained within the weathered rock as relic structure. The weathering process can be classified according to six grades from grade I – fresh rock to grade VI – residual soil (see Tables 3.1 and II.4 in Appendix II), and the geologic structure is retained in grades II to V.

Because most weathered rock, except the surficial residual soil, retains the relic geologic structure, it is important that these features are accounted for in the design of rock cuts in weathered rock. The orientation of the structure in the weathered rock will be identical to that of the fresh host rock because weathering is an *in situ* process. However, the shear strength of the discontinuities will be less than that in the fresh rock because the reduced strength on the joint surfaces will diminish the roughness component of the friction angle. Furthermore, as the weathering process develops, the rock on the joint surfaces may be converted to clays that form a low-shear-strength clay infilling, and the persistence of the discontinuities may increase as incipient joints are opened and lengthened. As a consequence of these conditions, structurally controlled instability in weathered rock may be more severe than in fresh rock.

If a cut is planned in weathered rock, it is important to investigate the relic geologic structure, with particular attention being paid to recovering low-strength infillings. Investigation methods may include test pits for shallow cuts, Standard penetration test (SPT) sampling in highly and completely weathered rock, or triple tube drilling, possibly with plastic core liners inside the split inner tube, in the less weathered rock. In cases where laboratory testing is to be carried out, the *in situ* moisture content of the sample should be maintained by sealing the core in plastic wrap and wax as soon as it is recovered.

EXAMPLE PROBLEM 2.1: STEREO PLOTS OF STRUCTURAL GEOLOGY DATA

Statement

A structural geology mapping programme for a proposed highway produced the following results for the orientation of the discontinuities (format: dip/dip direction):

40/080	45/090	20/160	80/310	83/312	82/305
23/175	43/078	37/083	20/150	21/151	39/074
70/300	75/305	15/180	80/010	31/081	

Required

1. Plot the orientation of each discontinuity as a pole on a stereonet using the equal area polar net and tracing paper.
2. Estimate and plot the position of the mean pole of each of the three sets of discontinuities.

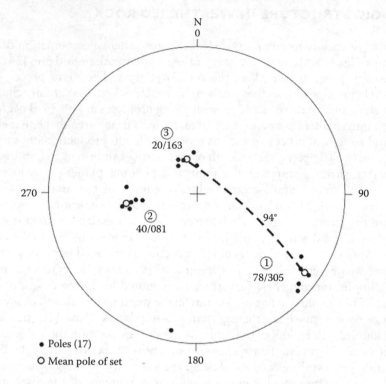

Figure 2.22 Example Problem 2.1 – plot of poles of structural mapping data (lower hemisphere projection).

3. Determine the angle between the mean poles with the steepest and shallowest dips.
4. Plot the great circles of the mean pole of each set on the equal area equatorial net.
5. Determine the plunge and trend of the line of intersection between the joint sets with the steepest and intermediate dip angles.
   Note: Stereonets are provided in Appendix I.

Solution

1. The poles of the 17 planes plotted in Figure 2.22 show that the discontinuities are oriented in three sets.
2. The mean pole of each discontinuity set is as follows:

   Set 1: 78/305;   *Set 2:* 40/081;   *Set 3:* 20/163

   The one pole oriented at 80/010 does not belong to any of the three discontinuity sets.
3. The angle between the mean poles of joint sets 1 and 3 is determined by rotating the stereonet until both mean poles lie on the same great circle. The number of divisions on this great circle is 94° as shown by the dotted line in Figure 2.22.
4. The great circles of the three mean poles are plotted in Figure 2.23.
5. The plunge and trend of the line of intersection of joint sets 1 and 2 are also shown in Figure 2.23. The values are as follows:

   Dip, $\psi_i = 27$;   dip direction, $\psi_i = 029$

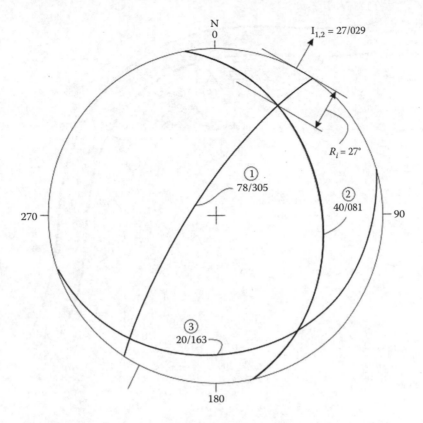

Figure 2.23 Example Problem 2.1 – great circles of mean poles plotted using equatorial net.

EXAMPLE PROBLEM 2.2: SLOPE STABILITY EVALUATION RELATED TO STRUCTURAL GEOLOGY

Statement

In order to make a 90° curve in a highway, a rock cut has been excavated that follows the curve of the highway, the face is cut at an angle of 50° (Figure 2.24).

The joints in the rock at the site form three sets with the following orientations:

Set 1: 78/305;   Set 2: 40/081;   Set 3: 20/163

Required

1. On a piece of tracing paper, draw a great circle representing the 50° slope face and a 25° friction circle.
2. Determine the most likely mode of failure, that is, plane, wedge or toppling, on the following slopes:
   - East-dipping slope
   - North-dipping slopes
3. State the joint set or sets on which sliding would occur on each slope.
4. Determine the steepest possible slope angle for these two slopes assuming that only the orientation of the discontinuities and the friction angle of the surfaces have to be considered.

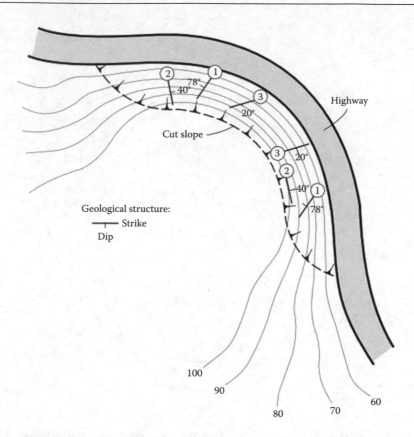

Figure 2.24 Cut slope and geological structures in cut face.

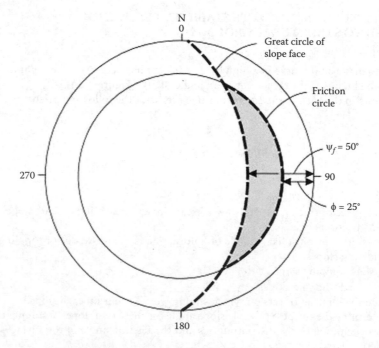

Figure 2.25 Great circle of slope dipping at 50° and 25° friction circle.

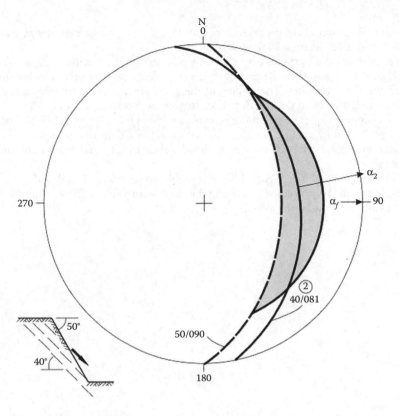

Figure 2.26 Stability conditions on east-dipping slope – plane failure on joint set 2.

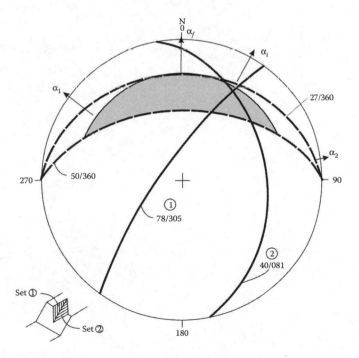

Figure 2.27 Stability conditions on north-dipping slope – wedge failure on joint set 1 and joint set 2.

Solution

1. Figure 2.25 shows the great circle of the slope dipping to the east at a face angle of 50°, and a 25° friction cone plotted as a circle.
2. The stability evaluation is carried out by placing first the tracing of the great circles (Figure 2.23) and then the tracing of the slope face and friction circle (Figure 2.25) on the equal area net. The tracing of the slope great circle is rotated to the corresponding orientation of the slope face to give the following results:
3. East-dipping slope: Plane failure possible on joint set 2 (Figure 2.26). Sliding could be prevented by cutting the slope at 40° coincident with the joint surfaces. If the friction angle had been greater than 40°, then sliding may not occur on these planes.
4. North-dipping slope: Wedge failure possible on joint sets 1 and 2 (Figure 2.27). Sliding could be prevented by cutting the slope at an angle of 27° so that the wedges are not undercut by the slope.

Chapter 3

Weathered rock and slope stability

3.1 WEATHERING PROCESSES AND WEATHERED ROCK OCCURRENCE

This chapter discusses the processes that cause rock to weather, the classification of weathered rock and its occurrence around the world. Also discussed are typical profiles of weathered rock in a variety of geologic conditions, and how slope stability is influenced by the extent of weathering. The importance of addressing the design of slopes in weathered rock is that these conditions occur throughout the tropical regions of the world, and in these areas, it is necessary to use slope designs that are specifically suited to the properties of weathered rock. While these designs share the same principles as those for fresh rock that are discussed in this book, the design methods for fresh and weathered rock are sufficiently different to warrant the separate treatment in this chapter.

Chapter 2 discusses structural geology, that is, the orientation and characteristics of discontinuities in strong rock, because it is these features that have a significant influence on slope design for civil excavations where the rock strength is generally greater than the state of stress in the slope. These conditions usually occur in areas that have been previously glaciated such that the weak, surficial rock has been removed by the scraping action of ice to expose fresh rock. However, in tropical areas with intense seasonal rainfall events and high temperatures, physical, chemical and biological processes acting over millions of years cause weathering of the rock that progressively changes fresh rock to residual soil. Since this is a progressive process, the degree of weathering decreases with depth and the rock properties change accordingly.

3.1.1 Influence of rock weathering on slope design

The design of slopes in weathered rock must take into account changes in the weathering profile with depth, and the corresponding properties of the rock in each class of weathering; weathering classes are defined in Table 3.1. This condition means that slope designs may also have to change with depth such that flatter slopes may be required in the upper part of the slope compared with steeper angles at greater depths. Furthermore, the designs must make allowance for the gradational contacts between weathering classes, and variations over short distances of the elevation of these contacts. Another issue to consider is that weathered rock, except for highly weathered residual soil, often contains remnant geological structure that can result in structurally controlled instability similar to that in fresh, jointed rock. Furthermore, if the rock will continue to weather during the life of the project, the design may have to consider that the rock mass strength will decrease with time.

Figure 3.1 shows an excavated slope in a weathered rock profile in which the upper face is in highly weathered rock with little evidence of the original rock, while the lower face

Table 3.1 Weathering and alteration grades

Grade	Term	Description
I	Fresh	No visible sign of rock material weathering; perhaps slight discolouration on major discontinuity surfaces
II	Slightly weathered	Discolouration indicates weathering of rock material and discontinuity surfaces. All the rock material may be discoloured by weathering and may be somewhat weaker externally than in its fresh condition
III	Moderately weathered	Less than half of the rock material is decomposed and/or disintegrated to a soil. Fresh or discoloured rock is present either as a continuous framework or as corestones
IV	Highly weathered	More than half of the rock material is decomposed and/or disintegrated to a soil. Fresh or discoloured rock is present either as a discontinuous framework or as corestones
V	Completely weathered	All rock material is decomposed and/or disintegrated to soil. The original rock mass structure is still largely intact
VI	Residual soil	All rock material is converted to soil. The mass structure and material fabric are destroyed. A large change (decrease) in volume occurs, but the soil has not been significantly transported

Source: Adapted from International Society for Rock Mechanics (ISRM). 1981a. Suggested methods for the quantitative description of discontinuities in rock masses. In: E.T. Brown, ed., Pergamon Press, Oxford, UK, 211pp.

Figure 3.1 Cut in weathered rock showing core stones of slightly weathered rock in matrix of highly weathered rock (Recife, Brazil).

contains core stones of fresh, slightly weathered rock in a matrix of highly weathered rock. This cut face shows the typical variability of weathered rock profiles.

Since most weathered rock occurs in tropical areas where high-intensity rainfall occurs, the low-strength rock and residual soil are subject to erosion that can rapidly scour substantial gullies in the slopes, and cause slope failures. Therefore, an important component of slopes designs in weathered rock is the provision of lined ditches to collect and discharge surface runoff, as well as vegetation to protect the surface from erosion.

3.1.2 Residual soil/weathered rock in slope engineering

The sequence of weathering encompasses residual soil at the surface to progressively fresher rock at depth. In tropical areas, the most frequent occurrences of slope instability, sometimes resulting in property damage and loss of life, are in the residual soil to a depth of 2–3 m (6–10 ft). Consequentially, residual soil and slope stability problems in this material have been studied extensively by the soil mechanics community (Huat, Toll and Prasad, 2012).

Because engineering methods in residual soil are well established, and because it is a soil mechanics issue that is outside the scope of this book, it has been decided that residual soil will not be addressed in detail in this book. However, the book does address slope engineering in weathered rock, that is, weathering grades II–V as shown in Table 3.1, in which substantial cuts may be required for many engineering projects. An important design issue for these slopes is the differing shear strength of the weathered rock grades, and how these may decrease during the life of the project. The shear strength of rock masses, both fresh and weathered, is discussed in Sections 5.5 and 5.6.

3.1.3 Classification of weathered rock

For projects in weathered rock, it is helpful to have a method of describing both the extent of weathering, as well as the characteristics of each class of weathered rock that are useful for engineering purposes. Furthermore, the descriptions of the weathered rock should be based on field observations and simple tests, and use terminology that is both unambiguous and used elsewhere in the description of geological materials. Appendix II lists the standardised terms that have been adopted by the International Society of Rock Mechanics for these descriptions. For example, intact rock strength is defined by seven categories ranging from 'extremely strong' ($\sigma_{ucs} > 250$ MPa [36,000 psi]) to 'extremely weak' ($\sigma_{ucs} = 0.25$–1 MPa [36–145 psi]), while the fabric of the rock can be characterised as 'blocky' or 'tabular' (International Society for Rock Mechanics, 1981). These are unambiguous terms that are used throughout the geological engineering community. Use of these terms avoids ill-defined words such as 'sound' or 'broken'.

Another objective of the classification system is that it should be universally applicable, that is, to all, or most, geological conditions. This is achieved by the use of expressions that describe the rock mass in terms of engineering characteristics such as strength, rather than geological terms such as lithology. For example, a strong sandstone and a strong schist can perform similarly if they have comparable characteristics with respect to weathering and discontinuities.

Finally, the value of a standardised classification system for weathered rock is that it facilitates communication between professionals because they all have the same frame of reference in characterising site conditions. Furthermore, the classification can be used to compare conditions at different sites.

Many field geologic classifications of rock weathering have been developed for different areas of the world with tropical climates such as Brazil (Branner, 1895), Hong Kong (Brock, Grount and Swanson, 1943) and Nigeria (Thomas, 1966), and for different purposes such as geology studies and geological engineering. Deere and Patton (1971) summarise a total of 12 classification systems spanning 17 years between 1953 and 1969, and propose a classification system that incorporates both the earlier classification systems and engineering properties. Deere and Patton proposed three weathering classes: *I* – residual soil; *II* – weathered rock and *III* – unweathered rock, with sub-classes within each class. A further classification criterion has been proposed by Vaz (1998) that incorporates the excavation method – diggable with a 'blade', or requiring blasting.

In order to unify these classification systems into a single system that can be widely/universally applied, the International Society of Rock Mechanics (ISRM, 1981b) has proposed

(a)   (b)

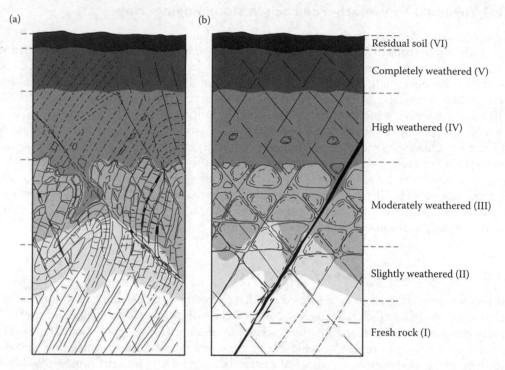

Residual soil (VI)

Completely weathered (V)

High weathered (IV)

Moderately weathered (III)

Slightly weathered (II)

Fresh rock (I)

Figure 3.2 Definition of weathering grades encompassing fresh, unweathered rock (Grade I) to completely weathered residual soil (Grade VI): (a) folded metamorphic formation; (b) granite with orthogonal jointing (Deere and Patton, 1971).

the classification system shown in Table 3.1, with the corresponding illustration of these grades depicted in Figure 3.2. The illustration shows how weathering of the fresh rock (Grade I) is initiated at discontinuities where water flow occurs (Grade II), with this progressing until more than half the rock has decomposed to soil and the original rock remains only as core stones (Grades III and IV). In Grade V, all the rock is decomposed to soil except for occasional core stones, but the original structure of the rock mass is retained, while Grade VI is entirely soil with no remnant of the rock remaining.

It is considered that the ISRM classification system is generally applicable to a wide range of geological conditions, and readers are encouraged to use the system in order to standardise descriptions of weathered rock.

3.1.4 Occurrence of weathered rock

Since the formation of weathered rock is related to climatic conditions that occur in the tropics, the occurrence of weathered rock is generally restricted to equatorial areas of the world between latitudes 20°N and 20°S of the equator as shown in Figure 3.3. That is, deep weathering is found in Central America (e.g. southern Mexico and Panama), northern South America (e.g. Venezuela and Brazil), Central Africa (e.g. Congo and Uganda) and South-east Asia (e.g. southern India, Thailand and northern Australia). An exception to this condition is the occasional existence, in present temperate climates, of remnant areas of ancient weathered rock that have escaped glaciation.

The tropical climatic conditions that cause rock weathering are temperatures that are consistently above 18°C (64°F), and precipitation levels of at least 600 mm (24 in.) per month

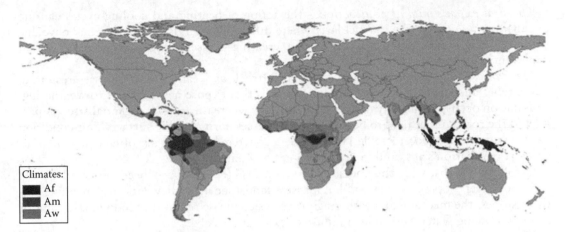

Figure 3.3 Locations of weathered rock, in tropical climate zones around the world (Peel et al., 2007).

related to warm, moist and unstable air masses and high humidity. The three classes of tropical climate related to the annual rainfall are as follows (Huat, Toll and Prasad, 2012):

1. *Tropical wet (Af)* – abundant rainfall in every month of the year
2. *Tropical monsoonal (Am)* – short dry season but a very rainy wet season
3. *Tropical savannah (Aw)* – longer dry season and prominent but not particularly rainy wet season

Another criterion for significant weathering is long-term stability that allows sufficient time for the progressive weathering processes to develop. As a result of this requirement, significant weathered rock does not generally occur in active volcanic areas where bedrock is buried in new formations before much weathering can develop.

As a consequence of the influence of climate and geological stability on rock weathering, the extent of weathering will be greatest in the hottest, wettest climates with low angle, stable slopes where the weathering process acts for an extended period of time. Under these conditions, the depth of weathering can extend to hundreds of metres.

3.2 WEATHERING PROCESSES

The essential components of weathered rock are first, that bedrock is exposed to long-term weathering processes that are sufficiently active to cause alteration of the rock, and second that these processes occur with no transportation of the weathered products such that the weathered rock remains *in situ*. Weathering is a combination of progressive processes in which physical and chemical changes act continuously so that the rock becomes increasingly weathered with time, and to a greater depth. The following sections discuss the processes that can cause weathering of fresh rock to the grades of weathered rock and residual soil shown in Figure 3.2.

3.2.1 Physical weathering

Physical weathering of the host rock involves mechanical processes that break the rock into smaller fragments, thus creating additional surfaces on which weathering can occur. Typical physical weathering processes are discussed below.

Thermal expansion and contraction – the action of heating and cooling cycles causing expansion and contraction, particularly where different minerals within the rock have dissimilar coefficients of expansion/contraction, creating stresses along grain boundaries that can eventually fracture the rock.

Mechanical exfoliation – removal of surficial rock as weathering progresses results in stress relief and the formation of exfoliation slabs that expose new surfaces to weathering. Exfoliation occurs commonly in granitic batholiths where the *in situ* tangential stresses parallel to the rock surface tend to be higher than stresses normal to the surface. This condition can be widely observed in Rio de Janeiro where concrete buttresses are often constructed to support exfoliation slabs, and in granite kopjes of Zimbabwe.

Exfoliation slabs, with thicknesses of 1–2 m (3–6 ft), have also been observed by the writer on steep slopes in very weak but massive sandstone in Stanley Park in Vancouver. On these slopes, the mechanical weathering processes comprise removal of individual grains of sand by flowing water, and sudden failure of exfoliation slabs.

Wetting and drying – the presence of water in the rock introduces water molecules between mineral grains causing swelling, and shrinkage when drying occurs that can develop tensional stresses in the rock mass. When the material dries, slaking of the loosened grains is possible.

Scour – the flow of water over the rock surface and through the rock mass removes loose particles of weathered rock, exposing new surfaces to weathering.

3.2.2 Chemical weathering

Chemical weathering involves chemical reactions between water and the minerals in the host rock resulting in either dissolution of the minerals, or the formation of new compounds. This type of weathering is, by far, the greatest cause of rock degradation.

The rate and type of weathering depends on temperature, with high temperatures in tropical regions increasing the rate of chemical reactions, and on the presence of dissolved compounds in the readily available water, such as carbon dioxide, to form carbonic acid, and ammonia, chlorides and sulphates. Chemical reactions also occur more readily with small grains that have a relatively large surface area, and with minerals that have formed at high temperatures and pressures and are now far from their equilibrium state, such as olivine and plagioclase (see Figure 3.5).

Chemical weathering reactions involve selective attack of constituent rock minerals that weaken the rock mass by creating new, weaker compounds, developing internal stresses where the reactions cause expansion or removing soluble components of the rock. The rate of weathering and the constituents of the resulting weathered rock are dependent on the type of host rock.

3.2.3 Biological weathering

Biological processes contribute to weathering in the upper part of the regolith (heterogeneous, loose surficial material) weathered profile. The processes include the formation of humic acids from the decay of vegetation, as well as the actions of plants and animals that aerate the soil and enhance the flow and circulation of water within the soil and rock mass. These conditions also promote chemical weathering.

3.2.4 Topography

The slope angle and its orientation can influence the degree of weathering. For example, on slopes steeper than about 20°, both erosion and slides in the upper 2–3 m (7–10 ft) of residual soil that can occur during heavy rainfall events, will remove the most weathered material and expose fresher rock to renewed weathering action. The weathering products produced by these events are deposited and accumulate on the flatter slopes below. In contrast, in flat terrain where little disturbance of the surficial material occurs, weathering will occur continuously, with the result that the depth of weathering in flat terrain may be greater than in steep terrain.

Furthermore, for slopes that are oriented such that they receive greater quantities of precipitation and sunshine, the depth of weathering will exceed slopes that are relatively sheltered from rainfall and sunshine. That is, relatively sheltered slopes are those in rainfall 'shadow' areas, and slopes that face south in the southern hemisphere, or north in the northern hemisphere.

3.2.5 Duration

While the rate of weathering is clearly related to processes discussed in the previous paragraphs, a guideline on the rate of weathering is that acid rocks (e.g. granite, rhyolite) weather more slowly than basic rocks (e.g. basalt), partly because acid rocks contain more silica that is highly resistant to weathering. As a guideline on the rate of weathering, measurements of the mean lifetime of 1 mm of fresh rock in humid, tropical climates is 20–70 years for acid rock, and 40 years for basic rock (Nahon, 1991).

3.3 LITHOLOGY: ROCK AND SOIL PRODUCTS OF WEATHERING

The weathering processes, and the products formed by these processes, are closely related to the chemical composition of minerals forming the host rock. The two most abundant elements in the solid crust of the earth are oxygen (47.3%) and silicon (27.7%). These two elements can combine in the form of a silica tetrahedron containing one silicon atom and four oxygen atoms, and the tetrahedra themselves combine such that oxygen atoms are shared with adjacent tetrahedra with the overall composition SiO_2 (Figure 3.4). The mineral formed by this configuration is quartz, which is an abundant mineral that forms 12% of igneous rock.

The two next most abundant minerals in the earth's crust are aluminium (7.8%) and iron (4.5%). Aluminium ions (Al^{3+}) are an important component of the crystal structure of many clays such as kaolinite and the smectites, while the weathering of iron produces the familiar red colour of weathered soils. Possible weathering products of iron are ferric oxide (Fe_2O_3) which is red coloured, and the less common hydroxyl form $Fe(OH)_2$ that is green coloured.

Silica tetrahedra shown in Figure 3.4 combine to create frameworks of stable crystal structures that form at a relatively low temperature of 573°C and for this reason quartz itself is a very stable mineral that is highly resistant to weathering. The fact that beach sand is composed primarily of quartz is a demonstration of the resistance of this mineral to degradation and abrasion.

Silica tetrahedra can also combine into other crystal structures in which cations (group of ions with positive charge) of new elements are incorporated with, or displace the silicon, to form a wide variety of minerals. Examples of these minerals and their corresponding structures and composition are as follows (Mitchell, 1976):

- *Olivine* – island structure containing Mg^{2+} and Fe^{3+} cations.
- *Bentonite and beryl* – ring structures containing Ba^+ and Ti^+ cations (bentonite) and Be^{3+} and Al^{2+} cations (beryl).

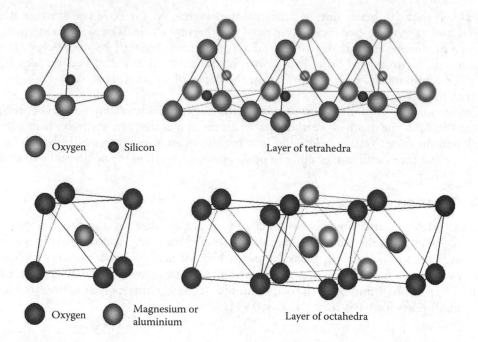

Figure 3.4 Silica tetrahedron, and tetrahedra arranged in a hexagonal network.

- *Pyroxenes* – chain structures containing Ca^+ and Mg^+ cations (e.g. diopside).
- *Amphiboles* – band structures containing Ca^+ and Mg^+ cations as well as hydroxyl (OH^{2+}) groups.
- *Mica* – sheet structures that form micas, chlorite, talc and clays. Crystals containing K^+ and Al^{2+} cations and hydroxyl groups form muscovite (white mica), while crystals that also contain Mg^{3+} and Fe^{3+} form biotite (dark coloured mica).

All these minerals occur widely in rock and their characteristics are altered by the physical, chemical and biological weathering processes that break down the rock mass as described in Section 3.2. The dominant weathering process is chemical which depends upon the presence of water to develop hydrolysis and the reaction between the ions of the minerals in the rock and the H^+ and $(OH)^-$ ions of water. The small size of the H^+ ion enables it to enter the lattice of minerals and replace existing cations. Hydrolysis will not take place in the presence of static water because the reaction is driven by the removal of soluble materials resulting in leaching, complexing, adsorption or precipitation, as well as the continual introduction of new H^+ ions.

Some of the products of these processes are the following three types of common sheet-type clay minerals, some of which have swelling characteristics. The general cause of swelling is the easy penetration of water between adjacent oxygen ions of the silica tetrahedra sheets, causing the individual layers of clay particles to separate and swell.

- *Kaolinite* is a very common clay formed from feldspars and micas by leaching of acid, quartz-rich rocks such as granite. Kaolinite is non-swelling.
- *Smectite* minerals are also very common, the parent materials for which are rocks high in alkaline earths such as basic and intermediate igneous rocks and volcanic ash; they are formed in arid and semiarid areas where evaporation exceeds precipitation.

Common smectites are montmorillonite and vermiculite, both of which are highly expansive, while illite (hydrous mica) is also common, but is non-swelling.

- *Chlorite* may form by alteration of smectite where Mg^{2+} cations are introduced into the lattice structure. Biotite from igneous and metamorphic rocks may alter to form chlorites and mixed layer chlorite–vermiculite. Chlorites are non-swelling.

This description of the weathering process is greatly simplified, particularly because progressive weathering can cause an already weathered product to be further weathered into a new mineral.

The susceptibility of minerals to weathering is related to the reaction series shown in Figure 3.5 in which olivine is the least stable, and quartz and muscovite are the most stable; minute flakes of muscovite can be seen in soils that have weathered from schists, for example. Olivine is formed at high temperatures in the magma and as the temperature is reduced below the crystalline temperature, the structure of the mineral changes to a less stable configuration. In general, acid rocks such as granite containing abundant quartz weather more slowly than basic rocks such as basalt. Furthermore, sedimentary rocks such as sandstone may be relatively resistant to weathering because they are formed from the stable products of earlier weathering processes.

Limestones – weathering of limestones differs from the quartz-rich igneous and metamorphic rock in that they are high in calcium that is dissolved by seeping water, particularly where the water is slightly acidic due to the presence of carbon dioxide or sulphides. The solution of limestone results in the formation of fissures and caverns with orientations related to the bedding and joint planes, while impurities such as quartz sand grains and clays form thin soil deposits.

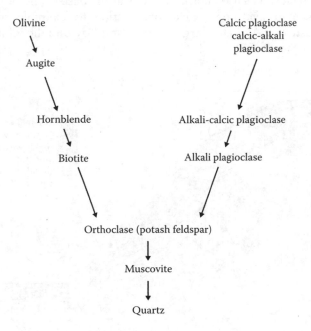

Figure 3.5 Mineral stability series closely parallels weathering stability series: olivine weathers rapidly, while quartz is stable (Goldich, 1938).

3.4 WEATHERING PROFILES RELATED TO GEOLOGY

This section describes typical weathering profiles for a number of rock types and geologic conditions – metamorphic rock, granites, basalt, shales, shale/sandstone interbedded formations and carbonate rock. For each rock type, the influence of lithology, structural geology and ground water on the weathering profile is discussed. The information for this section has been obtained primarily from an article by Deere and Patton (1971).

3.4.1 Intrusive igneous rock: Granite

Geology – Figure 3.6 shows a slope in weathered granite containing two orthogonal joints sets, with one set dipping at approximately the same angle as the slope face. The weathering profile encompasses the full range of weathered rock from residual soil at the surface (Class VI), overlying highly weathered rock with core stones (Class III) and then fresh rock at depth (Class I) (see Figure 3.7). As discussed in Section 3.3, a clay commonly formed by the weathering of acid, granitic rocks is kaolinite, a non-swelling mineral.

Ground water – shown in Figure 3.6 is a typical dry season water table located in the lower portion of the moderately weathered rock that has a relatively high-hydraulic conductivity, compared to the overlying clay-rich residual soil. The weathered rock (Grades II and III) has higher conductivity compared to the overlying residual soil and underlying fresh rock due to the opening of joints as the rock degrades preferentially along the discontinuity surfaces, and weathering products are removed by the ground water flow.

At times of high precipitation and infiltration of surface water, the ground water table would rise, possibly exacerbating slope instability.

Slope stability – a feature of this topography is the creek that can erode the surficial residual soil in the lower part of the valley resulting in the exposure of less weathered rock and core stones on the slope face. Ongoing erosion may then cause instability and slumping of this exposed material that then accumulates at the base of the slope. Removal of this weathered rock, due to excavation for road construction, for example, could result in slides in the residual soil (surface A–A), or larger-scale instability on relic joints extending to the crest of the slope (surface B–B).

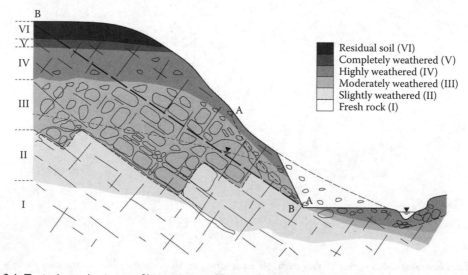

Figure 3.6 Typical weathering profile in granites (Deere and Patton, 1971).

Figure 3.7 (**See colour insert.**) Weathering sequence in granite showing irregular surface of fresh granite underlying weathered rock and residual soil (Sao Paulo, Brazil).

Stability of the overall slope is influenced by the orientation of the joint set dipping parallel to the slope face, and the reduced shear strength of the joints due to preferential weathering of these surfaces. As shown on surface B–B, this joint set dips at 35° which could exceed the friction angle of these degraded rock surfaces. Undercutting the base of the slope could result in failure of the overall slope.

3.4.2 Extrusive igneous rock: Basalt

Geology – Figure 3.8 shows a formation comprising interbedded basalt flows with residual soil layers buried by successive flows. The origin of the soil may be *in situ* weathering of basalt during long periods with no volcanic activity, by ash deposited between lava flows, and by sediments deposited on top of an existing flow. Regardless of the origin of the soil layers, the overall formation will be made up of relatively strong rock layers and much weaker soil layers. The upper portion of the slope shows the progressive weathering of the most recent flow where a surficial layer of residual soil is forming.

In Hawaii, lava is classified as either an 'aa' flow or a 'pahoehoe' flow, the characteristics of each are as follows*:

aa flows generally have a surface layer of blocky, breccia or clinker that forms a very rough and irregular surface, overlying an intermediate layer of relatively dense basalt, and finally a base layer of breccia that may be thinner than the upper layer; in some cases, the flow may comprise only breccia. The overall thickness of the flow may be between a few metres up to 20 m (65 ft).

Pahoehoe flows are fluid and more vesicular than aa flows, and form with a smooth surface. Flow thickness varies from a few centimetres to several metres. The interior of flows tends to be dense and the hydraulic conductivity is low if columnar jointing is absent (Figure 3.9).

---

* As an aid to remembering these names, 'aa aa' is the sound uttered when walking barefoot on the rough, clinker surface of an aa flow.

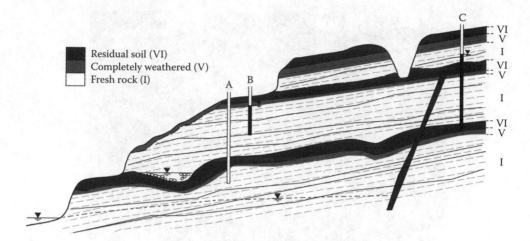

Figure 3.8 Typical weathering profile in multi-flow basalt formation (Deere and Patton, 1971).

Figure 3.9 (**See colour insert.**) Columnar basalt flow overlying weathered soil that has undermined the basalt (Sea to Sky Highway, British Columbia, Canada).

Another geologic feature shown in Figure 3.8 is a steeply dipping dyke that cuts through most of the lava flows. The formation of the dyke, that is, the injection of a seam of molten rock into the lava-soil sequence, creates a layer of baked and altered rock and soil, which may be 10–100 mm (4 in.) thick, on either side of the dyke. These baked layers tend to weather more rapidly than the surrounding rock and have relatively low-hydraulic conductivity.

Ground water – With respect to hydraulic conductivity, clay-rich soil, except for buried granular alluvium, will usually have low conductivity, while basalt may have either high or low conductivity depending on the degree of fracturing. Furthermore, horizontal conductivity tends to be greater than vertical conductivity, with the soil layers between the flows impeding the downward flow of ground water, which can result in a low overall water table as shown by the dry piezometer A.

The layered formation may also form a series of perched water tables with possible arte-sian water pressures as indicated by piezometer *B*, as well as springs at the slope surface where the ground water can exit the face. Other ground water features include possible high localised flows associated with buried stream channels C, and with lava tubes.

The dyke shown in Figure 3.8 can also have a significant effect on ground water, depend-ing on its conductivity. If it is a low-conductivity feature due to the layers of baked rock on the sides, then the dyke will inhibit flow as illustrated by the significant elevation differ-ence in the water table to the left and right of the dyke. Piezometer C shows the high-water table to the right of the dyke caused by the damming effect of this feature. Alternatively, the dyke could be closely fractured and have high conductivity in which case it could act as a drain. Another high-conductivity condition is where the structure has a high quartz content and the host rock is removed by weathering to leave a conduit of loose quartz fragments.

These ground water features in layered basalt formations can produce sudden flows of water if they are intersected in an excavation. The duration of these flows may be short lived if the source is an isolated pocket of water, or may flow indefinitely if the outlet in the face is connected to the overall ground water system by a high-conductivity conduit. The hazard with prolonged flows of water is that loose materials could be washed from within the slope forming cavities that may eventually collapse.

Stability – slope stability in multi-flow basalt formations is related to the dip of the flows and the shear strength of the soil deposits between the flows, as well as water pressures. Under adverse conditions where the beds dip out of the face at an angle that exceeds the friction angle of the soil, together with high confined water pressures in the soil layers, substantial blocks of rock can slide from the face. Anderson and Schuster (1970) found that the pyroclastic deposits (material generated by the explosive fragmentation of magma) weathered into soils with high montmorillonite clay content that had peak shear strengths of cohesion = 10 kPa (1.45 psi) and friction angle = 43.5°, and a residual friction angle of 10° with no cohesion.

3.4.3 Metamorphic rock

Geology – Figure 3.10 shows a weathering profile in a metamorphic formation, with the depth of weathering varying with the lithology. In both these rock types, remnants of the well-developed geologic structure are retained in the highly weathered rock with a resulting influence on stability for cuts made in these materials. Dykes, sills and other irregular intru-sions often occur in metamorphic rock that may weather faster or slower than the host rock depending on their mineralogy. The figure shows a relatively resistant quartzite dyke that has created a ridge on the slope surface. However, in different geology, weathering of dykes may extend to as much as 100 m (300 ft) below the depth of weathering of the surrounding rock as occurs in the Johannesburg area of South Africa.

The steeply dipping fault in the base of the valley may also contain weathering products derived from the original clay gouge formed during the faulting process.

The thickness of the weathered material in the valley bottom will depend on the hydro-logic conditions. Where the stream is eroding the channel and removing the weathered rock, the depth of weathering will be limited or absent, with good exposure of bedrock along the stream banks. However, in areas of low stream gradient and aggrading streams, and below steep slopes where colluvium material is deposited faster than it is removed by erosion, the valley floor will build up. This accumulated material will protect the underlying rock and allow the weathering processes to continue.

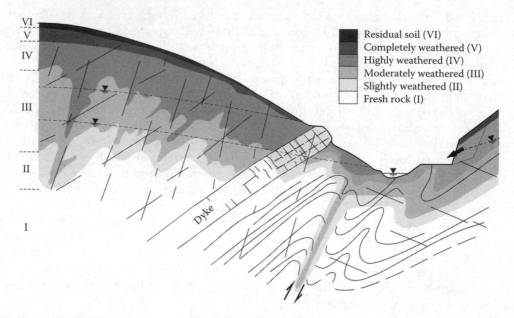

Figure 3.10 Typical slope in weathered metamorphic slope with relatively resistant quartzite dyke (Deere and Patton, 1971).

The weathering of acid metamorphic rocks with their high content of biotite and muscovite micas forms clays in which the biotite weathers to highly expansive vermiculite, and the more resistant muscovite weathers to non-swelling illite. Illite is a very common clay mineral, but the slower weathering muscovite can often be observed as fine, shiny flakes in the residual soil.

Ground water – the ground water table is likely to be within the weathered rock where the hydraulic conductivity is high compared to the overlying residual soil and the underlying fresh bedrock. Figure 3.10 shows the possible fluctuation in the water table between dry and wet seasons, with a possible spring developing during the wet season at the relatively less weathered and lower-conductivity dyke.

The steeply dipping fault in which preferential weathering has occurred is also likely to have relatively low conductivity and be a barrier to ground water flow.

Stability – slope stability in weathered metamorphic rock is usually related to the relic structure where the preferential weathering of joints reduces the shear strength of these surfaces, and high-water pressures develop in the low-conductivity materials. In Figure 3.10, the foliation dips into the slope on the left side of the valley, and dips out of the slope on the right side of the valley. Therefore, stability conditions are likely to be less favourable on the right side, as shown by the minor slide on relic foliation dipping out of the cut face.

3.4.4 Shale

Figure 3.11 shows a slope in a typical weathered shale formation, where shale includes the widest range of engineering characteristics from strong, cemented shales to the weaker compaction clay shales. Most of the minerals forming shales, that is, quartz, clays and iron oxides, are themselves derived from the weathering of other rocks. Therefore, shales tend to be more stable in a renewed weathering environment, and the depth of weathering tends to be less than that for igneous or metamorphic rock. For example, in Georgia, USA, shale

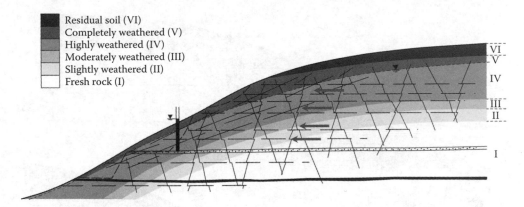

Residual soil (VI)
Completely weathered (V)
Highly weathered (IV)
Moderately weathered (III)
Slightly weathered (II)
Fresh rock (I)

Figure 3.11 Slope in weathered, horizontally bedded shale with slope movement along the bedding (Deere and Patton, 1971).

is weathered to a depth of about 5 m (16 ft) while granite in the same area is weathered to a depth of at least 16 m (50 ft) (Rodriguez, Castillo and Sowers, 1988). However, some minerals found in shales, such as iron pyrite and marcasite (iron sulphide), are less stable than the shales, and swell when subject to weathering. Also, calcite, if present, rapidly dissolves in tropical climates creating voids and open seams in the rock.

A feature of shale weathering is that the constituent minerals are generally resistant to chemical weathering as a result of their prior history, so the predominant weathering processes are physical – temperature changes, wetting and drying, and erosion. These processes fracture and break the rock into small fragments, commonly termed 'slaking', that removes the relatively weak shale and leaves prominent slabs or ridges of more resistant sandstone. Figures 3.11 and 3.12 show common shale occurrences in which shale and sandstone occur as an interbedded sequence.

The weathering profile for shale shown in Figure 3.11, which is similar to other weathering profiles, comprises a low-conductivity surficial soil layer that is underlain by jointed and fissured rock of higher conductivity; the transition from residual soil to weathered rock tends to be more gradual than in other rock types. The properties of the surficial soil are closely related to climatic conditions such as temperature changes, wetting and drying and chemical decomposition, the effects of which diminish with depth.

For slopes in shale where the rock at depth that is not subject to the climatic weathering, deep-seated strains and release of strain energy may occur in the rock forming the slope, producing fractures and fissures that are often slickensided; these fissures become more closely spaced towards the slope face. Movement on the horizontal bedding is enhanced by the high persistence and planarity of these planes, and the low-shear strength of clay infillings.

The cause of horizontal movement is often release of locked in horizontal stresses by the down cutting action of the river or stream; Figure 3.11 shows arrows indicating this movement. The presence of horizontal stresses at shallow depths in many parts of the world, due to vertical unloading, has been established from stress measurements made for the design of tunnels and other underground structures where it has been found that within the upper 100–300 m (330–1000 ft) of the earth's crust the ratio of horizontal to vertical stress (k value) can vary between 2 and 4 (Hoek, Kaiser and Bawden, 1995).

The presence of near-horizontal shear planes in shale formations, with the deepest planes at the base of the valley, has been observed in the Missouri River valley (Fleming, Spencer and Banks, 1970), the Panama Canal (Lutton and Banks, 1970) and at the Ardley dam site in Alberta, Canada (Brooker and Anderson, 1970).

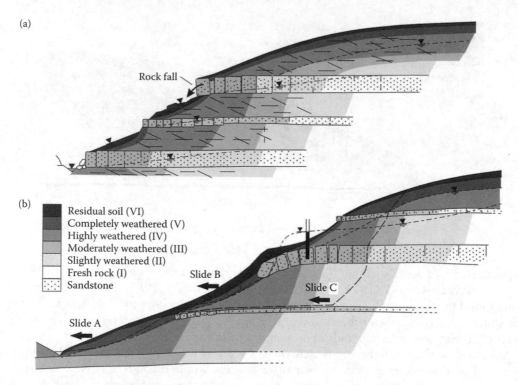

Figure 3.12 Typical slope in interbedded sandstone and shale with colluvium deposits at surface; (a) ground water flow concentrated in higher conductivity sandstone beds and (b) typical types of slope instability (Deere and Patton, 1971).

Ground water – ground water conditions in weathered shales are similar to those in the weathering profiles of other rock types. That is, the surficial residual soil has a relatively low-hydraulic conductivity, while the underlying fissured and open jointed shale has a somewhat higher conductivity. However, Figure 3.11 shows a typical water table in shale that is closer to the surface than that for slopes in granite (Figure 3.6) and metamorphic (Figure 3.10) as the result of the relatively low conductivity of shale bedrock. Also shown in Figure 3.11 is possible artesian pressure in a piezometer sealed into a relatively high-conductivity sandstone seam that is a conduit for ground water flow, but the outlet to surface is blocked by the low-conductivity surficial residual soil.

Stability – instability in weathered shale slopes may comprise shallow slides as indicated by A, B and C in Figure 3.12. Slide A occurs in the surficial soil and is usually associated with high rainfall. Slides B and C are related to specific geological features such as the sandstone seam in which high-water pressures have developed (slide B), or a low-shear strength bentonite clay seam (slide C). Slides B and C occur on near-horizontal features because of a combination of low-shear strength, high-water pressure on the base plane, and bearing capacity failure of the weak shale underlying the sandstone. The dimensions of slides that may develop on low-shear strength horizontal seams can be a significant part of the overall slope. The deep-seated horizontal movement is expressed as offsets and opening of the steep joints where they intersect the horizontal planes (Figure 3.11).

As deep-seated movement occurs and the rock fractures and opens, ground water infiltration into the bedrock is enhanced. However, the low-conductivity surficial materials inhibit drainage of this water resulting high-water pressures developing locally in the slope.

3.4.5 Interbedded sandstones and shales

Geology – Figure 3.12 shows two typical slopes in a horizontal, interbedded sandstone and shale sequence, in which the sandstone is both stronger and has higher-hydraulic conductivity than the shale. This is a common geological formation that occurs, for example, over a wide area of the east-central United States including Pennsylvania and Ohio. In some formations, the sandstone is substituted by siliceous siltstones, limestones, coals or occasionally basalt, all of which have similar characteristics with respect to strength and conductivity.

Jointing of the rock is generally orthogonal to the beds, with vertical joints oriented in the direction parallel to the slope face, being most prominent in the more brittle sandstone. These joints form in response to relief of the locked-in horizontal stress as the result of down cutting in river valleys. Another effect of stress relief is shearing within the relatively low-modulus shales forming low-shear strength seams on which sliding may occur.

Ground water – ground water conditions in interbedded sandstone and shale formations are related mainly to the relatively high-hydraulic conductivity of the sandstone. That is, flow occurs predominately in the sandstone both horizontally and vertically when vertical stress relief joints are opened. Comparatively little flow occurs through the shale beds. Consequently, the overall hydraulic conductivity of the formation is anisotropic with the conductivity being higher horizontally than vertically. Also, ground water discharge at the slope surface tends to be at the base of the sandstone beds, which can be observed as a series of horizontal lines of seepage along the face. This is more pronounced in the winter when icicles may form on the face. For reference, Figure 6.10 shows details of a flow net in a horizontally bedded, high- and low-conductivity formation.

A feature of these formations related to weathering is the accumulation of low-conductivity shale colluvium on the slope surface. This material can inhibit the discharge of ground water at the face resulting in increased water pressures within the slope, as shown by the artesian pressure in the sandstone in Figure 3.12b.

Stability – the stability of the interbedded formations shown in Figure 3.12 is related to the relatively low strength and more-rapid weathering of the shale (Slide A). Surficial instability also occurs where weathering and slaking of the shale undermines the stronger sandstone beds that can form substantial overhangs (Figure 3.13). Slides occur due to both tensile failure of the sandstone, and bearing capacity failure of the weathered shale supporting the overhang (Slide B). This instability can occur with no warning, which can be a hazard to facilities at the base of the slope.

Deep-seated instability can occur along low-shear strength beds in the shale in response to down cutting of the valley and horizontal stress relief (see Section 3.4.4), and particularly when high-water pressures develop within the slope as shown by surface C–C in Figure 3.12. Where tectonic action has caused folding or faulting and the beds are inclined, the potential for instability is significant if the beds dip out of the slope face.

3.4.6 Carbonate rocks

Geology – Figure 3.14 shows a typical weathering profile in carbonate rocks that include limestone, dolomite and marble. The weathering of these rock types differs from that of igneous rock as described earlier in this chapter in that carbonates are dissolved by ground-water, and the solutes are removed to create cavities within the rock. The residual soil that is formed by the weathering processes is the insoluble portion of the original rock, namely quartz, chert, clay and iron and manganese oxides. These weathering products form a residual soil that is usually mainly clay, but may also contain sand or pebbles. The volume of the residual soil represents only a small fraction of the original rock volume since most of the

Figure 3.13 Example of overhanging sandstone bed formed by the relatively rapid weathering of the underlying shale (Kentucky, USA).

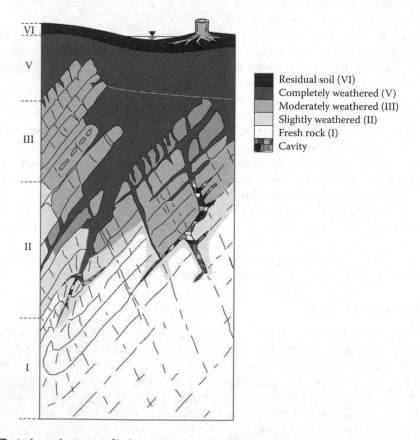

Figure 3.14 Typical weathering profile for carbonate rocks (Deere and Patton, 1971).

rock has been removed by solution. This is in contrast to residual soils formed from igneous rocks where about 50%–70% of the original rock constituents remain in place.

As shown in Figure 3.14, weathering is initiated on the discontinuities in the rock where the ground water flow is concentrated, and these openings are gradually widened as the solutes are removed. Eventually, cavities are formed with 'floaters' of intact rock remaining within the residual soil. This process is highly unpredictable with wide variations in the depth of weathering over short distances. The geologic formations formed by weathering of carbonate rocks is termed 'karstic'.

A feature of carbonate rock weathering in comparison with that of other rock types is that the contact between fresh rock and residual soil is usually abrupt with little intermediate, partially weathered rock. Figure 3.15 shows images of weathered limestone where, (a) on a macro scale, cavities have developed within the rock mass, and the ground surface is highly irregular, while (b) on a micro scale, a network of fine cavities has been formed.

Most of the cavities in the weathered rock become filled with clay that is soft, saturated and normally loaded, and often a deep red colour, with many stability problems being associated with these low-strength materials.

Ground water – in weathered limestones, the ground water table will be below the depth to which voids and high-hydraulic conductivity rock occurs. The downward flow of water in this rock will continue the weathering process of removing dissolved rock and extending the cavities, with high-flow volumes often occurring in the bedrock. In contrast to the high-conductivity rock, the residual soils have low conductivity and perched water tables may occur in the soil. The flow of water, with the accompanying scour of loose materials, from the surface to the sub-surface network of cavities is the primary cause of instability in these formations.

An important indication of karstic terrain is loss of drill circulation water in exploration drill holes, as well as voids into which the drill rods drop with little or no resistance.

Stability – Figure 3.16 shows common features of weathered carbonate rocks and the overlying residual soils. Stability conditions in these formations relate to a combination of the low-strength clay, cavities in the rock, the flow of water through these materials and the geological structure in the bedrock. In Figure 3.16, the bedding in the limestone dips at about 45° to the left and weathering has taken place preferentially along these discontinuities, and on the orthogonal joint set dipping to the right. This pattern of solution and cavity formation has created three depressions at the ground surface that are in different stages of instability.

First, for the cavity on the left side of Figure 3.16, the soils filling the cavity have begun to subside as the result of the downward flow of ground water washing the loose deposits into the underlying cavities. Eventually, when the rate at which the soils are removed exceeds the rate of accumulation of the soil, which may occur at times of high-precipitation levels, a sinkhole may develop, as shown by the central feature in Figure 3.16. The development of sinkholes can be slow as the soil is gradually removed, or sudden if the soil removal occurs at depth and undermines the upper soil that then collapses with little warning. If the ground water flow through the bedrock continues to remove soil, then the solution cavities will be open, which will in turn enhance ground water flow and accelerate solution of the rock.

The feature on the right side of Figure 3.16 shows a solution cavity that is currently filled with soil. This is an incipient sinkhole because the ground water flow into the underlying cavities may progressively remove soil to create a feature similar to that in the centre of the figure.

The sudden appearance of sinkholes in karstic terrain, which occurs in many parts of the world, can result in both property damage and loss of life if they are located in urban areas. The common feature of karstic terrain is the extreme irregularity of the ground surface comprising isolated peaks of rocks interspersed with deep cavities where sinkholes have formed.

(a)

(b)

Figure 3.15 (**See colour insert.**) Typical weathered limestone formation: (a) irregular ground surface and cavities of varying dimensions; (b) closeup of weathered rock showing network of fine cavities at Cottesloe Beach, Perth, Western Australia. (Image by I. McDougall.)

Another stability issue illustrated in Figure 3.16 is the bedrock slopes, both natural and excavated, where the geological structure has a significant influence on stability. For the sinkhole illustrated in the centre of the drawing, the natural slopes are steep in the left side where the bedding dips into the face and the slope has formed along the steeply dipping orthogonal joints. In contrast, the right-hand slope, which has formed along the bedding, has a much flatter dip. If an excavation were to be made in the bedrock in a direction parallel to the strike of the beds, the cut face angles would need to conform to the geologic structure. If the slopes were designed at a relatively steep angle of 76° (¼H:1 V), then planar failure may occur along the shallower dipping structure (see Chapter 7).

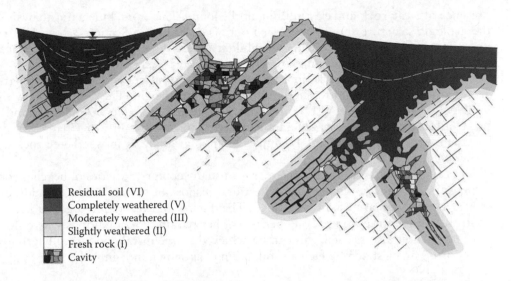

Residual soil (VI)
Completely weathered (V)
Moderately weathered (III)
Slightly weathered (II)
Fresh rock (I)
Cavity

Figure 3.16 Common features of weathered carbonate rocks and residual soils (Deere and Patton, 1971).

Another issue to consider in the design of rock cuts is the presence of cavities behind the face. These features could be detrimental to stability where collapses could occur that undermine the upper part of the slope, resulting in rock falls or larger-scale failures. Remedial work for this condition could include filling the cavities with shotcrete or concrete if their dimensions are less than 1–2 m (3–6 ft), or having to recut the slope in less weathered rock if the cavities are pervasive.

3.5 SLOPE DESIGN IN WEATHERED ROCK

As discussed in Section 3.1.2, the most common mode of instability in tropical regions with deep weathering profiles is surficial slides, with thicknesses of up to 3 m (10 ft) approximately, in residual soils. These slides are usually triggered by intense tropical rainfall when the rate of surface infiltration exceeds the rate of sub-surface drainage into the higher conductivity underlying weathered rock mass. These types of slides are an issue in soil mechanics that is extensively covered in the literature and is not repeated here (Huat, Toll and Prasad, 2012).

3.5.1 Slope design concepts for weathered rock

This book addresses the stability of slopes in weathered rock that lie beneath the residual soil and above the fresh rock, with the assumption that the slope design will incorporate measures to prevent surficial soil failures. The thickness of weathered rock may vary from a few metres to hundreds of metres, and its characteristics depend on the grade of weathering and the properties of the host rock.

The three basic issues to be addressed in the design of cuts in weathered rock are

1. *Remnant geological structure* – since weathering is an *in situ* process, the orientation of the discontinuities in the weathered rock will be the same as that in the fresh host rock, although the shear strength will be less in the weathered rock due to the

presence of weak rock and clay infilling of the joints. Therefore, kinematic analysis of the geologic structure should be carried out, as described in Section 2.6, to identify potential planar, wedge or toppling instability. Where slope stability is not influenced by structure, then circular failure analysis should be carried out.

2. *Slope angle* – the overall stable slope angle should be consistent with the rock strength, geologic structure, ground water conditions and slope height. The designs follow the limit equilibrium analysis (LEA) methods described in Chapters 7 (planar) and 8 (wedge) for slopes where the remnant structure controls stability, or Chapter 9 (circular) for rock mass instability. Examples of circular analyses in weathered rock are shown in Section 9.7.

3. *Slope configuration* – cuts in weathered rock usually comprise a series of benches containing lined drainage ditches that collect and dispose of surface water. In addition, the excavated faces are usually vegetated. The ditches and vegetation together help to control erosion of the slope during periods of heavy rain. For conditions where slope reinforcement is also needed, this can be achieved by the installation of fully grouted steel dowels and a shotcrete facing ('soil nailing') as shown in Figure 1.2.

3.5.2 Slope design in shallow weathered rock

Figure 3.17 shows a possible slope configuration used for cuts in weathered rock where the cut extends through the weathered profile and into fresh rock, and where no stabilisation measures are employed except for vegetation for erosion control. This design would be used where sufficient property is available to lay back the slope at angles that are stable without the use of support.

The weak and highly erodible residual soil (VI) and completely weathered rock (V) may be stripped back along the crest at a low angle so that shallow slides will not occur in these materials. The highly to slightly weathered rock can be excavated at a steeper angle than the VI and V materials, and if these materials are thicker than about 8–10 m (26–33 ft),

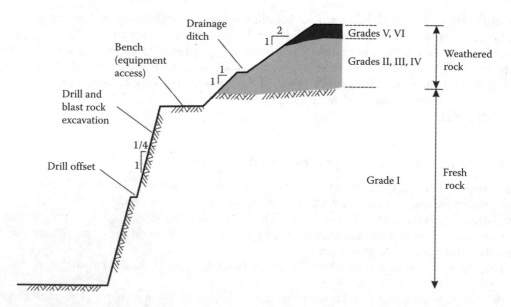

Figure 3.17 Example of rock cut geometry for geological conditions encompassing residual soil to fresh rock showing differing face angles and a catchment bench.

intermediate benches with drainage ditches to control surface water runoff would be incorporated into the slope.

Design of the slope angles in these materials would use circular failure analysis methods as discussed in Chapter 9, assuming that the rock mass is reasonably homogeneous with no significant relic structure. However, it is still advisable to review the orientation of the discontinuities in the host rock to identify possible structurally controlled instability such as planar and wedge failures (see Chapters 7 and 8, respectively). In carrying out planar and wedge failure analyses, reduced shear strengths on the sliding planes would be used that take into account the weathering that has occurred on these surfaces. The likely effect of weathering in the Class V and VI materials is to eliminate cohesion and to reduce the friction angle due to the loss of asperities and the reduced shear strength on the joint surfaces.

At the slightly weathered/fresh rock contact zone, it is usually prudent to include a bench to contain ravelling material and minor slides. The bench should always be machine accessible so that it can be cleaned to prevent accumulation of slide material that would otherwise fall down the rock face. The minimum bench width for machine access and safe operation is usually about 5 m (16 ft) to allow for loss of width along the crest, keeping in mind that just one slide that reduces the width significantly could cut off access to the bench. The use of appropriate blast designs to prevent damage to the crest of the cut face is discussed in Chapter 13.

An important design requirement for the cut shown in Figure 3.17 is that the elevation of the bench at the top of the rock must be known with reasonable accuracy before starting excavation so that the crest of the cut can be positioned correctly. If the top of rock elevation is lower than expected, it is very difficult (and expensive) to recut the slope starting at the crest. Obtaining adequate information on the depth of weathering, which is usually unpredictable, may require an investigation programme involving geophysics and drilling.

The other consideration for slopes in weathered rock is the control of erosion during intense rainfall that requires a lined ditch on each bench, and vegetation on the excavated faces (see Figures 1.2, 9.22 and 9.23). One method of establishing vegetation on slopes that was observed in Malaysia was to cut shallow, horizontal grooves in the excavated face that serve to trap the grass seeds when they are applied by hydroseeding. The grooves prevent the seeds from being washed off the slope before they have had time to germinate. A grass that is commonly used for erosion control in tropical climates is Vetiver that both forms a thick mat on the ground surface to limit scour, and has a root system up to a metre deep that helps to strengthen the surficial weathered rock.

Chapter 4

Site investigation and geological data collection

4.1 PLANNING AN INVESTIGATION PROGRAMME

The design of rock cuts is often an iterative process that proceeds from initial reconnaissance, through preliminary and final design, followed by construction. This process involves progressively collecting more detailed design data, specific to the site conditions and needs of the project. Typically, the three stages of a complete investigation are as follows:

- *Reconnaissance* – examination of published geological maps and reports, study of air photographs, gathering of local experience, field visits to examine, if possible, the performance of existing slopes in similar geological conditions, and geophysics studies if outcrops are limited.
- *Route selection/preliminary slope design* – if the project involves the evaluation of alternative routes, limited investigations could be carried out of each route comprising outcrop mapping, geophysics to find overburden thickness and index tests of rock properties. For a quarry, considerable geological information on the property will usually be generated during the exploration programme. This will often include mapping, geophysics and drilling from which geotechnical data can be obtained. It is beneficial to the design of the quarry slopes if geotechnical data can be collected as part of the exploration programme.
- *Detailed investigations* – final design would usually require detailed mapping of outcrops and existing cuts to study structural geology, test pits to obtain information on overburden thickness and properties, and diamond drilling to investigate rock conditions at depth. Components of the drilling could include core orientation to obtain structural geology information, and installation of piezometers to measure ground water levels, and possibly hydraulic conductivity. Rock strength testing could comprise laboratory testing of drill core to determine the friction angle of discontinuities, uniaxial compressive strength and slake durability.

Figure 4.1 shows diamond drilling in progress in a vertical hole, using a triple tube core barrel to investigate targeted geological structure and collect high-quality core in closely fractured rock. Some of the current (2016) requirements for drilling and sampling programmes may include collection of all circulation water and drill cuttings ('zero discharge drilling'), and complete restoration of the drill site and access roads.

Because of the wide variety of both site conditions and slope designs, it is not considered appropriate to draw up any rules on the types and quantity of investigation programmes. That is, every investigation is unique. The only general rule that applies to investigation for rock slope design is that information is required on geology, rock strength and ground water. These three sets of design parameters are discussed in the pages to come.

Figure 4.1 Typical diamond drilling equipment drilling vertical hole, with core logging being conducted on site. (Image by Norman Norrish.)

4.1.1 Geology

A distinguishing feature of many investigations for rock slopes is that it is particularly important to focus on details of the structural geology. For example, the characteristics of one clay filled fault that dips out of the face can make the difference between stability and instability. Structural geology data provided by surface mapping, where available, are usually more reliable than that obtained by diamond drilling because outcrops and cuts show larger-scale features and undisturbed *in situ* conditions compared to the very small volume of a drill core. Furthermore, the orientation of discontinuities in core is not known unless the core is oriented.

It is recommended wherever possible that the mapping be carried out by the same person or engineering group who will carry out the design so that the objectives of the mapping programme are clearly identified, and the data collected are relevant to the design. For example, a large number of short, non-persistent joints that have little influence on the rock mass strength or stability should be given less attention during mapping than a limited number of shears with continuous lengths equal to that of the slope height. A design engineer who is analysing the data and is not familiar with the site may be unable to distinguish on a contoured stereonet the relative importance between the many non-persistent joints and the more persistent shears (refer to Figure 2.3).

Alternative approaches to geological investigations are as follows. First, a number of existing slopes, either natural or excavated, may exist near the site where the geological conditions are similar to those on the project. In this case, strong reliance could be placed on extrapolating the performance of these slopes to the new design. In these circumstances, it may not be necessary to collect additional data, except to carefully document existing slope performance and assess how this may be applied to the proposed design. Alternatively, where little local experience on cut slope stability is available, it may be necessary to conduct an extensive investigation programme involving mapping, drilling and laboratory testing. As this programme develops, it should be modified to suit particular conditions at the site. For example, the drilling and mapping may show that although the rock is strong and the jointing is favourable to stability, a number of faults may exist that could control stability conditions. The investigation programme would then concentrate on determining the location and orientation of these faults, and their shear strength properties.

This chapter describes geological investigation methods.

4.1.2 Rock strength

The rock strength parameters that are used in slope design are primarily the shear strength of discontinuities and the rock mass, the weathering characteristics of the rock where applicable, and to a lesser extent the compressive strength of intact rock. The shear strength of discontinuities can be measured in the laboratory on samples obtained from drill core, or samples cut from lumps of rock that are intersected by a discontinuity. The shear strength of the rock mass can either be determined by back analysis of slope failures, or calculated by an empirical method that requires information on the intact rock strength, the rock type and the degree of fracturing. The compressive strength of rock can be measured on core samples, or from index tests applied to outcrops in the field. The susceptibility of rock to weathering can also be measured in the laboratory, or assessed by field index tests.

Details of rock strength testing methods are described in Chapter 5.

4.1.3 Ground water

The investigation of ground water plays an important part in any slope design programme. In climates with high precipitation levels, water pressures should always be included in the design. The design water pressures should account for likely peak pressures that may develop during intense rainfall events or snow melt periods, rather than the pressure due to the average seasonal water table. Furthermore, if drainage measures are installed, the design may account for the possible degradation of these systems over time due to lack of maintenance.

Similarly to geological investigations, the extent of the ground water investigation will also depend on site conditions. In most cases, it is sufficient to install piezometers to measure the position and variation in the water table so that realistic values of water pressure can be used in design. However, if it is planned to install extensive drainage measures such as a drainage adit, then measurements of hydraulic conductivity are beneficial to assess whether the adit will be successful in draining the slope, and then to determine the optimum location and layout of the adit and drain holes.

Details of ground water investigations are described in Chapter 6.

4.2 SITE RECONNAISSANCE

The following is a discussion on some of the reconnaissance techniques that may be used early in a project, mainly for the purpose of project evaluation. It is rare that the information gathered at this stage of a project would be adequate for use in final design, so these studies would be followed by more detailed investigations such as surface mapping and drilling once the overall project layout has been finalised.

Part of any site reconnaissance is the collection of all relevant existing data on the site ranging from published data from both government and private sources, to observations of the performance of existing natural slopes and cut faces. These sources will provide information such as rock types, depth of weathering, likely slope failure modes and the frequency and size of rock falls.

An important first step in the reconnaissance stage of a project is to define *zones*, in each of which the geological properties are uniform with regard to the requirements of the project (ISRM, 1981a). Typical boundaries between zones include rock type contacts, faults or major folds. The zoning of the rock mass should provide information on the location, orientation and type of boundary between zones, as well as information on the engineering properties of the rock mass in each. By defining the boundaries of each zone, it is possible to determine the extent to which stability conditions will vary along the alignment or across the site, and plan more detailed investigations in those zones in which the potential for instability exists.

Figure 4.2 shows an example of a reconnaissance-level geologic map for a highway project. This map shows basic location features such as Universal Transverse Mercator (UTM) co-ordinates, a river, the existing highway and railway and the proposed new alignment. Note that ground contours have been left off the map for the sake of clarity. The geologic data include the rock type, a landslide, thrust faults, and the strike and dip orientation of discontinuities mapped on outcrops. The numbers such as '10-3' are references linked to a table that provides more information on the characteristics of each discontinuity; the locations of the discontinuities were determined using a GPS (geographical positioning system) unit. Orientations of the bedding and two primary orthogonal joint sets are shown on the stereonets in Figures 2.10 to 2.12.

4.2.1 Aerial and terrestrial photography

The study of stereographic pairs of vertical aerial photographs or oblique terrestrial photographs provides much useful information on the larger-scale geological conditions at a site (Peterson, Sullivan and Tater, 1982). Often these, large features will be difficult to identify in surface mapping because they are obscured by vegetation, rock falls or more closely spaced discontinuities. Photographs commonly used in geotechnical engineering are black and white, vertical photographs taken at heights between 500 and 3000 m (1600–10,000 ft) with scales ranging from 1:10,000 to 1:30,000. On some projects, it is necessary to have both high- and low-level photographs, with the high-level photographs being used to identify landslides for example, while the low-level photographs provide more detailed information on geological structure.

The use of aerial photographs has now been somewhat replaced with Google Earth images that have coverage of the entire earth, and are regularly updated allowing changes in conditions over time to be examined. At present (2016), the generation of stereo pairs from Google earth images is not routine. However, detailed LiDAR scans of slopes can be interpreted to provide information comparable to that of aerial photographs (see Section 4.2.3).

One of the most important uses of aerial photographs is the identification of landslides that have the potential for causing movement, or even destruction of facilities. Landslide

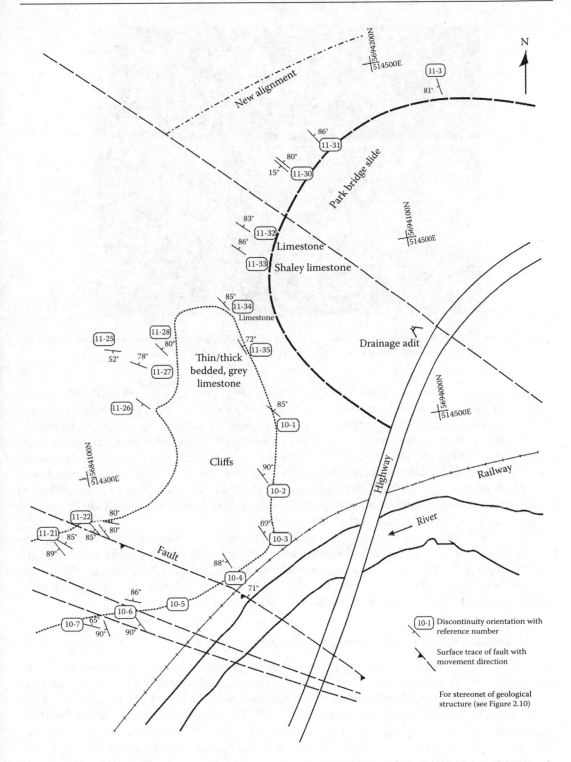

Figure 4.2 Typical reconnaissance-level geological map showing structural geology features. (Mapping by Dr. C. H. B. Leitch.)

Figure 4.3 Stereo pair aerial photographs of Hope Slide, British Columbia; slide had volume of 47 million cubic metres of rock and buried a 3.2-km length of highway.

features that are often readily apparent on vertical aerial photographs are scarps along the crest of the slide, hummocky terrain in the body of the slide and areas of fresh disturbance in the toe, including sudden changes in river direction. By comparing photographs taken over a number of years, it may be possible to determine the rate of movement of a slide, and whether it is growing in size. Figure 4.3 shows stereo pair aerial photographs of the Hope Slide in British Columbia that occurred in 1965. The volume of rock that failed was about 47 million cubic metres (60 million cubic yards); a 3.2 km (2 mile) length of the highway was buried to a depth of about 80 m (260 ft). The slide mass failed on a continuous foliation plane dipping at about 30° out of the slope face.

Other features that may be evident on aerial photographs are major geological structures such as faults, bedding planes and continuous joint sets. The photographs may provide information on the position, length and continuity of these features (Goodman, 1976).

As discussed in Section 4.2.3, the use of LiDAR scans to create digital terrain models (DTMs) is becoming more common than aerial photography.

4.2.2 Slope design in deep weathered rock

For geological conditions where the entire cut is in weathered rock, a multi-bench design will usually be required where the required overall face angle required for stability defines the combination of bench height, berm width and interberm face angle. In practice, the vertical spacing of benches may be about 6–8 m (20–26 ft), and the minimum width of benches required to incorporate a drainage ditch is about 3 m (10 ft). Furthermore, the intermediate slope angles that will be steeper than the overall slope must also be stable in order to maintain the drainage ditches on each bench.

Examples of circular stability analyses of slopes in weathered rock are shown in Section 9.7. These slope designs show both the overall slope angles, and the bench configurations.

4.2.3 Geophysics

Geophysical methods are often used in the reconnaissance or preliminary stages of a site investigation to provide information such as the depth of weathering, the bedrock profile, contacts between rock types of significantly different density, the location of major faults, and the degree of fracturing of the rock. The results obtained from geophysical measurements are usually not sufficiently accurate to be used in final design and should preferably be calibrated by putting down a number of test pits or drill holes to spot check actual properties and contact elevations. However, geophysical surveys provide a continuous profile of sub-surface conditions and this information can be used as a fill-in between drill holes. For rock slope engineering purposes, the most common geophysical investigation method is seismic refraction as described below.

Seismic refraction – seismic surveys are used to determine the approximate location and density of layers of soil and rock, a well-defined water table, or the degree of fracturing, porosity and saturation of the rock. Seismic velocities of a variety of rock types have also been correlated with their rippability, which is a useful guideline in the selection of rock excavation methods as shown in Figure 4.4. The seismic method is effective to depths in the range of tens of metres to a maximum of a few hundred metres. Discontinuities will not be detected by seismic methods unless shear displacement has occurred and a distinct elevation change of a layer with a particular density as a result of fault movement.

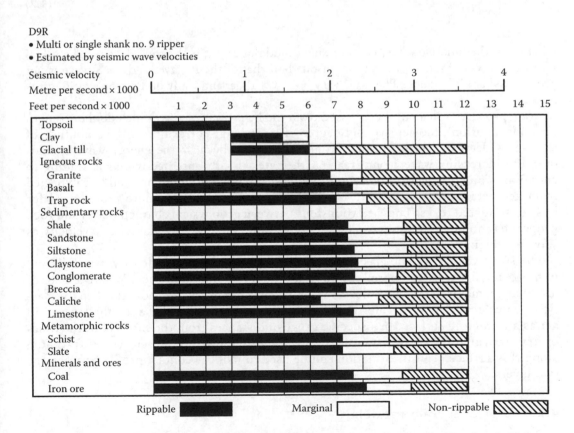

Figure 4.4 Approximate range of seismic velocities for common geological materials related to their excavation by ripping (Caterpillar Inc. 2015).

Seismic surveys measure the relative arrival times, and thus the velocity of propagation, of elastic waves travelling between a shallow energy source and a number of transducers set out in a straight line along the required profile. The energy source may be a hammer blow, an explosion of a propane–oxygen mixture in a heavy chamber (gas-gun), or a light explosive charge. In elastically homogeneous ground subject to a sudden stress near its surface, three elastic pulses travel outward at different speeds. Two are body waves that are propagated as spherical fronts affected to only a minor extent by the free surface of the ground. The third wave is a surface wave (Raleigh wave) that is confined to the region near the surface; its amplitude falling off rapidly with depth. The two body waves, namely the primary or 'P' wave and the secondary or 'S' wave, differ in both their direction of motion and speed. The P wave is a longitudinal compressive wave in the direction of propagation, while the S wave induces shear stresses in the medium. The velocities of the primary (V_p) and secondary (V_s) waves are related to the elastic constants and density of the medium by the equations:

$$V_p = \left(\frac{(K + (4G/3))}{\gamma_r} \right)^{1/2} \tag{4.1}$$

$$V_s = \left(\frac{G}{\gamma_r} \right)^{1/2} \tag{4.2}$$

where K is the bulk modulus, G is the shear modulus and γ_r is the rock density. The velocity of the S wave in most rocks is about one half that of the P wave, and the S wave is not propagated at all in fluids. The ratio V_p/V_s ($V_p/V_s > 1$) depends only on Poisson's ratio of the medium.

Rock properties that can be assessed from the behaviour of waves include the strength (or consolidation), the density and degree of fracturing. For example, the amplitude of the waves decreases with distance from their source because the energy wave spreads over the increasing wave front area. Earth materials are imperfectly elastic leading to energy loss and attenuation of the seismic waves that is greater than would be expected from geometric spreading alone. This reduction in amplitude is more pronounced for less consolidated rocks. These relationships between geophysical characteristics and rock properties can be utilised to correlate the rippability of rock with the seismic velocity as shown in Figure 4.4.

The sonic velocity of the elastic wave will be greater in higher-density material, and in more massive rock compared to low-density, closely fractured rock. Where a layer of low-density material overlies a denser layer, such as soil overlying bedrock, then the elastic wave velocity will be greater in the bedrock and the contact between the layers will act as a refracting surface. In a specific range of distances from the shot point, the times of first arrival at different distances from the shot point will represent waves travelling along this surface. This information can be used to plot the contact profile between the two layers.

4.2.4 Digital terrain models

As of 2016, the use of digital images and LiDAR scanning to create DTMs is a rapidly advancing and growing technology. These techniques can be used for applications such as remote geological mapping of rock faces (see Section 4.3.2), measuring shotcrete thickness, monitoring slope movement (see Chapter 15), determining overbreak/underbreak relative to design lines and measuring volumes of excavations or rock falls (Hiltunen, Hudyma and Quigley, 2007; Birch, 2008; Fekete, Diederichs, Lato et al., 2008). The precision of these measurements will depend on factors such as locating the camera/LiDAR scanner at stable, fixed and surveyed reference points, and whether single or multiple, overlapping images are obtained. It is also possible to obtain images from drones. The three-dimensional models can also be useful for keeping records of construction progress.

In applying DTM technology, it is necessary to match the sophistication, and cost, of the equipment with the required precision of the results. Images obtained using inexpensive digital cameras can be widely used for these applications, but a precise, complete scan of the face requires the use of LiDAR equipment. Some limitations of LiDAR scans are that vegetation, snow and heavy rainfall that cannot be easily penetrated by the laser beam will limit the value of the results, and, of course, only line-of-sight features can be obtained. However, for some conditions, it is possible to strip vegetation from a DTM to produce a model of the ground surface.

4.3 GEOLOGIC MAPPING

Geological mapping of surface outcrops or existing cuts, in similar geological formations to that in which the excavation will be made, usually furnishes the fundamental information on site conditions required for slope design. While mapping is a vital part of the investigation programme, it is also an inexact process because a certain amount of judgement is usually required to extrapolate the small amount of information available from surface outcrops to the overall cut slope.

In order to produce geological maps and descriptions of the engineering properties of the rock mass that can be used with confidence in design, it is important to have a well-defined process that produces comparable results obtained by different personnel working at several sites. To meet these requirements, standard mapping procedures have been drawn up which have the following objectives:

- To provide a language-enabling observers to transmit their general impression of a rock mass, particularly with regard to its anticipated mechanical behaviour. The language of the geological description must be unambiguous so that different observers of a given rock mass describe the rock mass in the same way.
- To contain as far as possible quantitative data of interest to the solution of definite practical problems.
- Whenever possible, to use simple measurements rather than visual observations alone.
- To provide a complete specification of the rock mass for engineering purposes.

4.3.1 Line and window mapping

Methods of structural mapping that will systematically examine all significant geological features are 'line' and 'window' mapping (see also Appendix II).

Line mapping comprises stretching a tape along the face and mapping every discontinuity that intersects the line; line lengths are normally between 50 and 100 m (160–330 ft). If the ends of the line are surveyed, then the location of all the discontinuities can be determined. Window mapping comprises mapping all discontinuities within a representative segment or 'window' of fixed size, spaced at regular intervals along the exposure. The intervening areas are examined for similarity of structure. The dimensions of a window would normally be about 10 m (33 ft). Either of these mapping techniques may be used in both the reconnaissance and final design stages of a project, depending on the extent of the face available for mapping. If the initial investigations identify a particular feature that is likely to have a significant effect on stability, then more detailed mapping, such as roughness and persistence measurements, could be carried out on these structures.

4.3.2 Stereogrammetric mapping of discontinuities

Circumstances may occur where it is not possible to directly access a rock face for mapping because, for example, rock falls are a hazard to geologists or the face is overhanging. Under these conditions, indirect methods of geological mapping are available using terrestrial photography. The basic principle involves obtaining the co-ordinates of at least three points on each surface, from which its orientation can be calculated.

One system that provides this facility is SIROJOINT (CSIRO, 2001). A digital image of the slope face is taken by a camera at a known location, and the image is converted into three-dimensional spatial data defining the surface of the rock face. Each spatial point has a position (x, y, z co-ordinates) in space, and each local set of three spatial points defines a triangle. From these co-ordinates, the orientation of the triangles can be defined in terms of the dip and dip direction, together with the co-ordinates of the centroid. The software allows a mouse to be used to outline the surface to be analysed, and the calculated dip and dip direction can then be imported directly to a stereonet.

4.3.3 Types of discontinuity

Geological investigations usually categorise discontinuities according to the manner in which they were formed. This is useful for geotechnical engineering because discontinuities within each category usually have similar properties as regards both dimensions and shear strength properties that can be used in the initial review of stability conditions of a site. The following are standard definitions of the most commonly encountered types of discontinuities:

1. *Fault* – discontinuity along which an observable amount of displacement has occurred. Faults are rarely single planar units; normally they occur as parallel or sub-parallel sets of discontinuities along which movement has taken place to a greater or lesser extent.
2. *Bedding* – surface parallel to the surface of deposition, which may or may not have a physical expression. Note that the original attitude of the bedding plane should not be assumed to be horizontal.
3. *Foliation* – parallel orientation of platy minerals, or mineral banding in metamorphic rocks.
4. *Joint* – discontinuity on which no observable relative movement has taken place. In general, joints intersect primary surfaces such as bedding, cleavage and schistosity. A series of parallel joints is called a *joint set*; two or more intersecting sets produce a *joint system*; two sets of joints approximately at right angles to one another are said to be *orthogonal*.
5. *Cleavage* – parallel discontinuities formed in incompetent layers in a series of beds of varying degrees of competency are known as cleavages. In general, the term implies that the cleavage planes are not controlled by mineral particles in parallel orientation.
6. *Schistosity* – foliation in schist or other coarse-grained crystalline rock due to the parallel arrangement of mineral grains of the platy or prismatic type, such as mica.

4.3.4 Definition of geological terms

The following is a summary of information that may be collected to provide a complete description of the rock mass, and comments on how these properties influence the performance of the rock mass. This information is based primarily on the procedures developed by ISRM (1981a). More details of the mapping data are provided in Appendix II which includes mapping field sheets, and tables relating descriptions of rock mass properties to quantitative measurements.

Figure 4.5a illustrates the 12 essential features of geological structure, each of which is described in more detail in this section. The diagram and photograph in Figure 4.5 show that sets of discontinuities often occur in orthogonal sets (mutually at right angles) in response to the stress field that has deformed the rock; the photograph shows three orthogonal joints in massive granite. Orthogonal structure is also illustrated in the stereonet in Figure 2.11. The value of recognising orthogonal structure on an outcrop or in a stereonet is that these features are often the most prevalent in a cut and are likely to control stability.

The following is a list, and a description of the parameters that define the characteristics of the rock mass as illustrated in Figure 4.5a:

A. *Rock type* – the rock type is defined by the origin of the rock (i.e. sedimentary, metamorphic or igneous), the mineralogy, the colour and grain size (Deere and Miller, 1966). The importance of defining the rock type is that wide experience has been acquired on the performance of different rock types (e.g. granite is usually stronger and more massive than shale), and this information provides a useful guideline on the likely behaviour of the rock.

B. *Discontinuity type* – discontinuity types range from clean tension joints of limited length, to faults containing several metres thickness of clay gouge and lengths of many kilometres; the shear strength of faults will generally be less than that of joints. Section 4.3.3 provides a definition of the six most common types of discontinuity.

C. *Discontinuity orientation* – the orientation of discontinuities is expressed as the dip and dip direction (or strike) of the surface. The dip of the plane is the maximum angle of the plane to the horizontal (angle ψ), while the dip direction is the direction of the horizontal trace of the line of dip, measured clockwise from north (angle α) (see Figure 2.4). For the plane shown in Figure 4.5 that dips to the north–east, the orientation of the plane can be completely defined by five digits: 30/045, where the dip is 30° and the dip direction is 45°. This method of defining discontinuity orientation facilitates mapping because the dip and dip direction can be read from a single compass reading (Figure 4.6). Also, the results can be plotted directly on a stereonet to analyse the structural geology (see Section 2.5). In using the compass shown in Figure 4.6, the dip is read off a graduated scale on the lid hinge, while the dip direction is read off the compass scale that is graduated from 0 to 360°.

Some compasses allow the graduated circle to be rotated to account for the magnetic declination at the site so that measured dip directions are relative to true north. If the compass does not have this feature, then the magnetic readings can be adjusted accordingly: for a magnetic declination of *20° east*, for example, *20° is added to the magnetic readings* to obtain true north readings.

D. *Spacing* – discontinuity spacing can be mapped in rock faces and in drill core, with the true spacing being calculated from the apparent spacing for discontinuities inclined to the face (see also Section 4.4.1); spacing categories range from extremely wide (>2 m) to very narrow (<6 mm). Measurement of discontinuity spacing of each set of discontinuities will define the size and shape of blocks and give an indication of stability modes

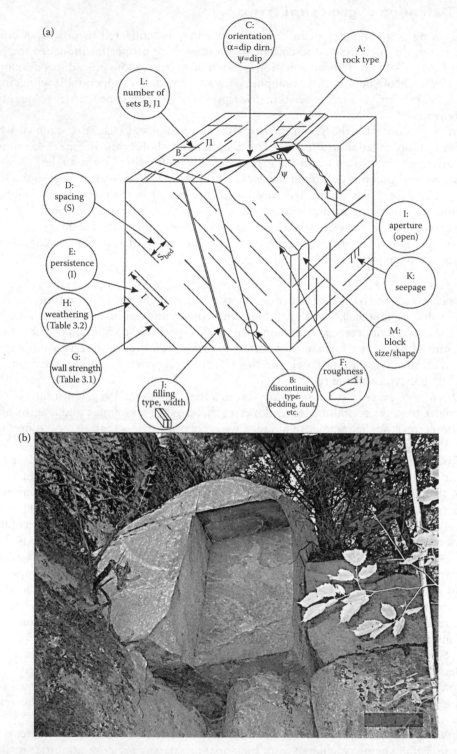

Figure 4.5 (**See colour insert.**) Characteristics of discontinuities in rock masses: (a) parameters describing the rock mass; letter ('A', etc.) refers to description parameters in text (Wyllie, 1999). (b) Photograph of blocky granite containing three joints orientated in orthogonal directions (near Hope, British Columbia).

Figure 4.6 Geological compass (Clar type) used to directly measure dip and dip direction of surfaces. Lid is placed on rock surface and body is rotated until it is horizontal, as indicated by the spirit level; the needle is released to indicate dip direction, and the dip is read off scale on the hinge (The Brunton Co., Riverton, Wyoming).

such as toppling failure. Also, the rock mass strength is related to spacing because in closely fractured rock, the individual discontinuities will more readily join to form a continuous zone of weakness. A method of calculating the strength of fractured rock masses taking into account discontinuity spacing is discussed in Section 5.5.

E. *Persistence* – persistence is the measure of the continuous length or area of the discontinuity; persistence categories range from very high (>20 m) to very low (<1 m). This parameter defines the size of blocks and the length of potential sliding surfaces, so the mapping should concentrate on measuring the persistence of the set of discontinuities that will have the greatest influence on stability. Where it is not possible to measure the length of discontinuities directly on the face, the procedure described in Section 4.4.2 can be used to estimate the average length of fractures that extend beyond the dimensions of the face.

F. *Roughness* – the roughness of a discontinuity surface is often an important component of the shear strength, especially where the discontinuity is not displaced and remains interlocked. Roughness becomes less important where the discontinuity is infilled, or displaced and interlock is lost. Roughness should be measured in the field on exposed surfaces with lengths of at least 2 m (6 ft) if possible, and in the anticipated direction of sliding, and can be described in terms of a combination of both the large- and small-scale features, namely:

Shape:	Stepped	Roughness:	Rough
	Undulating		Smooth
	Planar		Slickensided

The degree of roughness can be quantified in terms of the $i°$ value, which is a measure of the inclination of the irregularities (or *asperities*) on the surface (see detail F,

Figure 4.5a). The total friction angle of a rough surface is $(\phi + i)$. Values for i can be determined either by direct measurement of the surface, or by comparing the surface with standard profiles of irregular joint surfaces. These techniques are described in Section 4.4.3. Usual practice would be to use the standard roughness profiles during the preliminary mapping. If a discontinuity exists that will be critical to slope stability, then the values estimated from the profiles could be calibrated with a number of detailed *in situ* measurements of roughness.

G. *Wall strength* – the strength of the rock forming the walls of discontinuities will influence the shear strength of rough surfaces. Where high stresses, compared to the wall strength, are generated at local contact points during shearing, the asperities will be sheared off, resulting in a reduction of the roughness component of the friction angle. In the initial stages of weathering where water flow is concentrated in discontinuities, a reduction in rock strength on the discontinuity surfaces will often occur, resulting in a diminished roughness value. It is usually adequate to estimate the compressive strength from the simple field tests as shown in Table 4.1 (ISRM, 1981a), or if core or lump samples are available, by carrying out point load tests (see Section 5.7.2). The Schmidt hammer test is also a method of estimating the compressive strength of rock at discontinuity surfaces.

Table 4.1 Classification of rock and soil strengths

Grade	Description	Field identification	Approximate range of uniaxial compressive strength (MPa)
R6	Extremely strong rock	Specimen can only be chipped with geological hammer	>250
R5	Very strong rock	Specimen requires many blows of geological hammer to fracture it	100–250
R4	Strong rock	Specimen requires more than one blow of geological hammer to fracture it	50–100
R3	Medium strong rock	Cannot be scraped or peeled with a pocket knife, specimen can be fractured with single firm blow of geological hammer	25–50
R2	Weak rock	Can be peeled by a pocket knife with difficulty, shallow indentations made by firm blow with point of geological hammer	5.0–25
R1	Very weak rock	Crumbles under firm blows with point of geological hammer, can be peeled by a pocket knife	1.0–5.0
R0	Extremely weak rock	Indented by thumbnail	0.25–1.0
S6	Hard clay	Indented with difficulty by thumbnail	>0.5
S5	Very stiff clay	Readily indented by thumbnail	0.25–0.5
S4	Stiff clay	Readily indented by thumb but penetrated only with great difficulty	0.1–0.25
S3	Firm clay	Can be penetrated several inches by thumb with moderate effort	0.05–0.1
S2	Soft clay	Easily penetrated several inches by thumb	0.025–0.05
S1	Very soft clay	Easily penetrated several inches by fist	<0.025

Source: Adapted from International Society for Rock Mechanics (ISRM). 1981a. Suggested methods for the quantitative description of discontinuities in rock masses. In: E. T. Brown, ed., Pergamon Press, Oxford, UK, 211pp.

Note: Discontinuity wall strength will generally be characterised by grades R0–R6 (rock). Some rounding of strength values has been made when converting to S.I. units.

The strength of the intact rock is also one of the parameters used to calculate rock mass strength as discussed in Section 5.5.

H. *Weathering* – reduction of rock strength due to weathering will reduce the shear strength of discontinuities as described in (G) above. Weathering will also reduce the shear strength of the rock mass due to the diminished strength of the intact rock. Weathering categories range from fresh rock to residual soil. Weathering of rock takes the form of both physical disintegration and chemical decomposition. Physical weathering is the result of environmental conditions such as wetting and drying, freezing and thawing that break down the exposed surface layer. Physical weathering is most prevalent in sedimentary rocks such as sandstones and shales, particularly if they contain swelling clays, and in metamorphic rocks with a high mica content. Chemical weathering refers to changes in rock produced by chemical agents such as oxidation (e.g. yellow discolouration in rock containing iron), hydration (e.g. decomposition of feldspar in granite to kaolinite clay) and carbonation (e.g. solution of limestone). Table 4.2 lists weathering grades that categorise the rock mass according to the degree of disintegration and decomposition (see also Chapter 3).

I. *Aperture* – aperture is the perpendicular distance separating the adjacent rock walls of an open discontinuity, in which the intervening space is air or water filled; categories of aperture range from cavernous (>1 m), to very tight (<0.1 mm). Aperture is thereby distinguished from the 'width' of a filled discontinuity. It is important in predicting the likely behaviour of the rock mass, such as hydraulic conductivity and deformation under stress changes, to understand the reason that open discontinuities develop. Possible causes include scouring of infillings, solution of the rock forming the walls of a discontinuity, shear displacement and dilation of rough discontinuities, tension features at the head of landslides and relaxation of steep valley walls following glacial retreat or erosion. Aperture may be measured in outcrops or tunnels provided that care is taken to discount any blast-induced open discontinuities, in drill core if recovery is excellent, and in boreholes using a borehole camera if the walls of the hole are clean.

Table 4.2 Weathering grades

Term	Description	Grade
Fresh	No visible sign of rock material weathering; perhaps slight discolouration on major discontinuity surfaces.	I
Slightly weathered	Discolouration indicates weathering of rock material and discontinuity surfaces. All the rock material may be discoloured by weathering and may be somewhat weaker externally than in its fresh condition.	II
Moderately weathered	Less than half of the rock material is decomposed and/or disintegrated to a soil. Fresh or discoloured rock is present either as a continuous framework or as corestones.	III
Highly weathered	More than half of the rock material is decomposed and/or disintegrated to a soil. Fresh or discoloured rock is present either as a discontinuous framework or as corestones.	IV
Completely weathered	All rock material is decomposed and/or disintegrated to soil. The original mass structure is still largely intact.	V
Residual soil	All rock material is converted to soil. The mass structure and material fabric are destroyed. A large decrease in volume occurs, but the soil has not been significantly transported.	VI

Source: Adapted from International Society for Rock Mechanics (ISRM). 1981a. Suggested methods for the quantitative description of discontinuities in rock masses. In: E. T. Brown, ed., Pergamon Press, Oxford, UK, 211pp.

J. *Infilling/width* – infilling is the term for material separating the adjacent walls of discontinuities, such as calcite or fault gouge; the perpendicular distance between the adjacent rock walls is termed the *width* of the filled discontinuity. A complete description of filling material is required to predict the behaviour of the discontinuity including the following: mineralogy, particle size, over-consolidation ratio, water content/conductivity, wall roughness, width and fracturing/crushing of the wall rock. If the filling is likely to be a potential sliding surface in the slope, samples of the material (undisturbed if possible) should be collected for shear testing.

K. *Seepage* – the location of seepage from discontinuities provides information on aperture because ground water flow is confined almost entirely in the discontinuities (secondary conductivity); seepage categories range from very tight and dry, to continuous flow that can scour infillings. These observations will also indicate the position of the water table, or water tables in the case of rock masses containing alternating layers of low- and high-conductivity rock such as shale and sandstone, respectively. In dry climates, the evaporation rate may exceed the seepage rate and it may be difficult to observe seepage locations. In cold weather, icicles provide a good indication of even very low seepage rates. The flow quantities will also help anticipate conditions during construction such as flooding and pumping requirements of excavations.

L. *Number of sets* – the number of sets of discontinuities that intersect one another will influence the extent to which the rock mass can deform without failure of the intact rock. As the number of discontinuity sets increases and the block size diminishes, the greater is the opportunity for blocks to rotate, translate and crush under applied loads. Mapping should distinguish between systematic discontinuities that are members of joint sets, from less frequent, randomly oriented, discontinuities.

M. *Block size/shape* – the block size and shape are determined by the discontinuity spacing and persistence, and the number of sets. Block shapes include blocky, tabular, shattered and columnar, while block size ranges from very large (>8 m^3) to very small (<0.0002 m^3). The block size can be estimated by selecting several typical blocks and measuring their average dimensions.

Using the terms outlined in this section, a typical description for a *rock material* would be as follows:

> Moderately weathered, weak, fine-grained, dark grey to black carbonaceous shale.

Note that the rock name comes last since this is less important than the engineering properties of the rock.

An example of a *rock mass* description is as follows:

> Interbedded sequence of shale and sandstone; typically, the shale units are 200–400 mm (8–16 in.) thick and the sandstone units are 1000–5000 mm (3–20 ft) thick. In the shale, the bedding spacing is 100–200 mm (4–8 in.) and many contain a soft clay infilling with width up to 20 mm; the bedding is planar and often slickensided. The sandstone is blocky with the spacing of the beds and two conjugate joint sets in the range of 500–1000 mm (1.6–3 ft); the bedding and joint surfaces are smooth and undulating, and contain no infilling. The upper contacts of the shale beds are wet with some dripping; the sandstone is dry.

4.4 SPACING, PERSISTENCE AND ROUGHNESS MEASUREMENTS

A component of the geological mapping of rock exposures may involve detailed measurements of surface roughness, and discontinuity spacing and persistence. Usually these detailed measurements would only be carried out on discontinuity sets or specific features that have been identified to have a significant influence on stability because, for example, they are persistent and dip out the face.

4.4.1 Spacing of discontinuities

The spacing of discontinuities will determine the dimensions of blocks in the slope, which will influence the size of rock falls and the design of bolting patterns. A factor to consider in the interpretation of spacing measurements from surface mapping is the relative orientation between the face and the discontinuities. That is, the relative orientation introduces a bias to the number and spacing of discontinuities in which the actual spacing is less than the apparent spacing, and the actual number of discontinuities is greater than the number mapped. The bias arises because all discontinuities oriented at right angles to the face will be visible on the face at their true spacing, while few apparently widely spaced discontinuities oriented sub-parallel to the face will be visible (Figures 4.5a and 4.7). The bias in spacing can be corrected as follows (Terzaghi, 1965):

$$S = S_{app} \cdot \sin\theta \qquad (4.3)$$

where S is the true spacing between discontinuities of the same set, S_{app} is the measured (apparent) spacing, and θ is the angle between the face and the strike of discontinuities.

The number of discontinuities in a set can be adjusted to account for the relative orientation between the face and the strike of the discontinuity as follows:

$$N = \frac{N_{app}}{\sin\theta} \qquad (4.4)$$

where N is the adjusted number of discontinuities and N_{app} is the measured number of discontinuities. For example, a vertical drill hole will intersect few steeply dipping discontinuities, and a vertical face will intersect few discontinuities parallel to this face; the Terzaghi

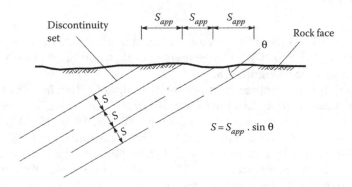

Figure 4.7 Relationship between apparent and true spacing for a set of discontinuities.

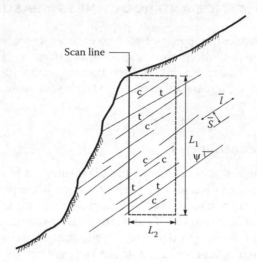

Figure 4.8 Measurement of average spacing and persistence of a set of discontinuities dipping at angle ψ in an outcrop.

correction will calculate an appropriate increase in the number of these surfaces. Some stereonet programs can apply the Terzaghi correction to increase the number of discontinuities to allow for the bias in sampling orientation, and more accurately represent the population of discontinuities.

In rock outcrops, the spacing between discontinuities in a set will be variable, and the following is a discussion on methods of calculating the average spacing of a joint set, and the application of the Terzaghi correction. Figure 4.8 shows a rock outcrop in which it is planned to make a steep cut; the rock contains a set of joints that dip out of the face at angle ψ. The characteristics of this joint set can be determined using line mapping by hanging a tape vertically down the face and recording the characteristics of each joint that intersects the tape. The tape can be termed a *scan line*, which in this case has a length of 15 m (50 ft) (see Section 4.3.1).

The average true spacing ($\bar{s}$) of joint set in Figure 4.8 is calculated using Equation 4.5, for the conditions where the scan line is vertical and is not at right angles to the joints. If N' joints with dip ψ intersect a scan line of length L_1, then the value of $\bar{s}$ is given by

$$\bar{s} = L_1 \cdot \cos \psi / N' \tag{4.5}$$

In Figure 4.8, nine joints ($N' = 9$) with an average dip of 35° intersect the scan line, which has a length of 15 m ($L_1 = 15$). The average spacing of these joints is 1.5 m (5 ft); this average spacing is plotted to scale in Figure 4.8.

One approach that may be taken to study the spacings of different sets of discontinuities is to make measurements along scan lines with different orientations; it is preferable that the scan lines are at right angles to each set (Hudson and Priest, 1979, 1983).

The Terzaghi correction can be applied to the joints measured on the scan line as follows. The scan line is intersected by nine joints ($N_{app} = 9$), and the average angle between the joints dipping at 35° and the vertical scan line is 55°. Therefore, Equation 4.4 shows that approximately 11 joints would have been intersected by a 15 m (50 ft) long scan line orientated at right angles to the joints.

4.4.2 Persistence of discontinuity sets

Persistence of discontinuities is one of the most important rock mass parameters because it defines, together with spacing, the size of blocks that can slide from the face. Furthermore, a small area of intact rock between low-persistence discontinuities can have a significant influence on stability because the strength of the rock will often be much higher than the shear stress acting in the slope. Unfortunately, persistence is one of the more difficult parameters to measure because often only a small part of the discontinuity is visible in the face. In the case of drill core, no information on persistence is available.

A number of procedures have been developed to calculate the approximate average persistence of a set of discontinuities by measuring their exposed trace lengths on a specified area of the face (Pahl, 1981; Priest and Hudson, 1981; Kulatilake and Wu, 1984).

The procedure developed by Pahl (1981) comprises, first, defining a mapping area on a face with dimensions L_1 and L_2 (Figure 4.8). Then the total number of discontinuities (N'') of a particular set (with dip ψ) in this area is counted, and the numbers of these discontinuities that are either contained within (N_c), or transect (N_t) the mapping area are identified. Contained discontinuities are short and have both ends visible within the area, while transecting discontinuities are relatively long and have neither end visible. The approximate average length ($\bar{l}$) of a set of discontinuities is calculated from Equations 4.6 to 4.8 that are independent of the assumed form of the statistical distribution of the lengths.

$$\bar{l} = H' \cdot \frac{(1+m)}{(1-m)} \tag{4.6}$$

where

$$H' = \frac{L_1 \cdot L_2}{(L_1 \cdot \cos \psi + L_2 \cdot \sin \psi)} \tag{4.7}$$

and

$$m = \frac{(N_t - N_c)}{(N'' + 1)} \tag{4.8}$$

As demonstrated by these equations, the basis of this method of estimating the average length of discontinuities is to count discontinuities on a face with a known area, and does not involve measuring the length of individual discontinuities, which can be a much more time-consuming task.

For the joints depicted in Figure 4.8, the average persistence can be calculated as follows. The total number of joints (dip, ψ ~35°) within the scan area is 15 ($N'' = 15$), of which five are contained within the scan area ($N_c = 5$), and four transect the scan area ($N_t = 4$). If the dimensions of the scan area are $L_1 = 15$ and $L_2 = 5$, then the value of m is −0.07 and the value of H' is 4.95. From Equation 4.6, the average persistence of the joints in this set is 4.3 m (14 ft). This average persistence is plotted to scale in Figure 4.8.

If a second set of discontinuities occurs in the scan area that would influence stability, then these could be counted using the same procedure to determine their average spacing and persistence.

4.4.3 Roughness of rock surfaces

The friction angle of a rough surface comprises two components – the friction of the rock material (ϕ), plus interlocking produced by the irregularities (asperities) of the surface (i). Because roughness can be a significant component of the total friction angle, measurement of roughness is often an important part of a mapping programme.

During the preliminary stages of an investigation, it is usually satisfactory to make a visual assessment of the roughness as defined by the joint roughness coefficient (JRC) (Barton, 1973). JRC varies from zero for smooth, planar and particularly slickensided surfaces, to as much as 20 for rough, undulating surfaces. The value of JRC can be estimated by visually comparing the surface condition with standard profiles based on a combination of surface irregularities (at a scale of several centimetres) and waviness (at a scale of a several metres) as shown in Figure 4.9.

JRC is related to the roughness of the surface i, and the strength of the rock on the discontinuity surface, by the following equation:

$$i = \text{JRC} \cdot \log_{10}\left(\frac{\text{JCS}}{\sigma'}\right) \tag{4.9}$$

where JCS (joint compressive strength) is the compressive strength of the rock on the discontinuity surface (see Table 4.1), and σ' is the effective normal stress on the surface due to the weight of the overlying rock less any uplift water pressure on the surface. Equation 4.9 shows that i diminishes as the asperities are ground off when the rock strength is low-strength compared to the applied normal stress. The application of Equation 4.9 to the determination of the shear strength of rock surfaces is discussed in more detail in Section 5.2.

In the final design stage of a project, a few discontinuities having a significant effect on stability may be identified, and a number of methods are available to accurately measure the surface roughness of these critical surfaces. A method developed by Fecker and Rengers (1971) consists of measuring the orientation of the discontinuity using a geological compass with a series of plates of different diameters attached to the lid. If the diameter of the larger plates is about the same dimensions as the wavelength of the roughness, then the measured

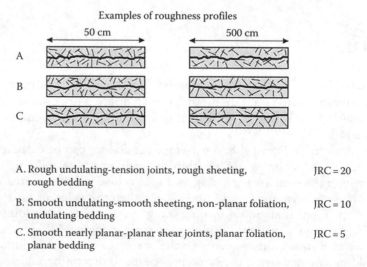

Examples of roughness profiles

A. Rough undulating-tension joints, rough sheeting, rough bedding — JRC = 20

B. Smooth undulating-smooth sheeting, non-planar foliation, undulating bedding — JRC = 10

C. Smooth nearly planar-planar shear joints, planar foliation, planar bedding — JRC = 5

Figure 4.9 Standard profiles defining joint roughness coefficient (Barton, 1973).

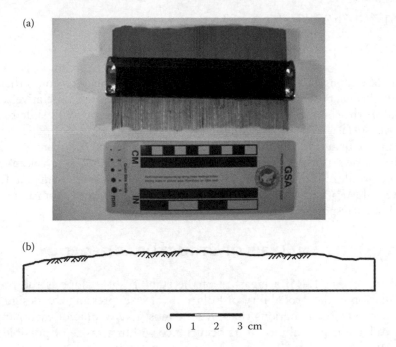

(a)

(b)

0   1   2   3 cm

Figure 4.10 Measurement of joint roughness: (a) photography of profilometer; (b) profile of joint with JRC of 11.4.

orientation will be approximately equal to the average orientation of the surface. However, the smaller-diameter plates will show a scatter in the orientation measurements as the plates lie on irregularities with shorter wavelengths. If the orientation measurements are plotted on a stereonet, the degree of scatter in the poles about the mean orientation is a measure of the roughness.

Quantitative methods of profile measurement have been established by Tse and Cruden (1979) using a mechanical profilometer, and Maerz, Franklin and Bennett (1990) have developed a shadow profilometer that records the shape of the surface with a video camera and image analyser.

The method developed by Tse and Cruden is illustrated in Figure 4.10. The profile is defined by measuring the distance (y_i) of the surface from a fixed reference line at specified equal intervals (Δx) over a length of M intervals. From these measurements, the coefficient $Z2$ is defined as

$$Z2 = \left[\frac{1}{M \cdot (\Delta x)^2} \sum_{i=1}^{M} (y_{i+1} + y_i)^2 \right]^{1/2}$$

(4.10)

A study has also been carried out to assess the effect of the size of the sampling interval (Δx) along the profile on the calculated value of JRC (Yu and Vayassde, 1991). This study found that the calculated value of JRC was dependent on the size of Δx, and that the most accurate results were obtained with small sampling intervals. The value of JRC can be calculated from the coefficient $Z2$ using one of the following equations for the appropriate sampling interval:

$$JRC = 60.32 \cdot (Z2) - 4.51 \quad \text{for } \Delta x = 0.25 \text{ mm}$$

(4.11)

$$JRC = 61.79 \cdot (Z2) - 3.47 \quad \text{for } \Delta x = 0.5 \text{ mm} \tag{4.12}$$

$$JRC = 64.22 \cdot (Z2) - 2.31 \quad \text{for } \Delta x = 1 \text{ mm} \tag{4.13}$$

One means of making profile measurements is to use a carpenter's comb that consists of a series of metal rods, positioned in a frame, such that they can slide relative to each other (Figure 4.10a). If the comb is pressed against a rock face, the rods will slide to conform to the shape of the surface. This profile can then be traced on to a piece of paper and the shape of the rock surface quantified by measuring the distance of each rod from a reference line.

Figure 4.10b shows a profile measured with a carpenter's comb of a rough, planar joint in granite. The profile of the surface was digitised, with 145 measurements made on a 1 mm (0.04 in.) interval ($M = 145$; $\Delta x = 1$), from which $Z2$ was calculated from Equation 4.13 to be 0.214 with a corresponding JRC value of 11.4.

4.5 PROBABILISTIC ANALYSIS OF STRUCTURAL GEOLOGY

As discussed in Section 1.4.5, a measure of the stability of a slope is the probability of failure. Calculation of the probability of failure involves expressing the design parameters in terms of probability distributions that give the most likely value of each parameter (e.g. average), as well as the probability of its occurrence within a range of possible values (e.g. standard deviation).

This section demonstrates techniques to determine the probability distributions of structural geology data. The distribution in orientation can be calculated from the stereonet, while the distributions of persistence and spacing are calculated from field measurements.

4.5.1 Discontinuity orientation

The natural variation in orientation of discontinuities results in scatter of the pole positions when plotted on the stereonet. It can be useful to incorporate this scatter into the stability analysis of the slope because, for example, a wedge analysis using the mean values of pair of discontinuity sets may show that the line of intersection of the wedge does not daylight in the face and that the slope is stable. However, an analysis using orientations other than the mean values may show that some unstable wedges can be formed. The risk of occurrence of this condition would be quantified by calculating the mean and standard deviation of the dip and dip direction as described below.

A measure of the dispersion, and from this the standard deviation, of a discontinuity set can be calculated from the direction cosines as follows (Goodman, 1980). The direction cosines of any plane with dip ψ and dip direction α are the unit vectors l, m and n, where

$$\left. \begin{array}{l} l = \sin \psi \cdot \cos \alpha \\ m = \sin \psi \cdot \sin \alpha \\ n = \cos \psi \end{array} \right\} \tag{4.14}$$

For a number of poles (i), the direction cosines (l_R, m_R and n_R) of the mean orientation of the discontinuity set is the sum of the individual direction cosines as follows:

$$l_R = \frac{\Sigma l_i}{|R|}; \quad m_R = \frac{\Sigma m_i}{|R|}; \quad n_R = \frac{\Sigma n_i}{|R|} \tag{4.15}$$

where $|R|$ is the magnitude of the resultant vector given by

$$|R| = [(\Sigma l_i)^2 + (\Sigma m_i)^2 + (\Sigma n_i)^2]^{1/2} \tag{4.16}$$

The dip ψ_R and dip direction α_R of the mean orientation are

$$\left. \begin{aligned} \psi_R &= \cos^{-1}(n_R) \\ \alpha_R &= +\cos^{-1}(l_R/\sin\psi_R) \quad \text{for } m_R \geq 0 \\ \alpha_R &= -\cos^{-1}(l_R/\sin\psi_R) \quad \text{for } m_R < 0 \end{aligned} \right\} \tag{4.17}$$

A measure of the scatter of a set of discontinuities comprising N poles can be obtained from the dispersion coefficient C_d which is calculated as follows:

$$C_d = \frac{N}{(N - |R|)} \tag{4.18}$$

If the scatter in the orientation of the discontinuities is limited, the value of C_d is large, and the value of C_d diminishes as the scatter increases.

From the dispersion coefficient, it is possible to calculate from Equation 4.19 the probability P that a pole will make an angle $\theta°$ or less than the mean orientation.

$$\theta = \cos^{-1}[1 + (1/C_d) \cdot \ln(1 - P)] \tag{4.19}$$

For example, the angle from the mean defined by one standard deviation occurs at a probability P of 0.16 (refer to Figure 1.13). If the dispersion is 20, one standard deviation lies at 7.6° from the mean.

Equation 4.19 is applicable when the dispersion in the scatter is approximately uniform about the mean orientation, which is the case of joint set A in Figure 2.11. However, in the case of the bedding in Figure 2.11, less scatter occurs in the dip than in the dip direction. The standard deviations in the two directions can be calculated approximately from the stereonet as follows. First, two great circles are drawn at right angles corresponding to the directions of dip and dip direction, respectively. Then the angles corresponding to the 7% and 93% levels (P_7 and P_{93}) are determined by counting the number of poles in the set and removing the poles outside these percentiles. The equation for the standard deviation along either of the great circles is as follows (Morriss, 1984):

$$SD = \tan^{-1}\{0.34 \cdot [\tan(P_{93}) - \tan(P_7)]\} \tag{4.20}$$

More precise methods of determining the standard deviation are described by McMahon (1982), but the approximate method given by Equation 4.20 may be sufficiently accurate considering the difficulty in obtaining a representative sample of the discontinuities in the set.

An important aspect of accurate geological investigations is to account for bias when mapping a single face or logging a single borehole when few of the discontinuities aligned parallel to the line of mapping are measured. This bias in the data can be corrected by applying the Terzaghi correction as described in Section 4.4.1.

4.5.2 Discontinuity length and spacing

The length and spacing of discontinuities determines the size of blocks that will be formed in the slope. Designs are usually concerned with persistent discontinuities that could form blocks with dimensions great enough to influence the overall slope stability. However, discontinuity dimensions have a range of values and it is useful to have an understanding of the distribution of these values in order to predict how the extreme values may be compared to values obtained from a small sample. This section discusses probability distributions for the length and spacing of discontinuities, and discusses the limitations of making accurate predictions over a wide range of dimensions.

The primary purpose of making length and spacing measurements of discontinuities is to estimate the dimensions of blocks of rock formed by these surfaces (Priest and Hudson, 1976; Cruden, 1977; Kikuchi, Kuroda and Mito, 1987; Dershowitz and Einstein, 1988; Kulatilake, 1988; Einstein, 1993). This information can then be used, if necessary, to design appropriate stabilisation measures such as rock bolts and rock fall barriers. Attempts have also been made to use these data to calculate the shear strength of 'stepped' surfaces comprising joints separated by portions of intact rock (Jennings, 1970; Einstein, Veneziano, Baecher et al., 1983). However, it has since been found that the Hoek–Brown method of calculating the shear strength of rock masses is more reliable (see Section 5.5).

4.5.3 Probability distributions of length and spacing

Discontinuities are usually mapped along a scan line, such as drill core, slope face or wall of a tunnel. Individual measurements are made of the properties of each fracture, including its visible length and the spacing between discontinuities in each set (Appendix II). The properties of discontinuities typically vary over a wide range and it is possible to describe the distribution of these properties by means of probability distributions. A normal distribution is applicable if a particular property has values in which the mean value is the most commonly occurring. This condition would indicate that the property of each discontinuity, such as its orientation, is related to the property of the adjacent discontinuities reflecting that the discontinuities were formed by stress relief. For properties that are normally distributed, the mean and standard deviation are given by Equations 1.16 and 1.17.

A negative exponential distribution is applicable for properties of discontinuities, such as their length and spacing, which are randomly distributed indicating that the discontinuities are mutually independent. A negative exponential distribution would show that the most commonly occurring discontinuities are short and closely spaced, while persistent, widely spaced discontinuities are less common. The general form of a probability density function $f(x)$ of a negative exponential distribution is (Priest and Hudson, 1981):

$$f(x) = \frac{1}{\bar{x}}(e^{-x/\bar{x}}) \tag{4.21}$$

and the associated cumulative probability $F(x)$ that a given spacing or length value will be less than dimension x will is given by

$$F(x) = (1 - e^{-x/\bar{x}}) \tag{4.22}$$

where x is a measured value of length or spacing and $\bar{x}$ is the mean value of that parameter. A property of the negative exponential distribution is that the standard deviation is equal to the mean value.

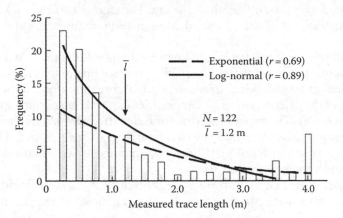

Figure 4.11 Histogram of joint trace lengths, with best-fit exponential and lognormal curves (Priest and Hudson, 1981).

From Equation 4.22 for a set of discontinuities in which the mean spacing is 2 m (6 ft), the probability that the spacing will be less than 1 and 5 m (3 and 16 ft), respectively, is

$$F(x) = (1 - e^{-1/2}) = 40\% \quad \text{and} \quad F(x) = (1 - e^{-5/2}) = 92\%$$

Equation 4.20 could be used to estimate the probability of occurrence of discontinuities with a specified length. This result could be used, for example, to determine the likelihood of a plane being continuous through a slope.

Another distribution that can often be used to describe the dimensions of discontinuities is the lognormal distribution which is applicable where the variable ($x = \ln y$) is normally distributed (Baecher, Lanney and Einstein, 1977). The probability density function for a lognormal distribution is (Harr, 1977)

$$f(x) = \frac{1}{y \cdot SD_x \cdot \sqrt{2\pi}} \cdot \exp\left[-\frac{1}{2}\left(\frac{\ln y - x}{SD_x}\right)^2\right] \tag{4.23}$$

where $\bar{x}$ is the mean value and SD_x is the standard deviation.

Figure 4.11 shows the measured lengths of 122 joints in a Cambrian sandstone for lengths of less than 4 m; the mean length $\bar{l}$ is 1.2 m (4 ft) (Priest and Hudson, 1981). To this data have been fitted both exponential and lognormal curves for which the correlation coefficients r are 0.69 and 0.89, respectively. While the lognormal curve has a higher correlation coefficient, the exponential curve has a better fit at longer discontinuity lengths. This demonstrates that for each set of data the most appropriate distribution should be determined.

4.6 DIAMOND DRILLING

On many projects, surface mapping is supplemented by diamond drilling to obtain core samples of the sub-surface rock. The extent of the drilling will depend on factors such as the soil cover, the availability of rock outcrops and the confidence with which surface data can be extrapolated over the full depth of the cut. For example, if the rock at the surface is weathered or disturbed by blasting, then drilling may be required to find rock conditions at depth.

The type of information that can be obtained by diamond drilling may be somewhat different from surfacing mapping information. While surface mapping is the primary means of obtaining information on geological structure, in drill core, no information can be obtained on persistence, and the orientation of discontinuities can only be obtained if the core is oriented (see Section 4.6.4). The types of information that are provided by drill core are *in situ* rock strength, fracture frequency and the characteristics of shear zones. The core can also be used for laboratory strength testing (Chapter 5), and instrumentation such as piezometers can be installed in the holes (Chapter 6).

A basic requirement of diamond drilling for geotechnical purposes is the recovery of the weakest portion(s) of the rock mass, such as faults, since these are the features that are most likely to influence stability. Therefore, the drilling techniques must be used that minimise the risk of loss of core in these zones. These required techniques, as discussed below, include the use of a triple tube core barrel, appropriate drilling fluid and moderating the advance rate.

4.6.1 Diamond drilling equipment

Figure 4.1 shows a typical diamond drill. The major components of the drill comprise a motor, usually gasoline or diesel powered, a head to generate torque and thrust to the drill rods, a mast to support the wire line equipment, and a string of drill rods at the lower end of which is a core barrel and a diamond impregnated bit. Drill rods are flush coupled, usually in 3 m (10 ft) lengths, and the diameters available for common North American equipment are listed in Table 4.3, together with the corresponding hole and core sizes. The rod diameters are designated by four letters (e.g. BQTT) – the first letter indicates the diameter, the second that it is used with wire line equipment and TT indicates a triple tube core barrel. Also listed on the table are casing diameters; casing is used in the upper part of holes, in soil or weathered rock, to prevent caving. The casing and rod diameters are arranged such that N rods fit inside N casing, and that B rods fit inside N rods, for example. This allows holes to be 'telescoped' down to a smaller diameter, if necessary, as the hole is advanced. It is usual in geotechnical drilling to use NQTT rods, or HQTT if the rock is highly broken because better core recovery is achieved with the larger-diameter hole.

Table 4.3 Dimensions of diamond drilling equipment triple tube core barrels

Drill size	AQTT	BQTT	NQTT	HQTT	PQTT
Hole diameter, mm (in.)	–	60 (2-23/64)	75.7 (2-63/64)	96 (3-25/32)	122.6 (4-53/64)
Core diameter, mm (in.)	–	33.5 (1-5/16)	45 (1-25/32)	61.1 (2-13/32)	83 (3-9/32)
Hole volume, l/100 m (US gal/100 ft)	–	282 (22.7)	451 (36.3)	724 (58.3)	1180 (95.1)
Casing ID, mm (in.)	48.4 (1-29/32)	60.3 (2-3/8)	76.2 (3.0)	101.6 (4.0)	127 (5.0)
Casing OD, mm (in.)	57.1 (2-1/4)	73 (2-7/8)	88.9 (3-1/2)	114.3 (4-1/2)	139.7 (5-1/2)
Casing weight kg/3 m length (lb/10 ft)	17 (38)	31.3 (70)	38.4 (86)	50.5 (113)	64.3 (144)
Drill rod ID, mm (in.)	34.9 (1-3/8)	46 (1-13/16)	60.3 (2-3/8)	77.8 (3-1/16)	103.2 (4-1/16)
Drill rod OD, mm (in.)	44.5 (1-3/4)	55.6 (2-3/16)	69.9 (2-3/4)	88.9 (3-1/2)	117.5 (4-5/8)
Drill rod weight kg/3 m length (lb/10 ft)	14 (31)	18 (40)	23.4 (52)	34.4 (77)	47.2 (106)

Source: Adapted from Christensen Boyles Corp. 2000. *Diamond Drill Products – Field Specifications.* CBC, Salt Lake City, UT.

Table 4.3 also lists the weight of the drill rods and casing that can be useful information where the equipment is transported by helicopter.

For geotechnical drilling where one of the objectives is to recover the weakest portion of the rock, recommended practice is to use a triple tube core barrel. The triple tube barrel comprises an outer tube coupled to the drill rods and bit, which turns with the rods while drilling. The middle tube locks into a roller bearing at the top of the core barrel such that it remains stationary while the drill rods rotate. The inner tube is split into two pieces longitudinally, and also remains stationary during drilling.

The drilling procedure involves rotating the rods and bit at high speed (up to about 1000 rpm), while applying a steady thrust to the bit and pumping water down the centre of the drill rods. In this way, the bit is advanced into the rock while the water cools the bit, and removes the cuttings in the annular space outside the rods. The core fills the core barrel as the bit advances, and the drilling is stopped when the core barrel is full – the core barrel is usually 3 m (10 ft) long. In holes with depths less than about 20 m (66 ft), it is possible to recover the core at the end of each run by removing the rods from the hole. For deeper holes, the usual practice is to lower a wireline down the hole with a core catcher on the end which locks into the core barrel and releases it from the drill rods. The core barrel is lifted from the hole without moving the rods. The core is extracted from the core barrel by pumping the inner split tube out of the barrel and then removing the upper half of the tube. The core can then be logged in the core barrel with minimal disturbance. In contrast, if a double tube barrel is used, the core has to be pumped or hammered out of the tube, which inevitably results in damage and disturbance, especially to the weaker portions of the rock that are the most important to recover.

In very poor quality rock, it is possible to modify the triple tube by inserting a clear plastic liner inside the split inner tube. The core is contained in the plastic tube so that it can be logged and then stored with minimal disturbance.

4.6.2 Diamond drilling operations

The following are factors that may need to be considered in a diamond drilling operation. First, the drill rig should be set up on level ground, or on a level platform if the ground surface is irregular. Furthermore, the platform should be robust and the rig should be securely attached to the ground or platform. If the rig is able to move during drilling, vibration in the rods will diminish core quality and may damage the rods.

It is essential that the drill bit be continuously flushed with a fluid, usually water, to cool the bit and remove the cuttings. The circulating water also lubricates the drill string to reduce the torque required to turn the rods, and reduces vibration of the rods. If the drill hole intersects permeable zones of rock in which the drill water is lost, then it is necessary to seal the hole with the use of drill muds or cement grout as described below.

Water is usually supplied to the site by pumping from a nearby river, or by a tanker truck. Factors to consider in the supply of water include the pumping head between the supply and the site, freezing of the pipeline, and available road access. It is usual that the return drill water is collected in a settling tank at the site to remove the cuttings, and is then recirculated down the hole. This reduces the quantity of supplied water, and eliminates environmental contamination by silt-laden water.

The addition of certain chemicals and solids to the drilling water can improve the properties of the circulating fluid, and can be essential for successfully drilling highly permeable or unstable formations (Australian Drilling Industry, 1996). The most common additive in diamond drilling is organic, long-chain polymers that have thixotropic properties. That is, they have low viscosity when they are stirred or pumped, but gel when allowed to stand.

The effect of these properties is that the mud can be readily circulated in the hole to remove the cuttings in the narrow annular space around the rods, but will gel to form a cake to seal and stabilise the walls of the hole.

The correct application of these dual properties of mud will greatly enhance diamond drilling operations. For example, if the hole intersects a zone of broken and weak rock, the mud can stabilise the walls, and it will not be necessary to extend the casing down to this level. Similarly, if the circulation fluid is lost in high-conductivity zones, these can be sealed with a mud cake. Ideally, the mud cake should be thin and have low permeability, and should penetrate the formation so that it is not broken up by the rotation of the drill rods. The fluid pressure in the hole then helps to keep the cake in place. Where the polymer muds are not sufficiently viscous to form a mud cake, possible additives to the mud include fine flakes of mica or paper to help seal fine openings in the rock on the wall of the hole. In the case of artesian flows, the density of the mud can be increased by bentonite–barytes mixtures so that the mud weight balances the upward pressure of the water in the artesian formation.

If muds are not effective in sealing or stabilising the hole, then it is often necessary to fill the hole to just above the zone of broken rock with cement grout, let this set and then drill down through the grouted zone.

For holes that are to be used for hydrologic testing, such as conductivity measurements and the installation of piezometers, it is necessary that the walls of the hole are free of mud cakes. For these conditions, biodegradable muds are available that break down with time to leave the walls of the hole clean.

At the completion of the hole and any associated down-hole testing, it is often a requirement to fill the hole with cement grout to prevent changes to the hydrological conditions due to the flow of water within the hole.

4.6.3 Core logging

Recording the properties of the recovered drill core involves making a detailed and complete log of the rock; an example of a diamond drill core log is shown in Figure 4.12. The log should be prepared using the same properties and descriptions of the rock mass discussed in Section 4.3 so that the surface and sub-surface data are consistent. These data will include the rock description, the properties of the discontinuities and their orientation with respect to the core axis. Measurements can also be made of the rock quality designation (RQD), fracture index and core recovery, which are indicative of the rock mass quality, as described below. Also, the log records all testing carried out in the hole such as conductivity measurements, the results of strength tests on core samples and the position of instrumentation such as piezometers. Finally, the core is photographed, complete with a colour reference and dimension scale.

Where rock types such as shales are recovered that are highly susceptible to degradation when exposed to the atmosphere, it may be necessary to preserve the rock as soon as it has been logged. Preservation of core samples would be necessary where they are to be transported to the laboratory for strength testing and it is necessary that they be tested close to their *in situ* state. Often, the lowest-strength rock will influence slope design, and these samples should not be allowed to break down and lose strength, or bake in the sun and gain strength.

One method of preservation is to wrap the core in plastic film and then dip it in molten wax to create a barrier to moisture loss. Finally, the sealed core can be embedded in a rigid foam so that is can be transported without damage. A further precaution that may be necessary is to prevent the sample from freezing because the formation of ice can fracture weak rock.

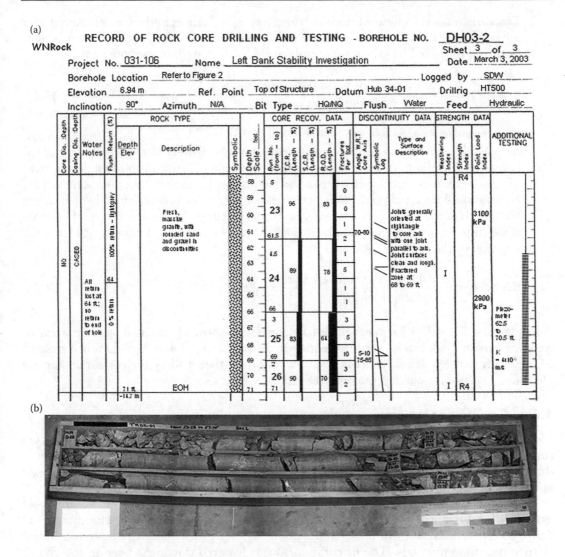

Figure 4.12 Diamond drill records: (a) typical core log with histograms indicating values for recovery and RQD; (b) photograph of drill core with colour reference chart.

The following are measurements of the core that are routinely made to assess the intact rock strength and the degree of fracturing.

1. RQD is an index related to the degree of fracturing of the core. The RQD is calculated by measuring the total length of all pieces of core in a drill run with lengths greater than 100 mm (4 in.), discounting fractures due to drilling. These lengths are then added together, and the total length is expressed as a percentage of the length of the drill run. A low RQD value would be indicative of a closely fractured rock, while an RQD of 100% means that all pieces of core are longer than 100 mm. RQD is calculated as follows:

$$RQD = \frac{\Sigma(\text{Length of core pieces with length} > 100 \text{ mm})}{\text{Total length of core run}} \tag{4.24}$$

2. *Fracture index* is a count of the number of natural fractures in the core measured over a fixed length of 0.5 m (1.5 ft). This parameter is related to the RQD value but is standardised to a fixed length so is not influenced by the length of the core runs.

$$\text{Fracture index} = \frac{\text{Number of natural discontinuities}}{0.5 \text{ m length of core}} \qquad (4.25)$$

3. *Core recovery* is a measure of how much rock has been lost during drilling. Core loss may result from weak zones being washed out by the drilling water, or grinding of the core during drilling, or the presence of an open cavity. Drillers can often detect areas of very weak rock or cavities where a sudden increase in the advance rate occurs; these zones should be noted on the log. The core recovery value is determined by measuring the length of recovered core compared to the total length drilled. If a length of lost core is identified when logging the core, it is good practice to install a wooden spacer at the location of the lost core.

$$\text{Recovery} = \frac{\text{Total length of recovered ore}}{\text{Length of drill run}} \times 100 \qquad (4.26)$$

In Figure 4.12, values for these three parameters are shown graphically so that areas of weak or broken rock can be readily identified when scanning down the log. When making RQD and fracture index measurements, it is important that drilling breaks in the core are identified and not included in the reported values.

4.6.4 Core orientation

For conditions where insufficient design data are available on discontinuity orientation from surface mapping, it may be necessary to obtain these data from drill core. This will require that the core be oriented, that is, a line is marked along the top of the core and the plunge and trend of this line is determined.

The first step in orienting core is to determine the plunge and trend of the drill hole using a down-hole survey tool. One such tool comprises an aluminium (non-magnetic) drill rod that contains a dip meter and a compass, both of which can be photographed at specified time intervals. The orientation tool is lowered down the hole on the end of the drill string, and is held stationary at the times specified for the camera operation. At each time interval a photograph is taken of the dip meter and compass, and the depth is recorded. When the tool is recovered from the hole, the film is developed to show the hole orientation at the recorded depths. Other hole orientation tools include the Tropari single shot instrument, and gyroscopes for use in magnetic environments (Australian Drilling Industry, 1996).

Most methods of orienting core involve marking a line down the core representing the top of the hole. Since the orientation of this line is known from the hole survey, the orientation of all discontinuities in the core can be measured relative to this line, from which their dip and strike can be calculated (Figure 4.13). Figure 4.13 shows that a plane intersected by the core has the shape of an ellipse, and the first step in the calculation process is to mark the down-hole end, major axis of this ellipse. The dip (δ) of this plane is then measured relative to the core axis, and a reference angle (α) is measured clockwise (looking down-hole) around the circumference of the core from the top of core line to the major axis of the ellipse. The dip and dip direction of the plane are calculated from the plunge and trend of the hole and the measured angles δ and α. The true dip and dip direction of a discontinuity

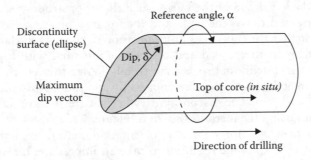

Reference angle, α

Discontinuity
surface (ellipse)

Dip, δ

Maximum
dip vector

Top of core (in situ)

Direction of drilling

Figure 4.13 Measurement of orientation of discontinuity in oriented drill core.

in the core can be determined by stereographic methods (Goodman, 1976), or by spherical/analytical geometry methods (Lau, 1983).

In a few cases, the core may contain a distinct and consistent marker of known orientation, such as bedding, which can be used to orient the core and measure the orientation of the other discontinuities. However, minor and unknown variations in the orientation of the marker bed are likely to lead to errors. Therefore, it is usually preferable to use one of the following three mechanical methods to determine the top of the core.

Clay impression core barrel – the clay impression method to orient core involves fabricating a wireline core barrel with one side weighted so that the barrel can rotate and position the weight at the bottom of the hole (Figure 4.14) (Call, Savely and Pakalnis, 1982). A piece

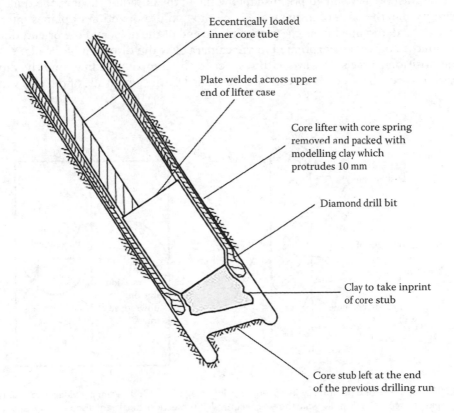

Eccentrically loaded
inner core tube

Plate welded across upper
end of lifter case

Core lifter with core spring
removed and packed with
modelling clay which
protrudes 10 mm

Diamond drill bit

Clay to take inprint
of core stub

Core stub left at the end
of the previous drilling run

Figure 4.14 Clay impression core barrel to orient drill core in an inclined hole (Call, Savely and Pakalnis, 1982).

of clay is placed at the lower end of the barrel such that it protrudes past the drill bit when the core barrel is lowered down the rods and locked into place. A light pressure is then applied to the rods so that the clay takes an impression of the rock surface at the end of the hole. The core barrel is then removed and the clay impression, with top-of-core reference line, is retrieved. The next drill run proceeds normally. When this length of core is removed, the top of the core is matched with the clay impression and the top-of-core line is transferred from the clay to the core run. The discontinuities in the core run are then oriented relative to the top-of-core line using the method shown in Figure 4.13.

The advantages of the clay impression core orientation method are its simplicity and low equipment cost. However, the time required to take an impression on each drill run slows drilling, and the method can only be used in holes inclined at angles flatter than about 70°. Also, the orientation line will be lost at any place where the core is broken and it is not possible to extend the top-of-core line past the break.

Christiensen–Hugel core orientation – a more sophisticated core orientation tool is the Christiensen–Hugel device that scribes a continuous line down the core during drilling; the orientation of the scribed line is determined by taking photographs of a compass in the head of the core barrel. The advantage of this method is that a continuous reference line is scribed so the orientation line is not lost in zones of broken core. However, this is a more expensive and sophisticated piece of equipment than the clay impression tool.

Scanning borehole camera – a scanning borehole camera developed by Colog Inc. takes a continuous 360° image of the wall of the hole as the camera is lowered down the hole. The image can then be processed to have the appearance of a piece of core that can be rotated and viewed from any direction, or can be 'unwrapped' (Figure 4.15). In the core view, planes intersecting the core have an elliptical shape, while in the unwrapped view, the trace of each discontinuity has the form of a sine wave. The dip and dip direction of planes intersecting the core can be determined from the plunge and trend of the hole, and the orientation of the image from the compass incorporated in the camera. The dip direction of the plane is found from the position of the sine wave with respect to the compass reading, and the dip of the plane with respect to the core axis can be determined by the amplitude of the sine wave. The

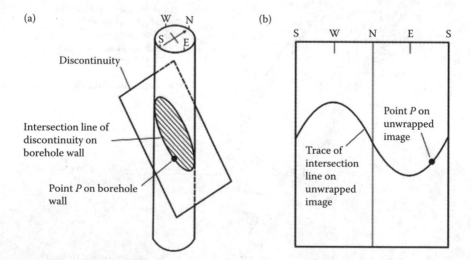

Figure 4.15 Orientation of discontinuity in borehole using image from 360° scanning video camera: (a) core-like image of drill hole wall showing elliptical intersection between discontinuity and core; (b) 'unwrapped' view of borehole wall with discontinuity displayed as a sine wave (Colog Inc. 1995).

software with the camera system allows orientation data indicated by the sine waves to be plotted directly on a stereonet.

The significant advantages of Colog camera system are that the camera is run down the hole at the completion of the hole so no interruption to drilling occurs. Also, the image provides a continuous record of rock conditions, including cavities and zones of broken rock that may be lost in the recovered core. The disadvantage of the system is the cost and the need for a stable hole with clean walls, that is either dry or is filled with clean water.

Reflex orientation system – this core orientation system is based on recovering the core barrel orientation at the conclusion of a drill run (Ureel, Momayez and Oberling, 2013). The reflex tool begins the orientation process by inserting the tool in the core barrel using a specially made shoe. The tool records the orientation of the core barrel every minute during the core run. The reflex sleeve that is attached to the upper drill rod measures the orientation of the top-of-hole using built-in accelerometers. On completion of the run, the drill string is left undisturbed while the communication tool, which is on the surface, counts down the time to the next reading, after which the barrel can be withdrawn. On the surface, the tool is inserted into the top of the core barrel and the barrel is rotated until the tool indicates that the barrel is in the same up–down position as it was in the hole. After the inner core tube is split, the top-of-core mark is transferred along the length of the recovered core.

While the reflex system is a reliable means of orienting the core, good results depend on the core recovery and the ability to mark the orientation line down the full length of the core run.

4.7 INVESTIGATION OF WEATHERED ROCK

Chapter 3 discusses the occurrence and properties of weathered rock, and Table 3.1 describes the characteristics of the weathering grades that range from residual soil at the surface (Grade VI) to slightly weathered at depth (Grade II). Field investigation of these materials is essentially identical to those described in this chapter for fresh rock, with the exceptions described below.

First, test pits can be excavated to depths of 2–3 m (6–10 ft) to investigate the characteristics of the residual soil if the stability of slopes in these surficial materials is of importance. This will also allow undisturbed block samples to be collected for laboratory testing, if required.

Second, weathering is an *in situ* process in which the geologic structure of the host rock is retained in the various grades of weathered rock so the orientation and shear strength properties of the discontinuities must often be considered in slope design. Therefore, the investigation should often include determination of discontinuity orientation within the weathered zones.

Third, weathering is initiated by the flow of water through the discontinuities causing chemical and physical alteration of the rock on the walls of the discontinuities. These actions result in the wall rock becoming weaker and being eventually converted to clays of various types related to the type of the host rock. These conditions mean that the shear strength is reduced as the rock becomes more weathered, and since weathering is a progressive process in which rock near the surface is more weathered than that at depth, shear strengths will also vary with depth.

To account for the change in shear strength with depth, it is necessary to obtain samples within the range of depths of interest for slope design and to determine the corresponding strength parameters. If the slope extends through several grades of weathering, it is likely that the strength properties of both the rock material and the discontinuities used in design will increase with depth.

Chapter 5

Rock strength properties and their measurement

5.1 INTRODUCTION

In analysing the stability of a rock slope, the most important factor to be considered is the geometry of the rock mass behind the face. As discussed in Chapter 4, the relationship between the orientation of the discontinuities and of the excavated face will determine whether parts of the rock mass are free to slide or topple. After geology, the next most important factor governing stability is the shear strength of the potential sliding surface, which is the subject of this chapter.

5.1.1 Scale effects and rock strength

The sliding surface in a slope may consist of a single plane continuous over the full area of the surface, or a complex surface made up of both discontinuities and fractures through intact rock. Determination of reliable shear-strength values is a critical part of slope design because, as will be shown in later chapters, small changes in shear strength can result in significant changes in the safe height or face angle of a slope. The choice of appropriate shear-strength values depends not only on the availability of test data, but also on a careful interpretation of these data in the light of the behaviour of the rock mass that makes up the overall slope. For example, it may be possible to use the results obtained from a shear test on a joint in designing a slope in which failure is likely to occur along a single joint similar to the one being tested. However, these shear test results could not be used directly in designing a slope in which a complex failure process involving several joints and some failure of intact rock is expected. In this book, the term *rock mass* is used for rock materials in which this complex failure process occurs.

This discussion demonstrates that the selection of an appropriate shear strength of a slope depends to a great extent on the relative scale between the sliding surface and structural geology. For example, in the deep, multi-bench excavation illustrated in Figure 5.1, the dimensions of the overall slope are much greater than the discontinuity spacing, so any failure surface will pass through the jointed rock mass, and the appropriate rock strength to use in design of the overall slope is that of the rock mass. In contrast, the bench height is about equal to the joint length so stability could be controlled by a single joint, and the appropriate rock strength to use in design of the benches is that of the joints set that dips out of the face. Finally, at a scale of less than the joint spacing, blocks of intact rock occur and the appropriate rock strength to use in the assessment of drilling and blasting methods, for example, would be primarily that of the intact rock.

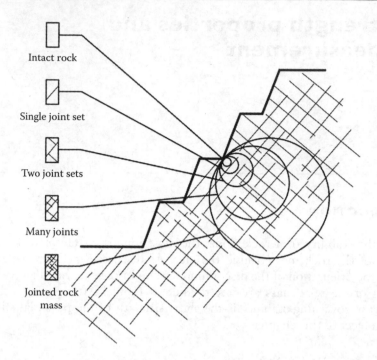

Figure 5.1 Idealised diagram showing transition from intact rock to jointed rock mass with increasing sample size.

Based on this relationship between sample size and rock strength characteristics, this chapter examines methods of determining the strength of the following three classes of rock:

1. *Discontinuities* – single bedding planes, joints or faults. The properties of discontinuities that influence shear strength include the shape and roughness of the surfaces, the rock on the surface which may be fresh or weathered, and infillings that may be low-strength or cohesive.
2. *Rock mass* – the factors that influence the shear strength of a jointed rock mass include the compressive strength and friction angle of the intact rock, and the spacing of discontinuities and the condition of their surfaces.
3. *Intact rock* – a factor to consider in measuring the strength of the intact rock is that the strength could diminish over the design life of the slope due to weathering.

In this book, the subscript 'i' is used to designate intact rock and the subscript 'm' to designate the rock mass; the respective compressive strengths are σ_{ci} and σ_{cm}.

5.1.2 Examples of rock masses

Figures 5.2 to 5.5 show four different geological conditions that are commonly encountered in the design of rock slopes. These are typical examples of rock masses in which the strength of laboratory-size samples may differ significantly from the shear strength along the overall sliding surface. In all four cases, instability occurs because of shear movement along a sliding surface that either lies along an existing fracture, or passes partially or entirely through

(a)

(b)

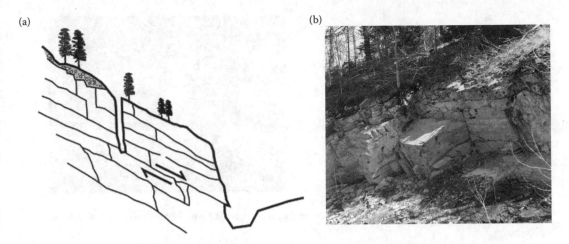

Figure 5.2 (a) Plane failure on continuous bedding plane dipping out of the slope; (b) strong, blocky lime-
    stone, Crowsnest Pass, Alberta, Canada.

intact rock. Figures 5.2 to 5.5 also show that, in fractured rock, the shape of the sliding
surface is influenced by the orientation and length of the discontinuities.

Figure 5.2 shows a strong, blocky limestone containing a set of continuous bedding
surfaces that dip out of the slope face. Because the near-vertical cut face is steeper than
the dip of the bedding, the bedding surfaces are exposed, or 'daylight' on the face and
sliding has occurred with a tension crack opening along the sub-vertical, orthogonal joint
set. Under these circumstances, the shear strength used in stability analysis is that of the
bedding surfaces. Section 5.2 of this chapter discusses the strength properties of discontinuity surfaces, and Section 5.3 describes methods of measuring the friction angle in the
laboratory.

(a)

(b)

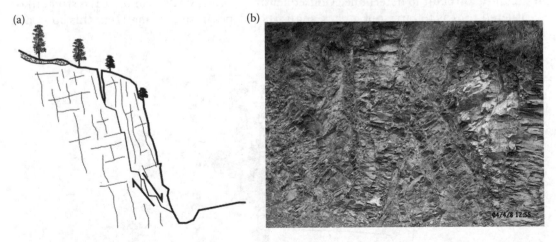

Figure 5.3 (a) Shallow slope movement in closely jointed, randomly orientated rock mass; (b) weathered
    basalt, Island of Oahu, Hawaii.

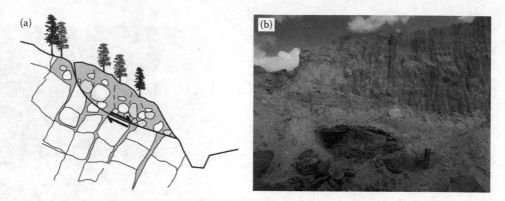

Figure 5.4 (a) Circular failure in residual soil and weathered rock; (b) weathered gneiss, Recife, Brazil.

Figure 5.3 shows a face cut in moderately weathered, medium weak basalt containing very low persistence, closely spaced joints that occur in a wide variety of orientations. Because the joints are discontinuous, no single joint controls the stability of the slope. However, if a sliding surface was to develop in this slope, it would follow a stepped path which would partly lie along joint surfaces, and partly pass through intact rock. The shear strength of this complex sliding surface cannot be determined analytically, so a set of empirical equations has been developed from which the cohesion and friction angle can be calculated with respect to the degree of fracturing, and the rock strength. This procedure is described in Section 5.5.

Figure 5.4 shows a slope cut in a weathered rock in which the degree of weathering varies from residual soil in the upper part of the slope, to intact core stones in a matrix of highly weathered rock in the lower part of the cut. For these conditions, the sliding surface will lie predominantly in the weaker materials in the upper part of the slope. Stability analysis of this formation would require the use of different strength parameters for the upper and lower portions of the slope. Because the degree of degradation of weathered rock tends to be highly variable, the strength of the rock mass will also be variable and can be difficult to determine. Consequently, a means of determining the strength of weathered rock is to carry out a back analysis of slopes in similar material; this approach is described in Section 5.4.

Figure 5.5 (a) Shallow failure in very weak, massive rock containing no discontinuities; (b) Volcanic tuff, Trans European Highway, near Ankara, Turkey.

A fourth geological condition that may be encountered is that of a very weak but intact rock containing essentially no discontinuities. Figure 5.5 shows a cut face in tuff, a rock formed by the consolidation of volcanic ash. A geological hammer could be embedded in the face with a few blows, indicating the low strength of this rock. However, because this rock contains no discontinuities, it has a significant cohesive strength in addition to a moderate friction angle. Therefore, it was possible to cut a stable, near-vertical face to a height of at least 20 m (65 ft) in this material, provided water pressures and erosion were controlled.

5.1.3 Classes of rock strength

Based on the scale effects and geological conditions discussed in the previous sections, it can be seen that sliding surfaces can form either along discontinuity surfaces, or through the rock mass, as illustrated in Figure 5.6. The importance of the classification shown in Figure 5.6 is that in essentially all slope stability analyses it is necessary to use the shear-strength properties of either the discontinuities, or of the rock mass, and the different procedures for determining the strength properties are as follows:

- *Discontinuity shear strength* can be measured in the field and the laboratory as described in Sections 5.2 and 5.3.

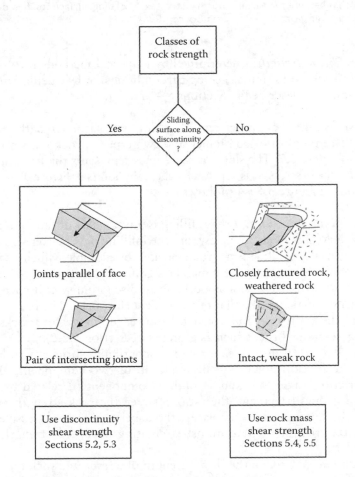

Figure 5.6 Relationship between geology and classes of rock strength.

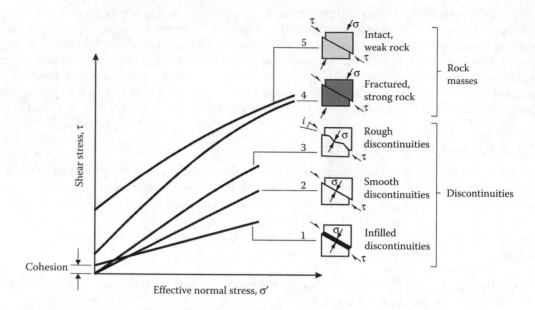

Figure 5.7 Relationship between shear and normal stresses on sliding surface for five different geological conditions (Transportation Research Board, 1996).

- *Rock mass shear strength* is determined by empirical methods involving either back analysis of slopes cut in similar geological conditions, or by calculation involving rock strength indices as described in Sections 5.4 and 5.5.

As a further illustration of the effects of geology on shear strength, typical strength parameters for three types of discontinuity and two types of rock mass are shown on the Mohr diagram in Figure 5.7. The slope of these lines represents the friction angle, and the intercept with the shear stress axis represents the cohesion (see Figure 1.9a). A description of these conditions in Figure 5.7 is as follows:

Curve 1. *Infilled discontinuity*: if the infilling is a weak clay or fault gouge, the infilling friction angle (ϕ_{inf}) is likely to be low, but the infilling may be cohesive if undisturbed. Alternatively, if the infilling is a strong calcite, for example, which produces a healed surface, then the cohesive strength may be significant (see Section 5.2.5).

Curve 2. *Smooth discontinuity*: a smooth, clean discontinuity will have zero cohesion, and the friction angle will be that of the rock surfaces (ϕ_r). The friction angle of rock is related to the grain size, and is generally lower in fine-grained rocks such as shale, than in coarse-grained rocks such as granite (see Section 5.2.2).

Curve 3. *Rough discontinuity*: clean, rough discontinuity surfaces will have zero cohesion, and the friction angle will be made up of two components. First, the rock material friction angle (ϕ_r), and second, a component (i) related to the roughness (asperities) of the surface and the ratio between the rock strength and the normal stress. As the normal stress increases, the asperities are progressively sheared off and the total friction angle diminishes creating a curved strength envelope (see Section 5.2.4).

Curve 4. *Fractured rock mass*: the shear strength of a fractured rock mass, in which the sliding surface lies partially on discontinuity surfaces and partially passes through intact rock, can be expressed as a curved envelope. At low-normal stresses where the confinement of the fractured rock is limited and the individual fragments may move and rotate, the cohesion is low but the friction angle is high. At higher normal stresses, crushing of the rock fragments begins to take place with the result that the friction angle diminishes. The shape of the strength envelope is related both to the degree of fracturing, and the strength of the intact rock (see Section 5.5).

Curve 5. *Weak intact rock*: rock such as the tuff shown in Figure 5.5 is composed of fine-grained volcanic ash in which the ash particles are bound (cemented) together. Therefore, while the rock has a low-friction angle, because it contains no discontinuities, the cohesion can be higher than that of a strong, but closely fractured rock.

The range of shear-strength conditions that may be encountered in rock slopes as illustrated in Figures 5.2 to 5.5 clearly demonstrates the importance of examining both the characteristics of the discontinuities and the rock strength during the site investigation.

5.2 SHEAR STRENGTH OF DISCONTINUITIES

If geological mapping and/or diamond drilling identify discontinuities on which shear failure could take place, it will be necessary to determine the friction angle and cohesion of the sliding surface in order to carry out stability analyses. The investigation programme should also obtain information on characteristics of the sliding surface that may modify the shear-strength parameters. Important discontinuity characteristics include continuous length, surface roughness, and the thickness and characteristics of any infilling, as well as the effect of water on the properties of the infilling.

The following sections describe the relationship between the shear strength and the properties of the discontinuities.

5.2.1 Definition of cohesion and friction

In rock slope design, rock is assumed to be a Coulomb material in which the shear strength of the sliding surface is expressed in terms of the cohesion (c) and the friction angle (ϕ) (Coulomb, 1773). The application of these two strength parameters to rock is discussed in the following paragraphs.

Assume a number of test samples were cut from a block of rock containing a smooth, planar discontinuity. Furthermore, the discontinuity contains a cemented infilling material such that a tensile force would have to be applied to the two halves of the sample in order to separate them. Each sample is subjected to a force at right angles to the discontinuity surface (normal stress, σ), and a force is applied in the direction parallel to the discontinuity (shear stress, τ) while the shear displacement (δ_s) is measured (Figure 5.8a).

For a test carried out at a constant normal stress, a typical plot of the shear stress against the shear displacement is shown in Figure 5.8b. At small displacements, the specimen behaves elastically and the shear stress increases linearly with displacement. As the force resisting movement is overcome, the curve becomes non-linear and then reaches a maximum that represents the *peak* shear strength of the discontinuity. Thereafter, the stress required to cause displacement decreases and eventually reaches a constant value termed the *residual* shear strength.

If the peak shear-strength values from tests carried out at different normal stress levels are plotted, a relationship shown in Figure 5.8c is obtained; this is termed a Mohr diagram (Mohr, 1900). The features of this plot are first that it is approximately linear and

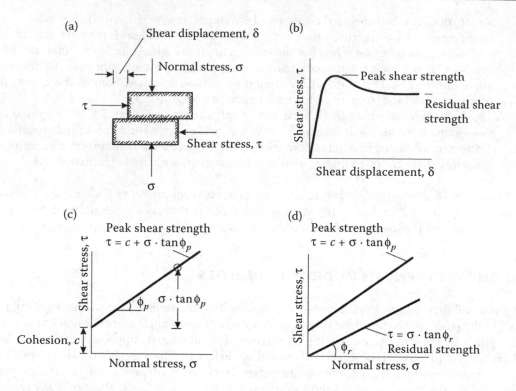

Figure 5.8 Definition of shear strength of discontinuity surface: (a) shear test of discontinuity; (b) plot of shear displacement versus shear stress; (c) Mohr plot of peak strength; (d) Mohr plot of peak and residual strength.

the slope of the line is equal to the peak friction angle ϕ_p of the rock surface. Second, the intercept of the line with the shear stress axis represents the cohesive strength c of the cementing material. This cohesive component of the total shear strength is independent of the normal stress, but the frictional component increases with increasing normal stress. Based on the relationship illustrated in Figure 5.8c, the peak shear strength is defined by the equation

$$\tau = c + \sigma \cdot \tan \phi_p \tag{5.1}$$

If the residual shear stress values at each applied normal stress are plotted on the Mohr diagram, the residual shear-strength line is obtained as shown in Figure 5.8d, and is defined by the equation

$$\tau = \sigma \cdot \tan \phi_r \tag{5.2}$$

where ϕ_r is the residual friction angle. For the residual strength condition, the cohesion is lost once displacement has broken the cementing action; on the Mohr diagram, this is represented by the strength line passing through the origin of the graph. Also, the residual friction angle is less than the peak friction angle because the shear displacement grinds the minor irregularities on the rock surface and produces a smoother, lower friction surface.

Table 5.1 Typical ranges of friction angles for a variety of rock types

Rock class	Friction angle range	Typical rock types
Low friction	20–27°	Schists (high mica content), shale, marl
Medium friction	27–34°	Sandstone, siltstone, chalk, gneiss, slate
High friction	34–40°	Basalt, granite, limestone, conglomerate

5.2.2 Friction angle of rock surfaces

For a planar, clean (no infilling) discontinuity, the cohesion will be zero and the shear strength will be defined solely by the friction angle. The friction angle of the rock material is related to the size and shape of the grains exposed on the fracture surface. Thus, a fine-grained rock such as shale, and rock with a high mica content aligned parallel to the surface such as a phyllite, will tend to have a low-friction angle, a course-grained rock such as a granite will have a high friction angle. Table 5.1 shows typical ranges of friction angles for a variety of rock types (Barton, 1973; Jaeger and Cook, 1976).

The friction angles listed in Table 5.1 should be used as a guideline only because actual values will vary widely with site conditions. Laboratory testing procedures to determine friction angle are described in Section 5.3.

5.2.3 Shearing on an inclined plane

In the previous section, it was assumed that the discontinuity surface along which shearing occurs is exactly parallel to the direction of the shear stress, τ. Consider now a discontinuity surface inclined at an angle i to the shear stress direction (Figure 5.9). In this case, the shear and normal stresses acting on the sliding surface, τ_i and σ_i respectively, are given by

$$\tau_i = \tau \cdot \cos^2 i - \sigma \cdot \sin i \cdot \cos i \tag{5.3}$$

$$\sigma_i = \sigma \cdot \cos^2 i + \tau \cdot \sin i \cdot \cos i \tag{5.4}$$

If it is assumed that the discontinuity surface has zero cohesion and that its shear strength is given by

$$\tau_i = \sigma_i \cdot \tan \phi \tag{5.5}$$

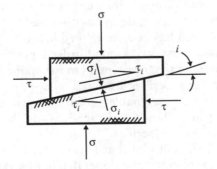

Figure 5.9 Shear displacement on an inclined plane.

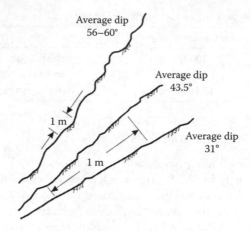

Figure 5.10 Patton's observations of bedding plane traces in unstable limestone slopes (Patton, 1966).

then Equations 5.3 and 5.4 can be substituted in Equation 5.5 to give the relationship between the applied shear stress and the normal stress as

$$\tau = \sigma \cdot \tan(\phi + i) \tag{5.6}$$

This equation was confirmed in a series of tests on models with regular surface projections carried out by Patton (1966) who must be credited with having emphasised the importance of this simple relationship in the analysis of rock slope stability.

Patton convincingly demonstrated the particular significance of this relationship by measuring the average value of the angle i from photographs of bedding plane traces in unstable limestone slopes. Three of these are reproduced in Figure 5.10 showing that the rougher the bedding plane trace, the steeper the slope angle. Patton found that the inclination of the bedding plane trace was approximately equal to the sum of the friction angle of the rock ϕ found from laboratory tests on planar surfaces, and the average roughness i angle.

5.2.4 Surface roughness

All natural discontinuity surfaces exhibit some degree of roughness, varying from polished and slickensided sheared surfaces with very low roughness, to rough and irregular tension joints with considerable roughness. These surface irregularities are given the general term *asperities*, and because they can have a significant effect on the stability of a slope, they should be accounted for appropriately in design as discussed in this section.

The discussion in Section 5.2.3 has been simplified because Patton found that asperities can be divided into two classes: first- and second-order asperities as shown in Figure 5.11. The first-order asperities are those that correspond to the major undulations on the bedding surfaces, while the second-order asperities are small bumps and ripples on the surface have higher i values. In order to obtain reasonable agreement between field observations of the dip of the unstable bedding planes shown in Figure 5.10 and the $(\phi + i)$ values, it was necessary to measure only the first-order asperities.

Later studies by Barton (1973) showed that Patton's results were related to the normal stress acting across the bedding planes in the slopes that he observed. At low-normal stresses, the second-order projections come into play and Barton quotes $(\phi + i)$ values in the range of

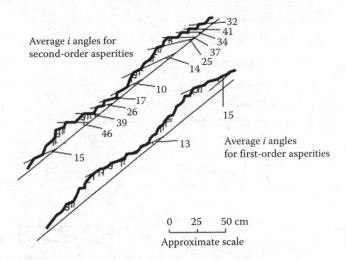

Figure 5.11 Measurement of roughness angles i for first- and second-order asperities on rough rock surfaces (Patton, 1966).

69–80° for tests conducted at low-normal stresses ranging from 20 to 670 kPa (3–100 psi) (Goodman, 1970; Paulding, 1970; Rengers, 1971). Assuming a friction angle for the rock of 30°, these results show that the effective roughness angle i varies between 40° and 50° for these low-normal stress levels.

The actual shear performance of discontinuity surfaces in rock slopes depends on the combined effects of the surface roughness, the rock strength at the surface, the applied normal stress and the amount of shear displacement. This is illustrated in Figure 5.12 where the asperities are sheared off, with a consequent reduction in the friction angle with increasing normal stress. That is, transition occurs from dilation to shearing of the rock. The extent to which the asperities are sheared will depend on both the magnitude of the normal force in relation to the compressive strength of the rock on the fracture surface, and the displacement distance. A rough surface that is initially undisturbed and interlocked will have a peak friction angle of $(\phi + i)$. With increasing normal stress and displacement, the asperities will be sheared off, and the friction angle will progressively diminish to a minimum value of the basic, or residual, friction angle of the rock. This dilation-shearing condition is represented on the Mohr diagram as a curved strength envelope with an initial slope equal to $(\phi + i)$, reducing to ϕ_r at higher normal stresses.

The shear stress–normal stress relationship shown in Figure 5.12 can be quantified using a technique developed by Barton (1973) based on the shear-strength behaviour of artificially produced rough, clean 'joints'. The study showed that the shear strength of a rough rock surface depends on the relationship between the roughness, the rock strength and the normal stress, and can be defined by the following empirical equation:

$$\tau = \sigma' \cdot \tan\left(\phi + JRC \cdot \log_{10}\left(\frac{JCS}{\sigma'}\right)\right) \tag{5.7}$$

where JRC is the joint roughness coefficient, JCS is the compressive strength of the rock at the fracture surface and σ' is the effective normal stress. The value of JRC is determined using the techniques discussed in Section 4.4.3 that involve either comparing the roughness

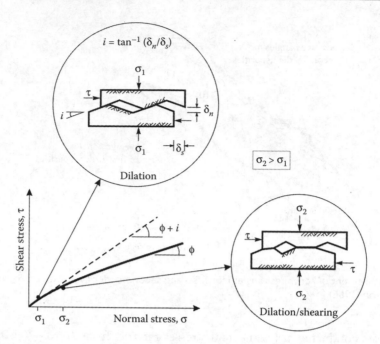

Figure 5.12 Effect of surface roughness and normal stress on friction angle of discontinuity surface (Transportation Research Board, 1996).

of the surface with standard roughness profiles, or making measurements of the surface. The compressive strength of the rock (JCS) can be estimated using the simple field observations described in Table 5.1, or with a Schmidt hammer to measure the rock on the joint surface. The normal stress σ_N, acting on the surface, is the component acting normal to the surface, of the product of the height of rock (H) above the sliding surface and the unit weight of the rock ($\sigma_N = H \cdot \gamma_r \cdot \cos(\psi_p)$).

In Equation 5.7, the term ($JRC \cdot \log_{10}(JCS/\sigma')$) is equivalent to the roughness angle i in Equation 5.6. At high stress levels, relative to the rock strength, when (JCS/σ') = 1 and the asperities are sheared off, the term ($JRC \cdot \log_{10}(JCS/\sigma')$) equals zero. At low stress levels, the ratio (JCS/σ') tends to approach infinity and the roughness component of the shear strength becomes very large. In order that realistic values of the roughness component are used in design, the term ($\phi + i$) should not exceed about 50° and the useful range for the ratio (JCS/σ') is between about 3 and 100.

It is found that both JRC and JCS values are influenced by scale effects, that is, as the discontinuity size increases, a corresponding decrease occurs in JRC and JCS values. The reason for this relationship is that small-scale roughness of a surface becomes less significant compared to the dimensions of the discontinuity, and eventually large-scale undulations have more significance than the roughness (Barton and Bandis, 1983; Bandis, 1993). These effects are consistent with properties of first- and second-order asperities shown in Figure 5.11. The scale effect can be quantified by the following two equations:

$$JRC_n = JRC_0 \cdot \left(\frac{L_n}{L_0}\right)^{-0.02 \cdot JRC_0} \quad ; \quad \text{and} \quad JCS_n = JCS_0 \cdot \left(\frac{L_n}{L_0}\right)^{-0.03 \cdot JRC_0} \tag{5.8}$$

where L_0 is the dimension of the surface used to measure JRC such as the wire comb (Figure 4.10), and L_n is the dimension of the sliding surface.

An example of the application of Equations 5.7 and 5.8 is as follows. Consider a discontinuity dipping at $\psi_p = 35°$, and the dimension of this surface is 10 m (L_n) (32.8 ft). If the average depth (H) of the discontinuity is 20 m (66 ft) below the crest of the slope, and the rock unit weight (γ_r) is 26 kN/m³ (165 lb/ft³), then the effective normal stress (σ') for a dry slope is

$$\sigma' = \gamma_r \cdot H \cdot \cos(\psi_p)$$
$$= 26 \cdot 20 \cdot \cos(35)$$
$$= 426 \text{ kPa (61.8 psi)}$$

If the JRC_0 value measured with a wire comb ($L_0 = 0.2$ m) is 15, and the rock is strong with a JCS_0 value of 50,000 kPa (7250 psi), then the scaled values are

$$JRC_n = 15\left(\frac{10}{0.2}\right)^{-0.02 \cdot 15} \approx 5; \quad \text{and} \quad JCS_n = 50,000\left(\frac{10}{0.2}\right)^{-0.03 \cdot 15} \approx 8600 \text{ kPa (1247 psi)}$$

and the roughness angle is

$$JRC \cdot \log_{10}(JCS/\sigma') = 5 \cdot \log_{10}(3050/426)$$
$$\approx 4°$$

The concepts of shear strength of rough joints and scale effects discussed in this section can be applied to rock slope design as illustrated in Figure 5.13. For example, stress relief, and possibly blast damage during construction, can cause shear movement along discontinuities and dilation of the rock at the face. Also, the stress levels on the sliding surface may be high enough to cause some shearing of the asperities (Figure 5.13a). For these conditions, the first effect is that any cohesion existing on the undisturbed surface is diminished. The second effect is loss of interlock on the rough surface so that the second-order asperities

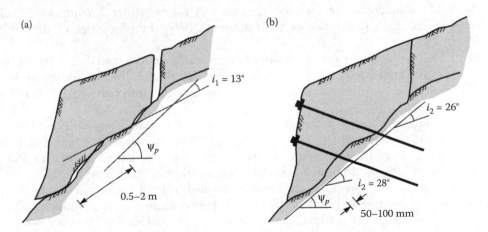

Figure 5.13 Effect of asperities on stability of sliding blocks: (a) shear strength of displaced block controlled by first-order asperities (i_1); (b) tensioned rock bolts prevent dilation along potential sliding surface and produce interlock along second-order asperities s (i_2).

have a diminished effect on the shear strength. The consequence of this dilation is that the roughness angle corresponding to undulating first-order asperities would be used in design.

In contrast to the displaced block shown in Figure 5.13a, the rock mass can be prevented from movement and dilation by the use of careful blasting (see Chapter 13), and installation of tensioned rock anchors or passive support such as dowels and buttresses (see Chapter 14). Under these conditions, interlock along the sliding surface is maintained and the rougher, second-order asperities contribute to the shear strength of the potential sliding surface.

The difference in the total friction angle between the displaced and undisturbed rock will have a significant effect on stability and design of stabilisation measures. This demonstrates the value of using construction measures that minimise relaxation and dilation of rock masses.

5.2.5 Discontinuity infilling

The preceding section discussed rough, clean discontinuity surfaces with rock-to-rock contact and no infilling, in which the shear strength is derived solely from the friction angle of the rock material. However, if the discontinuity contains an infilling, the shear-strength properties of the fracture are often modified, with both the cohesion and friction angle of the surface being influenced by the thickness and properties of the infilling. For example, for a clay-filled fault zone in granite, it would be assumed that the shear strength of the discontinuity would be that of the clay and not the granite. In the case of a healed, calcite-filled fracture, a high cohesion would be used in design, but only if it were certain that the discontinuity would remain healed after any disturbance caused by blasting when excavating the slope.

The presence of infillings along discontinuity surfaces can have a significant effect on stability. It is important that infillings be identified in the investigation programme, and that appropriate strength parameters be used in design. For example, one of the contributing factors to the massive landslide into the Vajont Reservoir in Italy that resulted in the death of about 2000 people was the presence of low-shear strength clay along the bedding surfaces of the shale (Trollope, 1980).

The effect of the infilling on shear strength will depend on both the thickness and strength properties of the infilling material. With respect to the thickness, if it is more than about 25%–50% of the amplitude of the asperities, little or no rock-to-rock contact will occur, and the shear-strength properties of the fracture will be the properties of the infilling (Goodman, 1970).

Figure 5.14 is a plot of the results of direct shear tests carried out to determine the peak friction angle and cohesion of filled discontinuities (Barton, 1973). Examination of the test results shows that the infillings can be divided approximately into two groups, as follows:

- *Clays* – montmorillonite and bentonitic clays, and clays associated with coal measures have friction angles ranging from about 8–20° and cohesion values ranging from 0 to about 200 kPa (29 psi). Some cohesion values were measured as high as 380 kPa (55.1 psi), which would probably be associated with very stiff clays.
- *Faults, shears and breccias* – the material formed in fault zones and shears in rocks such as granite, diorite, basalt and limestone may contain clay in addition to granular rock fragments. These materials have friction angles ranging from about 25–45°, and cohesion values ranging from 0 to about 100 kPa (14.5 psi). Fault gouge derived from coarse-grained rocks such as granites tends to have higher friction angles than those from fine-grained rocks such as limestones.

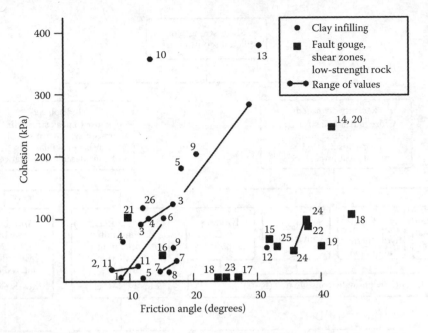

Figure 5.14 Shear strength of filled discontinuities (modified from Barton, 1974).

1. Bentonite shale
2. Bentonite seams in chalk
3. Bentonite; thin layers
4. Bentonite; triaxial tests
5. Clay, over consolidated
6. Limestone, 10–20 mm clay infillings
7. Lignite and underlying clay contact
8. Coal measures; clay mylonite seams
9. Limestone; <1 mm clay infillings
10. Montmorillonite clay
11. Montmorillonite; 80 mm clay seam in chalk
12. Schists/quartzites; stratification, thick clay
13. Schists/quartzites; stratification, thick clay
14. Basalt; clayey, basaltic breccia
15. Clay shale; triaxial tests
16. Dolomite, altered shale bed
17. Diorite/granodiorite; clay gauge
18. Granite; clay-filled faults
19. Granite; sandy-loam fault fillings
20. Granite; shear zone, rock and gauge
21. Lignite/marl contact
22. Limestone/marl/lignites; lignite layers
23. Limestone; marlaceous joints
24. Quartz/kaolin/pyrolusite; remolded triaxial
25. Slates; finely laminated and altered
26. Limestone; 10–20 mm clay infillings

Some of the tests shown in Figure 5.14 also determine residual shear-strength values. It was found that the residual friction angle was only about 2–4° less than the peak friction angle, while the residual cohesion was essentially zero.

Shear strength–displacement behaviour is an additional factor to consider regarding the shear strength of filled discontinuities. In analysing the stability of slopes, this behaviour will indicate whether a reduction in shear strength with displacement is likely. In conditions where a significant decrease in shear strength occurs with displacement, slope failure can occur suddenly following a small amount of movement.

Filled discontinuities can be divided into two general categories, depending on whether previous displacement of the discontinuity has occurred (Barton, 1973). These categories are further subdivided into either normally consolidated (N–C) or over-consolidated (O–C) materials (Figure 5.15):

1. *Recently displaced discontinuities* – these discontinuities include faults, shear zones, clay mylonites and bedding-surface slips. In faults and shear zones, the infilling is

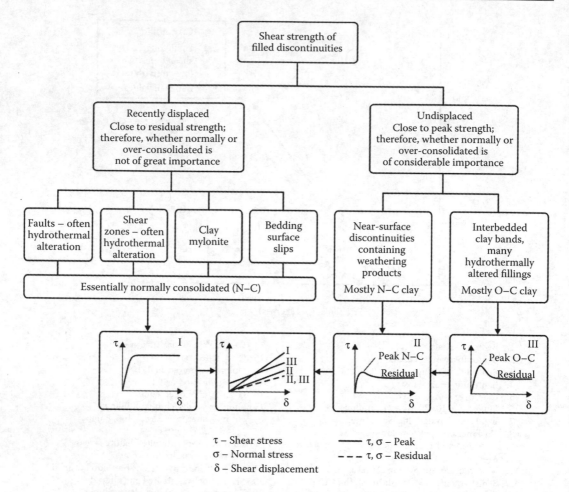

Figure 5.15 Division of filled discontinuities into displaced and undisplaced, and N–C and O–C categories (modified from Barton, 1974).

formed by the shearing process which may have occurred many times and produced considerable displacement. The gouge formed in this process may include both clay-size particles, and breccia with the particle orientation and striations of the breccia aligned parallel to the direction of shearing. In contrast, the mylonites and bedding-surface slips are discontinuities that were originally clay bearing, and along which sliding occurred during folding or faulting.

For these types of discontinuities their shear strength will be at, or close to, the residual strength (Figure 5.15, graph I). Any cohesive bonds that existed in the clay due to previous over-consolidation will have been destroyed by shearing and the infilling will be equivalent to the normally consolidated state. In addition, strain softening may occur, with any increase in water content resulting in a further strength reduction.

2. *Undisplaced discontinuities* – infilled discontinuities that have undergone no previous displacement include igneous and metamorphic rocks that have weathered along the discontinuity to form a clay layer. For example, diabase weathers to amphibolite and eventually to clay. Other undisplaced discontinuities include thin beds of clay

and weak shales that are found with sandstone in interbedded sedimentary formations. Hydrothermal alteration is another process that forms infillings that can include low-strength materials such as montmorillonite, and high-strength materials such as quartz and calcite.

The infillings of undisplaced discontinuities can be divided into normally consolidated (N–C) and over-consolidated (O–C) materials that have significant differences in peak strength values. This strength difference is illustrated in Figure 5.15, graphs II and III. While the peak shear strength of over-consolidated clay infillings may be high, a significant loss of strength can occur due to softening, swelling and pore pressure changes on unloading. Unloading occurs when rock is excavated for a slope or foundation, for example. Strength loss also occurs on displacement in brittle materials such as calcite.

5.2.6 Influence of water on shear strength of discontinuities

The most important influence of water in a discontinuity is the diminished shear strength resulting from the reduction of the effective normal–shear stress acting on the surface. The effective normal stress is the difference between the weight of the overlying rock and the uplift pressure produced by the water pressure. The effect of water pressure, U, on the shear strength can be incorporated into the shear-strength equation as follows:

$$\tau = c + (\sigma - U) \cdot \tan \phi_p - \text{peak strength} \qquad (5.9)$$

or

$$\tau = (\sigma - U) \cdot \tan \phi_r - \text{residual strength} \qquad (5.10)$$

These equations assume that the cohesion and friction angle are not changed by the presence of water on the surface. In most hard rock and in many sandy soils and gravels, the strength properties are not altered by water. However, many clay, shale and mudstones and similar materials will exhibit significant reduction in strength with changes in moisture content. It is important therefore that the moisture content of the test samples be as close as possible to that of the *in situ* rock (see Section 4.6.3 regarding the preservation of rock samples).

5.3 LABORATORY TESTING OF SHEAR STRENGTH

The friction angle of a discontinuity surface can be determined in the laboratory using a direct shear box of the type shown in Figure 5.16. This is portable equipment that can be used in the field if required, and is suited to testing samples with dimensions up to about 75 mm, such as NQ and HQ drill core. The most reliable values are obtained if a sample with a smooth, planar surface is used because it is found that, with an irregular surface, the effect of surface roughness can make the test results difficult to interpret.

The test procedure consists of using plaster of Paris or molten sulphur to set the two halves of the sample in a pair of steel boxes (ISRM, 1981a,b). Particular care is taken to ensure that the two pieces of core are in their original, matched position, and the discontinuity surface is exactly parallel to the direction of the shear force. A constant normal load is then applied using the cantilever, and the shear load gradually increased until sliding failure occurs. Measurement of the vertical and horizontal displacement of the upper block relative

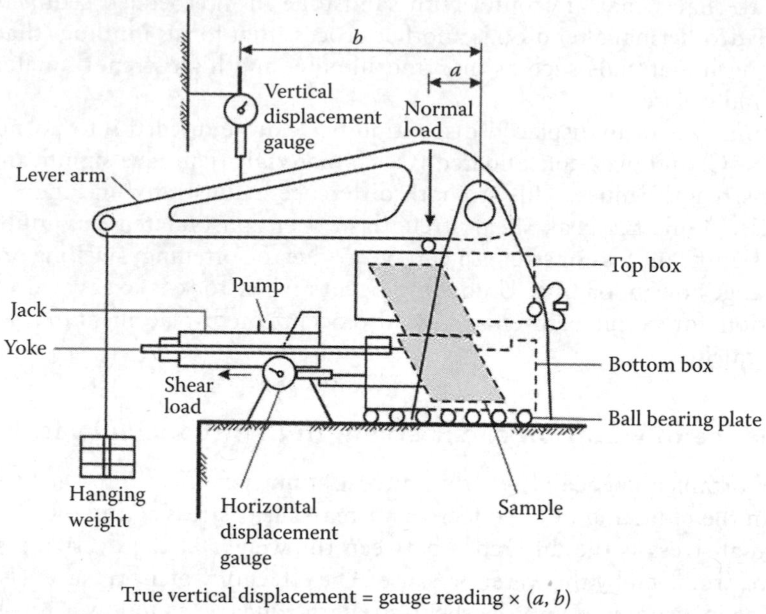

True vertical displacement = gauge reading × (a, b)

Figure 5.16 Simple equipment for performing direct shear tests on rock samples up to about 75 mm (3 in.) dimensions.

to the lower block can be made most simply with dial gauges, while more precise, continuous displacement measurements can be made with linear variable differential transformers (LVDTs) (Hencher and Richards, 1989).

Each sample is usually tested three or four times, at progressively higher normal loads. That is, when the residual shear stress has been established for a normal load, the sample is reset, the normal load increased and another shear test conducted. The test results are expressed as plots of shear displacement against shear stress from which the peak and residual shear stress values are determined. Each test will produce a pair of shear stress–normal stress values, which are plotted to determine the peak and residual friction angles of the surface.

Figure 5.17 shows a typical result of a direct shear test on a surface with a 4 mm (0.16 in.) thick sandy silt infilling. The curves on the upper right are shear stress–shear displacement plots showing an approximate peak shear stress, as well as a slightly lower, residual shear stress. The sample was initially undisplaced, and so exhibited a difference between peak and residual strengths (see Figure 5.8). The normal stresses at the peak and residual shear stress values are calculated from the applied normal load and the contact area. When calculating the contact area, an allowance is made for the decrease in area as shear displacement takes place. For diamond drill core in an inclined hole, the fracture surface is in the shape of an ellipse, and the formula for calculating the contact area is as follows (Hencher and Richards, 1989):

$$A = \pi \cdot a \cdot b - \left[\frac{\delta_s \cdot b \, (4a^2 - \delta_s^2)^{1/2}}{2a} \right] - 2 \cdot a \cdot b \cdot \sin^{-1}\left(\frac{\delta_s}{2a} \right) \tag{5.11}$$

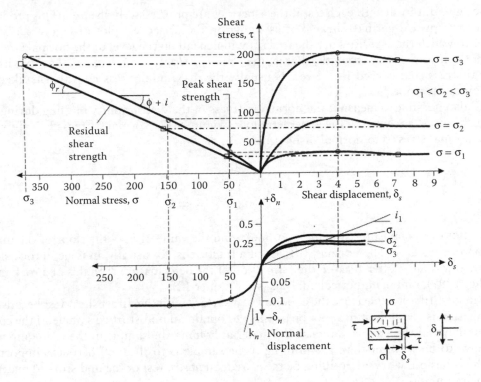

Figure 5.17 Results of direct shear test of filled discontinuity showing measurements of shear strength, roughness (*i*) and normal stuffiness (k_n) (modified from Erban and Gill, 1988).

where *A* is the gross area of contact, *2a* the major axis of ellipse, *2b* the minor axis of ellipse and δ_s the relative shear displacement.

The increase in the normal stress with displacement for a constant normal load is shown in the upper left diagram of Figure 5.17, where the normal stress for the residual shear stress is greater than that for the peak shear stress.

The measured friction angle is the sum of the friction angle of the rock (ϕ_r), and the roughness of the surface (*i*). The roughness of the surface is calculated from the plots of shear and normal displacement, (δ_s and δ_n, respectively, on the lower right side of Figure 5.17) as follows:

$$i = \tan^{-1}\left(\frac{\delta_n}{\delta_s}\right) \tag{5.12}$$

This value of *i* is then subtracted from the friction angle calculated from the plot of shear and normal stresses at failure to obtain the friction angle of the rock. While the shear test can be conducted on a sawed sample on which the roughness component is nil, the saw may polish the surface resulting in a low value of the friction angle compared to a natural surface.

As shown in Figure 5.17, it is usual to test each sample at a minimum of three normal stress levels, with the sample being reset to its original position between tests. When the tests are run at progressively higher normal stress levels, the total friction angle of the surface will diminish with each test if the asperities are progressively sheared. This produces a concave upward normal–shear plot as shown in the upper left plot of Figure 5.17. The extent to which the asperities are sheared off will depend on the level of the normal stress in comparison to the rock strength, that is, the ratio (JCS/σ) in Equation 5.7. The maximum normal stress that is used in the test is usually the maximum stress level that is likely to develop in the slope.

It is also possible to measure the normal stiffness of the discontinuity infilling during the direct shear test as shown in the lower left plot in Figure 5.17. Normal stiffness k_n is the ratio of the normal stress σ to normal displacement δ_n, or

$$k_n = \left(\frac{\sigma}{\delta_n} \right) \tag{5.13}$$

The plot of σ against δ_n is highly non-linear and the value of k_n is the slope of the initial portion of the curve. The normal stiffness of a fracture is not usually an issue in rock slope design, and is more often used in the estimation of the deformation modulus of a rock mass (Wyllie, 1999), and in numerical analysis (see Chapter 12).

It can be difficult to measure the cohesion of a surface with the direct shear test because, if the cohesion is very low, it may not be possible to obtain an undisturbed sample. If the cohesion is high and the sample is intact, the material holding the sample in the test equipment will have to be stronger than the infilling if the sample is to shear. Where it is important that the cohesion of a weak infilling be measured, an *in situ* test of the undisturbed material may be required.

5.4 SHEAR STRENGTH OF ROCK MASSES BY BACK ANALYSIS OF SLOPE FAILURES

For the geological conditions shown in Figure 5.3 where a cut has been made in fractured rock, no distinct discontinuity surface exists on which sliding can take place. A sliding surface in this rock mass will comprise both natural discontinuities aligned on the sliding surface, together with some shear failure through intact rock. It is generally difficult and expensive to sample and test large samples (~1 m diameter) of fractured rock. Consequently, two empirical methods of determining the friction angle and cohesion of rock masses have been developed – first, back analysis that is described in this section, and second, the Hoek–Brown strength criterion that is discussed in Section 5.5. In both methods, it is necessary to categorise the rock mass in terms of both the intact rock strength and the characteristics of the discontinuities. This may require some judgement and it is advisable, where possible, to compare the strength values obtained by both methods to improve the reliability of values used in design.

Probably, the most reliable method of determining the strength of a rock mass is to back analyse a failed, or failing slope. This involves carrying out a stability analysis using available information on the position of the sliding surface, the groundwater conditions at the time of failure and any external forces such as foundation loads and earthquake ground motions, if applicable. With the factor of safety set at 1.0, the stability analysis is used to calculate the friction angle and cohesion. This section describes the back analysis of a failed slope in a limestone quarry, and the use of the calculated shear-strength values to design an appropriate slope angle for a deeper pit (Roberts and Hoek, 1971).

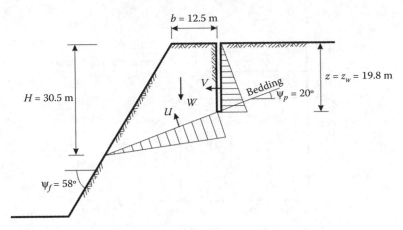

Figure 5.18 Cross-section of quarry slope failure showing geometry and water forces.

Figure 5.18 shows the geometry of the slope failure in which sliding occurred on bedding planes striking parallel to the face, and dipping out of the face at an angle of 20°. These conditions are applicable to plane failure conditions as shown in Figure 2.16 and described in more detail in Chapter 7.

At the time of failure, a tension crack had opened on the upper, horizontal bench of the quarry and the dimensions of the failed mass were defined by the dimensions H, b, ψ_f and ψ_p. From these dimensions and a rock unit weight of 25.1 kN/m³, the weight of the sliding mass was calculated to be 12.3 MN/m. Immediately prior to failure a heavy rainstorm flooded the upper bench so that the tension crack was full of water ($z_w = z$). It was assumed that water forces acting in the tension crack and on the bedding plane could be represented by triangular distributions with magnitudes of $V = 1.92$ MN/m and $U = 3.26$ MN/m, respectively.

In back analysis, both the friction angle and the cohesion of the sliding surface are unknown, and their values can be estimated by the following method. The likely range of the friction angle can usually be estimated by inspection (see Table 5.1), or by laboratory testing if samples containing bedding planes are available. In this case where the limestone was fine-grained and the bedding planes smooth, it was estimated that the friction angle was in the range of 15–25°. The next step was to carry out a number of stability analyses with a range of cohesion values and a factor of safety of 1.0. The results of this analysis show that, at a friction angle of 20°, the corresponding cohesion value is about 110 kPa (16 psi), while for friction angles of 15° and 25°, the corresponding cohesion values are 130 kPa (19 psi) and 80 kPa (12 psi), respectively (Figure 5.19).

The shear-strength values calculated in this manner can be used to design slopes excavated in this limestone, provided that careful blasting is used to maintain the cohesion on the bedding planes. Figure 5.20 shows the relationship between the factor of safety and the face angle for a 64 m (210 ft) high cut, assuming plane failure on a bedding plane dipping at 20° out of the face. If the slope is drained, the shear strength is sufficient for the face to stand vertically, but if the slope is saturated, the steepest stable slope is about 50°.

In many cases, it may not be feasible to carry out a back analysis of a slope in similar geological conditions to that in which the new slope is to be excavated. In these circumstances, published results of rock mass shear strength can be used in design. Figure 5.21 shows the results of back analyses of slope failures in a variety of geological conditions (as described in Table 5.2), and the shear-strength parameters (ϕ/c values) calculated at failure. By adding

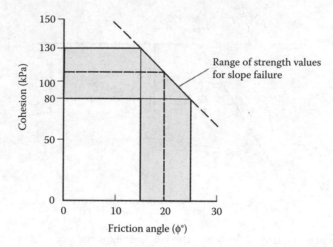

Figure 5.19 Shear strength mobilised on bedding plane for slope failure shown in Figure 5.18.

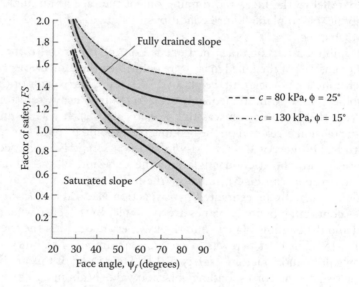

Figure 5.20 Relationship between factor of safety and face angle of dry and saturated slope for slope shown in Figure 5.18.

additional points to Figure 5.21 for local geological conditions, it is possible to draw up a readily applicable rock mass strength chart for shear failures.

5.5 HOEK–BROWN STRENGTH CRITERION FOR FRACTURED ROCK MASSES

As an alternative to back analysis to determine the strength of fractured rock masses, an empirical method has been developed by Hoek and Brown (1980a,b) in which the shear strength is represented as a curved Mohr envelope. This strength criterion was derived from the Griffith crack theory of fracture in brittle rock (Hoek, 1968), as well as from

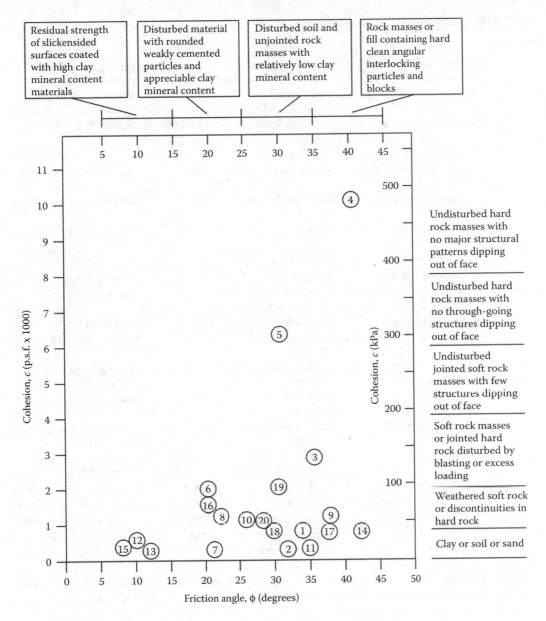

Figure 5.21 Relationship between friction angles and cohesive strength mobilised at failure for slopes analysed in Table 5.2.

observations of the behaviour of rock masses in the laboratory and the field (Marsal, 1967, 1973; Jaegar, 1970; Brown, 1970).

Hoek and Brown introduced their failure criterion to provide input data for the analyses required for the design of underground excavations in hard rock. The criterion started from the properties of intact rock, and then introduced factors to reduce these properties based on the characteristics of joints in a rock mass. The authors sought to link the empirical criterion to geological observations by means of one of the available rock mass classification schemes and, for this purpose, they chose the Rock Mass Rating proposed by Bieniawski (1976).

Table 5.2 Sources of shear-strength data plotted in Figure 5.21

Point number	Material	Location	Slope height (m)	References
1	Disturbed slates and quartzites	Knob Lake, Canada	–	Coates, Gyenge and Stubbins (1965)
2	Soil	Any location	–	Whitman and Bailey (1967)
3	Jointed porphyry	Rio Tinto, Spain	50–110	Hoek (1970)
4	Ore body hanging wall in granite rocks	Grangesberg, Sweden	60–240	Hoek (1974)
5	Rock slopes with slope angles of 50–60°	Any location	300	Ross-Brown (1973)
6	Bedding planes in limestone	Somerset, England	60	Roberts and Hoek (1971)
7	London clay, stiff	England	–	Skempton and Hutchinson (1948)
8	Gravelly alluvium	Pima, Arizona	–	Hamel (1970)
9	Faulted rhyolite	Ruth, Nevada	–	Hamel (1971b)
10	Sedimentary series	Pittsburgh, Pennsylvania	–	Hamel (1971a)
11	Kaolinised granite	Cornwall, England	75	Ley (1972)
12	Clay shale	Fort Peck Dam, Montana	–	Middlebrook (1942)
13	Clay shale	Gardiner Dam, Canada	–	Fleming, Spencer and Banks (1970)
14	Chalk	Chalk Cliffs, England	15	Hutchinson (1970)
15	Bentonite/clay	Oahe Dam, South Dakota	–	Fleming, Spencer and Banks (1970)
16	Clay	Garrison Dam, North Dakota	–	Fleming, Spencer and Banks (1970)
17	Weathered granites	Hong Kong	13–30	Hoek and Richards (1974)
18	Weathered volcanics	Hong Kong	30–100	Hoek and Richards (1974)
19	Sandstone, siltstone	Alberta, Canada	240	Wyllie and Munn (1979)
20	Argillite	Yukon, Canada	100	Wyllie (project files)

Because of the lack of suitable alternatives, the criterion was soon adopted by the rock mechanics community and its use quickly spread beyond the original limits used in deriving the strength reduction relationships. Consequently, it became necessary to reexamine these relationships and introduce new elements from time to time to account for the wide range of practical problems to which the criterion was being applied (Hoek, Carranza-Torres and Corkum, 2002). Typical of these enhancements were the introduction of the idea of 'undisturbed' and 'disturbed' rock masses (Hoek and Brown, 1988), and of a modified criterion to force the rock mass tensile strength to zero for very poor quality rock masses (Hoek, Wood and Shah, 1992).

The original Hoek–Brown strength criterion was defined in terms of the principal stress relationships by the equation

$$\sigma_1' = \sigma_3' + \sigma_{ci}' \left(m \cdot \frac{\sigma_3'}{\sigma_{ci}} + s \right)^{0.5} \tag{5.14}$$

where σ_1' and σ_3' are, respectively, the major and minor effective principal stresses at failure, σ_{ci} is the uniaxial compressive strength of the intact rock material and m and s are material constants; $s = 1$ for intact rock.

However, many geotechnical problems, particularly slope stability analysis, are more conveniently dealt with in terms of shear and normal stresses, and an exact relationship between Equation 5.14 and the normal and shear stresses at failure was derived by J. W. Bray (reported by Hoek [1983]) and later by Ucar (1986), as discussed in Section 5.5.3.

Hoek (1990) discussed the derivation of equivalent friction angles and cohesive strengths for various practical situations. These derivations were based upon tangents to the Mohr envelope derived by Bray. Hoek (1994) suggested that the cohesive strength determined by fitting a tangent to the curvilinear Mohr envelope is an upper bound value and may give optimistic results in stability calculations. Consequently, an average value, determined by fitting a linear Mohr–Coulomb relationship by least squares methods, may be more appropriate. In the 1994 paper, Hoek also introduced the concept of the generalised Hoek–Brown criterion in which the shape of the principal stress plot or the Mohr envelope could be adjusted by means of a variable coefficient a in place of the 0.5 power term in Equation 5.14.

Hoek and Brown (1997) attempted to consolidate all the previous enhancements into a comprehensive presentation of the failure criterion and they gave a number of worked examples to illustrate its practical application.

In addition to the changes in the equations, it was also recognised that the Rock Mass Rating of Bieniawski was no longer adequate as a vehicle for relating the failure criterion to geological observations in the field, particularly for very weak rock masses. This resulted in the introduction of the geological strength index (*GSI*) by Hoek, Wood and Shah (1992), Hoek (1994) and Hoek, Kaiser and Bawden (1995). This index was subsequently extended for weak rock masses in a series of articles by Hoek, Marinos and Benissi (1998), Marinos and Hoek (2000), Marinos and Hoek (2001) and Hoek and Marinos (2000).

The *GSI* provides a system for estimating the reduction in rock mass strength for different geological conditions. Values of *GSI* are related to both the degree of fracturing and the condition of fracture surfaces, as shown in Figures 5.22 and 5.23, respectively, for blocky rock masses and schistose metamorphic rock masses. The strength of a jointed rock mass depends on the properties of the intact rock pieces, as well as the freedom of the rock pieces to slide and rotate under different stress conditions. This freedom is controlled by the geometrical shape of the intact rock pieces and the condition of the surfaces separating the pieces. Angular rock pieces with clean, rough surfaces will result in a much stronger rock mass than one that contains rounded particles surrounded by weathered and altered material.

An important limitation of the Hoek–Brown strength criterion is that it applies to isotropic rock masses, and it is not applicable to highly anisotropic masses such as interbedded shale and sandstone sequences, and strongly foliated metamorphic rock. In these anisotropic rocks, the structure is likely to have a dominant effect on stability and the rock mass strength will be of less importance.

5.5.1 Generalised Hoek–Brown strength criterion

The generalised Hoek–Brown strength criterion is expressed in terms of the major and minor principal stresses, and is modified from Equation 5.14 as follows (Figure 5.24):

$$\sigma_1' = \sigma_3' + \sigma_{ci}\left(m_b \cdot \frac{\sigma_3'}{\sigma_{ci}} + s\right)^a \tag{5.15}$$

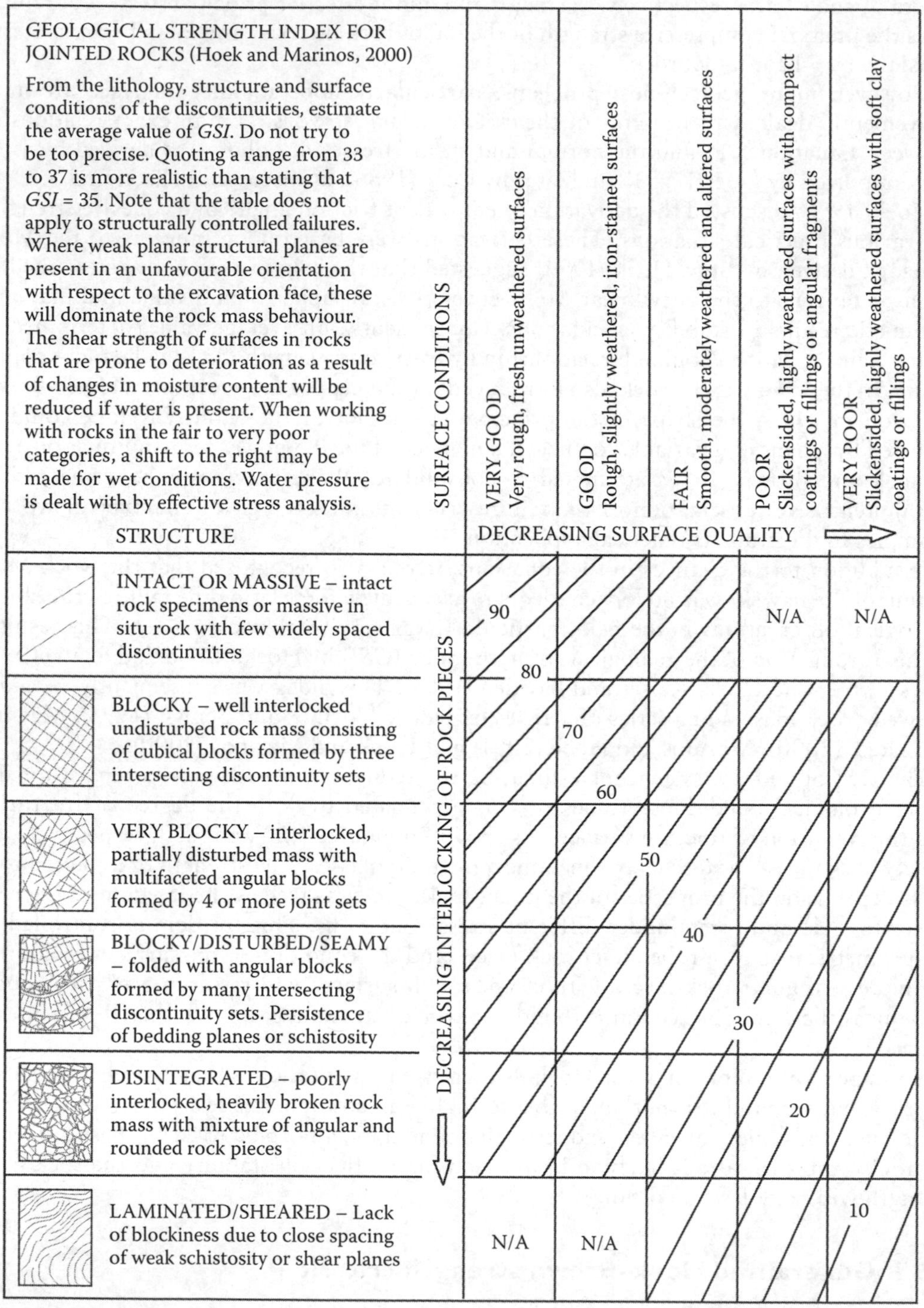

GEOLOGICAL STRENGTH INDEX FOR JOINTED ROCKS (Hoek and Marinos, 2000)

From the lithology, structure and surface conditions of the discontinuities, estimate the average value of *GSI*. Do not try to be too precise. Quoting a range from 33 to 37 is more realistic than stating that *GSI* = 35. Note that the table does not apply to structurally controlled failures. Where weak planar structural planes are present in an unfavourable orientation with respect to the excavation face, these will dominate the rock mass behaviour. The shear strength of surfaces in rocks that are prone to deterioration as a result of changes in moisture content will be reduced if water is present. When working with rocks in the fair to very poor categories, a shift to the right may be made for wet conditions. Water pressure is dealt with by effective stress analysis.

SURFACE CONDITIONS

DECREASING SURFACE QUALITY ⇒

STRUCTURE	VERY GOOD Very rough, fresh unweathered surfaces	GOOD Rough, slightly weathered, iron-stained surfaces	FAIR Smooth, moderately weathered and altered surfaces	POOR Slickensided, highly weathered surfaces with compact coatings or fillings or angular fragments	VERY POOR Slickensided, highly weathered surfaces with soft clay coatings or fillings
INTACT OR MASSIVE – intact rock specimens or massive in situ rock with few widely spaced discontinuities	90 80			N/A	N/A
BLOCKY – well interlocked undisturbed rock mass consisting of cubical blocks formed by three intersecting discontinuity sets		70 60			
VERY BLOCKY – interlocked, partially disturbed mass with multifaceted angular blocks formed by 4 or more joint sets			50		
BLOCKY/DISTURBED/SEAMY – folded with angular blocks formed by many intersecting discontinuity sets. Persistence of bedding planes or schistosity			40	30	
DISINTEGRATED – poorly interlocked, heavily broken rock mass with mixture of angular and rounded rock pieces				20	
LAMINATED/SHEARED – Lack of blockiness due to close spacing of weak schistosity or shear planes	N/A	N/A			10

DECREASING INTERLOCKING OF ROCK PIECES ↓

Figure 5.22 GSI values characterising blocky rock masses on the basis of particle interlocking and discontinuity condition.

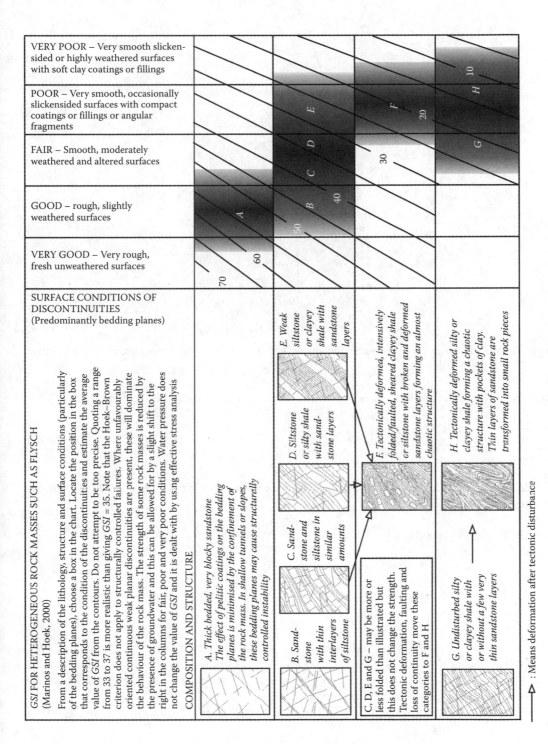

Figure 5.23 GSI values characterising heterogeneous rock masses on the basis of foliation and discontinuity condition.

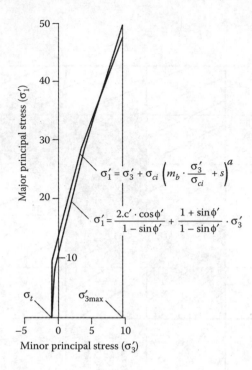

The figure contains, along its curve, the equations:

$$\sigma_1' = \sigma_3' + \sigma_{ci}\left(m_b \cdot \frac{\sigma_3'}{\sigma_{ci}} + s\right)^a$$

$$\sigma_1' = \frac{2.c' \cdot \cos\phi'}{1 - \sin\phi'} + \frac{1 + \sin\phi'}{1 - \sin\phi'} \cdot \sigma_3'$$

Figure 5.24 Relationships between major and minor principal stresses for Hoek–Brown and equivalent Mohr–Coulomb criteria.

where m_b is a reduced value of the material constant m_i for intact rock and is given by

$$m_b = m_i \cdot \exp\left(\frac{GSI - 100}{28 - 14D}\right) \tag{5.16}$$

Table 5.3 gives values of m_i for a wide variety of rock types, and s and a are constants for the rock mass given by

$$s = \exp\left(\frac{GSI - 100}{9 - 3D}\right) \tag{5.17}$$

$$a = \frac{1}{2} + \frac{1}{6}(e^{-GSI/15} - e^{-20/3}) \tag{5.18}$$

D is a factor that depends upon the extent of disturbance to which the rock mass has been subjected by blast damage and stress relaxation. It varies from 0 for undisturbed *in situ* rock masses, to 1 for very disturbed rock masses; guidelines for the selection of appropriate values for D are discussed in Section 5.5.6.

The uniaxial compressive strength of the rock mass, σ_c, is obtained by setting $\sigma_3' = 0$ in Equation 5.15, giving

$$\sigma_c = \sigma_{ci} \cdot s^a \tag{5.19}$$

Table 5.3 Values of constant m_i for intact rock, by rock type (values in parenthesis are estimates)

Rock type	Class	Group	Texture			
			Coarse	Medium	Fine	Very fine
Sedimentary	Clastic		Conglomerates[a]	Sandstones 17±4	Siltstones 7±2	Claystones 4±2
			Breccias[a]		Greywackes (18±3)	Shales (6±2)
						Maris (7±2)
	Non-clastic	Carbonates	Crystalline Limestone (12±3)	Sparitic Limestones (10±2)	Micritic Limestones (9±2)	Dolomites (9±3)
		Evaporites		Gypsum 8±2	Anhydrite 12±2	
		Organic				Chalk 7±2
Metamorphic	Non-foliated		Marble 9±3	Hornfels (19±4) Metasandstone (19±3)	Quartzites 20±3	
	Slightly foliated		Migmatite (29±3)	Amphibolites 26±6	Gneiss 28±5	
	Foliated[b]			Schists 12±3	Phyllites (7±3)	Slates 7±4
Igneous	Plutonic	Light	Granite 32±3 Granodiorite (29±3)	Diorite 25±5		
		Dark	Gabbro 27±3 Norite 20±5	Dolerite (16±5)		
	Hypabyssal		Porphyries (20±5)		Diabase (15±5)	Peridotite (25±5)
	Volcanic	Lava		Rhyolite (25±5) Andesite 25±5	Dacite (25±3) Basalt (25±5)	
		Pyroclastic	Agglomerate (19±3)	Breccia (19±5)	Tuff (13±5)	

[a] Conglomerates and breccias may present a wide range of m values depending on the nature of the cementing material and the degree of cementation, so they may range from values similar to sandstone, to values used for fine-grained sediments (even under 10).

[b] These values are for intact rock specimens tested normal to bedding or foliation. The value of m_i will be significantly different if failure occurs along a weakness plane.

The tensile strength of a rock mass s is given by the following equation:

$$\sigma_t = -\frac{s \cdot \sigma_{ci}}{m_b} \tag{5.20}$$

Equation 5.20 is obtained by setting $\sigma_1' = \sigma_3' = \sigma_t$ in Equation 5.15. This represents a condition of biaxial tension. Hoek (1983) showed that, for brittle materials, the uniaxial tensile strength is equal to the biaxial tensile strength.

Note that the 'switch' at $GSI = 25$ for the coefficients s and a (Hoek and Brown, 1997) has been eliminated in Equations 5.17 and 5.18, which give smooth continuous transitions for the entire range of GSI values.*

Normal and shear stresses are related to principal stresses by the equations published by Balmer (1952)[†]

$$\sigma_n' = \frac{\sigma_1' + \sigma_3'}{2} - \frac{\sigma_1' - \sigma_3'}{2} \cdot \frac{d\sigma_1'/d\sigma_3' - 1}{d\sigma_1'/d\sigma_3' + 1} \tag{5.21}$$

$$\tau = (\sigma_1' - \sigma_3') \frac{\sqrt{d\sigma_1'/d\sigma_3'}}{d\sigma_1'/d\sigma_3' + 1} \tag{5.22}$$

where

$$d\sigma_1'/d\sigma_3' = 1 + a \cdot m_b (m_b \cdot \sigma_3'/\sigma_{ci} + s)^{a-1} \tag{5.23}$$

5.5.2 Modulus of deformation

The Hoek–Brown failure criterion also allows the rock mass modulus of deformation to be calculated as follows:

$$E_m = \left(1 - \frac{D}{2}\right) \cdot \sqrt{\frac{\sigma_{ci}}{100}} \cdot 10^{(GSI-10)/40} \quad \text{(units: GPa)} \tag{5.24}$$

Note that the original equation proposed by Hoek and Brown (1997) has been modified, by the inclusion of the factor D, to allow for the effects of blast damage and stress relaxation.

The principal use of the rock mass modulus of deformation is in numerical analysis to calculate strain in rock slopes (see Chapter 12), and in foundation design.

5.5.3 Mohr–Coulomb criterion

The analysis of slope stability involves examination of the shear strength of the rock mass, expressed by the Mohr–Coulomb failure criterion, on the sliding surface. Therefore, it is

---

* Note that the numerical values of s and a, given by Equations 5.17 and 5.18, are very close to those given by the previous equations in Hoek and Brown (1997) and it is not necessary to revisit and make corrections to old calculations.

[†] The original equations derived by Balmer contained errors that have been corrected in Equations 5.21 and 5.22.

necessary to determine friction angles and cohesive strengths that are equivalent between the Hoek–Brown and Mohr–Coulomb criteria. These strengths are required for each rock mass and stress range along the sliding surface. This is done by fitting an average linear relationship to the curve generated by solving Equation 5.15 for a range of minor principal stress values defined by $\sigma_t < \sigma_3 < \sigma_{3max}$, as illustrated in Figure 5.24. The fitting process involves balancing the areas above and below the Mohr–Coulomb plot. This results in the following equations for the angle of friction ϕ' and cohesive strength c':

$$\phi' = \sin^{-1}\left[\frac{6a \cdot m_b(s + m_b \cdot \sigma'_{3n})^{a-1}}{2(1+a)(2+a) + 6a \cdot m_b(s + m_b \cdot \sigma'_{3n})^{a-1}}\right] \tag{5.25}$$

$$c' = \frac{\sigma_{ci}\left[(1+2a)s + (1-a)m_b \cdot \sigma'_{3n}\right](s + m_b \cdot \sigma'_{3n})^{a-1}}{(1+a)(2+a)\sqrt{1 + \left(6a \cdot m_b(s + m_b \cdot \sigma'_{3n})^{a-1}\right)/((1+a)(2+a))}} \tag{5.26}$$

where $\sigma_{3n} = \sigma'_{3max}/\sigma_{ci}$.

Figure 5.25 shows a curved Hoek–Brown strength envelope defined by the parameters listed in the caption, and the fitted straight line defining the Mohr–Coulomb cohesion and friction angle shear-strength parameters. The figure also shows points on each of the lines defined by normal and shear stress values.

Note that the value of σ'_{3max}, the upper limit of confining stress over which the relationship between the Hoek–Brown and the Mohr–Coulomb criteria is considered, has to be determined for each individual case. Guidelines for selecting these values for slopes are presented in Section 5.5.5.

The Mohr–Coulomb shear-strength τ, for a given normal stress σ, is found by substitution of these values of c' and ϕ' into the equation

$$\tau = c' + \sigma \cdot \tan\phi' \tag{5.27}$$

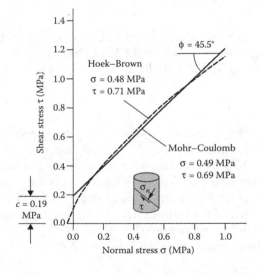

Figure 5.25 Non-linear Mohr envelope for fractured rock mass defined by Equations 5.25 and 5.26; best fit line shows cohesion and friction angle for applicable slope height. Rock mass parameters: $\sigma_c = 30$ MPa (4350 psi), $GSI = 50$, $m_i = 10$, $D = 0.7$, $H = 20$ m, $\gamma_r = 0.026$ MN/m³.

The equivalent plot, in terms of the major and minor principal stresses, is defined by

$$\sigma_1' = \frac{2c' \cdot \cos\phi'}{1 - \sin\phi'} + \frac{1 + \sin\phi'}{1 - \sin\phi'} \cdot \sigma_3' \tag{5.28}$$

5.5.4 Rock mass strength

The uniaxial compressive strength of the rock mass σ_c is given by Equation 5.19. For underground excavations, instability initiates at the boundary of the excavation when the compressive strength σ_c is exceeded by the stress induced on that boundary. The failure propagates from this initiation point into the biaxial stress field and it eventually stabilises when the local strength, defined by Equation 5.15, is higher than the induced stresses σ_1' and σ_3'. Most numerical models can follow this process of fracture propagation and this level of detailed analysis is very important when considering the stability of underground excavations in rock, and designing support systems.

However, for slope stability, failure is initiated along a sliding surface within the slope where the rock is subject to a biaxial stress field and it is useful to consider the overall behaviour of a rock mass rather than the detailed failure propagation process described above. This leads to the concept of a global 'rock mass strength' and Hoek and Brown (1997) proposed that this could be estimated from the Mohr–Coulomb relationship

$$\sigma_{cm}' = \frac{2c' \cdot \cos\phi'}{1 - \sin\phi'} \tag{5.29}$$

with c_c' and ϕ' determined for the stress range $\sigma_t < \sigma_3' < \sigma_{ci}/4$ giving the following value for the rock mass strength σ_{cm}':

$$\sigma_{cm}' = \sigma_{ci} \cdot \frac{(m_b + 4s - a(m_b - 8s)) \cdot (m_b/4 + s)^{a-1}}{2(1+a)(2+a)} \tag{5.30}$$

For the rock mass strength parameters listed in Figure 5.25, the corresponding rock mass strength calculated using Equation 5.30 is 3.16 MPa (460 psi), compared to the compressive strength of the intact rock of 30 MPa (4350 psi).

5.5.5 Determination of σ_{3max}'

The issue of determining the appropriate value of σ_{3max}' for use in Equations 5.25 and 5.26 depends upon the specific application. For the case of slopes, it is necessary that the calculated factor of safety and the shape and location of the failure surface be equivalent. Stability studies of rock slopes using Bishop's circular failure analysis for a wide range of slope geometries and rock mass properties have been carried out for both generalised Hoek–Brown and Mohr–Coulomb criteria to find the value of σ_{3max}' that gives equivalent characteristic curves. These analyses gave the following relationship between σ_{3max}', the rock mass strength σ_{cm}' and the stress level on the sliding surface, σ_0 (Figure 5.26):

$$\frac{\sigma_{3max}'}{\sigma_{cm}'} = 0.72 \left(\frac{\sigma_{cm}'}{\sigma_0} \right)^{-0.91} \tag{5.31}$$

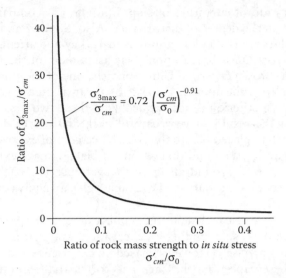

Figure 5.26 Relationship for the calculation of σ'_{3max} for equivalent Mohr–Coulomb and Hoek–Brown parameters for slopes.

The stress level on the sliding surface is related to the slope height H and the unit weight of the rock γ_r is given by

$$\sigma_0 = H \cdot \gamma_r \tag{5.32}$$

For the rock mass strength and slope parameters listed in Figure 5.25, the ratio $(\sigma'_{cm}/\sigma_0) = (3.16/0.52) = 6.1$, and the value for $\sigma'_{3max} = 0.44$ MPa (64 psi).

To illustrate the influence of the slope height on the rock mass strength, the shear-strength values for the parameters shown in Figure 5.25 were calculated for a slope height of 100 m (330 ft), for which $\sigma_0 = 2.6$ MPa. For the 100 m (330 ft) high slope, the rock mass strength is essentially unchanged, but the cohesion increases from 0.19 to 0.44 MPa (28–64 psi), while the friction angle decreases from 45.5° to 33.7°, that is, the tangent line at a higher normal stress level on the curved strength envelope. These changes in shear-strength parameters arise from the greater confining stresses in the higher slopes that results in the greater interlock between the rock fragments (higher cohesion), but increasing crushing of the fragments (lower friction angle).

5.5.6 Estimation of disturbance factor D

Experience in the design of slopes in very large open pit mines has shown that the Hoek–Brown criterion for undisturbed *in situ* rock masses ($D = 0$) results in rock mass properties that are too optimistic (Pierce, Brandshaug and Ward, 2001; Sjöberg, Sharp and Malorey, 2001). The effects of heavy blast damage as well as stress relief due to removal of the overburden result in disturbance of the rock mass (Hoek and Brown, 1988). It is considered that the 'disturbed' rock mass properties using $D = 1$ in Equations 5.15 and 5.16 are more appropriate for these rock masses.

A number of other studies to assess the extent of disturbance of the rock mass have been carried out by observing the performance of surface and underground excavations. For example, Lorig and Varona (2001) showed that factors such as the lateral confinement produced by different radii of curvature of slopes (in plan) as compared with their height also have an influence on the degree of disturbance. Also, Sonmez and Ulusay (1999) back-analysed five slope failures in open pit coal mines in Turkey and attempted to assign disturbance factors to each rock mass based upon their assessment of the rock mass properties predicted by the Hoek–Brown criterion. Unfortunately, one of the slope failures appears to be structurally controlled while another consists of a transported waste pile. Hoek considers that the Hoek–Brown criterion is not applicable to these two cases. In addition, Cheng and Liu (1990) report the results of very careful back analysis of deformation measurements, from extensometers placed before the commencement of excavation, in the Mingtan power cavern in Taiwan. It was found that a zone of blast damage extended for a distance of approximately 2 m (6.6 ft) around all large excavations. The back-calculated strength and deformation properties of the damaged rock mass give an equivalent disturbance factor $D = 0.7$.

From these references, it is clear that a large number of factors can influence the extent of disturbance in the rock mass surrounding an excavation, and that it may never be possible to quantify these factors precisely. However, based on experience and on the analysis of the details contained in these papers, Hoek, Carranza-Torres and Corkum (2002) have drawn up a set of guidelines for estimating the factor D (Table 5.4).

The influence of this disturbance factor can be large. This is illustrated by a typical example in which $\sigma_{ci} = 50$ MPa (7250 psi), $m_i = 10$ and $GSI = 45$. For an undisturbed *in situ* rock mass surrounding a tunnel at a depth of 100 m (328 ft), with a disturbance factor $D = 0$, the equivalent friction angle is $\phi' = 47.16°$ while the cohesive strength is $c' = 0.58$ MPa (84 psi). A rock mass with the same basic parameters but in highly disturbed slope of 100 m (328 ft) height, with a disturbance factor of $D = 1$, has an equivalent friction angle of $\phi' = 27.61°$ and a cohesive strength of $c' = 0.35$ MPa (51 psi).

Note that these are guidelines only, and the reader would be well advised to apply the values given with caution, and to compare the calculated results with those obtained by back analysis as shown in Figure 5.21. It is considered that the Hoek–Brown calculations can be used to provide a realistic starting point for any design and, if the observed or measured performance of the excavation turns out to be better than predicted, the disturbance factors can be adjusted downwards.

These methods have all been implemented in a program called RocData that is available from RocScience (www.rocscience.com). This program includes tables and charts for estimating the uniaxial compressive strength of the intact rock elements (σ_{ci}), material constant m_i, GSI and disturbance factor D.

5.6 SHEAR STRENGTH OF WEATHERED ROCK MASSES

Chapter 3 describes the occurrence, formation and properties of weathered rock. In summary, these materials occur in tropical climates where high temperatures (>18°C) and heavy rainfall occur frequently and the rock is modified, *in situ*, by chemical and physical processes that reduce the strength of the rock mass. Weathering is a gradational process that can range from fresh rock at depth to residual soil at the surface, and is quantified in six grades as described in Table 3.1.

For the design of slopes in weathered rock, it is necessary to have information on shear-strength properties corresponding to both the grade of weathering and the type of the host rock. For the higher grades of weathering – completely weathered rock and residual soil – it is possible to carry out laboratory testing to determine the strength if undisturbed samples are available.

Table 5.4 Guidelines for estimating disturbance factor D

Appearance of rock mass	Description of rock mass	Suggested value of D
	Excellent quality controlled blasting or excavation by tunnel boring machine results in minimal disturbance to the confined rock mass surrounding a tunnel	$D = 0$
	Mechanical or hand excavation in poor quality rock masses (no blasting) results in minimal disturbance to the surrounding rock mass Where squeezing problems result in significant floor heave, disturbance can be severe unless a temporary invert, as shown in the photograph, is placed	$D = 0$ $D = 0.5$ No invert
	Very poor quality blasting in a hard rock tunnel results in severe local damage, extending 2 or 3 m, in the surrounding rock mass	$D = 0.8$
	Small-scale blasting in civil engineering slopes results in modest rock mass damage, particularly if controlled blasting is used as shown on the left-hand side of the photograph. However, stress relief results in some disturbance.	$D = 0.7$ Good blasting $D = 1.0$ Poor blasting

(Continued)

Table 5.4 (Continued) Guidelines for estimating disturbance factor *D*

Appearance of rock mass	Description of rock mass	Suggested value of D
	Very large open pit mine slopes suffer significant disturbance due to heavy production blasting and also due to stress relief from overburden removal In some softer rocks, excavation can be carried out by ripping and dozing and the extent of damage to the slopes is less	$D = 1.0$ Production blasting $D = 0.7$ Mechanical excavation

However, for lower grades of weathering where the material is a variable composition of rock and soil, it is difficult to obtain representative samples for testing, and to have large enough equipment to carry out the tests. For these conditions, it is found that the Hoek–Brown rock mass strength criterion described in Section 5.5 can incorporate weathered rock by the use of appropriate *GSI* values. Examination of Figure 5.22 shows that the surface conditions describing the columns in the table range from 'Very rough, fresh, unweathered surfaces' to 'Highly weathered surfaces with soft clay coatings or infillings'. The five categories of surface conditions correspond approximately to the five grades of weathering described in Table 3.1 from fresh rock (grade I) to completely weathered rock (grade V), if grade VI for residual soil is neglected.

The second set of parameters defining the *GSI* of the rock mass on Figure 5.22 is the structure that ranges from 'intact or massive' to 'laminated/sheared'. Selection of the appropriate structure to use for the rock mass is based upon an examination of the *in situ* rock to assess the degree of fracturing related to both the discontinuity spacing and the number of discontinuity sets. The structure of the rock mass is independent of the grade of weathering because weathering is an *in situ* process that alters the properties of the intact rock but does not cause any additional fracturing because no deformation occurs.

The shear strength of a weathered rock mass can be determined using the Hoek–Brown strength criterion by selecting the appropriate value for the *GSI* from Figure 5.22, together with the value of the parameter m_i for the rock type (Table 5.3), and the compressive strength of the weathered rock fragments. The compressive strength will diminish with increasing weathering and if it is not possible to measures the strength, then information provided on Table 4.1 can be used to estimate the strength using the strengths corresponding to the listed field test result.

Figures 5.27 and 5.28 are modified versions of Figures 5.22 and 5.23, respectively, showing how weathering grades and geologic structure can be incorporated into the *GSI* system. The four sets of symbols show *GSI* values for weathering grades II, III, IV and V, corresponding to approximate rock compressive strengths of 20, 5, 1 and 0.5 MPa (2900, 725, 145 and 73 psi), respectively. In addition, three geologic conditions are illustrated – blocky, jointed (e.g. granite); very blocky, disturbed, moderately spaced joints (e.g. schist) and blocky, disturbed, many intersecting joints (e.g. basalt). For each rock type, the value for the constant m_i is provided from Table 5.3.

The shear strength of weathered rock masses can be calculated with the program RocData using the parameter values shown in Figures 5.27 and 5.28. In carrying out these calculations for slope applications, it is necessary to define a slope height since this determines the normal stress level on the sliding plane. For these examples, slope heights of 10 and 30 m (30 and 100 ft) have been used since this range covers many civil engineering applications. Figure 5.29 shows the results of the RocData calculations to determine the cohesion and friction angle of the weathered rock masses (grades II–IV) plotted on a $(c-\Phi)$ similar to Figure 5.21. Figure 5.29 shows the $(c-\Phi)$ relationship for four rock types – granite, basalt, schist and sandstone/shale – that encompass a wide range of geological conditions in which rock slopes may be designed.

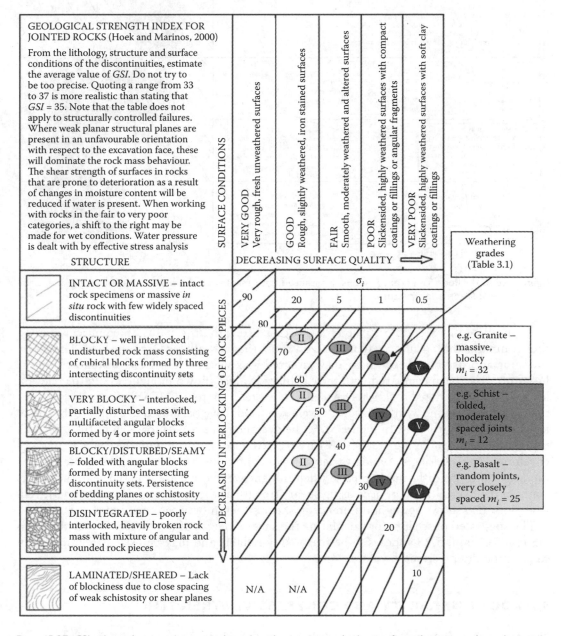

Figure 5.27 *GSI* values characterising weathered rock masses on the basis of weathering grade and particle interlocking.

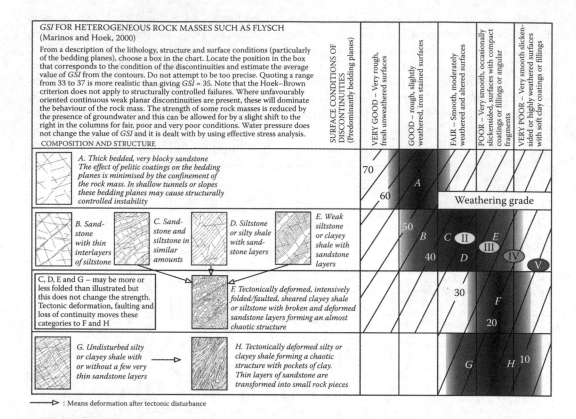

Figure 5.28 GSI values for heterogeneous, weathered rock masses on the basis of weathering grade, composition and structure.

As discussed in Section 5.5, the Hoek–Brown strength criterion is an empirical method that compares the calculated strengths with those that actually occur in the field. Therefore, the same approach has been taken for weathered rock where shear strengths reported in the literature have been compared with the calculated values shown in Figure 5.29; the relevant literature on weathered rock is referenced in Chapter 3. In total, 117 values of $(c - \phi)$, where the rock type and the weathering grade were reported, have been used to calibrate the calculated shear-strength data. In Figure 5.29, the average value $(c - \phi)$ for each of the four rock types and each weathering grade (II–V) are plotted in comparison with the calculated $(c - \phi)$ envelopes. Considering the wide variety of site conditions for the strengths quoted in the literature, and the difficulty in precisely defining the weathering grade in the field, the results confirm the overall trend of decreasing shear strength as the rock becomes more weathered. The plots also show the slight difference in shear strength between rock types.

The suggested procedure for calculating the shear strength of weathered rock as shown in Figures 5.27 and 5.28 appears to be reasonably reliable and may be used as a guideline in calculating design strength values.

5.7 ROCK DURABILITY AND COMPRESSIVE STRENGTH

The strength parameters that are usually of most significance to the analysis of slope stability are the cohesion and friction angle on the sliding surface, as discussed previously in

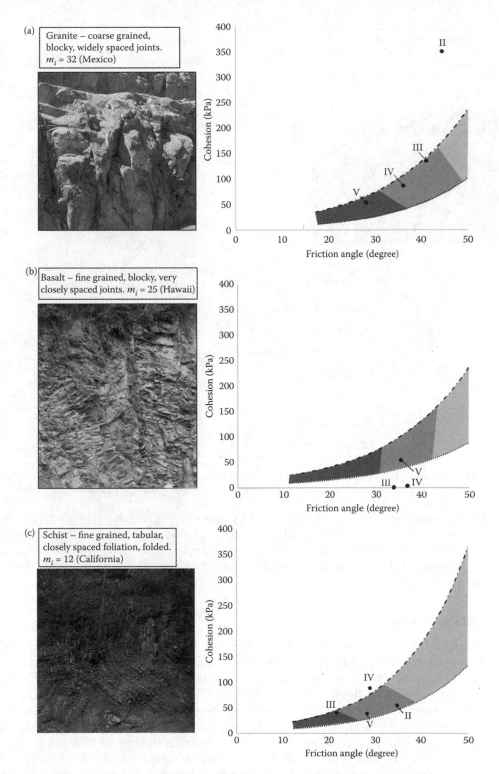

Figure 5.29 Plots of cohesion versus friction angle on the basis of rock type and weathering grade, with shear-strength values from the literature. (a) Granite, (b) basalt, (c) schist.   (*Continued*)

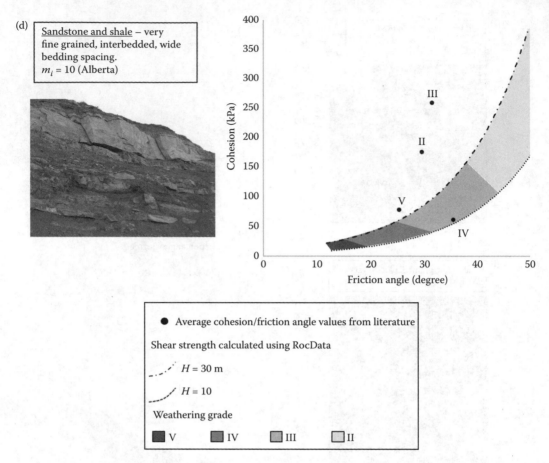

Figure 5.29 (Continued) Plots of cohesion versus friction angle on the basis of rock type and weathering grade, with shear-strength values from the literature. (d) sandstone and shale.

this chapter. However, both the durability and compressive strength of the rock may be of importance depending on the geological and stress conditions at the site, as discussed below. The durability and compressive strength test procedures are index tests best used to classify and compare one rock with another; if necessary, the index measurements can be calibrated by more precise laboratory tests.

5.7.1 Slake durability

Widely occurring rock materials are prone to degradation when exposed to weathering processes such as wetting and drying, and freezing and thawing cycles. Rock types that are particularly susceptible to degradation are shale and mudstone which usually have a high clay content. The degradation can take the form of swelling, and the time over which weakening and disintegration can occur after exposure may range from minutes to years. The effect of degradation on slope stability can range from surficial sloughing and gradual retreat of the face, to slope failures resulting from the loss of strength with time (Wu et al., 1981). In sedimentary formations comprising alternating beds of resistant sandstone and relatively degradable shale, the weathering process can develop overhangs in the sandstone and produce a rock fall hazard due to sudden failure of the sandstone (see Figure 3.13).

Figure 5.30 Slake durability testing equipment showing two wire mesh drums that are partially submerged in water and then rotated.

A common remediation measure for rocks susceptible to weathering is to apply a layer of shotcrete to protect the surface material from degradation (see Section 14.4.4).

A simple index test of the tendency of rock to weather and degrade is the slake durability test (International Society for Rock Mechanics, 1981a,b) (Figure 5.30). It is important that undisturbed samples are used that have not been excessively broken in the sampling procedure, or allowed to freeze. The test procedure comprises placing the sample in the wire mesh drum, drying it in an oven at 105°C for 2–6 hours, and then weighing the dry sample. The drum is then partially submerged in water and rotated at 20 revolutions per minute for a period of 10 minutes. The drum is then dried a second time and the loss of weight recorded. The test cycle is then repeated and the slake durability index is calculated as the percentage ratio of final to initial dry sample masses. A low slake durability index will indicate that the rock is susceptible to degradation when exposed. For highly degradable rocks it is useful to carry out soil classification tests such as Atterberg limits, as well as x-ray diffraction tests to identify clay mineral types and determine whether swelling clays such as bentonite and montmorillonite are present.

5.7.2 Compressive strength

In many slopes in moderately strong to strong rock, the stress level due to gravity loads will be less than the strength of the rock. Therefore, the intact rock within the slope is unlikely to fracture, and consequently compressive strength is a less important design parameter than shear strength. The compressive strength of the rock on the sliding surface is only used indirectly in stability analysis when determining the roughness of a fracture (*JCS* term in Equation 5.7), and in the application of the Hoek–Brown strength criteria (Section 5.5). For both these applications, it is satisfactory to use an estimate of the compressive strength because the results are not particularly sensitive to the value of this parameter. The compressive strength is also used in evaluating excavation methods and costs. For example, the progress rate for blast hole drilling and excavation by ripping, as well as selection of explosive types and design of blasts are all influenced by the compressive strength of the rock.

(a)

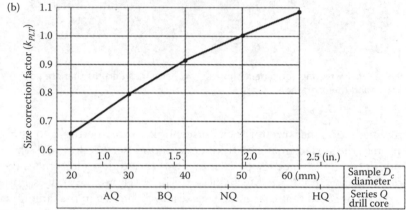

(b)

Figure 5.31 Point load testing: (a) point load test equipment and (b) relationship between sample equivalent core diameter D_e, and size correction factor k_{PLT}.

The point load test is an appropriate method to estimate the compressive strength for rock slope design (Figure 5.31a). The equipment is portable, and tests can be carried out quickly and inexpensively in the field on both core and lump samples. Because the point load test provides an index value for the strength, usual practice is to calibrate the results with a limited number of uniaxial compressive tests on prepared core samples.

The test procedure involves placing the sample between the platens and applying a load with the hydraulic jack to break the sample in tension. If P is the point load breaking strength, then the point load index, I_s given by

$$I_s = \frac{P}{D_e^2} \qquad\qquad (5.33)$$

where:

D_e is the equivalent core diameter, defined as
$D_e^2 = D^2 -$ diametral tests where D is the core diameter

or

$D_e^2 = (4W \cdot D)/\pi -$ axial, block or lump tests

where W is the specimen width and D is the distance between the platens. The term $(W \cdot D)$ is the minimum cross-sectional area of a lump sample for the plane through the platen contact points.

The size-corrected point load strength index $I_{s(50)}$ of a rock specimen is defined as the value of I_s that would have been measured by a diametral test with $D = 50$ mm. For tests conducted on samples with dimensions different from 50 mm, the results can be standardised to a size-corrected point load strength index by applying a correction factor k_{PLT} as follows:

$$I_{s(50)} = I_s \cdot k_{PLT} \tag{5.34}$$

The value of the size correction factor k_{PLT} is shown in Figure 5.31b and is given by

$$k_{PLT} = \left(\frac{D_e}{50}\right)^{0.45} \tag{5.35}$$

It has been found, on average, that the uniaxial compressive strength is about 20–25 times the point load strength index. However, tests on many different types of rock show that the ratio can vary between 15 and 50, especially for anisotropic rocks. Consequently, the most reliable results are obtained if uniaxial calibration tests are carried out.

Point load test results are not acceptable if the failure plane lies partially along a pre-existing fracture in the rock, or is not coincident with the line between the platens. For tests in weak rock where the platens indent the rock, the test results should be adjusted by measuring the amount of indentation and correcting the distance D.

If no equipment is available to measure the compressive strength, simple field observations can be used to estimate the strength with sufficient accuracy for most purposes. Table 4.1 describes a series of field index tests and observations of rock behaviour, and gives the corresponding range of approximate compressive strengths.

EXAMPLE PROBLEM 5.1: ANALYSIS OF DIRECT SHEAR-STRENGTH TEST RESULTS

Statement

The following table of results was obtained from a direct shear box test on a planar discontinuity in a sample of slightly weathered granite. The average normal pressure on the sample was 200 kPa (29 psi).

Shear stress (kPa)	Shear displacement (mm)
159	0.05
200	1.19
241	3.61
228	4.50
214	8.51
207	9.40
200	11.61
193	12.60
179	17.09
179	19.81

Required

1. Plot a graph of shear stress against shear displacement with shear stress on the vertical axis; from the graph determine the peak and residual shear strengths of the surface.
2. Plot a graph of the peak and residual shear strengths (on the vertical axis) against the average normal stress on the surface; from the graph determine the peak and residual friction angles of the surface.

Solution

1. The graph of shear stress against shear displacement is shown in Figure 5.31a; the peak strength is 241 kPa (42 psi) and the residual strength is 179 kPa (26 psi).
2. The graph of shear stress against normal stress is shown in Figure 5.31b; the peak friction angle is 50.3° and the residual friction angle is 41.8° (Figure 5.32).

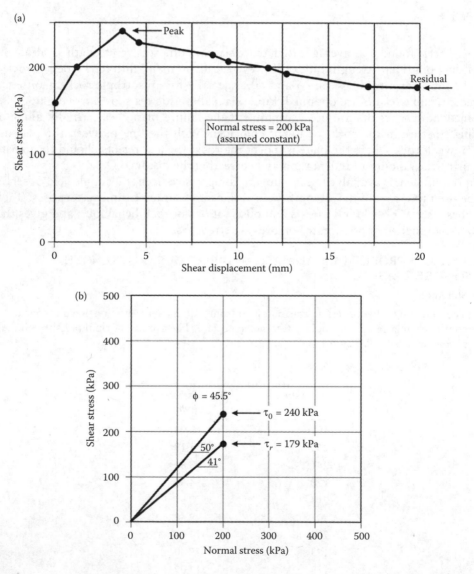

Figure 5.32 Analysis of direct shear-strength test, Example Problem 5.1: (a) plot of shear stress against shear displacement; (b) plot of shear stress against normal stress (Mohr diagram).

EXAMPLE PROBLEM 5.2: ANALYSIS OF POINT LOAD TEST RESULTS

Statement

A series of point load tests on pieces of NQ core 48 mm in diameter gave an average point load breaking strength (P) of 17.76 kN (4 kips) when the core was loaded diametrically.

Required

Determine the approximate average uniaxial compressive strength of the samples.

Solution

The point load strength index (I_s) is calculated from the point load breaking strength by

$$I_s = \frac{P}{D^2}$$

where D is the core diameter of 48 mm.

$$I_s = 17.76E3/(0.048)^2$$
$$= 7.71 \text{ MPa}$$

Correction factor $k_{PLT} = (48/50)^{0.48}$
$$= 0.98$$

The size-corrected point load strength, $I_{s(50)} = I_s \cdot k_{PLT}$
$$= 7.71 \cdot 0.98$$
$$= 7.56 \text{ MPa}$$

The approximate compressive strength of the rock samples σ_{ci}
$$\approx 24 \cdot 7.56$$
$$\approx 180 \text{ MPa}$$

Chapter 6

Groundwater

6.1 INTRODUCTION

The presence of ground water in a rock slope can have a detrimental effect upon stability for the following reasons:

1. *Water pressure* reduces the stability of the slopes by diminishing the shear resistance of potential failure surfaces as described in Chapter 1. Water pressure in tension cracks or similar near vertical fissures reduces stability by increasing the forces that induce sliding.
2. Changes in *moisture content* of some rock, particularly shales, can cause accelerated weathering and a decrease in shear strength.
3. *Freezing* of ground water can cause wedging in water-filled fissures due to temperature-dependent volume changes in the ice. Also, freezing of surface water on slopes can block drainage paths resulting in a buildup of water pressure in the slope with a consequent decrease in stability.
4. *Erosion* of weathered rock by surface water, and of low strength infillings by ground water can result in local instability where the toe of a slope is undermined, or a block of rock is loosened.
5. *Excavation costs* can be increased when working below the water table. For example, wet blast holes require the use of water-resistant explosives that are more expensive than non-water-resistant ANFO. Also, flow of ground water into the excavation or pit will require pumping and possibly treatment of the discharge water, and equipment trafficability may be poor on wet haul roads.

By far the most important effect of ground water in a rock mass is the reduction in stability resulting from water pressures within the discontinuities. Methods for including these water pressures in stability calculations and designing drainage systems are dealt with in later chapters of this book. This chapter describes the hydrologic cycle (Section 6.2), methods that are used to analyse the flow of water through fractured rock and the pressures developed by this flow (Sections 6.3 and 6.4). Sections 6.5 and 6.6 discuss, respectively, methods of making hydraulic conductivity and pressure measurements in the field.

Figure 6.1 shows a series of drain holes drilled into the base of a rock cut where significant flow is occurring in only one of the holes. This is a typical condition in jointed rock where ground water flow is concentrated in the discontinuities, and flow occurs in only those drains that intersect water-bearing discontinuities. However, other drain holes, not currently flowing, may have intersected joints that initially contained water under pressure, but the water was confined such that flow ceased once the pressure was relieved.

Figure 6.1 Horizontal drain holes, lined with perforated plastic pipes, in rock face; flow only occurs in holes that intersect water-bearing discontinuities (Sea to Sky Highway, British Columbia, Canada). (Image by J. K. McDonnell.)

In examining rock or soil slopes, it may be a mistake to assume that ground water is not present if no seepage appears on the slope face. The seepage rate may be lower than the evaporation rate, and hence the slope surface may appear dry and yet the water may be at a significant pressure within the rock mass. It is water pressure, and not rate of flow, which is responsible for instability in slopes and it is essential that measurement or calculation of this water pressure form part of site investigations for stability studies. Drainage, which is discussed in Section 14.4.7, is one of the most effective and economical means available for improving the stability of rock slopes. Rational design of drainage systems is only possible if the water flow pattern within the rock mass is understood, and measurement of hydraulic conductivity and water pressure provides the key to this understanding.

A useful means of assessing ground water conditions in a slope is to make observations during periods of freezing temperatures. At these times, even minor seeps on the face may form icicles that can show both the location of the water table and the set(s) of discontinuities in which flow is occurring.

6.2 THE HYDRAULIC CYCLE

A simplified hydrologic cycle illustrated in Figure 6.2 shows typical sources of ground water, and emphasises that ground water can travel considerable distances through a rock mass. It is important, therefore, to consider the regional geology of an area when starting a rock slope design programme. In general, ground water flows from recharge areas to discharge areas. A recharge area is one in which the net saturated flow of ground water is directed away from the water table, while in a discharge area the net saturated flow is directed towards the water table. In Figure 6.2, discharge areas occur at the rock cut and quarry, while the ocean and the lake are recharge areas to the quarry.

Clearly, precipitation in the catchment area is the most important source of ground water, and Figure 6.3 illustrates the typical relationship between the precipitation and

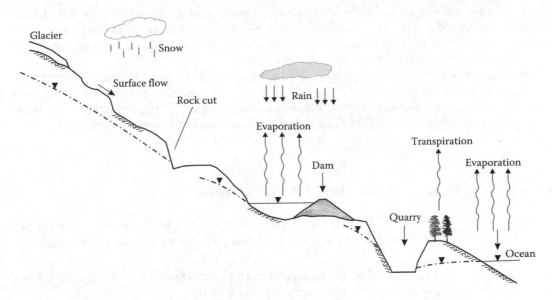

Figure 6.2 Simplified representation of a hydrologic cycle showing typical sources of ground water (modified from Davis and DeWiest, 1966).

ground water levels in three climatic regions. In tropical and desert climates, the ground water table is usually more predictable and consistent than temperate climates where the precipitation levels are more variable. In assessing the relationship between climate and ground water levels in the slope, both the average precipitation and the peak events should be considered because the peak events are those that usually cause instability. Examples of peak precipitation events that can lead to high-infiltration rates include typhoons, intense rainstorms and rapid snowmelt. If these climatic conditions exist at the site, then it is advisable to use correspondingly high-water pressures in design, or to design high-capacity drainage systems.

Sources of ground water in addition to precipitation may include recharge from adjacent rivers, reservoirs, lakes or oceans as shown in Figure 6.2. For example, substantial quarries

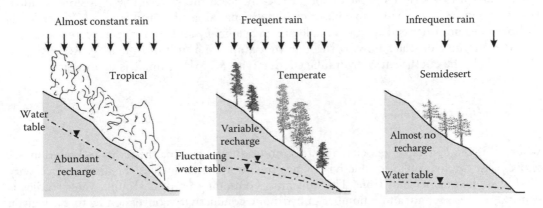

Figure 6.3 Relationship between water table level and precipitation (modified from Davis and DeWiest, 1966).

and open pit mines such as Dutra Minerals in California, and Granisle Copper and Island Copper in Canada have successfully operated below, and close to, substantial bodies of water. However, in such operations, significant seepage may develop into the pit, as well as slope instability resulting from high-water pressures.

Another important factor influencing ground water within a slope is distribution of rock types, and details of the structural geology such as fault infillings, persistence of joint sets and the presence of solution cavities. These features can result in regions of low- and high-hydraulic conductivity within the slope that are termed aquitards and aquifers, respectively. These matters are discussed in more detail in Section 6.4.

6.3 HYDRAULIC CONDUCTIVITY AND FLOW NETS

Where ground water effects are to be included in slope design, two possible approaches to obtaining data on distributions of the water pressures within a rock mass are as follows:

1. Deduction of the ground water flow pattern from consideration of the hydraulic conductivity of the rock mass and sources of ground water.
2. Direct measurement of water levels in boreholes or wells, or of water pressure by means of piezometers installed in boreholes.

Because of the important influence of water pressure on slope stability, it is essential that the best possible estimates of the likely range of pressures should be available before a detailed stability analysis is attempted. A large number of factors control ground water flow in jointed rock masses, and it is only possible in this book to highlight the general principles that may apply. If detailed studies of ground water conditions are required, it is advisable to obtain additional data from such sources as Freeze and Cherry (1979) and Cedergren (1989) on ground water analysis, and Dunnicliff (1993) on instrumentation.

6.3.1 Hydraulic conductivity

The basic parameter defining the flow of ground water, and the distribution of water pressure in geologic media, is hydraulic conductivity. This parameter relates the flow rate of water through the material to the pressure gradient applied across it (Scheidegger, 1960; Morgenstern, 1971).

Consider a cylindrical sample of soil or rock beneath the water table in a slope as illustrated in Figure 6.4. The sample has a cross-sectional area of A and length l. Water levels in boreholes at either end of this sample are at heights h_1 and h_2 above a reference datum and the quantity of water flowing through the sample in a unit of time is Q. According to Darcy's law, the coefficient of hydraulic conductivity K of this sample is defined as

$$K = \frac{Q \cdot l}{A(h_1 - h_2)} = \frac{V \cdot l}{(h_1 - h_2)} \tag{6.1}$$

where V is the discharge velocity. Substitution of dimensions for the terms in Equation 6.1 shows that the hydraulic conductivity K has the same dimensions as the discharge velocity V, that is, length per unit time. The unit most commonly used in ground water studies is centimetre per second, and a number of hydraulic conductivity conversion factors are given in Table 6.1.

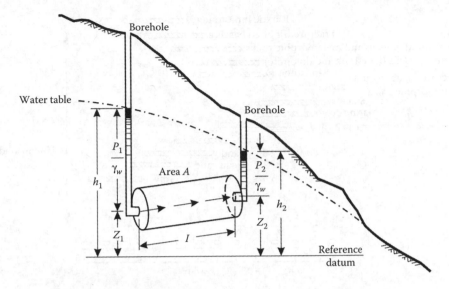

Figure 6.4 Illustration of Darcy's Law for definition of hydraulic conductivity.

Table 6.1 Hydraulic conductivity conversion table

To convert cm/s to	Multiply by
m/s	1.00×10^{-2}
ft/s	3.28×10^{-2}
US gal/day·ft^2	1.89×10^{4}
ft/year	1.03×10^{6}
m/year	3.14×10^{5}

Equation 6.1 can be rearranged to show the volume of water, Q flowing through the sample shown in Figure 6.4 under a specified head, as follows:

$$Q = \frac{K \cdot A(h_1 - h_2)}{l} \qquad (6.2)$$

In most rock types, flow through intact rock is negligible (defined by $K_{primary}$), and essentially all flow occurs along the discontinuities (defined by $K_{secondary}$). For example, the primary hydraulic conductivity for intact granite and basalt is about 10^{-10} cm/s, while for some course grained, poorly indurated sandstones the primary hydraulic conductivity may be as high as 10^{-4} cm/s. The term secondary hydraulic conductivity refers to flow in the rock mass and encompasses flow in both the intact rock and any discontinuities that are present. These conditions result in secondary hydraulic conductivities having a wide range of values depending on the persistence, width and infilling characteristics of the discontinuities. For example, granite which has very low-primary hydraulic conductivity usually contains tight, clean, low-persistence joints so the secondary hydraulic conductivity is also low. In contrast, sandstone may have some primary conductivity, and the presence of persistent bedding planes may result in high-secondary conductivity in the direction parallel to the bedding. For further discussion on flow in fractured rock, see Section 6.4.

Typical ranges of secondary hydraulic conductivity for variety rock types, as well as unconsolidated deposits, are shown in Figure 6.5. The range of hydraulic conductivities for

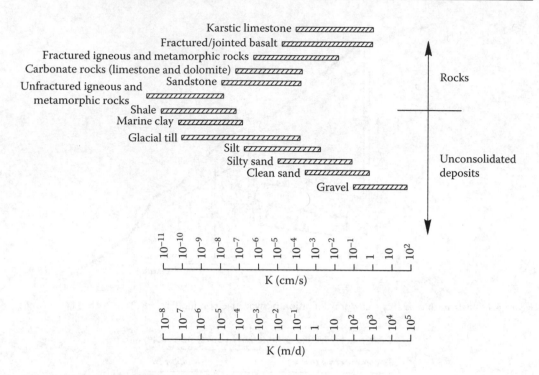

Figure 6.5 Hydraulic conductivity of various geologic materials (Atkinson, 2000).

geological materials covers 13 orders of magnitude, and for any single rock type, the range can be four orders of magnitude. This shows the difficulty in predicting water inflow quantities and pressures within slopes.

Figure 6.4 also shows that the total head h at any point can be expressed in terms of the pressure P and the height z above a reference datum. The relationship between these parameters is

$$h = \frac{P}{\gamma_w} + z \tag{6.3}$$

where γ_w is the density of water. The total head h represents the level to which water will rise in a borehole standpipe.

Darcy's law is applicable to porous media and so can be used to study ground water flow in both intact rock, and rock masses on a macroscopic scale. However, it is required that the flow be laminar, so Darcy's law is not applicable in the event of non-linear or turbulent flow in an individual fracture.

6.3.2 Porosity

The total volume V_T of a rock or soil is made up of the volume of solid portion V_s and the volume of the voids V_v. The porosity, n, of a geologic material is defined as the ratio

$$n = \frac{V_v}{V_T} \tag{6.4}$$

In general, rocks have lower porosities than soil. For example, the porosity of gravel is about 20%–40%, and clay has a porosity in the range of 51%–58%. In comparison, fractured basalt and karstic limestone may have porosities in the range of 5%–50%, while the porosity of dense crystalline rock is usually in the range of 0%–5%.

The significance of the porosity to rock slopes is in the design of drainage systems. For example, drains installed in a low-porosity granite need only discharge a small amount of water to lower the water pressures, while drains in karstic limestone may discharge large volumes of water with little effect on the ground water table.

6.3.3 Flow nets

The graphical representation of ground water flow in a rock or soil mass is known as a flow net and a typical example is illustrated in Figure 6.6. A flow net comprises two sets of intersecting lines as follows:

1. *Flow lines* are paths followed by the water in flowing through the saturated rock or soil.
2. *Equipotential lines* are lines joining points at which the total head h is the same. As shown in Figure 6.6, the water level is the same in standpipes that terminate at points A and B on the same equipotential line. Water pressures at points A and B are not the same since, according to Equation 6.3, the total head h is given by the sum of the pressure head P/γ_w, and the elevation z of the measuring point above the reference datum. The water pressure increases with depth along an equipotential line as shown by the horizontal hatching in Figure 6.6.

The following are characteristics of flow nets that are applicable under all conditions and must be used in drawing flow nets. First, equipotential lines must meet impermeable boundaries at right angles and be parallel to constant head boundaries. Second, a uniform head loss occurs between adjacent equipotential lines. Third, equipotential and flow lines intersect at right angles to form curve-linear squares in rock with isotropic hydraulic conductivity. For a flow net such as that shown in Figure 6.6, the equipotential lines illustrate how the

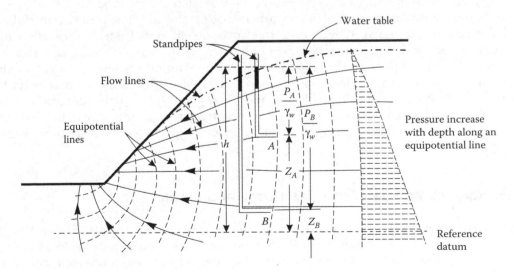

Figure 6.6 2D flow net in a slope.

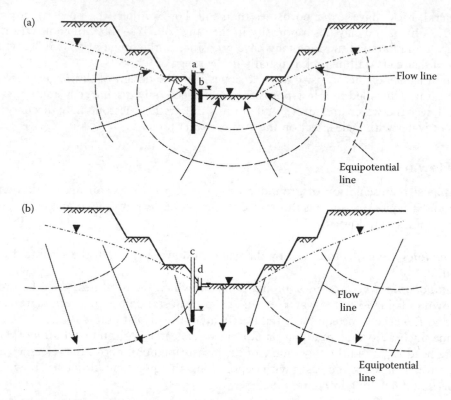

Figure 6.7 Ground water conditions for cut slopes in regional (a) discharge and (b) recharge areas (Patton and Deere, 1971).

ground water pressure varies within the slope, and that flow quantity is equal between the adjacent flow lines.

An example of the application of flow nets to study the distribution of pressures in rock slopes is illustrated in Figure 6.7. If the excavation is located in a recharge area, flow occurs towards the excavation, and artesian pressures can be developed below the excavation floor (a, b). In contrast, for excavations in discharge areas, flow is away from the excavation and the pressure below the excavation floor will be low (c, d).

A complete discussion on the construction or computation of flow nets exceeds the scope of this book and the interested reader is referred to the comprehensive texts by Cedergren (1989), Haar (1962) and Freeze and Cherry (1979) for further details. The use of graphical methods for constructing flow nets is often an important step in understanding how geology and drainage systems influence possible ground water conditions within a slope.

6.4 GROUND WATER FLOW IN FRACTURED ROCK

The flow of ground water in fractured rock is of great importance to the design of such facilities as nuclear waste storage sites and much important work is done in this field. However, for rock slopes detailed studies of fracture rock flow are probably only required where significant drainage systems such as tunnels and drain hole arrays are planned. For most

Figure 6.8 Rock mass with persistent vertical joints and relatively high-vertical hydraulic conductivity (modified from Atkinson, 2000).

rock cuts, it is adequate to study the relationship between geology and ground water flow as illustrated in Figures 6.8 to 6.12. These relationships will provide such information as likely ground water pressures within the slope, and the optimum orientation of drain holes to intersect water-bearing discontinuities.

As discussed in Section 6.2 on hydraulic conductivity, ground water flow in fractured rock masses occurs predominately along the discontinuities because of the very low-primary hydraulic conductivity of most intact rock. Therefore, the conductivity of rock masses will be influenced by the characteristics of the discontinuities, with a necessary condition required for flow being that the persistence of the discontinuities is greater than the spacing. Figure 6.8 shows a rock mass containing two vertical joint sets and one horizontal set in which the persistence of the vertical joints is much greater than the spacing, but the persistence of the horizontal set is less than the spacing. For these conditions, the hydraulic conductivity would be significantly greater in the vertical direction than that in the horizontal direction.

The analysis of flow in fractured rock can be carried out either assuming that the rock is a *continuum*, as has been assumed in the derivation of Darcy's equation and drawing flow nets, or that the rock is a *non-continuum* in which laminar flow occurs in individual discontinuities. Rock can be assumed to be a continuum if the discontinuity spacing is sufficiently close that the fractured rock acts hydraulically as a granular porous media so that the flow occurs through a number of discontinuities.

6.4.1 Flow in clean, rough discontinuities

The flow of water through fissures in rock has been studied in detail by Huitt (1956), Snow (1968), Louis (1969), Sharp (1970), Maini (1971) and others, and builds on the Navier–Stokes

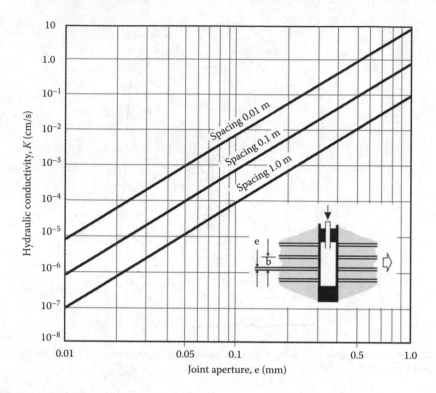

Figure 6.9 Influence of joint aperture, e and spacing b on hydraulic conductivity K along a set of smooth parallel joints in a rock mass.

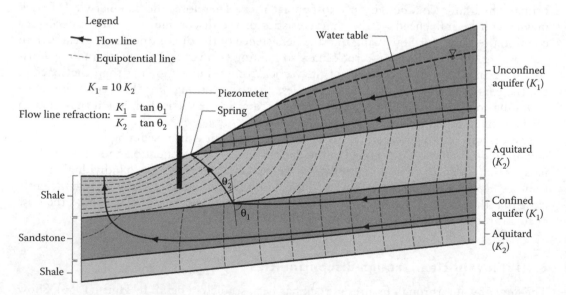

Figure 6.10 Water flow and pressure distribution in aquifers and aquitards formed by dipping sandstone and shale beds (Dr. P. Ward, plot by W. Zawadzki).

equation defining flow of viscous fluids. Subsequent to this, extensive research has been carried out on water flow in fractures in relation to such projects as underground nuclear waste storage facilities, contaminant transport and reservoir engineering (Bishop and Bjerrum, 1960). This work shows that the flow $Q(m^3/s)$ between two parallel smooth plates, not in contact at any point, is a cubic law expressed as

$$Q = \frac{e^3(\Delta p)}{12 \cdot \mu \cdot L} \tag{6.5}$$

where e is the distance (aperture) between the parallel plates (m), Δp the pressure drop along the flow path (Pa), μ the coefficient of kinematic viscosity (1.01×10^{-6} m^2/s for pure water at 20°C) and L the distance between the inlet and outlet boundaries. In order to apply Equation 6.5 to real discontinuities, the term e can be replaced with an equivalent aperture e_H (μm) that incorporates the surface roughness in terms of the joint roughness coefficient (JRC) (Barton and Bandis, 1983):

$$e_H = \frac{e_m^3}{JRC^{2.5}} \tag{6.6}$$

where e_m is the mean distance between parallel rough surfaces, and JRC quantifies the surface roughness as described in Section 4.4.3. Modelling of water flow through rough joints shows that the pressure drop along the flow path increases linearly with the JRC value (Hosseinian, Rasouli and Utikar, 2010).

The equivalent conductivity of a parallel array of discontinuities in relation to aperture and spacing is shown in Figure 6.9, where an increase in the aperture of one order of magnitude results in an increase in the conductivity of three orders of magnitude. Since the ground water flow quantity, and hydraulic conductivity, is proportional to the third power of the aperture, small changes in the aperture due, for example, to increasing stress in the rock will significantly decrease the flow. This condition could develop at the toe of a steep slope where high stresses decrease the aperture and result in a buildup of water pressure in the slope.

The relationship between discontinuity aperture and hydraulic conductivity was studied for the construction of the ship locks at the Three Gorges Project in China, which involved making parallel excavations with depths up to 170 m (560 ft) in strong, jointed granite (Zhang, Powrie, Harness et al., 1999). The excavations caused relaxation of the rock in the walls of the locks and the opening of the joints, which resulted in the hydraulic conductivity increasing by a factor of 18. The application of a support pressure of 2 MPa (290 psi) on the vertical walls of the excavation resulted in the hydraulic conductivity only increasing by a factor of 6 from the *in situ* condition.

6.4.2 Flow in filled discontinuities

Equation 6.5 applies only to laminar flow in planar, smooth, parallel discontinuities and represents the highest equivalent hydraulic conductivity for fracture systems. The lowest equivalent hydraulic conductivity occurs for infilled discontinuities, and is given by

$$K \approx \frac{e \cdot K_f}{b} + K_r \tag{6.7}$$

where K_f is the hydraulic conductivity of the filling and K_r is that of the intact rock. The term K_r is included in Equation 6.7 to account for the condition where flow occurs in both the intact rock and along discontinuities.

While Equations 6.5 and 6.7 illustrate the principles of water flow along discontinuity planes, this simple model cannot be used to calculate hydraulic conductivity of actual fractured rock masses. Methods of modelling ground water flow in rock have been developed using probabilistic techniques to simulate, in three dimensions, the likely ranges of discontinuity characteristics that may occur. One such modelling technique is termed FRACMAN (Dershowitz, Lee, Geier et al., 1994; Wei, Egger and Descoeudres, 1995).

6.4.3 Heterogeneous rock

Figure 6.10 shows a shallow dipping sequence of sandstone and shale beds. The shale, which is a fine grained rock with few persistent discontinuities, has low-hydraulic conductivity and is termed an *aquitard*. In contrast, the sandstone, which is course grained, has a relatively high-hydraulic conductivity and is termed an *aquifer*. Because of the significant difference between the hydraulic properties of the shale and sandstone, this is a heterogeneous rock mass. Flow nets in heterogeneous rock are modified from the simple net shown in Figure 6.6 because flowlines preferentially use the high-conductivity formations as conduits and traverse the low-conductivity formations by the shortest possible route. The equipotentials tend to lose a greater proportion of the head in the low-conductivity formation than in the higher-conductivity formation. This behaviour results in refraction of flow lines at the formation boundary, depending on the relative conductivities, according to the following relationship:

$$\frac{K_{\text{sandstone}}}{K_{\text{shale}}} = \frac{\tan\theta_1}{\tan\theta_2} \tag{6.8}$$

The refraction angles θ_1 and θ_2 are defined in Figure 6.10 for a conductivity ratio of $K_1/K_2 = 10$.

Features of the flow conditions shown in Figure 6.10 are as follows. First, flow in the upper, unconfined aquifer tends to flow down-dip in the sandstone and exits the slope at the sandstone/shale contact. This seepage line on the valley wall would be an indication of the location of the contact. Second, flow in the lower, confined aquifer is recharged from a source up-dip from the valley that develops artesian pressure in the sandstone. This condition could be demonstrated by completing a piezometer in the lower sandstone, in which the water would rise above the ground surface to the level of the equipotential in which it is sealed. Third, flow in the confined aquifer is refracted at the boundary and flows upwards in the shale to exit in the valley floor.

6.4.4 Anisotropic rock

In formations such as that shown in Figure 6.8, in which the conductivity of one set or sets of discontinuities is higher than another set, the rock mass will exhibit *anisotropic* hydraulic conductivity. For the rock shown in Figure 6.8, the vertical hydraulic conductivity will be considerably more than that in the horizontal direction. On the flow net, anisotropic hydraulic conductivity is depicted by squares formed by the flow lines and equipotentials being elongated in the direction of the higher-hydraulic conductivity. In general, the aspect ratio of the flow line/equipotential squares is equal to $(K_1/K_2)^{1/2}$.

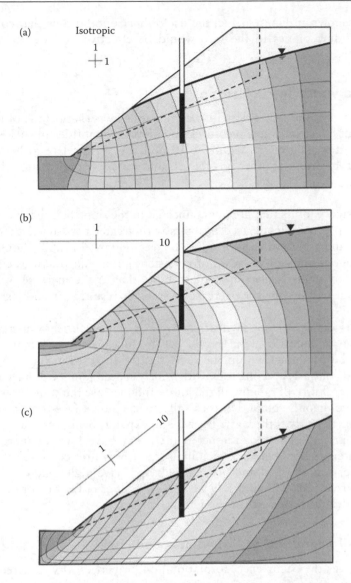

Figure 6.11 Flow nets and water pressures in slopes with isotropic and anisotropic hydraulic conductivity with piezometers showing water pressures at equivalent locations in the slope: (a) isotropic rock; (b) $K_{horizotal} = (10 \cdot K_{vertical})$; (c) $K_{parallel\ to\ slope} = (10 \cdot K_{perpendicular\ to\ slope})$ (plot by W. Zawadzki).

Examples of flow nets in isotropic and anisotropic rock are shown in Figure 6.11. The significance of these conditions to slope stability is as follows. First, in rock with high-hydraulic conductivity in the horizontal direction, such as horizontally bedded sandstone, the ground water can readily drain from the slope (Figure 6.11b). For these conditions, water pressures on potential sliding surfaces will be relatively low compared to the isotropic case. Second, in rock with high-hydraulic conductivity parallel to the face such as a slope cut parallel to bedding, flow to the face will be inhibited and high-water pressures will develop in the slope (Figure 6.11c). The water pressure in the slope for each condition is illustrated by piezometers at identical locations in these slopes. The piezometers show that the lowest pressure is in the slope with horizontal bedding (Figure 6.11b), while the highest pressure is in the slope with the bedding parallel to the slope face (Figure 6.11c).

For the slope shown in Figure 6.11c, the use of horizontal drains that connect the high-conductivity bedding planes to the face would be effective in lowering-water pressures within the slope.

6.4.5 Ground water in rock slopes

The discussion on ground water flow in rock masses shows that details of the geology can have a significant effect on water pressures and seepage quantities in rock slopes. In addition to the conditions shown in Figures 6.10 and 6.11 that relate to heterogeneous and anisotropic rock, a variety of other possible ground water conditions are related to geology as follows:

1. Low-persistence joints that are not connected to the slope face may develop high-transient water pressures, compared to joints with greater persistence that are connected to the face and allow water to drain at the face (Figure 6.12a). It should be noted that blast damage is one of the causes of persistent joints and fractures close to the face. However, any improvement in slope stability due to the increase in conductivity is probably out-weighed by the decrease in stability resulting from blast damage to the rock.
2. The porosity of the rock mass will affect the level of the transient water table in response to the same precipitation event (Figure 6.12b). In a porous rock mass, the infiltrating water will be contained within the rock and the increase in the water table will be minimal. In contrast, a strong rock with widely spaced joints will have low-porosity so ground water flow will rapidly fill the joints and increase the water pressure within the slope. It is commonly found that rock falls on steep rock faces occur soon after heavy rainfalls, particularly if the water freezes and expands behind the face.
3. Faults comprising clay and weathered rock may have low conductivity and act as ground water barriers behind which high-water pressures could develop. In contrast, faults comprising crushed and broken rock may have high conductivity and act as a drain (Figure 6.12c). Measurement of water pressures either side of the fault will indicate the hydraulic properties of these features.

6.4.6 Hydraulic conductivity of weathered rock

Chapter 3 describes the occurrence, formation and properties of weathered rock. In summary, these materials occur in tropical climates where temperatures are high (>18°C) and heavy rainfall occurs frequently such that the rock is modified, *in situ*, by chemical and physical processes that reduce the strength of the rock mass. Weathering is a gradational process that can range from fresh rock at depth to residual soil at the surface, and is quantified in six grades as described in Table 3.1.

The weathering process is initiated by the flow of water through discontinuities, resulting in the progressive alteration of the rock at the surface of these fractures. As the weathering develops, more of the rock is altered to clay minerals, or is lost to solution, until the entire rock mass becomes a residual soil.

The effect of weathering on the hydraulic conductivity is that the conductivity tends to increase, compared to the fresh rock, for weathering grades II (slightly weathered) to IV (highly weathered) where the discontinuities are widened by the weathering process. For grades V (completely weathered) and VI (residual soil), the conductivity tends to be lower because of the predominant clay content of the rock mass. Measurements of conductivity in

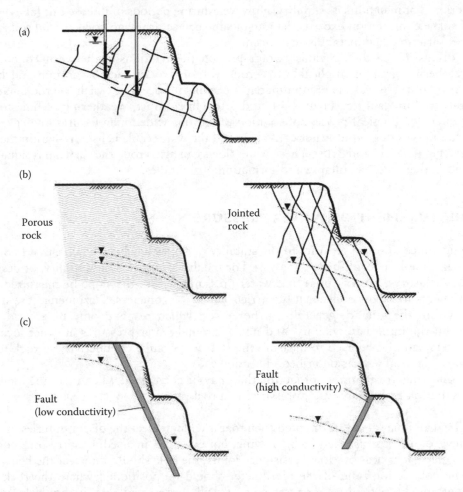

Figure 6.12 Relationship between geology and ground water in slopes: (a) variation in water pressure in joints related to persistence; (b) comparison of water tables in slopes excavated in porous and jointed rock; (c) faults as low-conductivity ground water barrier, and high-conductivity subsurface drain (Patton and Deere, 1971).

granite in Singapore gave the following values in relation to the weathering grades (Sharma, Chu and Zhao, 1999):

Residual soil (Grade VI)	$k = 1.0 \times 10^{-3}$ cm/s
Highly weathered (Grade IV)	$k = 5.1 \times 10^{-3}$ cm/s
Moderately weathered (Grade III)	$k = 1.8 \times 10^{-3}$ cm/s
Slightly weathered (Grade II)	$k = 1.6 \times 10^{-3}$ cm/s
Fresh granite (Grade I)	$k = 0.6 \times 10^{-3}$ cm/s

As a result of the relatively high conductivity in the grades II–IV rock, drainage occurs in these materials and the level of the water table tends to fluctuate with the seasons within these rock grades. This condition results in the residual soil also being drained which generates soil suction that maintains its stability. However, during periods of intense rainfall when the rate of surface infiltration exceeds the rate of sub-surface drainage, the surficial soil rapidly becomes saturated and instability can occur.

The discussion in the previous paragraphs on the relationship between conductivity and weathering grade is applicable to strong volcanic rocks such as granite and basalt. Weathering of these rocks is predominately a chemical process in which the rock materials are weathered to clays (see Figures 3.6 and 3.10). In contrast, weathering of limestones is predominately a physical process of solution that forms wide, open fissures and caverns in which high flows of ground water can occur and the water table is usually within the fresh rock (see Figures 3.14 and 3.16). These water flows can also erode the surficial residual soils resulting in their sudden collapse and formation of sinkholes.

6.5 MEASUREMENT OF WATER PRESSURE

The importance of water pressure to the stability of slopes has been emphasised in Section 1.4.2 regarding limit equilibrium analysis. For reliable study of slope stability, or design of drainage measures, it is essential that water pressures within the slope be measured. Such measurements are most conveniently carried out by piezometers. Piezometers are devices sealed within the ground, generally in boreholes, which respond only to ground water pressure in the immediate vicinity, and not to ground water pressures at other locations. Piezometers can also be used to measure the *in situ* hydraulic conductivity of rock masses using variable head tests as described in Section 6.6.

The following are a number of factors that may be considered when planning a piezometer installation to measure water pressures in a rock slope:

1. The drill hole should be oriented such that it will intersect the discontinuities in which the ground water is likely to be flowing. For example, in a sedimentary rock containing persistent beds but low-persistence joints, the hole should intersect the beds.
2. The completion zone of the piezometer should be positioned where the rock mass contains discontinuities. For example, if drill core is available, it should be studied to locate zones of fractured or sheared rock where ground water flow is likely to be concentrated. Positioning the completion zone in massive rock with few discontinuities may provide limited information on ground water pressures. The length of the completion zone in rock is usually longer than that in soil because of the need to intersect discontinuities.
3. Other geological features that may be considered in piezometer installations are fault zones. These may act as conduits for ground water if they contain crushed rock, or they may be barriers to ground water flow if they contain clay gouge (e.g. Figure 6.12c). In the case of high-hydraulic conductivity faults, the completion zone may be located in the fault, and for low-hydraulic conductivity faults, the completion zones may be located either side of the fault to determine any pressure differential.
4. The number of piezometers, or the number of completion zones in a single piezometer, may be determined by the geology. For example, in a sedimentary deposit containing low-hydraulic conductivity shale and relatively high-hydraulic conductivity sandstone, it may be necessary to install completion zones in each rock unit.
5. The hydrodynamic time lag is the volume of water required to register a head fluctuation in a piezometer standpipe. The time lag is dependent primarily on the type and dimensions of the piezometer and can be significant in rock with low-hydraulic conductivity. Standpipe piezometers have a greater hydrodynamic time lag than

diaphragm piezometers because a greater movement of pore or joint water is required to register a change in water pressure. The term 'slow response time' is used to describe a long-hydrodynamic time lag.

6. If the piezometer is being used to measure joint water pressure where pressure fluctuations are not likely to be significant, a standpipe piezometer is likely to be suitable. However, if the purpose of the piezometer is to measure the response of the ground water pressures to a drainage system such as a series of horizontal drains, or to detect transient water pressures in response to precipitation, then a diaphragm piezometer with a much shorter time lag would be more appropriate.

7. The filter material in the completion zone should be suited to the rock type. Installations in clay shales or weathered micaceous rocks should use fine-grained filter material that will not be clogged by rock weathering products washed in from the walls of the hole.

8. Cost and reliability are other factors to consider in selecting piezometer types. Standpipe piezometer are simple to install and can be read with inexpensive well sounders, while pneumatic and vibrating wire piezometers are expensive and require more costly read-out units. In situations where the slope is moving and piezometers may be lost, it would be preferable to install standpipe piezometers for reasons of economy.

The following is a brief description of piezometer types and the conditions in which they may be used (Dunnicliff, 1993):

1. *Observation wells* – ground water pressures may be monitored in open holes if the conductivity of the rock mass is greater than about 10^{-4} cm/s, such as coarse grained sandstones and highly fractured rock. The major limitation of observation wells is that they create a vertical connection between strata so their only application is in consistently permeable rock in which the ground water pressure increases continuously with depth. Therefore, observation wells are rarely utilised in monitoring ground water pressures in rock.

2. *Standpipe piezometers* – a standpipe piezometer consists of a length of plastic pipe, with a perforated or porous section at the lower end which is encased in clean gravel or sand to provide a good hydraulic connection with the rock (Figure 6.13). This perforated section of the piezometer, which is the point where the water pressure is measured, is isolated from the rest of the hole with a bentonite seal and a filter layer to prevent contamination of the clean sand around the perforated section. The bentonite is usually placed in the form of compacted pellets that will fall a considerable depth down a water-filled hole before they expand. In very deep holes, the balls can be first soaked in oil to form a protective layer that delays their expansion. However, cement is preferred as a seal for holes with depths greater than about 300 m (1000 ft).

   The water level in a standpipe piezometer can be measured with a well sounder consisting of a graduated electrical cable, with two bared ends, connected to an electrical circuit consisting of a battery and an ammeter. When the bared ends contact water, the circuit is closed and a current is registered on the ammeter. The advantages of this type of piezometer are that it is simple and reliable, but the disadvantages are that access to the top of the hole must be available, and the time lag in low-conductivity rock can be significant.

3. *Pneumatic piezometers* – a rapid response time can be achieved using pneumatic piezometers that comprise a valve assembly and a pair of air lines that connect the valve to the surface. The valve is placed in the sealed section of the piezometer to measure the water pressure at that point. The operating principle is to pump air down the supply line until the air pressure equals the water pressure acting on a diaphragm in

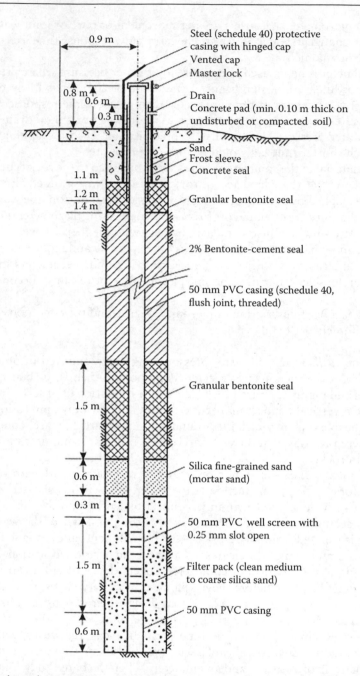

Figure 6.13 Typical standpipe piezometer installation.

the sealed section and the valve opens to start air flowing in the return tube. The pressure required to open the valve is recorded on a pressure gauge at the surface.

Pneumatic piezometers are suitable for low-conductivity rock installations, and because readings can be made at a remote location, they are particularly useful where no access is possible to the collar of the hole. The disadvantages of this type of piezometer are the risk of damage to the air lines either during construction or operation, and the need to maintain a calibrated readout unit.

4. *Electronic transducers* – water pressure measurements with electrical transducers allow very rapid response time and the opportunity to record and process the results at a considerable distance from the slope. Common types of electrical transducers include strain gauges and vibrating wire gauges that measure pressure with a high degree of accuracy. One advantage of these instruments is that they can be sealed in cement and do not require porous completion zones. It is recommended that all transducers be thoroughly tested and calibrated before installation. It should also be kept in mind that the long-term reliability of these sensitive electrical instruments may not equal the design life of the slope and provision should be made for their maintenance and possible replacement.

5. *Multi-completion piezometers* – for slopes excavated in rock types with differing hydraulic conductivities, it is possible that zones of high ground water pressure exist within a generally depressurised area. In such circumstances, it may be desirable to measure the ground water pressure at a number of points in a drill hole. This can be achieved by installing multiple piezometers in a single drill hole with bentonite or cement seals between each section of perforated pipe. The maximum number of such standpipes that can be installed in an NX borehole is three; and in more pipes, placement of filters and effective seals becomes very difficult.

An alternative method of measuring water pressures at a number of different points in a drill hole is to use a multi-port (MP) system that also allows measurement of hydraulic conductivity and retrieval of water samples (Black, Smith and Patton, 1986) (Figure 6.14). The MP system is a modular multiple-level monitoring device employing a single, closed access tube with valved ports. The valved ports provide access to several different levels of a drill hole in a single well casing, and the modular design permits as many monitoring zones as desired to be established in a drill hole. The system consists of casing components that are permanently installed in the drill hole with valves in the casing and zones of porous backfill located where water pressure measurements are required. Pressure measurements and water sampling are made by lowering-pressure transducers, sampling probes and specialised tools down the hole to the valve locations; two types of valved port couplings are used with capabilities to either measure pressure or take samples. The port assemblies can be isolated in the drill hole by sealing the annulus between the monitoring zones using either pairs of packers, or by filling the annulus with a cement grout or bentonite seal. The MP system has been used in drill holes up to 1200 m (4000 ft) deep.

6.6 FIELD MEASUREMENT OF HYDRAULIC CONDUCTIVITY

Determination of the hydraulic conductivity of a rock mass is necessary if estimates are required of ground water discharge from a slope, or in the design of a drainage system.

For evaluation of the stability of the slopes, it is the water pressure rather than the volume of ground water flow in the rock mass that is important. The water pressure at any point is independent of the hydraulic conductivity of the rock mass at that point, but it does depend upon the path followed by the ground water in arriving at the point (Figures 6.10 and 6.11). Hence, the heterogeneity and anisotropy of the rock mass with respect to the distribution of hydraulic conductivity is of interest in estimating the water pressure distribution in a slope.

Because ground water flow in fractured rock takes place predominately in the discontinuities, it is necessary that hydraulic conductivity measurements be made *in situ*; it is not possible to simulate water flow in a fractured rock mass in the laboratory. The following is

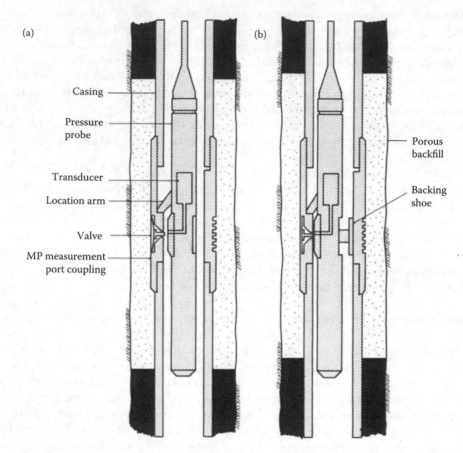

(a)

Casing

Pressure
probe

Transducer

Location arm

Valve

MP measurement
port coupling

(b)

Porous
backfill

Backing
shoe

Figure 6.14 Multiple completion piezometer installation (MP system, Westbay Instruments) with probe positioned to make pressure measurements. (a) Probe located at measurement port coupling; (b) probe measuring fluid pressure outside coupling (Black, Smith and Patton, 1986).

a brief description of the two most common methods of *in situ* conductivity testing, namely variable head tests and pumping tests. Detailed procedures for hydraulic conductivity tests are described in the literature, and the tests themselves are usually conducted by specialists in the field of hydrogeology.

6.6.1 Variable head tests

In order to measure the hydraulic conductivity at a 'point' in a rock mass, it is necessary to change the ground water conditions at that point and to measure the time taken for the original conditions to be reestablished. These tests are most conveniently carried out in a borehole, and the test length may represent the general rock mass properties in the slope, or the test may be located in a specific geologic feature such as a fault. An essential requirement of the test is that the borehole wall be clean and no clogging occurs of the discontinuities by drill cuttings or drilling mud. This will require flushing of the hole and the use of polymer muds that break down some time after drilling to leave a clean hole (see Section 4.6.2).

Test configurations: a number of borehole test configurations are possible. A piezometer installed in a drill hole will isolate a section at the end of the hole (Figure 6.15a), or at some point up the hole. It is also possible to conduct hydraulic conductivity measurements in open

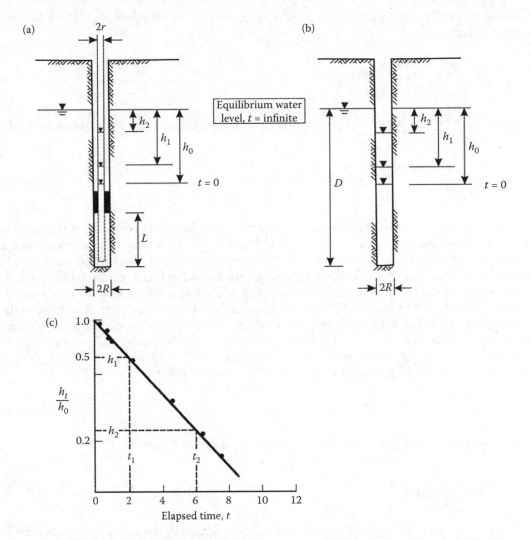

Figure 6.15 Method of calculating hydraulic conductivity for variable head test: (a) piezometer with completion length L; (b) open hole to depth D below water table; (c) typical plot of head increase (recovery) against time.

holes (Figure 6.15b), although it is necessary that the geologic conditions are consistent over the test length.

Test procedure: the procedure for the variable head test is first to establish the rest water level, which is the static equilibrium level of the water table at the drill hole location (Figure 6.15a and b). Pumping of circulation water during drilling will disturb this equilibrium and the conductivity results will be in error if insufficient time is allowed for equilibrium to be reestablished. Once equilibrium has been established, water can either be removed from (*bail test*) or added to (*slug test*) the standpipe piezometer to change the water level. If the test is conducted above the water table, it is necessary to perform a slug test, while for a test below the water table, a bail test is preferred because the flow of water out of the formation minimises clogging of the fractures.

Water is added or removed from the hole to change the water level by about 1–2 m (3–7 ft) and the rate at which the water level recovers to the equilibrium level is measured. For the test shown in Figure 6.14a, the hydraulic conductivity K is calculated from the following general relationship:

$$K = \frac{A}{F(t_2 - t_1)} \cdot \ln\left(\frac{h_1}{h_2}\right) \quad \text{for } \frac{L}{R} > 8 \tag{6.9}$$

where F is the shape factor. For a drill hole with radius R and a test zone of length L, the shape factor is given

$$F = \frac{2\pi \cdot L}{\ln(L/R)} \tag{6.10}$$

In Equation 6.9, A is the cross-section area of the standpipe ($A = \pi \cdot r^2$) and r is the internal radius of the standpipe. For a bail test, t_1 and t_2 are the times at which the water levels are at depths h_1 and h_2, respectively, below the equilibrium water level. The differential heads h_1 and h_2, as well as the initial equilibrium head h_0, are defined in Figure 6.15a, while a typical semilog plot of the rise in water level in the casing with time is shown in Figure 6.15c.

The types of test shown in Figure 6.15 are suitable for measurement of the hydraulic conductivity of reasonably uniform rock. Anisotropic hydraulic conductivity coefficients cannot be measured directly in these tests but allowance can be made for this anisotropy in the calculations as follows. If an estimate of the ratio of vertical and horizontal hydraulic conductivities K_v and K_h, respectively, is made, the ratio m is given by

$$m = \sqrt{K_h/K_v} \tag{6.11}$$

and the shape factor F given by Equation 6.10 is modified as follows:

$$F = \frac{2\,\pi \cdot L}{\ln(L \cdot (m/R))} \tag{6.12}$$

When this value of F is substituted in Equation 6.9, then the calculated value of the hydraulic conductivity, K is the mean hydraulic conductivity given by $\sqrt{(K_v \cdot K_h)}$.

Packer tests – where conductivity tests are required at specific locations within a drill hole, a test zone can be isolated using packers positioned at any location in the hole, and over any required length (Figure 6.16). Packer tests can be made during diamond drilling using a triple packer system that is lowered through the rods so that the test is conducted in a portion of the hole below the drill bit. The packer system consists of three inflatable rubber packers, each 1 m (3 ft) long that is sufficient to minimise the risk of leakage past the packer. The lower two packers are joined by a perforated steel pipe, the length of which depends on the required test length, while the top and middle packers are joined by a solid pipe. The whole packer assembly is lowered down the drill hole on the wire line through the drill rods and the lower two packers extend through the bit into the open hole, while the upper packer is located in the lower end of the core barrel. The three packers are then inflated with nitrogen through a small diameter plastic tube that runs down the hole inside the rods. The inflated packers seal the packer assembly into the rods, and isolate a length of drill hole below the bit. If water is removed from the drill rods, formation water will flow from the rock, over the test interval isolated by the two lower packers, and through the perforated pipe to restore the water level in the rods. This flow of water is measured by

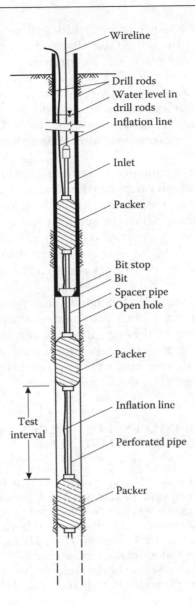

Wireline

Drill rods
Water level in
drill rods
Inflation line

Inlet

Packer

Bit stop
Bit
Spacer pipe
Open hole

Packer

Inflation line

Test
interval

Perforated pipe

Packer

Figure 6.16 Triple packer arrangement for making variable head hydraulic conductivity tests in conjunction with diamond drilling (Wyllie, 1999).

monitoring the change of water level in the drill rods with time. A plot of the results is a semilog head recovery plot such as that shown in Figure 6.15c. Cedergren (1989) provides a comprehensive discussion of variable head hydraulic conductivity testing.

6.6.2 Pumping test

The main limitations of conductivity tests carried out in drill holes are that only a small volume of rock in the vicinity of the hole is tested, and it is not possible to determine the anisotropy of the rock mass. Both these limitations are overcome by conducting pump tests, as described briefly below.

A pump test arrangement consists of a vertical well equipped with a pump, and an array of piezometers in which the water table elevation can be measured in the rock mass surrounding the well. The piezometers can be arranged so that the influence of various geologic features on ground water conditions can be determined. For instance, piezometers could be installed on either side of a fault, or in directions parallel and perpendicular to sets of persistent discontinuities such as bedding planes. Selection of the best location for both the pumped well and the observation wells requires considerable experience and judgement, and should only be carried out after thorough geological investigations have been carried out.

The test procedure consists of pumping water at a steady rate from the well and measuring the drop in water level in both the pumped well and the observation wells. The duration of the test can range from as short as 8 h to as long as several weeks, depending on the conductivity of the rock mass. When the pumping is stopped, the water levels in all the wells are measured until a static water level is determined – this is known as the recovery stage of the test. Plots of draw down (or recovery) against time can be used to calculate conductivity values using methods described by Cedergren (1989), Todd (1959), Jacob (1950) and Theis (1935).

Because of the cost and time required for conducting a pump test, they are rarely carried out for rock slope engineering. An example of a situation where a pump test may be justified would be to assess the feasibility of driving a drainage adit for stabilisation of a landslide. Generally, installation of piezometers to measure the ground water table and conduct variable head tests to measure hydraulic conductivity in boreholes provide sufficient information on ground water conditions for slope design purposes.

EXAMPLE PROBLEM 6.1: INFLUENCE OF GEOLOGY AND WEATHER CONDITIONS ON GROUND WATER LEVELS

Statement

Figure 6.17 shows a slope, cut in isotropic, fractured rock, under a variety of operational and climatic conditions; in all cases ground water is infiltrating the horizontal ground surface behind the crest of the slope (Terzaghi, 1962). In Figure 6.17a, the slope has been recently excavated, and prior to excavation, the water table was horizontal and at a shallow depth below the ground surface. In Figure 6.17a and b, the slope has been open for a sufficiently long time for ground water equilibrium conditions to be established, but the climatic conditions vary. The objective of the exercise is to sketch in the ground water table on each slope based on the general behaviour of ground water in slopes as governed by Darcy's Law.

Required

1. On the cross-section in Figure 6.17a, draw the approximate position of the ground water table after excavation of the slope.
2. On the cross-section in Figure 6.17b, draw the approximate positions of the ground water table for the following conditions:
   a. Large infiltration, low-hydraulic conductivity
   b. Small infiltration, high-hydraulic conductivity
3. On the cross-section in Figure 6.17c, draw the approximate positions of the ground water table for the following conditions:
   a. Joints on slope face plugged with ice
   b. Immediately following a heavy rainstorm
   c. Wet season
   d. Dry season

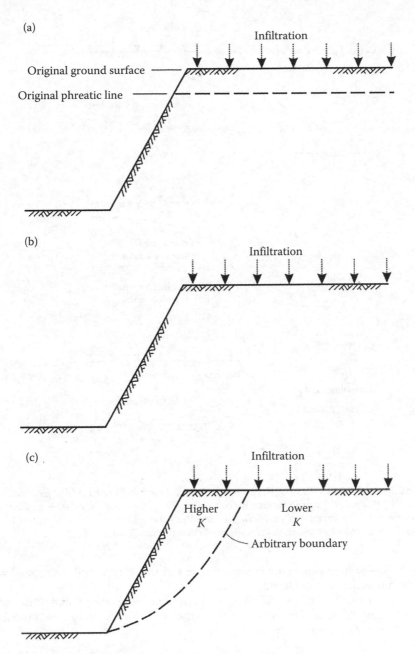

Figure 6.17 Cut slope in rock with water infiltration on ground surface behind crest for Example 6.1: (a) position of the ground water surface before excavation; (b) slope with variety of surface infiltration and rock conductivity condition; (c) slope with higher-conductivity rock close to face, and various climatic condition.

Solution

Figure 6.18a–c shows the positions of the ground water table under the various conditions shown in Figure 6.17.

In general, the rock near the slope face has been disturbed by blasting and has undergone stress relief so it will have a higher-hydraulic conductivity than the undisturbed rock.

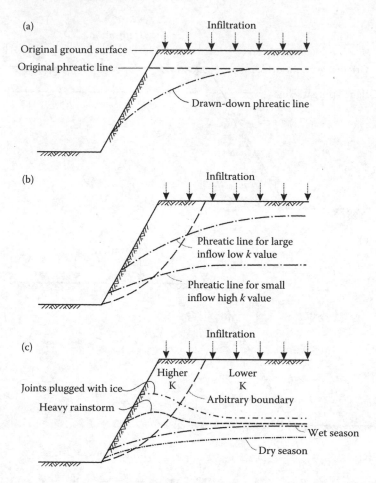

Figure 6.18 Positions of ground water surface for conditions shown in Figure 6.17 for Example 6.1: (a) position of the ground water surface before and after excavation; (b) relative positions of ground water table for variations of inflow and conductivity; (c) hypothetical positions of the ground water surface in jointed rock for variety of climatic conditions.

When the hydraulic conductivity is high, the rock drains readily and the ground water table has a relatively flat gradient.

If the face freezes and the water cannot drain from the slope, the ground water surface will rise behind the face. The same situation arises when heavy infiltration exceeds the rate at which the rock will drain.

Chapter 7

Plane failure

7.1 INTRODUCTION

A plane failure is a comparatively rare sight in rock slopes because it is only occasionally that all the geometric conditions required to produce such a failure occur in an actual slope. However, it would not be right to ignore the two-dimensional case because many valuable lessons can be learned from a consideration of the mechanics of this simple failure mode. Plane failure is particularly useful for demonstrating the sensitivity of the slope to changes in shear strength and ground water conditions – changes which are less obvious when dealing with the more complex mechanics of a three-dimensional slope failure. This chapter describes the method of analysis for plane failure (Sections 7.2 and 7.3), discusses the stability of 'dip slopes' (non-daylighting sliding plane) and demonstrates the design of reinforced slopes (Section 7.4) and probabilistic design methods (Section 7.6). Stability analysis methods for slopes subject to seismic ground motion – pseudo-static and Newmark displacement – are discussed in Chapter 11.

Two case studies describing the stabilisation of plane failures are described in Chapter 16.

7.2 GENERAL CONDITIONS FOR PLANE FAILURE

Figure 7.1 shows a typical plane failure in a rock slope where a block of rock has slid on a single plane dipping out of the face. In order for this type of failure to occur, the following geometrical conditions must be satisfied (Figure 7.2):

1. The plane on which sliding occurs must strike parallel or nearly parallel (within approximately ±20°) to the slope face.
2. The sliding plane must 'daylight' in the slope face, which means that the dip of the plane must be less than the dip of the slope face, that is, $\psi_p < \psi_f$.
3. The dip of the sliding plane must be greater than the angle of friction of this plane, that is, $\psi_p > \phi$.
4. The upper end of the sliding surface either intersects the upper slope, or terminates in a tension crack.
5. Release surfaces that provide negligible resistance to sliding must be present in the rock mass to define the lateral boundaries of the slide. Alternatively, failure can occur on a sliding plane passing through the convex 'nose' of a slope.

Figure 7.1 (**See colour insert.**) Plane failure on smooth, persistent bedding planes in shale (Interstate 40, near Newport, Tennessee).

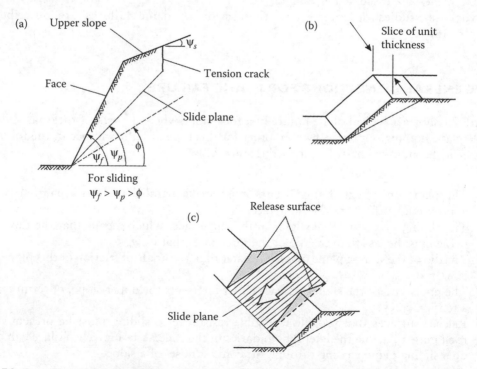

Figure 7.2 Geometry of slope exhibiting plane failure: (a) cross-section showing planes forming a plane failure; (b) unit thickness slide used in stability analysis; (c) release surfaces at the ends of plane failure.

7.3 PLANE FAILURE ANALYSIS

The slope geometries and ground water conditions considered in this analysis are defined in Figure 7.3, which shows two geometries as follows:

1. Slopes having a tension crack in the upper surface
2. Slopes with a tension crack in the face

When the upper surface is horizontal ($\psi_s = 0$), the transition from one condition to another occurs when the tension crack coincides with the slope crest, that is, when

$$\frac{z}{H} = (1 - \cot\psi_f \cdot \tan\psi_p) \tag{7.1}$$

where z is the depth of the tension crack, H is the slope height, ψ_f is the slope face angle and ψ_p is the dip of the sliding plane.

The following assumptions are made in plane failure analysis:

1. Both sliding surface and tension crack strike parallel to the slope.
2. The tension crack is vertical and is filled with water to a depth z_w.
3. Water enters the sliding surface along the base of the tension crack and seeps along the sliding surface, escaping at atmospheric pressure where the sliding surface daylights in the slope face. The pressure distributions induced by the presence of water in the tension crack and along the sliding surface are illustrated in Figure 7.3.

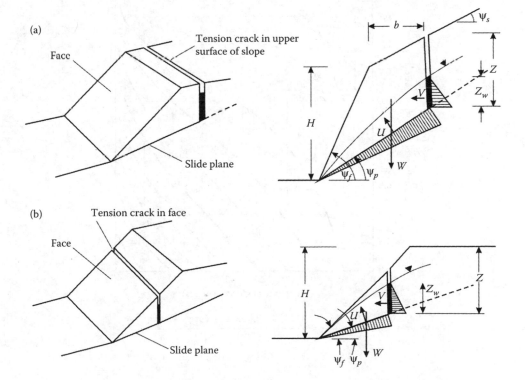

Figure 7.3 Geometries of plane slope failure: (a) tension crack in the upper slope; (b) tension crack in the face.

4. The forces W (the weight of the sliding block), U (uplift force due to water pressure on the sliding surface) and V (force due to water pressure in the tension crack) all act through the centroid of the sliding mass. In other words, it is assumed that the forces produce no moments that would tend to cause rotation of the block, and hence failure is by sliding only. While this assumption may not be strictly true for actual slopes, the errors introduced by ignoring moments are small enough to neglect. However, in steep slopes with steeply dipping discontinuities, the possibility of toppling failure should be kept in mind (see Chapter 10).

5. The shear strength τ of the sliding surface is defined by cohesion c and friction angle ϕ that are related by the equation $(\tau = c + \sigma \tan \phi)$ – as discussed in Chapter 5. In the case of a rough surface or a rock mass having a curvilinear shear strength envelope, the apparent cohesion and apparent friction angle are defined by a tangent that takes into account the normal stress acting on the sliding surface. The normal stress σ acting on a sliding surface can be determined from the curves given in Figure 7.4.

6. It is assumed that release surfaces are present so that no resistance to sliding is generated at the lateral boundaries of the failing rock mass.

7. In analysing two-dimensional slope problems, it is usual to consider a slice of unit thickness taken at right angles to the slope face. This means that on a vertical section through the slope, the area of the sliding surface can be represented by the length of the surface, and the volume of the sliding block is represented by the cross-section area of the block (Figure 7.2b).

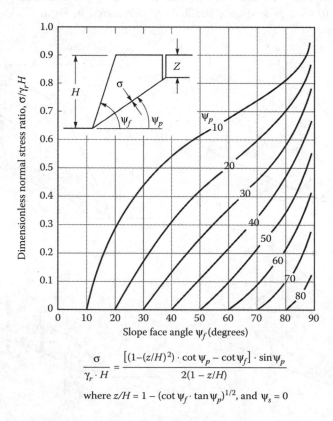

$$\frac{\sigma}{\gamma_r \cdot H} = \frac{[(1-(z/H)^2) \cdot \cot \psi_p - \cot \psi_f] \cdot \sin \psi_p}{2(1 - z/H)}$$

where $z/H = 1 - (\cot \psi_f \cdot \tan \psi_p)^{1/2}$, and $\psi_s = 0$

Figure 7.4 Normal stress acting on slide plane in a rock slope.

The factor of safety for plane failure is calculated by resolving all forces acting on the slope into components parallel and normal to the sliding plane. The vector sum of the shear forces ΣS acting down the plane is termed the *driving force*. The product of the total normal forces ΣN and the tangent of the friction angle ϕ plus the cohesive force is termed the *resisting force* (see Section 1.4.2). The factor of safety FS of the sliding block is the ratio of the resisting forces to the driving forces, and is calculated as follows:

$$FS = \frac{\text{Resisting force}}{\text{Driving force}} \tag{7.2}$$

$$= \frac{c \cdot A + \Sigma N \cdot \tan\phi}{\Sigma S} \tag{7.3}$$

where c is the cohesion and A is the area of the sliding plane.

Based on the concept illustrated in Equations 7.2 and 7.3, the factor of safety for the slope configurations shown in Figure 7.3 is given by

$$FS = \frac{c \cdot A + (W \cdot \cos\psi_p - U - V \cdot \sin\psi_p)\tan\phi}{W \cdot \sin\psi_p + V \cdot \cos\psi_p} \tag{7.4}$$

where A is given by

$$A = (H + b \cdot \tan\psi_s - z)\operatorname{cosec}\psi_p \tag{7.5}$$

The slope height is H, the tension crack depth is z and it is located at a distance b behind the slope crest. The dip of the slope above the crest is ψ_s.

When the depth of the water in the tension crack is z_w, the forces acting on the sliding plane U, and in the tension crack V, are given by

$$U = \frac{1}{2}\gamma_w \cdot z_w (H + b \cdot \tan\psi_s - z)\operatorname{cosec}\psi_p \tag{7.6}$$

$$V = \frac{1}{2}\gamma_w \cdot z_w^2 \tag{7.7}$$

where γ_w is the unit weight of water.

The weights of the sliding block W for the two geometries shown in Figure 7.3 are given by Equations 7.8 and 7.9.

For the tension crack in the inclined upper slope surface (Figure 7.3a)

$$W = \gamma_r \left[(1 - \cot\psi_f \cdot \tan\psi_p) \cdot \left(b \cdot H + \frac{1}{2}H^2 \cdot \cot\psi_f \right) + \frac{1}{2}b^2(\tan\psi_s - \tan\psi_p) \right] \tag{7.8}$$

and, for the tension crack in the slope face (Figure 7.3b)

$$W = \frac{1}{2}\gamma_r \cdot H^2 \left[\left(1 - \frac{z}{H} \right)^2 \cdot \cot\psi_p(\cot\psi_p \cdot \tan\psi_f - 1) \right] \tag{7.9}$$

where γ_r is the unit weight of the rock.

Figure 7.3 and Equations 7.4 to 7.9 illustrate that the geometry of a plane failure and the ground water conditions can be completely defined by four dimensions (H, b, z and z_w) and by three angles (ψ_f, ψ_p and ψ_s). These simple models, together with the ground water, rock bolting and seismic ground motion concepts discussed in the following sections allow stability calculations to be carried out for a wide variety of conditions.

7.3.1 Influence of ground water on stability

In the preceding discussion, it has been assumed that the influence of ground water on the stability of the slope is only the water present in the tension crack and along the sliding surface. This is equivalent to assuming that the rest of the rock mass is impermeable, an assumption that is certainly not always justified. Therefore, consideration must be given to water pressure distributions other than those presented in this chapter. Under some conditions, it may be possible to construct a flow net from which the ground water pressure distribution can be determined from the intersection of the equipotentials with the sliding surface (see Figure 6.10). Information that would assist in developing flow nets includes the rock mass hydraulic conductivity (and its anisotropy), the location of seepage on the face and recharge above the slope, and any piezometric measurements.

In the absence of actual ground water pressure measurements within a slope, the current state of knowledge in rock engineering does not permit a precise definition of the ground water flow patterns in a rock mass. Consequently, slope design should assess the sensitivity of the factor of safety to a range of realistic ground water pressures, and particularly the effects of transient pressures due to rapid recharge (see Figure 6.12).

The following are four possible ground water conditions that may occur in rock slopes, and the equations that can be used to calculate the water forces U and V. In these examples, the pressure distributions in the tension crack and along the sliding plane are idealised and judgement is required to determine the most suitable condition for any particular slope.

1. Ground water level is above the base of tension crack so water pressures act both in the tension crack and on the sliding plane. If the water discharges to the atmosphere where the sliding plane daylights on the slope face, then it is assumed that the pressure decreases linearly from the base of the tension crack to zero at the face. This condition is illustrated in Figure 7.3 and the method of calculating forces U and V is given by Equations 7.6 and 7.7, respectively.

2. Water pressure may develop in the tension crack only in conditions, for example, where a heavy rainstorm after a long dry spell results in surface water flowing directly into the crack. If the remainder of the rock mass is relatively impermeable, or the sliding surface contains a low-conductivity clay filling, then the uplift force U could also be zero or nearly zero. In either case, the factor of safety of the slope for these transient conditions is given by Equation 7.4 with $U = 0$ and V given by Equation 7.7.

3. Ground water discharge at the face may be blocked by freezing (Figure 7.5a). Where the frost penetrates only a few metres behind the face, water pressures can build up in the slope and the uplift pressure U can exceed that shown in Figure 7.3. For the idealised rectangular pressure distribution shown in Figure 7.5a, the uplift force U is given by

$$U = A \cdot p \tag{7.10}$$

where A is the area of the sliding plane given by Equation 7.5 and p is the pressure in the plane (and at the base of the tension crack) given by

$$p = \gamma_w \cdot z_w \tag{7.11}$$

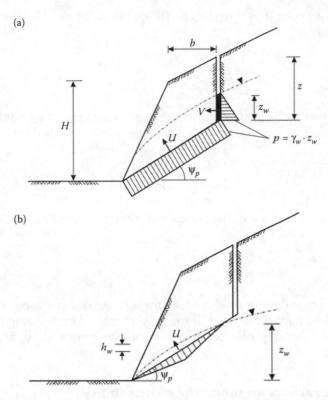

Figure 7.5 Possible ground water pressures in plane failure: (a) uniform pressure on slide plane for drainage blocked at toe; (b) triangular pressure on slide plane for water table below the base of tension crack.

The condition shown in Figure 7.5a may only occur rarely, but could result in a low factor of safety; a system of horizontal drains may help to limit the water pressure in the slope.

4. Ground water level in the slope is below the base of the tension crack so water pressure acts only on the sliding plane (Figure 7.5b). If the water discharges to the atmosphere where the sliding plane daylights on the face, then the water pressure can be approximated by a triangular distribution, for which the uplift force U is given by

$$U = \frac{1}{2} \cdot \frac{z_w}{\sin \psi_p} \cdot h_w \cdot \gamma_w \qquad (7.12)$$

where h_w is the estimated depth of water at the mid-point of the saturated portion of the sliding plane.

7.3.2 Critical tension crack depth and location

In the analysis it has been assumed that the position of the tension crack is known from its visible trace on the upper surface or on the face of the slope, and that its depth can be established by constructing an accurate cross-section of the slope. However, the tension crack position may be unknown due, for example, to the presence of soil above the slope crest, or an assumed location may be required for design. Under these circumstances, it becomes necessary to consider the most probable position of a tension crack.

When the slope is dry or nearly dry $(z_w/z = 0)$, Equation 7.4 for the factor of safety can be modified as follows:

$$FS = \frac{c \cdot A}{W \cdot \sin \psi_p} + \cot \psi_p \cdot \tan \phi \qquad (7.13)$$

The critical tension crack depth z_c for a dry slope can be found by minimising the right-hand side of Equation 7.13 with respect to z/H. This gives the critical tension crack depth as

$$\frac{z_c}{H} = 1 - \sqrt{\cot \psi_f \cdot \tan \psi_p} \qquad (7.14)$$

and the corresponding position of the critical tension crack b_c behind the crest is

$$\frac{b_c}{H} = \sqrt{\cot \psi_f \cdot \cot \psi_p - \cot \psi_f} \qquad (7.15)$$

Critical tension crack depths and locations for a range of dimensions for dry slopes are plotted in Figure 7.6a and b. However, if the tension crack forms during heavy rain or if it is located on a pre-existing geological feature such as a vertical joint, Equations 7.14 and 7.15 no longer apply.

7.3.3 Tension crack as an indicator of instability

Anyone who has examined excavated rock slopes cannot have failed to notice the occasional tension cracks behind the crest (Figure 7.7). Some of these cracks have been visible for tens of years and, in many cases, do not appear to have had any adverse influence on the stability of the slope. It is interesting, therefore, to consider how such cracks are formed and whether they can give any indication of slope instability.

In a series of very detailed model studies on the failure of slopes in jointed rocks, Barton (1971) found that the tension crack resulted from small shear movements within the rock mass. Although these individual movements were very small, their cumulative effect was significant displacement of the slope surfaces – sufficient to cause separation of vertical joints behind the slope crest and to form 'tension' cracks. The fact that the tension crack is caused by shear movements in the slope is important because it suggests that, when a tension crack becomes visible in the surface of a slope, it must be assumed that shear failure has initiated within the rock mass.

It is difficult to quantify the significance tension cracks since their formation is only the start of a complex progressive failure process within the rock mass; such failure mechanisms can be studied with numerical models as discussed in Chapter 12. It is probable that, in some cases, the improved drainage resulting from dilation of the rock structure, combined with the interlocking of individual blocks within the rock mass, could result in an increase in stability. However, where the failure surface comprises a single discontinuity surface such as a bedding plane daylighting in the slope face, initial movement could be followed by a very rapid decrease in stability because a small amount of movement could result in a reduction in the shear strength from the peak to residual value.

In summary, the presence of a tension crack should be taken as an indication of potential instability and that, in the case of an important slope, this should signal the need for a detailed investigation of stability.

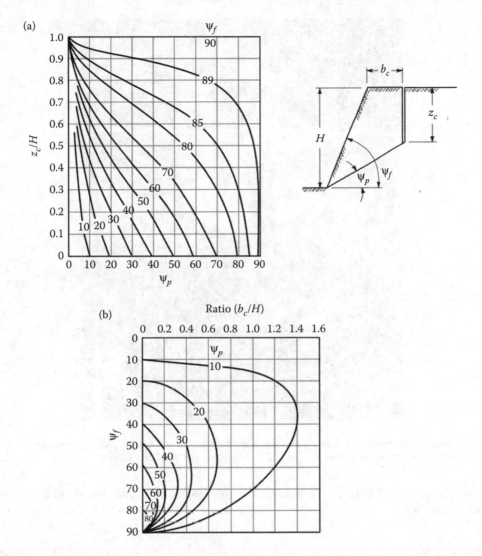

Figure 7.6 Critical tension crack locations for a dry slope: (a) critical tension crack depth relative to crest of cut; (b) critical tension crack location behind crest of cut.

7.3.4 Critical slide plane inclination

When a persistent discontinuity such as a bedding plane exists in a slope and the inclination of this discontinuity is such that it satisfies the conditions for plane failure defined in Figure 7.2, the stability of the slope will be controlled by this feature. However, where no such feature exists and a sliding surface, if it were to occur, would follow minor geological features and, in some places, pass through intact material, how can the inclination of such a failure path be determined?

The first assumption that must be made concerns the shape of the slide surface. For a slope cut in weak rock, or a soil slope with a face angle less than about 45°, the slide surface would have a circular shape. The analysis of such a failure surface is discussed in Chapter 9.

In steep rock slopes, the slide surface is approximately planar and the inclination of such a plane can be found by partial differentiation of Equation 7.4 with respect to ψ_p and by

Figure 7.7 (**See colour insert.**) A tension crack behind a sliding rock mass in which significant horizontal displacement has occurred (above: Kooteney Lake, British Columbia).

equating the resulting differential to zero. For dry slopes, this gives the critical slide plane inclination ψ_{pc} as

$$\psi_{pc} = \frac{1}{2}(\psi_f + \phi) \tag{7.16}$$

The presence of water in the tension crack will cause the slide plane inclination to be reduced by as much as 10%, but in view of the uncertainties associated with the inclination of this slide surface, the added complication of including the influence of ground water is not considered justified. Consequently, Equation 7.16 can be used to obtain an estimate of the critical slide plane inclination in steep slopes that do not contain through-going discontinuities.

7.3.5 Analysis of failure on a rough plane

The stability analyses discussed so far in this section have used shear strength parameters that are constant throughout the slope. However, as discussed in Section 5.2.4 on the shear strength of rough rock surfaces, the friction angle that will be mobilised in the slope may depend on the normal stress acting on the surface. That is, the friction angle will decrease with increasing normal stress as the asperities on the surface are ground off, as defined by

Equation 5.7. The significance of this relationship between friction angle and normal stress is illustrated below.

Consider a plane slope failure with the geometry as shown in Figure 7.3a. For a dry slope $(U = V = 0)$, the normal stress σ acting on the sliding surface is given by

$$\sigma = \frac{W \cdot \cos \psi_p}{A} \tag{7.17}$$

where W is the weight of the sliding block, ψ_p is the dip of the sliding surface and A is the area of this surface. If the sliding plane contains no cohesive infilling so that the shear strength comprises only friction, then the factor of safety can be calculated using Equations 1.2 to 1.6 for limit equilibrium analysis, Equation 5.7 to define the shear strength of the rough surface, and Equation 7.17 to define the normal stress on this surface. For these conditions, the factor of safety is given by

$$FS = \frac{\tau \cdot A}{W \cdot \sin \psi_p} \tag{7.18}$$

$$= \frac{\sigma \cdot \tan(\phi + JRC \cdot \log_{10}(JCS/\sigma)) \cdot A}{W \cdot \sin \psi_p} \tag{7.19}$$

$$= \frac{\tan(\phi + JRC \cdot \log_{10}(JCS/\sigma))}{\tan \psi_p} \tag{7.20}$$

$$= \frac{\tan(\phi + i)}{\tan \psi_p} \tag{7.21}$$

The application of these equations and the effect of a rough surface on the factor of safety can be illustrated by the following example. Consider a slope with dimensions $H = 30$ m (98 ft), $z = 15$ m (49 ft), $\psi_p = 30°$ and $\psi_f = 60°$, in which the properties of the clean rough joint forming the sliding surface are $\phi = 25°$, $JRC = 15$ and $JCS = 5000$ kPa (725 psi). From Figure 7.4, the normal stress ratio $(\sigma/\gamma_r \cdot H)$ is 0.36, and the value of σ is 281 kPa (40.7 psi) if the rock density γ_r is 26 kN/m³ (165 lb/ft³). The value of the σ calculated from Figure 7.4 is the average normal stress acting on the sliding surface. However, the maximum stress acting on this surface is below the crest of the slope where the depth of rock is 20 m (66 ft). The calculated maximum stress is

$$\begin{aligned} \sigma_{max} &= 20 \cdot 26 \cdot \cos 30 \\ &= 450 \text{ kPa (62.3 psi)} \end{aligned}$$

Using Equation 5.7 and the roughness properties quoted in the previous paragraph, the shear strength of the sliding surface and the corresponding factors of safety at the average and maximum normal stresses can be calculated as

$\sigma = 281$ kPa (40.7 psi); $\tau = 269$ kPa (39 psi); $(\phi + i) = 44°$ and $FS = 1.66$
$\sigma = 450$ kPa (62.3 psi); $\tau = 387$ kPa (56.3 psi); $(\phi + i) = 41°$ and $FS = 1.49$

These results indicate that the effect of increasing normal stress level on the sliding surface is to diminish the friction angle (due to the asperities being ground off) and that the corresponding decrease in the factor of safety is 10%.

7.4 DIP SLOPE STABILITY

A special case of plane failure is the 'dip slope' where a set of planar discontinuities is oriented parallel to the face to form a series of thin slabs over the full height of the slope. This configuration means that these discontinuities do not 'daylight' on the face such that instability requires some form of displacement or breakout through the slab to the slope face. Figure 7.8 shows typical dip slopes with possible modes of instability involving flexure, buckling or ploughing in the lower part of the slab.

This section describes occurrences of dip slope instability, failure mechanisms and methods of stability analysis.

7.4.1 Examples of dip slopes

Dip slopes occur most frequently in dipping sedimentary formations such as interbedded sandstone and shale formations, where the natural slope, or an excavated face, is parallel to the beds (Fisher, 2009). These conditions are common in coal mines where seams are mined to form a continuous footwall slope that may be planar, convex or concave depending on local folding of the beds (Brawner, Pentz and Sharp, 1971; Stead and Eberhardt, 1997). For example, at the Smoky River coal mine in Alberta, a 250 m (900 ft) long section of the 97 m (320 ft) high footwall failed in 1987. The failure occurred on a coal seam, dipping at an angle of 60–65°, located at a depth of 10–12 m (30–40 ft) behind the face (Dawson, Roth and Drescher, 1999). Vectors of slope movement measured prior to failure showed down-dip movement in the upper slope and outward movement in the lower slope that are indicative of a dip slope failure mechanism.

Buckling-type instability has been observed at Telegraph Hill in San Francisco, California where rock faces with an angle of 65–70° have been excavated in a sandstone quarry. At one location where a steeply dipping joint containing soil and tree roots was located at a distance of about 2 m (6 ft) behind the face, bulging of the toe was observed and the crest of the slope dropped about 0.3 m (1 ft) prior to sudden failure of the slope (Wallace and Marcum, 2015).

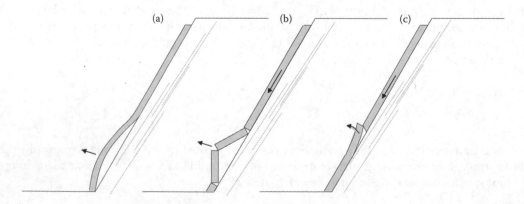

Figure 7.8 Possible modes of instability for dip slopes. (a) Flexure with Euler-type buckling of thin, unjointed slab; (b) three hinge buckling of jointed slab; (c) ploughing mechanism of dip slope movement.

The slope was about 20 m (66 ft) high so the ratio between the slope height (H) and the slab thickness (D) was about 0.1.

Larger-scale dip slope failures can occur in landslides where the base sliding plane comprises two linear segments where the upper segment is parallel to the slope, and the lower segment is at a shallower angle that daylights in the face. Examples of such landslides include the Vajont slide in Italy (Sitar, MacLaughlin and Doolin, 2005) (see Figure 7.9), and the Mitchell Creek slide in British Columbia, Canada (Clayton and Stead, 2015). The Mitchell Creek slide has a volume of about 75 million cubic metres with the upper plane dipping at 30°, at a depth of 140 m (460 ft), and the lower plane being near horizontal; the total length of the sliding plane from the crown to the toe is about 800 m (2600 ft). Field observations and computer modelling show that both these landslides comprise the basic components of bi-planar instability – an active upper zone that slides on the surface parallel to the slope surface, a transition zone where rock fracturing occurs and passive zone at the base of the slide where displacement occurs on the face.

Stability analysis of the Mitchell Creek slide is also discussed in Section 12.5.8 on numerical analysis.

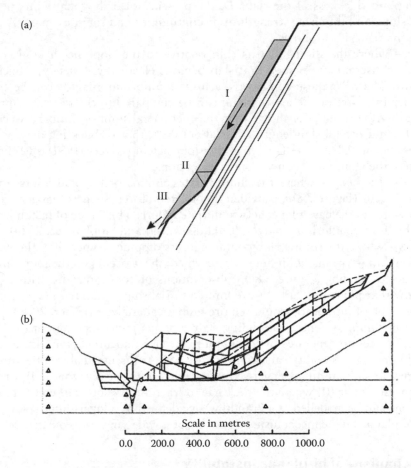

Figure 7.9 Examples of bi-planar dip slope failures showing rock failure in transition zones. (a) Bi-planar instability for steep slope with downslope movement of relatively thin active slab (I), rock fracture in Prandtl wedge transition zone (II) and outward movement of passive slab on toe release joint (III) (adapted from Fisher, 2009). (b) Analysis of Vajont Slide in Italy using DDA with upper base plane dipping at 33° at a depth of about 150 m (500 ft) (Sitar, MacLaughlin and Doolin, 2005).

7.4.2 Types of dip slope failure

Dip slope failures can occur by a variety of mechanisms that depend on details of site topography and geology as shown in Figures 7.8 and 7.9. In order for this type of slope instability to occur, it is first necessary that the discontinuity (usually bedding) forming the base of the upper slab is persistent and that the dip of this surface is greater than the friction angle, and/or high-water pressures act on this surface, so that a driving force is generated by the component of the weight of the slab acting down dip. This upper slab is generally termed the 'active' slab. The driving force of the active slab will increase at steeper slope angles because of the increasing down slope component of the weight, and the reduced the shear resistance as the result of the reduction of the normal force on the base of the slab.

The thrust produced by the driving force of the active slab can cause movement in the lower part of the slope, termed the 'passive' slab, by a number of mechanisms as shown in Figures 7.8 and 7.9. A common feature of these mechanisms is that, for slope failure to occur, displacement and/or fracturing of rock has to take place in the transition zone between the upper, active slab and the lower, passive slab.

Slope conditions that influence the failure mechanism include the slope height H, the bedding spacing and thickness of the slabs D, the presence of orthogonal joints in the lower part of the slope and the shear strength of discontinuities and the rock mass. A discussion of these mechanisms is as follows.

Buckling – where the sliding slab is thin relative to the slope height and contains no through-going discontinuities such that slab behaves elastically, Euler-type buckling may occur (Figure 7.8a). For these conditions, a limit equilibrium analysis can be carried out where the driving force of the active block can be compared to the resistance provided by elastic buckling. This analysis shows that the theoretical limiting ratio between the slab thickness (D) and overall slab length (L) is about 0.01, that is, buckling may occur where the thickness of the slab is less than 1% of the slope height (Cavers, 1981). Buckling is exacerbated where the slope has a convex shape in section.

Three hinge buckling – where the sliding slab contains orthogonal joints, three hinge buckling can occur (Figure 7.8b), provided that a lateral force, such as that generated by the presence of water in the base plane, acts to initiate failure. This mode of instability can also be studied by limit equilibrium analysis (Cala, Kowalski and Stopkowicz, 2014).

Ploughing – where the sliding slab contains joints dipping steeper than the normal into the face, ploughing-type instability may occur where shear occurs on the joints and the slab fractures due to bending (Figure 7.8c). Observations of coal mines in British Columbia, Canada show that ploughing failures are limited to slabs up to about 5 m (16 ft) thick. The study of this type of instability would require numerical analysis (Fisher, 2009).

Bi-planar instability – where the sliding slab contains joints dipping out of the face, or where these surfaces are generated by shear failure through intact rock, toe release surfaces are formed and bi-planar instability can occur (Figure 7.9). As discussed above, the three components of bi-planar slide are an upper active zone (I) and a lower passive zone (III), with an intermediate transition zone (II) where the rock has to fracture for slope movement to occur; the transition zone can be modelled as a Prandtl wedge defined by a logarithmic spiral. Figure 7.9 shows the bi-planar failure mechanism can occur for a wide range of geometric conditions.

7.4.3 Mechanism of bi-planar instability

For a bi-planar failure, the location of the interslab shear planes can be defined by frictional plasticity theory and the Mohr–Coulomb failure hypothesis, assuming that the maximum principal stress σ_1 is aligned parallel to the slope surface (Figure 7.10). For these conditions,

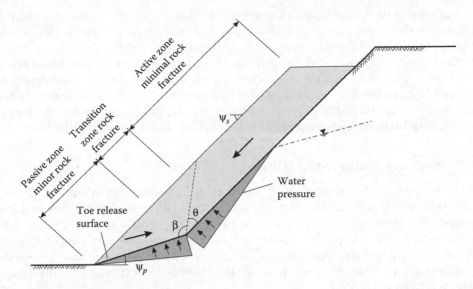

Figure 7.10 Dip slope instability involving fracture of rock within the transition zone between the active and passive zones, with angles defining the interblock shear surface and the toe release surface (adapted from Fisher 2009).

the following equations define the approximate inclinations of the upper interslab shear surface and the toe release joint that form the transition zone (Havaej, Stead, Eberhardt et al., 2014a):

$$\psi_p = \left(\psi_s + \frac{\phi}{2} - 45 \right) \tag{7.22}$$

$$\beta = (90 + \theta) \tag{7.23}$$

$$\theta = \left(45 - \frac{\phi}{2} \right) \tag{7.24}$$

If the dip of the toe release joint is known from Equation 7.22, or is a pre-existing discontinuity, then limit equilibrium analysis of the slope can be carried out, making the conservative assumption that the shear resistance in the transition zone is neglected; the analysis is independent of orthogonal jointing because no shear occurs on these surfaces. If the slope D/H ratio is not defined by the slope characteristics, then sensitivity analyses can be carried out to find the critical value of the D/H ratio for which the factor of safety is a minimum (Brawner, Pentz and Sharp, 1971). This analysis can incorporate water pressures as shown in Figure 7.10, and bolting forces using the procedures shown in Section 7.5.

Bi-planar failures are the most common type of dip slope instability because they can occur for a wide variety of slope characteristics. For example, the depth of the base plane may be several hundred metres as in the case of the Mitchell Creek and Vajont slides, and the dip of the plane may be as flat as 30° or be near vertical. Stability analysis of these slides requires the use of numerical analyses such as the distinct element method (DEM) (see Chapter 12), or discontinuous deformation analysis (DDA) (Goodman and Shi, 1985) that allow rock fracture and displacement to occur within the sliding mass. Numerical analysis is discussed in Chapter 12.

The analyses of dip slope failures show how the slope geometry influences the relative importance, with respect to stability, of various slope parameters. For example, where the upper base plane is less than about 40°, the most important strength parameter is that of the upper plane because significant shear resistance is generated on this plane. However, when the dip of the upper plane is greater than about 60°, little shear resistance is generated on the upper plane because the normal stress is low, and most of the shear resistance is generated on the lower, toe release surface. These concepts provide guidelines on how to focus field investigation and laboratory testing programmes (Fisher and Eberhardt, 2012).

7.4.4 Triggering of dip slope failures

Dip slope failures can be triggered by water pressures, particularly where the upper base plane contains a low-conductivity infilling and water is dammed behind this surface. For these conditions, the installation of drain holes through the plane would be effective. Ground motions due to earthquakes may also trigger instability; however experience at Telegraph Hill in San Francisco shows that water pressures are a more common cause of instability; no significant slope failures occurred in either the 1906 or 1989 (Loma Prieta) earthquakes (Wallace and Marcum, 2015).

It is likely that a potentially dangerous triggering mechanism is excavation at the base of the slope. As excavation progresses and the slope height increases, the driving force increases and eventually the excavation may expose a toe release joint dipping out of the face resulting in sudden instability. Therefore, for active mining operations, it is important to recognise potential dip slope conditions and take appropriate remedial measures as excavation progresses.

7.4.5 Remedial measures for dip slopes

Remedial measures for dip slopes follow the usual approach for slope stabilisation involving either reduction of the driving force or increase in the resisting force, consistent with the principles of limit equilibrium analysis as described in this chapter. It may also be possible, if site conditions are suitable, to provide protection against instability by locating facilities at sufficient set-back distances from the slope.

The following is a brief description of remedial measures that may be considered for dip slopes (see Figure 7.11):

1. *Drainage* – where water pressure occurs on the sliding planes (see Figure 7.10), drains holes could be drilled from the face to provide pressure relief. This depressurisation would be most effective if a pattern of holes is drilled in the face as the slope is excavated, rather than just along the base of the slope.
2. *Toe buttress* – construction of a toe buttress at the base of the slope can be an effective method of preventing movement of the passive zone. A rock fill buttress is usually an economic construction method that also provides for drainage of the slope. However, a rock fill buttress does require that sufficient space be available below the slope for a buttress with the required dimensions, taking into account that the density of clean rock fill is about $17 \, kN \cdot m^3$ ($108 \, lb/ft^3$) compared to $26 \, kN \cdot m^3$ ($165 \, lb/ft^3$) for solid rock.
3. *Bolting* – the installation of rock bolts anchored into stable rock behind the potential sliding planes can be effective in decreasing the driving force in the active zone, and increasing the resisting force in the passive zone. If a pattern of bolts are installed as the slope is excavated, then it is probably acceptable to install fully grouted, untensioned bolts. However, if the bolts are installed in the base of a marginally unstable slope, then

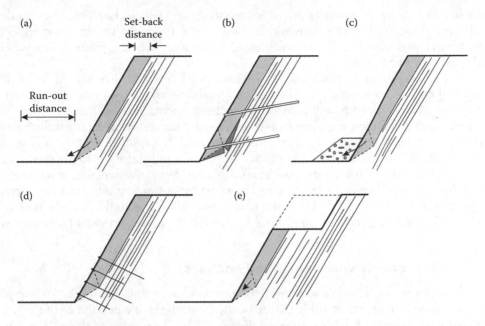

Figure 7.11 Possible remedial measures for dip slope stability. (a) Set-back and run-out distance defining development exclusion zones; (b) slope drainage; (c) rock fill toe buttress; (d) rock bolting; (e) crest unloading.

tensioned bolts may be required to provide an active resisting force (see Figure 14.5). It is also advisable to install bolts with staggered lengths because bolts with a uniform length may form a defined plane of weakness at the end of the bolts.

4. *Unloading* – excavation of the crest of the slope can be effective in reducing the driving force of the active slab. However, excavation should only be carried out if the blasting forces below the base of the excavation will not reduce the shear strength of the sliding plane. That is, ground vibrations produced by blasting may be sufficient to reduce the cohesion component of the shear strength. Also, the crest should not be unloaded without first reducing the water pressure in the slope because a reduction in the weight of the active slab with no change in the ground water pressure acting on the sliding plane is likely to result in the factor of safety being reduced significantly.

5. *Set-back distance* – where it is not possible to carry out any remedial work because, for example, access to the slope is not feasible, it may be necessary to define set-back distances where development is excluded at the crest and/or base of the slope. The set-back distance at the crest could be determined by the geology, or analysis methods described in this section. For safety at the base of the slope, the rock fall run-out distance described in Section 14.6 could be used as a guideline to define the set-back distance.

7.5 REINFORCEMENT OF A SLOPE

When it has been established that a slope is potentially unstable, reinforcement may be an effective method of improving the factor of safety. Methods of reinforcement include the installation of tensioned anchors or fully grouted, untensioned dowels, or the construction of a toe buttress. Factors that will influence the selection of an appropriate system for the site include the site geology, the required capacity of the reinforcement force, drilling equipment availability and access, and time required for construction. This section describes the design methods for slope reinforcement, while Section 14.4.2 discusses construction aspects of slope reinforcement, and Section 16.2 discusses a case study where a plane failure was reinforced with tensioned cables.

If rock anchors are to be installed, it is necessary to decide if they should be anchored at the distal end and tensioned, or fully grouted and untensioned. Untensioned dowels are less costly to install, but they will provide less reinforcement than tensioned anchors of the same dimensions, and their capacity cannot be tested. One technical factor influencing the selection is that if a slope has relaxed and loss of interlock has occurred on the sliding plane, then it is advisable to install tensioned anchors to apply normal and shear forces on the sliding plane. However, if the reinforcement can be installed before excavation takes place, then fully grouted dowels are effective in reinforcing the slope by preventing relaxation on potential sliding surfaces (see Figure 14.5). Untensioned dowels can also be used where the rock is randomly jointed and it is necessary to reinforce the overall slope, rather than a particular plane.

7.5.1 Reinforcement with tensioned anchors

A tensioned anchor installation involves drilling a hole extending below the sliding plane, installing a rock bolt or strand cable that is bonded into the stable portion of the slope, and then tensioning the anchor against the face (Figure 7.12). The tension in the anchor T modifies the normal and shear forces acting on the sliding plane, and the factor of safety of the anchored slope is given by

$$FS = \frac{c \cdot A + (W \cdot \cos\psi_p - U - V \cdot \sin\psi_p + T \cdot \sin(\psi_T + \psi_p))\tan\phi}{W \cdot \sin\psi_p + V \cdot \cos\psi_p - T \cdot \cos(\psi_T + \psi_p)} \qquad (7.25)$$

where T is the tension in the anchor inclined at an angle ψ_T below the horizontal. Equation 7.25 shows that the normal component of the anchor tension ($T \cdot \sin(\psi_p + \psi_T)$) is added to the normal force acting on the sliding plane, which has the effect of increasing the shear resistance to sliding. Also, the shear component of the anchor tension ($T \cdot \cos(\psi_p + \psi_T)$) acting up the sliding plane is subtracted from the driving forces, so the combined effect of the anchor force is to improve the factor of safety (if $(\psi_p + \psi_T) < 90°$).

The factor of safety of a slope reinforced with tensioned rock anchors varies with the inclination of the bolt (Figure 7.12). It can be shown that the most efficient angle ($\psi_{T(opt)}$) for a tensioned rock bolt is when

$$\phi = (\psi_{T(opt)} + \psi_p), \quad \text{or} \quad \psi_{T(opt)} = (\phi - \psi_p) \qquad (7.26)$$

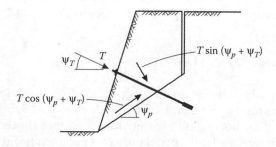

Figure 7.12 Reinforcement of a slope with tensioned rock bolt.

This relationship shows that the optimum installation angle for a tensioned bolt is flatter than the normal to the sliding plane. In practice, cement grouted anchors are installed at about 10–15° below the horizontal to facilitate grouting, while resin grouted anchors may be installed in up-holes. It should be noted that bolts installed at an angle steeper than the normal to the sliding plane (i.e. $(\psi_p + \psi_T) > 90°$) can be detrimental to stability because the shear component of the tension, acting down the plane, increases the magnitude of the displacing force.

Since the stability analysis of plane failures is carried out on a 1-m-thick slice of the slope, the calculated value of T for a specified factor of safety has units kN/m. The procedure for designing a bolting pattern using the calculated value of T is as follows. For example, if the tension in each bolt is T_B, and a pattern of bolts is installed with n bolts in each vertical row, then the total bolting force in each vertical row is $(T_B \cdot n)$. Since the required bolting force is T, then the horizontal spacing S between each vertical row is given by

$$S = \frac{T_B \cdot n}{T} \left(\frac{kN}{kN/m} \right) \tag{7.27}$$

This design method is illustrated in the worked example at the end of this chapter.

7.5.2 Reinforcement with fully grouted untensioned dowels

Fully grouted, untensioned dowels comprise steel bars installed in holes drilled across the potential sliding plane, which are then encapsulated in cement or resin grout. The dowels are also termed soil nails. The steel acts as a rigid shear pin across any plane of weakness in the rock. A method of calculating the reinforcement provided by dowels, developed by Spang and Egger (1990), is discussed below.

Figure 7.13 shows the results of a finite element analysis of a steel bar grouted into a drill hole in a rock containing a joint surface; the angle α between the axis of the bolt and the normal to the joint is 30°. Shear displacement on the joint causes deformation of the bolt that takes place in three stages as follows:

1. *Elastic stage* – after overcoming the cohesion of the joint, the blocks begin to slide relative to each other. The shear resistance of the dowelled joint comprises the shear strength due to friction on the joint, and the elastic response of the steel dowel, grout and rock.
2. *Yield stage* – at displacements of less than about 1 mm (0.04 in.) in installations where the rock is deformable and the grout thickness is at least equal to the dowel radius, the steel is deformed in order to mobilise shear resistance. As a result of the deformation, the yield strengths of the steel and grout are reached in bending and compression, respectively.
3. *Plastic stage* – all the materials in the grouted dowel installation yield at an early stage of shear displacement and at low-shear forces. Therefore, the shear resistance of the dowelled joint depends on the shear force–displacement relationship of the plasticised materials. The contribution of the dowel to the total shear strength of the joint is a function of the friction angle (ϕ) and roughness (i) of the joint, the dowel inclination (α), the compressive strength of the rock and grout (σ_{ci}) and the tensile strength of the steel bar ($\sigma_{t(s)}$). In general, the shear resistance is enhanced where the joint has a high-friction angle and is rough so that dilation occurs during shearing, the inclination angle α is between about 30° and 45°, and the rock is deformable but not so soft that the dowel cuts into the rock (Figure 7.13).

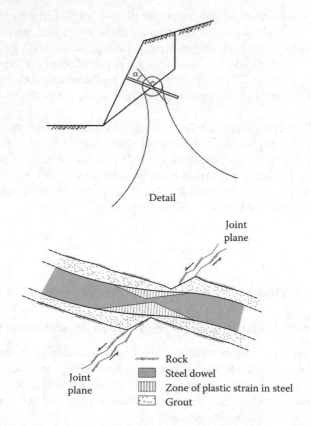

Figure 7.13 Strain in fully grouted steel dowel due to shear movement along joint (modified from Spang and Egger, 1990).

Based on the tests conducted by Spang and Egger, the shear resistance R_b (kN) of a dowelled joint is given by

$$R_b = \sigma_{t(s)}\left[1.55 + 0.011 \cdot \sigma_{ci}^{1.07} \cdot \sin^2(\alpha + i)\right] \cdot \sigma_{ci}^{-0.14}[0.85 + 0.45 \cdot \tan\phi] \tag{7.28}$$

where the units of σ_{ci} are MPa and of $\sigma_{t(s)}$ are kN. The corresponding displacement δ_s of a dowelled joint is given by

$$\delta_s = (15.2 - 55.2\ \sigma_{ci}^{-0.14} + 56.2\ \sigma_{ci}^{-0.28}) \cdot (1 - \tan\alpha \cdot (70/\sigma_{ci})^{0.125} \cdot (\cos\alpha)^{-0.5}) \tag{7.29}$$

In the stability analysis of a plane failure in which fully grouted dowels have been installed across the sliding plane, Equation 7.4 for the factor of safety is modified as follows:

$$FS = \frac{c \cdot A + N \cdot \tan\phi + R_b}{S} \tag{7.30}$$

or

$$FS = \frac{c \cdot A + N \cdot \tan\phi_b}{S - R_b} \tag{7.31}$$

Equation 7.30 refers to the case where the bolt force increases the resisting force, while Equation 7.31 is for the case where the bolt force reduces the driving force. While no rule has been established for which equation to use, it would seem reasonable to use Equation 7.30 for the situation where bolts are installed to increase the shear strength across the sliding plane. For the situation where a passive buttress is installed at the base of a sliding plane as discussed in Section 7.5.3, Equation 7.31 may be used because the buttress is essentially reducing the driving force.

Equations 7.30 and 7.31 produce different values for the factor of safety for the same value of R_b, so it is worthwhile to calculate the value of R_b for a required FS by both equations to determine what level of conservatism should be included in the design.

7.5.3 Reinforcement with buttresses

The two previous sections discussed reinforcement by installing anchors across the potential sliding surface. An alternate method is to construct a buttress at the toe to provide external support to the slope, using the methods shown in Figure 7.14. In both cases, the factor of safety is calculated using Equation 7.31 with the appropriate value for R_b for the system installed.

Near the crest of the slope where the bedding forms a series of slabs, steel bars can be grouted into holes drilled into the rock and then encased in shotcrete or concrete. The steel provides shear resistance to movement, while the concrete provides continuous support between the dowels and keeps small fragments of rock in place. These buttresses are particularly applicable where the rock along the crest is slightly weathered, and if rock anchors were installed, the ongoing weathering would eventually expose the bolts. It is likely that the maximum thickness of slab that can be supported in this manner is about 2 m (6.6 ft).

Larger-scale support can be provided by placing a waste rock buttress at the toe of the slope. The support provided by such a buttress depends on the buttress weight (density of clean fill is about 17 kN · m³) (108 lb/ft³), and the shear resistance generated along the base of the buttress that is a function of the weight of the rock, and the roughness and inclination of the base. This method can only be used, of course, if sufficient space is available at the toe to accommodate the required volume of rock. It is also important that the waste rock be free draining so that water pressures do not build up behind the buttress. It is usual that the height of a waste rock toe buttress needs to be at least one-third of the slope height to provide effective support.

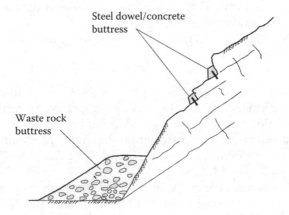

Steel dowel/concrete
buttress

Waste rock
buttress

Figure 7.14 Reinforcement of slope with buttresses.

The design of a buttress could be carried out using either Equations 7.30 or 7.31, as discussed in Section 7.5.2.

7.6 PROBABILISTIC DESIGN – PLANE FAILURE

The design procedures discussed so far in this chapter all use, for each design parameter, single values that are assumed to be the average or best estimate values. In reality, each parameter has a range of values that may represent natural variability, changes over time, and the degree of uncertainty in measuring their values. Therefore, the factor of safety can be realistically expressed as a probability distribution, rather than a single value. In design, this uncertainty can be accounted for by applying judgement in using a factor of safety consistent with the variability/uncertainty in the data. That is, a high factor of safety would be used where the values of the parameters are not well known. Alternatively, the uncertainty can be quantified using probabilistic analysis, such as Monte Carlo analysis, to calculate the probability of failure (see Section 1.4.5).

Figure 7.14 shows the results of a probabilistic stability analysis of the slope described in Section 5.4 (see Figures 5.18 to 5.20). Section 5.4 describes the calculation of the shear strength properties of the bedding planes, assuming that the factor of safety was 1.0 when the tension crack filled with water and the slope failed. The purpose of the probabilistic analysis described in this section is to show the range in factor of safety that is likely to exist in practice due to the variability in the slope parameters.

The first step in the analysis is to select probability distributions for each parameter for which the values are variable or have a range of values. The distribution may be based on the actual measurements such as the friction angle from laboratory testing of core samples. For parameters such as cohesion and water pressures where no measurements are available, judgement and experience will be required to select realistic distributions.

One criterion used to define these distributions is that they are more likely to have finite minimum and maximum values, rather than extend to infinity as is the case with a normal distribution (see Figure 1.13). Examples of distributions that meet this criterion are triangular and beta distributions that have maximum and minimum values, and a most likely value that can be either symmetrical or skewed to represent the expected most likely value. For joint persistence and spacing that is often expressed as a lognormal distribution that extends to infinity (see Section 4.5.3), the distribution can be cut off at a value that is expected to be the maximum that may occur in the field. An advantage of using distributions that have finite limits rather than extending to infinity is that this limits the range of the calculated factor of safety distribution.

For the slope shown in Figure 7.15a, Figure 7.15b shows the probability distributions for each of the following parameters:

1. *Friction angle* ϕ – skewed beta distribution with the most likely value of 18°, and maximum and minimum values of 15° and 25°, respectively; the mean value is 19° and the standard deviation is 2.3°.
2. *Cohesion c* – skewed triangular distribution with most likely value of 90 kPa (13 psi) and maximum and minimum values of 80 and 125 kPa (11.6–18.1 psi), respectively; the mean value is 98 kPa (14.2 psi).
3. *Water pressure* – expressed as per cent filling of the 19 m deep tension crack, skewed triangular distribution ranging from 5 m (26%) to full (100%), with the most likely value being 15 m (80%); the mean depth is 13 m (68%).

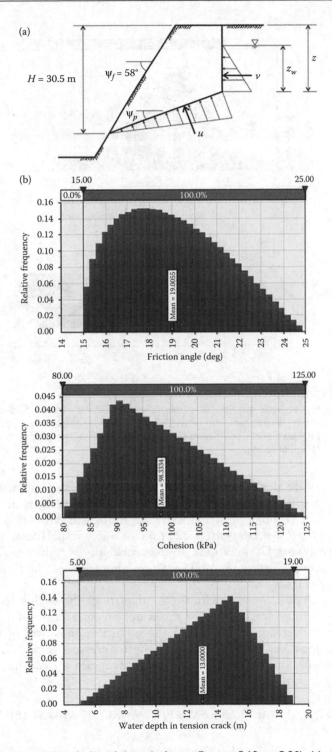

Figure 7.15 Probabilistic analysis of plane failure (refer to Figures 5.18 to 5.20): (a) slope model showing water pressures *U* and *V*; (b) probability distributions of cohesion, friction angle and depth of water in tension crack.   (*Continued*)

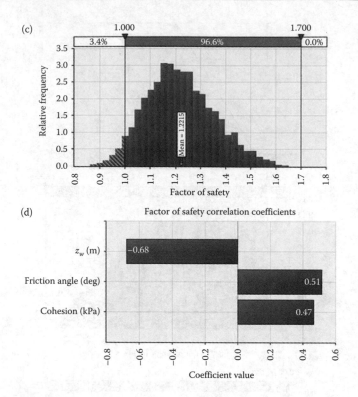

Figure 7.15 (Continued) Probabilistic analysis of plane failure (refer to Figures 5.18 to 5.20): (c) probability distribution of factor of safety showing 3.4% of probability of failure; (d) correlation coefficients showing effect of z_w (negative), ϕ and c (positive) on factor of safety.

Figure 7.14c shows the distribution of the factor of safety generated, using the Monte Carlo method (see Section 1.4.5), as the result of 5000 iterations with values randomly selected from the input parameter distributions. The histogram shows that the mean, maximum and minimum factors of safety are 1.22, 1.65 and 0.87, respectively. Also, the factor of safety was less than 1.0 for 170 iterations, so the probability of failure is 3.4%. If the mean values of all the input parameters are used in the stability analysis, the calculated deterministic factor of safety is 1.22.

The sensitivity analysis associated with these calculations shows that the factor of safety is most strongly influenced (negatively) by the water pressure in the slope, that is, by the depth of water in the tension crack, while the influences (positive) of the friction and cohesion are about equal but less than that of the water pressure (Figure 7.14d).

This analysis was performed using the computer program @Risk that is an add-on to Excel (Palisades Corporation, 2012).

EXAMPLE PROBLEM 7.1: PLANE FAILURE – ANALYSIS AND STABILISATION

Statement

A 12 m (40 ft) high-rock slope has been excavated at a face angle of 60°. The rock in which this cut has been made contains persistent bedding planes that dip at an angle of 35° into the excavation. The 4.35 m (14.3 ft) deep tension crack is 4 m (13.1 ft) behind the crest, and is filled with water to a height of 3 m (9.8 ft) above the sliding surface (Figure 7.16). The strength parameters of the sliding surface are as follows:

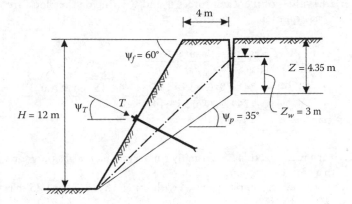

Figure 7.16 Plane failure geometry for Example Problem 7.1.

Cohesion, $c = 25$ kPa (3.6 psi)
Friction angle, $\phi = 37°$

The unit weight of the rock is 26 kN/m³ (165 lb/ft³), and the unit weight of the water is 9.81 kN · m³ (62.4 lb/ft³).

Required

Assuming that a plane slope failure is the most likely type of instability, analyse the following stability conditions.

Factor of safety calculations

1. Calculate the factor of safety of the slope for the conditions given in Figure 7.16.
2. Determine the factor of safety if the tension crack were completely filled with water due to run-off collecting on the crest of the slope.
3. Determine the factor of safety if the slope was completely drained.
4. Determine the factor of safety if the cohesion were to be reduced to zero due to excessive vibrations from nearby blasting operations, assuming that the slope was still completely drained.
5. Determine whether the 4.35-m-deep tension crack is the critical depth (use Figure 7.6).

Slope reinforcement using rock bolts

1. It is proposed that the drained slope with zero cohesion be reinforced by installing tensioned rock bolts anchored into sound rock beneath the sliding plane. If the rock bolts are installed at right angles to the sliding plane, that is, $\psi_T = 55°$, and the total load on the anchors per lineal metre of slope is 400 kN, calculate the factor of safety.
2. Calculate the factor of safety if the bolts are installed at a flatter angle so that the ψ_T is decreased from 55° to 20°.
3. If the working load for each bolt is 250 kN, suggest a bolt layout, that is, the number of bolts per vertical row, and the horizontal and vertical spacing between bolts to achieve a bolt load of 400 kN/m of slope length.

Solution

Factor of safety calculations

1. The factor of safety is calculated using Equations 7.4 to 7.9: The weight W of the block is 1241 kN/m (Equation 7.8), and the area A of the sliding plane is 13.34 m²/m (Equation 7.5). For water in the tension crack to depth, $z_w = 4.35$ m

(14.3 ft), the values of the water forces U and V acting on the block are 196.31 and 44.15 kN/m (Equations 7.6 and 7.7).

$$FS = \frac{c \cdot A + (W \cdot \cos \psi_p - U - V \cdot \sin \psi_p) \tan \phi}{W \cdot \sin \psi_p + V \cdot \cos \psi_p}$$

$$= \frac{(25 \cdot 13.34) + (1241.70 \cdot \cos 35 - 196.31 - 44.15 \cdot \sin 35) \tan 37}{1241.70 \cdot \sin 35 + 44.15 \cdot \cos 35}$$

$$= 1.25$$

In civil applications, this FS is usually a marginal factor of safety for a permanent slope with a high consequence of failure.

2. If the tension crack is completely filled with water, that is, $z_w = 4.35$ m (14.3 ft), and the new factor of safety is

$$FS = \frac{333.50 + (1017.14 - 284.57 - 53.23) \tan 37}{712.21 + 76.02}$$

$$= 1.07$$

This FS value indicates that the slope is close to failure.

3. If the slope were drained so no water pressures act in the tension crack or on the sliding plane, that is, $z_w = V = U = 0$, then the new factor of safety is

$$FS = \frac{333.50 + 1017.14 \cdot \tan 37}{712.21}$$

$$= 1.54$$

This is usually an adequate factor of safety.

The relationship between the calculated factor of safety and the depth of the water in the tension crack is plotted in Figure 7.17.

4. If the slope is drained and the cohesion on the sliding plane is reduced from 25 kPa (3.6 psi) to zero by blast vibrations, then the new factor of safety is

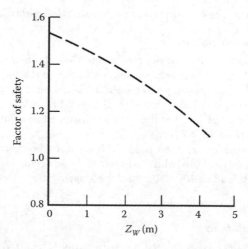

Figure 7.17 Plot of factor of safety against depth of water in tension crack for Example Problem 7.1.

$$FS = \frac{0 + 1017.14 \cdot \tan 37}{712.21}$$
$$= 1.08$$

The loss of cohesion reduces the factor of safety from 1.54 to 1.08, which illustrates the sensitivity of the slope to the cohesion on the sliding plane.

5. Figure 7.6a shows that the critical tension crack depth is 4.32 m (i.e. $z/H = 0.36$), which is close to the position of the tension crack (i.e. $4.35/12 = 0.36$).

Slope reinforcement with rock bolts

1. The factor of safety of plane slope failure reinforced with rock bolts is calculated using Equation 7.25. In this case, the slope is drained and the cohesion is zero, that is,

$$c = U = V = 0$$

Therefore, for a reinforcement force of T of 400 kN/m installed at a dip angle ψ_T of 55°, the factor of safety is

$$FS = \frac{[W \cdot \cos \psi_p + T \cdot \sin(\psi_T + \psi_p)] \cdot \tan \phi}{W \cdot \sin \psi_p - T \cdot \cos(\psi_T + \psi_p)}$$
$$= \frac{[1241.70 \cdot \cos 35 + 400 \cdot \sin(55 + 35)] \cdot \tan 37}{1241.70 \cdot \sin 35 - 400 \cdot \cos(55 + 35)}$$
$$= \frac{1067.90}{712.21}$$
$$= 1.5$$

2. If the bolts are installed at a flatter angle, $\psi_T = 20°$, then the factor of safety is

$$FS = \frac{[1241.70 \cdot \cos 35 + 400 \cdot \sin(20 + 35)] \cdot \tan 37}{1241.70 \cdot \sin 35 - 400 \cdot \cos(20 + 35)}$$
$$= \frac{1013.38}{482.78}$$
$$= 2.10$$

This shows the significant improvement that can be achieved by installing bolts at an angle flatter than the normal to the sliding surface. The optimum angle is when (see Equation 7.26)

$$\psi_{T opt} = (\phi - \psi_p)$$
$$= (37 - 35°)$$
$$= 2°, \text{ and factor of safety} = 2.41$$

3. The rock bolt pattern should be laid out so that the distribution of bolts on the slope is as even as possible. If four bolts are installed in each vertical row, the horizontal spacing S of the vertical rows is calculated as follows: (see Equation 7.27)

$$S = \frac{T_B \cdot n}{T} \left(\frac{kN}{kN/m} \right)$$
$$= \frac{240 \cdot 4}{400}$$
$$= 2.5 \text{ m}$$

Chapter 8

Wedge failure

8.1 INTRODUCTION

Chapter 7 dealt with slope failure resulting from sliding on a single planar surface dipping out of the face, and striking parallel or nearly parallel to the slope face. It was stated that the plane failure analysis is valid if the strike of the failure plane is within ±20° of the strike of the slope face. This chapter is concerned with the failure of slopes containing discontinuities striking obliquely to the slope face where sliding of a wedge of rock takes place along the line of intersection of two such planes (Figure 8.1). Wedge failures can occur over a much wider range of geologic and geometric conditions than plane failures. Therefore, the study of wedge stability is an important component of rock slope engineering. The analysis of wedges has been extensively discussed in geotechnical literature, and this chapter draws heavily upon the work of Goodman (1964), Wittke (1965), Londe (1965), Londe, Vigier and Vormeringer (1969, 1970) and John (1970).

In this chapter, the structural geological conditions that result in the formation of a wedge formed by two intersecting planes are defined, and the method of identifying wedges on the stereonet is illustrated. The stereonet defines the shape of the wedge, the orientation of the line of intersection and the direction of sliding, but not the dimensions. This information can be used to assess the potential for the wedge to slide from the cut face. The procedure is termed *kinematic analysis*, the purpose of which is to identify potentially unstable wedges, although it does not provide precise information on their factor of safety.

The chapter presents design charts that can be used to find the factor of safety of wedges for which friction is the only component of the shear strength, and no external forces such as water pressures or bolting act on the wedge. In addition, equations are presented that can be used to calculate the factor of safety of wedges where the shear strength on the two slide planes is defined by cohesion and friction angle, and each plane can have different shear strengths. The analysis can also incorporate water pressure.

The presence of a tension crack, and the influence of external forces due to water pressures, tensioned anchors, seismic accelerations or bridge foundations results in a significant increase in the complexity of the equations. Appendix III presents the complete solution for the wedge analysis.

8.2 DEFINITION OF WEDGE GEOMETRY

Typical wedge failures, illustrated in Figures 8.1 and 8.2, show the conditions that are normally assumed for the analytical treatment of wedges. Figure 8.1 shows a cut slope in granitic rock where a wedge is formed by two continuous, planar discontinuities and the line of intersection of these two planes daylights just at the toe of the rock face. That is, the

Figure 8.1 **(See colour insert.)** Typical wedge failure involving sliding on two persistent joints with line of intersection of joints daylighting at toe of rock face, and an upper plane that formed a tension crack (strong, volcanic rock on Interstate 5, near Grants Pass, Oregon).

trend of the line of intersection and the dip direction of the face are approximately equal. Furthermore, the plunge of the line of intersection is about 50–55°, while the friction angle of these joints is in the range of 35–40°. That is, the line of intersection dips steeper than the friction angle. These conditions meet the kinematic requirements for failure of the wedge. Figure 8.1 also illustrates how a slight change in the site conditions would result in a stable slope. For example, if the line of intersection had been slightly behind the face, or just one of the joints had been discontinuous, then no failure would have occurred.

The wedge in Figure 8.2 is formed by bedding on the left and a conjugate joint set on the right. As seen in Figure 8.1, the line of intersection daylights in the slope face and failure occurred. However, in this wedge, sliding occurred almost entirely on the bedding with the joint acting as a release surface. Therefore, the shear strength of the joint has little effect on stability.

The geometry of the wedge for analysing the basic mechanics of sliding is defined in Figure 8.3. Based on this geometry, the general conditions for wedge failure are as follows:

1. Two planes will always intersect in a line (Figure 8.3a). On the stereonet, the line of intersection is represented by the point where the two great circles of the planes intersect, and the orientation of the line is defined by its trend (α_i) and its plunge (ψ_i) (Figure 8.3b).

Figure 8.2 Wedge formed by bedding (left) and a conjugate joint set (right); sliding occurred on bedding with joints acting as a release surface (bedded shale, near Helena, Montana).

2. The plunge of the line of intersection must be flatter than the dip of the face, and steeper than the average friction angle of the two slide planes, that is, $\psi_f > \psi_i > \phi$ (Figure 8.3b and c). The inclination of the slope face ψ_f is measured in the view at right angles to the line of intersection. Note that the ψ_{fi} would only be the same as ψ_f, the true dip of the slope face, if the trend of the line of intersection was the same as the dip direction of the slope face.

3. The line of intersection must dip in a direction out of the face for sliding to be feasible; the possible range in the trend of the line of intersection is between α_i and α_i' (Figure 8.3d).

In general, sliding may occur if the intersection point between the two great circles of the sliding planes lies within the shaded area in Figure 8.3b. That is, the stereonet will show whether wedge failure is kinematically feasible. However, the actual factor of safety of the wedge cannot be determined from the stereonet, because it depends on the details of the geometry of the wedge, the shear strength of each plane and water pressure, as described in the following sections.

The trend α_i and plunge ψ_i of the line of intersection of planes A and B can be determined on the stereonet, or calculated using Equations 8.1 and 8.2 as follows:

$$\alpha_i = \tan^{-1}\left(\frac{\tan\psi_A \cdot \cos\alpha_A - \tan\psi_B \cdot \cos\alpha_B}{\tan\psi_B \cdot \sin\alpha_B - \tan\psi_A \cdot \sin\alpha_A}\right) \tag{8.1}$$

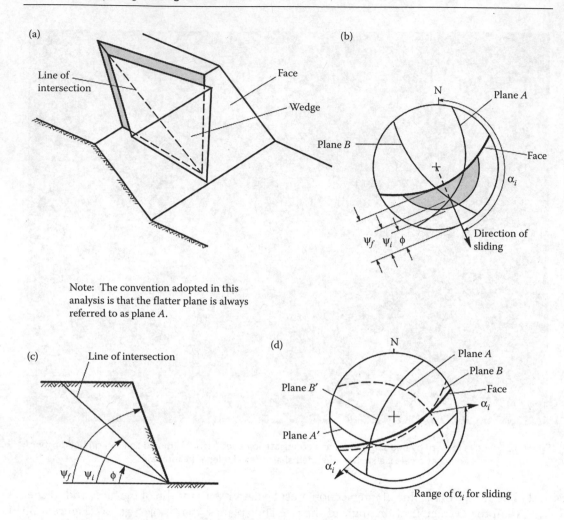

Figure 8.3 Geometric conditions for wedge failure: (a) pictorial view of wedge failure; (b) stereoplot showing the orientation of the line of intersection, and the range of the plunge of the line of intersection ψ_i where failure is feasible; (c) view of slope at right angles to the line of intersection; (d) stereonet showing the range in the trend of the line of intersection α_i where wedge failure is feasible.

$$\psi_i = \tan^{-1}(\tan\psi_A \cdot \cos(\alpha_A - \alpha_i)) = \tan^{-1}(\tan\psi_B \cdot \cos(\alpha_B - \alpha_i)) \tag{8.2}$$

where α_A and α_B are the dip directions and ψ_A and ψ_B are the dips of the two planes. Equation 8.1 gives two solutions 180° apart; the correct value lies between α_A and α_B.

8.3 ANALYSIS OF WEDGE FAILURE

The factor of safety of the wedge defined in Figure 8.3, assuming that sliding is resisted only by friction and that the friction angle ϕ is the same for both planes, is given by

$$FS = \frac{(R_A + R_B) \cdot \tan\phi}{W \cdot \sin\psi_i} \tag{8.3}$$

where R_A and R_B are the normal reactions provided by planes A and B as illustrated in Figure 8.4, and the component of the weight acting down the line of intersection is $(W \cdot \sin \psi_i)$. The forces R_A and R_B are found by resolving them into components normal and parallel to the direction along the line of intersection as follows:

$$R_A \cdot \sin\left(\beta - \frac{1}{2}\xi\right) = R_B \cdot \sin\left(\beta + \frac{1}{2}\xi\right) \tag{8.4}$$

$$R_A \cdot \cos\left(\beta - \frac{1}{2}\xi\right) + R_B \cdot \cos\left(\beta + \frac{1}{2}\xi\right) = W \cdot \cos\psi_i \tag{8.5}$$

where the angles ξ and β are defined in Figure 8.4a. Angles ξ and β are measured on the great circle containing the pole to the line of intersection and the poles of the two slide planes. In order to meet the conditions for equilibrium, the normal components of the reactions are equal (Equation 8.4), and the sum of the parallel components equals the component of the weight acting down the line of intersection (Equation 8.5).

The values of R_A and R_B are found from Equations 8.4 and 8.5 by solving and adding as follows:

$$R_A + R_B = \frac{W \cdot \cos\psi_i \cdot \sin\beta}{\sin(\xi/2)} \tag{8.6}$$

Hence,

$$FS = \frac{\sin\beta}{\sin(\xi/2)} \cdot \frac{\tan\phi}{\tan\psi_i} \tag{8.7}$$

In other words,

$$FS_W = K \cdot FS_P \tag{8.8}$$

where FS_W is the factor of safety of a wedge supported by friction only, and FS_p is the factor of safety of a plane failure in which the slide plane, with friction angle ϕ, dips at the same angle as the line of intersection ψ_i.

K is the wedge factor that, as shown by Equation 8.7, depends upon the included angle of the wedge ξ and the angle of tilt β of the wedge. Values for the wedge factor K, for a range of values of ξ and β, are plotted in Figure 8.5.

The method of calculating the factor of safety of wedges as discussed in this section is, of course, simplistic because it does not incorporate different friction angles and cohesions on the two slide planes, or ground water pressures. When these factors are included in the analysis, the equations become more complex. Rather than developing these equations in terms of the angles ξ and β, which cannot be measured directly in the field, the more complete analysis is presented in terms of directly measurable dips and dip directions. The following section gives equations for the factor of safety of a wedge with cohesion and friction acting on the slide planes and water pressure. The complete set of equations for stability analysis of a wedge is shown in Appendix III; this analysis includes parameters to define the shape and dimensions of the wedge, different shear strengths on each slide surface, water pressures and two external loads.

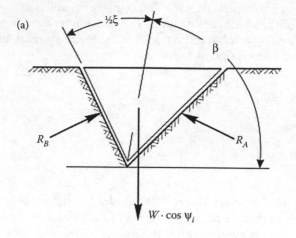

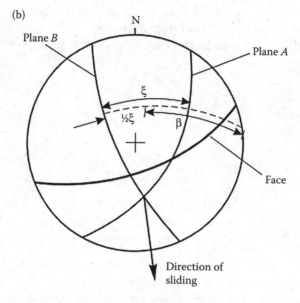

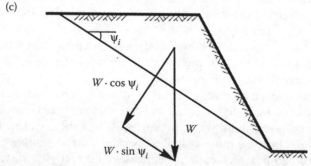

Figure 8.4 Resolution of forces to calculate factor of safety of wedge: (a) view of wedge looking at face showing definition of angle β and ξ, and reactions on sliding planes R_A and R_B; (b) stereonet showing measurement of angles β and ξ; (c) cross-section of wedge showing resolution of wedge weight W on the line of intersection.

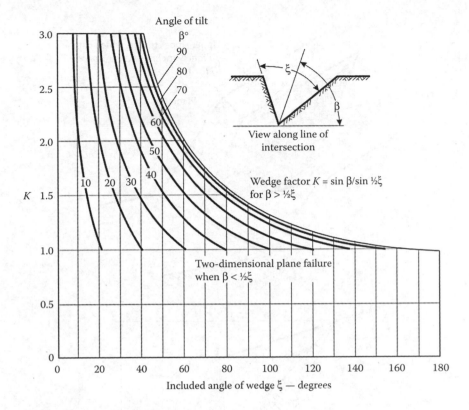

Figure 8.5 Wedge factor *K* as a function of wedge geometry.

This section shows the important influence of the wedging action as the included angle of the wedge decreases below 90°. The increase by a factor of 2 or 3 on the factor safety determined by plane failure analysis is of great practical importance because, as shown in Figure 8.5, the factor of safety of a wedge can be significantly greater than that of a plane failure (i.e., *K* > 1). Therefore, where the structural features that are likely to control the stability of a rock slope do not strike parallel to the slope face, the stability analysis should be carried out by means of 3D methods discussed in this chapter.

8.4 WEDGE ANALYSIS INCLUDING COHESION, FRICTION AND WATER PRESSURE

Section 8.3 discussed the geometric conditions that could result in a wedge failure, but this kinematic analysis provides limited information of the factor of safety because the dimensions of the wedge were not considered. This section describes a method to calculate the factor of safety of a wedge that incorporates the slope geometry, different shear strengths of the two slide planes and ground water (Hoek, Bray and Boyd, 1973). However, the limitations of this analysis are that it does not include a tension crack, or external forces such as bolting.

Figure 8.6a shows the geometry and dimensions of the wedge that will be considered in the following analysis, with the five numbered lines that define the shape of the wedge. Note that the upper slope surface in this analysis can be obliquely inclined with respect to the slope face, thereby removing a restriction that has been present in the stability analyses that

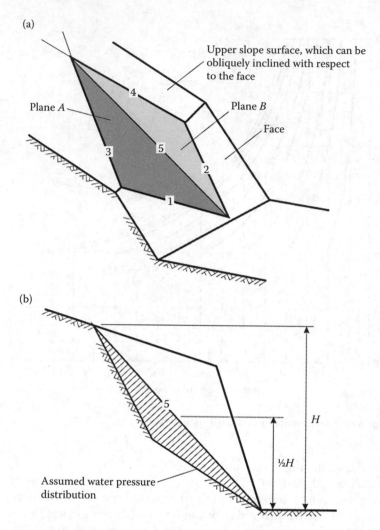

Figure 8.6 Geometry of wedge used for stability analysis including the influence of friction and cohesion, and of water pressure on the slide surfaces: (a) pictorial view of wedge showing the numbering of the intersection lines and plane; (b) view normal to the line of intersection (5) showing wedge height and water pressure distribution.

have been discussed so far in the book. The total height of the slope H is the difference in vertical elevation between the upper and lower extremities of the line of intersection along which sliding is assumed to occur. The water pressure distribution assumed for this analysis is based upon the hypothesis that the wedge itself is impermeable and that water enters the top of the wedge along lines of intersections 3 and 4 and discharges to the atmosphere from the slope face along lines of intersections 1 and 2. The resulting pressure distribution is shown in Figure 8.6b – the maximum pressure occurring along the line of intersection 5 and the pressure being zero along lines 1, 2, 3 and 4. This is a triangular pressure distribution with the maximum value occurring at the mid-height of the slope, with the estimated maximum pressure being equal to $(1/2 \cdot \gamma_w \cdot H)$. This water pressure distribution is believed to be representative of the extreme conditions that could occur during heavy rain and the slope is saturated.

The two planes on which sliding occurs are designated A and B, with plane A having the shallower dip. The numbering of the five lines of intersection of the four planes defining the wedge is as follows:

Line 1. Intersection of plane A with the slope face
Line 2. Intersection of plane B with the slope face
Line 3. Intersection of plane A with upper slope surface
Line 4. Intersection of plane B with upper slope surface
Line 5. Intersection of planes A and B

It is assumed that sliding of the wedge always takes place along the line of intersection numbered 5, and its factor of safety is given by (Hoek, Bray and Boyd, 1973)

$$FS = \frac{3}{\gamma_r \cdot H}(c_A \cdot X + c_B \cdot Y) + \left(A - \frac{\gamma_w}{2\gamma_r} \cdot X\right) \cdot \tan\phi_A + \left(B - \frac{\gamma_w}{2\gamma_r} \cdot Y\right) \cdot \tan\phi_B \tag{8.9}$$

where c_A and c_B are the cohesive strengths, and ϕ_A and ϕ_B are the angles of friction, respectively, on planes A and B; γ_r is the unit weight of the rock, γ_w the unit weight of the water and H the total height of the wedge measured along the line of intersection 5. The dimensionless factors X, Y, A and B depend upon the geometry of the wedge.

The values of parameters X, Y, A and B are given in Equations 8.10 to 8.13

$$X = \frac{\sin\theta_{24}}{\sin\theta_{45} \cdot \cos\theta_{2.na}} \tag{8.10}$$

$$Y = \frac{\sin\theta_{13}}{\sin\theta_{35} \cdot \cos\theta_{1.nb}} \tag{8.11}$$

$$A = \frac{\cos\psi_a - \cos\psi_b \cdot \cos\theta_{na.nb}}{\sin\psi_5 \cdot \sin^2\theta_{na.nb}} \tag{8.12}$$

$$B = \frac{\cos\psi_b - \cos\psi_a \cdot \cos\theta_{na.nb}}{\sin\psi_5 \cdot \sin^2\theta_{na.nb}} \tag{8.13}$$

where ψ_a and ψ_b are the dips of planes A and B, respectively, and ψ_5 is the dip of the line of intersection, line 5. The angles required for the solution of these equations can be measured most conveniently on a stereoplot that defines the geometry of the wedge and the slope (Figure 8.7).

The application of the equations discussed in this section is illustrated in the following example, using the parameters shown in Table 8.1.

The total height of the wedge H is 40 m (131 ft), the unit weight of the rock is 25 kN/m^3 (165 lb/ft^3) and the unit weight of the water 9.81 kN/m^3 (62.5 lb/ft^3).

The stereoplot of the great circles representing the four planes involved in this example is presented in Figure 8.7, and all the angles required for the solution of Equations 8.10 to 8.13 are marked in this figure.

Determination of the factor of safety is most conveniently carried out on a spread sheet such as that presented in Table 8.2. Setting the calculations out in this manner not only enables the

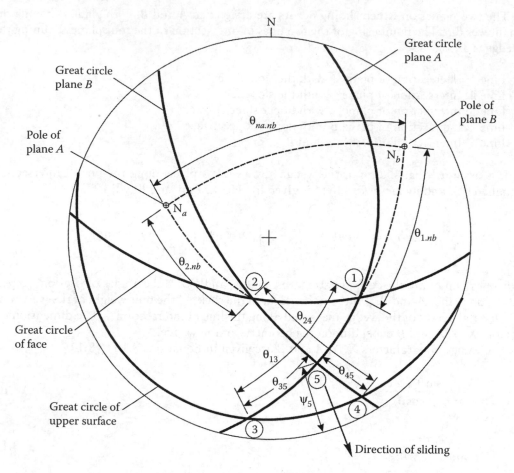

Figure 8.7 Stereoplot of data required for wedge stability analysis.

user to check all the data, but also shows how each variable contributes to the overall factor of safety. Hence, if it is required to check the influence of the cohesion on both planes falling to zero, this can be done by setting the two groups containing the cohesion values c_A and c_B to zero, giving a factor of safety of 0.62. Alternatively, the effect of drainage can be checked by varying the water density to simulate the effect of reducing the water pressure. In this example, with cohesive strength, the factor is 1.98 when the slope is completely drained, and the factor of safety is 1.24 for friction only, drained.

As has been emphasised in previous chapters, this ability to check the sensitivity of the factor of safety to changes in material properties or water pressures is important because the value of these parameters is difficult to define precisely.

Table 8.1 Parameters defining properties of wedge

Plane	Dip	Dip direction	Properties
A	45	105	$\phi_A = 30°$, $c_A = 24$ kPa (3.5 psi)
B	70	235	$\phi_B = 20°$, $c_B = 48$ kPa (7 psi)
Slope face	65	185	
Upper surface	12	195	

Table 8.2 Wedge stability calculation sheet

Input data	Function value	Calculated values
$\psi_a = 45°$	$\cos \psi_a = 0.707$	$A = \dfrac{\cos \psi_a - \cos \psi_b \cdot \cos \theta_{na.nb}}{\sin \psi_5 \cdot \sin^2 \theta_{na.nb}} = \dfrac{0.707 + 0.342 \cdot 0.191}{0.518 \cdot 0.964} = 1.548$
$\psi_b = 70°$	$\cos \psi_b = 0.342$	
$\psi_5 = 31.2°$	$\sin \psi_5 = 0.518$	
$\psi_{na.nb} = 101°$	$\cos \psi_{na.nb} = -0.19$	$B = \dfrac{\cos \psi_b - \cos \psi_a \cdot \cos \theta_{na.nb}}{\sin \psi_5 \cdot \sin^2 \theta_{na.nb}} = \dfrac{0.342 + 0.707 \cdot 0.191}{0.518 \cdot 0.964} = 0.956$
	$\sin \psi_{na.nb} = 0.982$	
$\theta_{24} = 65°$	$\sin \theta_{24} = 0.906$	$X = \dfrac{\sin \theta_{24}}{\sin \theta_{45} \cdot \cos \theta_{2.na}} = \dfrac{0.906}{0.423 \cdot 0.643} = 3.336$
$\theta_{45} = 25°$	$\sin \theta_{45} = 0.423$	
$\theta_{2.na} = 50°$	$\cos \theta_{2.na} = 0.643$	
$\theta_{13} = 62°$	$\sin \theta_{13} = 0.883$	$Y = \dfrac{\sin \theta_{13}}{\sin \theta_{35} \cdot \cos \theta_{1.nb}} = \dfrac{0.883}{0.515 \cdot 0.5} = 3.429$
$\theta_{35} = 31°$	$\sin \theta_{35} = 0.515$	
$\theta_{1.nb} = 60°$	$\cos \theta_{1.nb} = 0.5$	
$\phi_A = 30°$	$\tan \phi_A = 0.577$	$FS = \dfrac{3}{\gamma_r \cdot H}(c_A \cdot X + c_B \cdot Y) + \left(A - \dfrac{\gamma_w}{2\gamma_r} \cdot X\right)\tan \phi_A + \left(B - \dfrac{\gamma_w}{2\gamma_r} \cdot Y\right)\tan \phi_B$
$\phi_B = 20°$	$\tan \phi_B = 0.364$	
$\gamma_r = 25 \text{ kN/m}^3$	$\gamma_w/2 \cdot \gamma_r = 0.196$	
$\gamma_w = 9.81 \text{ kN/m}^3$	$(3 \cdot c_A/\gamma_r) \cdot H = 0.072$	$FS = 0.241 + 0.494 + 0.893 - 0.376 + 0.348 - 0.244$
$c_A = 24 \text{ kPa}$	$(3 \cdot c_B/\gamma_r) \cdot H = 0.144$	$\quad = 1.36$
$c_B = 48 \text{ kPa}$		
$H = 40 \text{ m}$		

8.5 WEDGE STABILITY CHARTS FOR FRICTION ONLY

A rapid check of the stability of a wedge can be made if the slope is drained and cohesion is zero on both slide planes A and B. Under these conditions, Equation 8.9 reduces to

$$FS = A \cdot \tan \phi_A + B \cdot \tan \phi_B \qquad (8.14)$$

The dimensionless factors A and B are found to depend upon the dips and dip directions of the two planes. The values of these two factors have been computed for a range of wedge geometries, and the results are presented as a series of charts (Figures 8.8 to 8.15).

Note that the factor of safety calculated from Equation 8.14 is independent of the slope height, the angle of the slope face and the inclination of the upper slope surface. This rather surprising result arises because the weight of the wedge occurs in both the numerator and denominator of the factor of safety equation and, for the friction-only case, this term cancels out, leaving dimensionless ratios that define the factor of safety. As discussed in Section 1.4, the factor of safety of a plane failure is also independent of the slope dimensions if the slope is drained and the cohesion is zero. This simplification is very useful in that it enables the user of these charts to carry out a quick check on the stability of a slope based on the dips and dip directions of the two discontinuities that form the slide planes of the wedge. An example of such an analysis is presented later in this chapter.

Many trial calculations have shown that a wedge having a factor of safety in excess of 2.0, as obtained from the friction-only stability charts, is unlikely to fail under even the most severe combination of conditions to which the slope is likely to be subjected. Consider the example discussed in Section 8.4 in which the factor of safety for the worst conditions (zero cohesion and maximum water pressure) is 0.62. This is 50% of the factor of safety of 1.24 for the friction-only case. Hence, had the factor of safety for the friction-only case been 2.0, the factor of safety for the worst conditions would have been 1.0, assuming that the ratio of the factors of safety for the two cases remains constant.

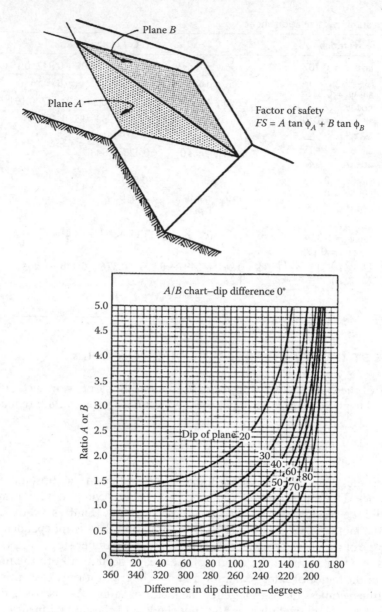

Factor of safety
$$FS = A \tan \phi_A + B \tan \phi_B$$

Figure 8.8 Wedge stability charts for friction only: A and B charts for a dip difference of 0°. Note: The flatter of the two planes is always called plane A.

On the basis of such trial calculations, it is suggested that the friction-only stability charts can be used to define those slopes that are adequately stable and can be ignored in subsequent analyses, that is, slopes having a factor of safety in excess of 2.0. Slopes with a factor of safety, based upon friction only, of less than 2.0 must be regarded as potentially unstable and require further detailed examination as discussed in Section 8.6 and Appendix III.

In the design of the cut slopes on many projects, it will be found that these friction-only stability charts provide all the information that is required for preliminary design and planning stages of a slope project. This information will help identify potentially unstable wedges before excavation of the slope is started. During construction, the charts can be used

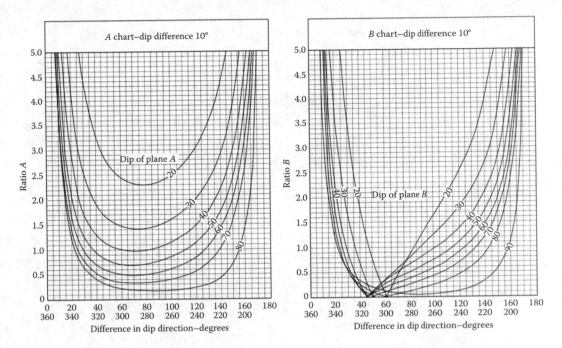

Figure 8.9 Wedge stability charts for friction only: A and B charts for a dip difference of 10°.

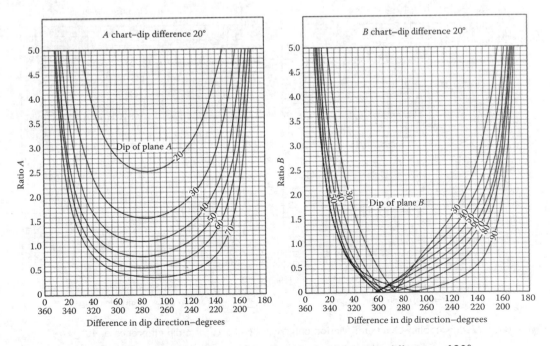

Figure 8.10 Wedge stability charts for friction only: A and B charts for a dip difference of 20°.

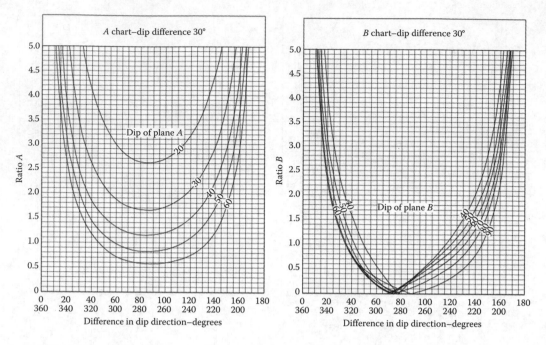

Figure 8.11 Wedge stability charts for friction only: A and B charts for a dip difference of 30°.

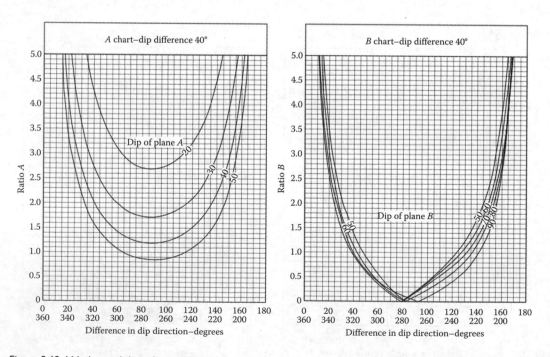

Figure 8.12 Wedge stability charts for friction only: A and B charts for a dip difference of 40°.

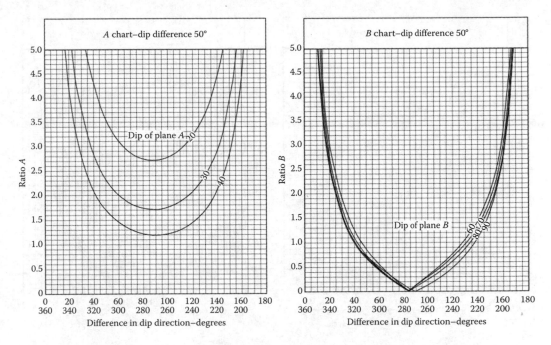

Figure 8.13 Wedge stability charts for friction only: A and B charts for a dip difference of 50°.

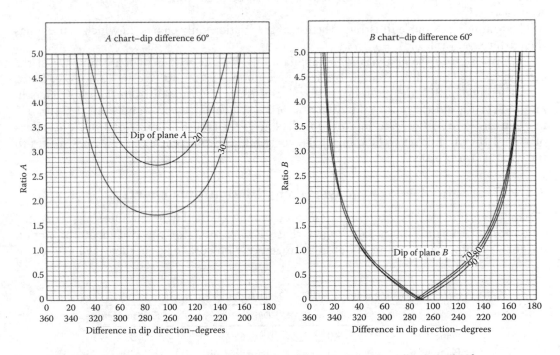

Figure 8.14 Wedge stability charts for friction only: A and B charts for a dip difference of 60°.

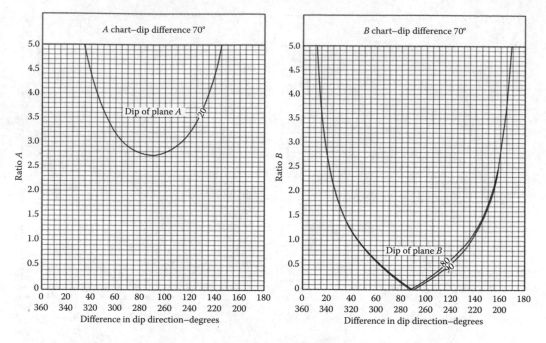

Figure 8.15 Wedge stability charts for friction only: A and B charts for a dip difference of 70°.

to make a rapid check of stability conditions when, for example, faces are being mapped as the excavation is proceeding and decisions are required on the need for support. If the factor of safety is less than 2.0, the detailed analysis can be used to design a bolting pattern (see Appendix III).

The following example illustrates the use of the friction-only charts, in which plane A has the flatter dip (Table 8.3).

The first step in the analysis is to calculate the absolute values of the difference in the dip angles, and the difference in the dip direction angles (third line in Table 8.3). For a dip difference of 30°, the values of ratios A and B are determined from the two charts in Figure 8.11 for a difference in dip direction of 120°. The values of A and B are 1.5 and 0.7, respectively, and substitution in Equation 8.14 gives the factor of safety of 1.30. The values of A and B give a direct indication of the contribution which each of the planes makes to the total factor of safety.

8.5.1 Example of wedge analysis using friction-only charts

During the route location study for a proposed highway, the layout engineer has requested guidance on the maximum safe angles that may be used for the design of the slopes. Extensive geological mapping of outcrops, together with core logging, identified five sets of

Table 8.3 Wedge stability analysis for friction only

	Dip	Dip direction	Friction angle
Plane A	40	165	35
Plane B	70	285	20
Differences	30	120	

Table 8.4 Orientations of discontinuity sets in wedge analysis example

Discontinuity set	Dip	Dip direction
1	66 ± 2	298 ± 2
2	68 ± 6	320 ± 15
3	60 ± 16	360 ± 10
4	58 ± 6	76 ± 6
5	54 ± 4	118 ± 2

discontinuities in the rock mass through which the road will pass. The dips and dip directions of these discontinuities are shown in Table 8.4, together with the measured variation in these measurements.

Note that, because the mapping covers the entire alignment that extends over several kilometres, the scatter in the dip and dip direction measurements can be taken into account in the analysis.

Figure 8.16 shows the pole locations for these five sets of discontinuities (inverted triangle). Also shown in this figure are the extent of the scatter in the pole measurements, and the great circles corresponding to the average pole positions. The dashed figure surrounding the great circle intersections is obtained by rotating the stereoplot to find the extent to which the intersection point is influenced by the scatter around the pole points. The intersection of great circles 2 and 5 has been excluded from the dashed figure because it defines a line of intersection dipping flatter than 20°, which is less than the estimated angle of friction.

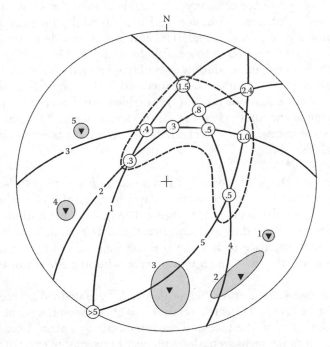

Figure 8.16 Stereoplot of geological data for preliminary design of highway slopes. Black triangles mark most likely position of poles of five sets of discontinuities present in rock mass; shaded area surrounding pole position defines extent of scatter in measurements; factors of safety for each combination of discontinuities are shown in circle over corresponding intersection of great circles; dashed line surrounds area of potential instability (FS < 2).

The factors of safety for each of the discontinuity intersections are determined from the wedge charts, assuming a friction angle of 30° (some interpolation is necessary), and the values are given in the circles over the intersection points. Because all of the planes are steep, some of the factors of safety are dangerously low. Since it is likely that slopes with a factor of safety of less than 1.0 will fail as the slope is excavated, the only practical solution is to cut the slopes at an angle that is coincident with the dip of the line of intersection. For marginally stable wedges with factors of safety between 1.0 and 2.0, a detailed analysis can be carried out to determine the bolting force required to increase the factor of safety to an acceptable level.

The stereoplot in Figure 8.16 can be used to find the maximum safe slope angle for the slopes of any dip direction. This stereographic analysis involves positioning the great circle representing the slope face for a particular dip direction in such a way that the unstable region (shaded) is avoided. The maximum safe slope angles are marked around the perimeter of this figure and their positions correspond to the orientation of the slope face (Figure 8.17a). For example, if a through cut for a highway is planned in this rock, the slope on the south side of the cut would be an angle of 30° and the slope on the north side would be at 85° (Figure 8.17b).

8.6 COMPREHENSIVE WEDGE ANALYSIS

8.6.1 Data for comprehensive analysis

If the friction-only wedge stability charts show that the factor of safety is less than 2.0, then a comprehensive stability analysis may be required to calculate, for example, the bolting force to achieve a required factor of safety. This analysis takes into account the dimensions and shape of the wedge, different cohesion and friction angles on each slide plane, water pressure and a number of external forces. The external forces that may act on the wedge include earthquake ground motion, tensioned bolts and possible loads generated by structures that may include a bridge foundation or building located on the wedge.

Figure 8.18 shows the features of the slope involved in the comprehensive wedge analysis, and a complete list of the equations used in the analysis is included in Appendix III. A fundamental assumption in the analysis is that all the forces act through the centre of gravity of the wedge and so no moments are generated. The following is a description of the components of the analysis:

1. *Wedge shape* – the shape of the wedge is defined by five surfaces: the two slide planes (1 and 2) with their line of intersection daylighting in the face, the upper slope (3) and face (4), and a tension crack (5) (Figure 8.18a). The orientations of these surfaces are each defined by their dips and dip directions. The range of orientations that the analysis can accommodate include an overhanging face, different dip directions for the upper slope and the face, and a tension crack dipping either towards or away from the face.
2. *Wedge dimensions* – the dimensions of the wedge are defined by the two dimensions $H1$ and L (Figure 8.18a). $H1$ is the vertical height between the point where the line of intersection daylights on the face and the intersection of plane 1 with the crest of the slope, L is the distance measured along plane 1 between the crest of the slope face (4) and the tension crack (5).
3. *Wedge weight* – the orientations of the five planes, and the two dimensions can be used to calculate the volume of the wedge, and the weight is determined from the unit weight of the rock (γ_r).

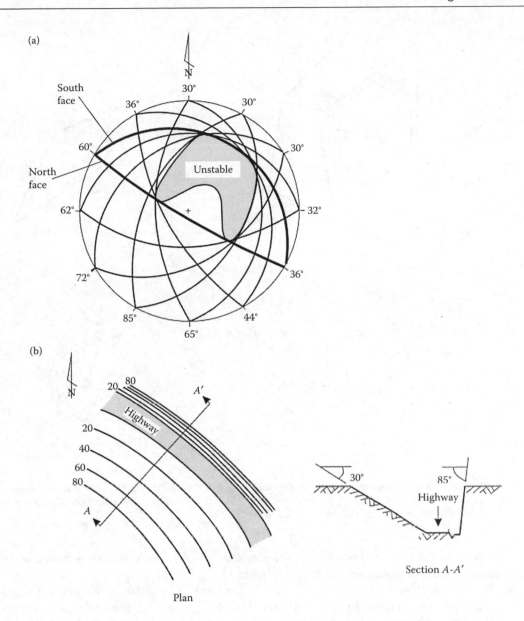

Figure 8.17 Slope design based on wedge stability analysis: (a) stereoplot of great circles representing stable slopes in a rock mass containing the five sets of discontinuities defined in Figure 8.16; (b) stable cut slopes angles based on wedge stability analysis shown in (a).

4. *Water pressures* – if it is assumed that the tension crack (5) is filled with water and that the water discharges to the atmosphere where planes 1 and 2 intersect the face (plane 4), then triangular water pressures act on planes 1, 2 and 5 (Figure 8.18b). The water pressure p at the base of the tension crack (and the top of the line of intersection) is equal to $(h_5 \cdot \gamma_w)$, where h_5 is the average vertical depth below the top of the tension crack. The water forces U_1, U_2 and V are calculated by integrating the pressures over the areas of planes 1, 2 and 5, respectively.

In the analysis, the effect of slope drainage is simulated by reducing the unit weight of the water to less than 9.81 kN/m³ (62.5 lb/ft³). This approach simplifies the calculation

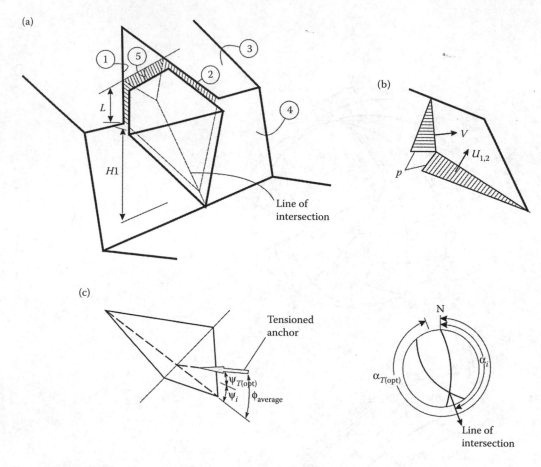

Figure 8.18 Comprehensive wedge analysis: (a) dimensions and surfaces defining size and shape of wedge; (b) water pressures acting in tension crack and along the line of intersection; (c) optimum anchor orientation for reinforcement of a wedge.

of areas of the saturated portions of the planes 2, 3 and 5, and was a convenience when the calculations were being carried out by hand.

5. *Shear strengths* – the slide planes (1 and 2) can have different shear strengths defined by the cohesion (c) and friction angle (ϕ). The shear resistance is calculated by multiplying the cohesion by the area of slide plane, and adding the product of the effective normal stress and the friction angle. The normal stresses are found by resolving the wedge weight in directions normal to each slide plane (see Equations 8.3 to 8.5).

6. *External forces* – external forces acting on the wedge are defined by their magnitude and orientation (plunge ψ, and trend α). The equations listed in Appendix III can accommodate a total of two external forces; if three or more forces are acting, the vectors are added as necessary. One external force that may be included in the analysis is the pseudo-static force used to simulate seismic ground motion (see Section 11.6). The horizontal component of this force would act in the same direction as the line of intersection of planes 1 and 2.

7. *Bolting forces* – if tensioned anchors are installed to stabilise the wedge, they are considered to be an external force. Note that the calculated bolting force is the required

force to support the entire wedge, in comparison to plane failure where all forces are calculated per linear metre of slope.

The orientation of the anchors can be optimised to minimise the anchor force required to produce a specified factor of safety. The optimum anchor plunge $\psi_{T(opt)}$ and trend $\alpha_{T(opt)}$, with respect to the line of intersection (ψ_i/α_i), are as follows (Figure 8.18c):

$$\psi_{T(opt)} = (\phi_{average} - \psi_i) \tag{8.15}$$

and

$$\alpha_{T(opt)} = (180 + \alpha_i) \quad \text{for} \quad \alpha_{T(opt)} \leq 360 \tag{8.16}$$

where $\phi_{average}$ is the average friction angle of the two slide planes.

8.6.2 Computer programs for comprehensive analysis

Appendix III lists equations that can be used to carry out a comprehensive stability analysis of a wedge using the input parameters discussed in Section 8.6.1. These equations were originally developed by Dr. John Bray and were included in the third edition of *Rock Slope Engineering* (1981). This method of analysis has been used in a number of computer programs that allow rapid and reliable analysis of wedge stability. However, the following limitation to this analysis should be noted.

Wedge geometry – the analysis procedure is to calculate the dimensions of a wedge that extends from the face to the point where planes 1, 2 and 3 intersect. The next step is to calculate the dimensions of a second wedge formed by sliding planes 1 and 2, the upper slope (plane 3) and the tension crack (plane 5). The dimensions of the wedge in front of the tension crack are then found by subtracting the dimensions of the wedge in front of the tension crack from the overall wedge (see Equations III.54 to III.57, Appendix III). Prior to performing the subtractions, the programme tests to see whether a wedge is formed and a tension crack is valid (Equations III.48 to III.53, Appendix III). The programme will terminate if the dip of the upper slope (plane 3) is greater than the dip of the line of intersection of planes 1 and 2, or if the tension crack is beyond the point where planes 1, 2 and 3 intersect.

While these tests are mathematically valid, they do not allow for a common geometric condition that may exist in steep mountainous terrain. That is, if plane 3 is steeper than the line of intersection of planes 1 and 2, a wedge made up of five planes can still be formed if a tension crack is located behind the slope face to create a valid plane 5. Where this physical condition exists in the field, an alternative method of analysis is to use key block theory in which the shape and stability condition of removal wedges can be completely defined (Goodman and Shi, 1985; PanTechnica Corporation, 2002) or other commercial programmes (Rocscience Inc).

8.6.3 Example of comprehensive wedge analysis

The following is an example of the comprehensive stability analysis of a wedge using the programme Swedge (Rocscience Inc). Consider the wedge formed by joint sets 3 and 5 in Figure 8.16. This has a friction-only factor of safety of 1.0, so water pressures and seismic ground motion may result in instability, depending on the cohesions on the slide surfaces.

Table 8.5 Orientation of planes forming wedge

Plane	Dip	Dip direction	Shear strength
1	60	360	$\phi_1 = 30°, c_1 = 50$ kPa (7.3 psi)
2	54	118	$\phi_2 = 30°, c_2 = 0$
Slope face	76	060	
Upper surface	15	070	
Tension crack	80	060	

This analysis shows the bolting force required to raise the factor of safety to 1.5, based on the input parameters shown in Table 8.5.

The other input parameters are as follows:

Wedge height, $H1 = 28$ m
Distance of tension crack from crest measured along plane 3, $L = 15$ m
Rock density, $\gamma_r = 26$ kN/m³ (62.5 lb/ft³)
Water density, $\gamma_w = 9.81$ kN/m³ (165 lb/ft³)
Seismic coefficient (horizontal), $k_H = 0.1 \cdot g$

This wedge has a line of intersection of 38° (plunge) and 063° (trend). This plunge angle is greater than the friction angle of either sliding surface and so sliding of the wedge is possible. The weight of the wedge is 121.3 MN, which indicates the magnitude of the bolting force that is required to significantly improve the factor of safety.

The stability analysis of the wedge using these parameters gave the following results:

1. Dry, static, $c_1 = c_2 = 0$, $T = 0$ $FS = 1.05$
   Case (1) corresponds to the simplified analysis using the friction-only charts for a dry, static slope that gives a factor of safety of 1.05.
2. $z_w = 50\%$, static, $c_1 = c_2 = 0$, $T = 0$ $FS = 0.97$
   Case (2) shows that for the friction-only condition and the tension crack half filled with water, the factor of safety reduces from 1.05 to 0.97.
3. $z_w = 50\%$, static, $c_1 = 50$ kPa (7.3 psi), $c_2 = 0$, $T = 0$ $FS = 1.24$
   Case (3) shows that the effect of the cohesion of only 50 kPa (7.3 psi) acting on the 491 m² area of plane 1 is to increase the factor of safety to 1.24.
4. $z_w = 50\%$, $k_H = 0.1 \cdot g$, $c_1 = 50$ kPa (7.3 psi), $c_2 = 0$, $T = 0$ $FS = 1.03$
   Case (4) shows that the effect of a pseudo-static force $k_H = 0.1 \cdot g$ for the same water pressure ($z_w = 50\%$) is to reduce the factor of safety from 1.24 to 1.03.
5. $T = 24{,}000$ kN, $\alpha_T = 243°$, $\psi_T = -8°$, $z_w = 50\%$, $k_H = 0.1 \cdot g$, $c_1 = 50$ kPa (7.3 psi), $c_2 = 0$ $FS = 1.50$
   Case (5) shows that if tensioned bolts with a capacity of 24,000 kN (5400 kips) are installed at the optimum angle as defined by Equations 8.15 and 8.16, that is, parallel to line of intersection and 8° above the horizontal, the factor of safety is increased to 1.5.
6. $T = 25{,}500$ kN, $\alpha_T = 249°$, $\psi_T = +10°$, $z_w = 50\%$, $k_H = 0.1 \cdot g$, $c_1 = 50$ kPa (7.3 psi), $c_2 = 0$ $FS = 1.50$
   Case (6) shows that if the bolts are installed at right angles to the face and at a plunge angle of 10° below the horizontal to facilitate grouting, then the bolting force has to be increased from 24,000 kN to 25,500 kN (5732 kips) for the factor of safety to be the same as that for the bolts installed at the optimum orientation.

Chapter 9

Circular failure

9.1 INTRODUCTION

Although this book is concerned primarily with the stability of rock slopes containing well-defined sets of discontinuities, it is also necessary to design cuts in weak materials such as highly weathered or closely fractured rock, and rock fills. In such materials, failure occurs along a surface that approaches a circular shape (Figure 9.1), and this chapter is devoted to a discussion on the stability analysis of slopes in these materials.

In a review of the historical development of slope stability theories, Golder (1972) traced the subject back almost 300 years. Much of the development of circular failure analysis methods was carried out in the 1950s and 1960s, and these techniques have since been used to prepare computer programs that have the versatility to accommodate a wide range of geologic, geometric, ground water and external loading conditions. This chapter discusses the principles of the theoretical work, and demonstrates their application in design charts and in the results of computer analyses. During the past half century, a vast body of literature on the subject of circular failure has accumulated, and no attempt will be made to summarise the material in this chapter. Standard soil mechanics textbooks such as those by Taylor (1937), Terzaghi (1943) and Lambe and Whitman (1969), and papers by Skempton (1948), Bishop (1955), Janbu (1954), Morgenstern and Price (1965), Nonveiller (1965), Peck (1967), Spencer (1967, 1969) and Duncan (1996) all contain excellent discussions on the stability of soil slopes.

The approach adopted in this chapter is to present a series of slope stability charts for circular failure. These charts enable the user to carry out a rapid check on the factor of safety of a slope, or upon the sensitivity of the factor of safety to changes in ground water conditions, slope angle and material strength properties. These charts should only be used for the analysis of circular failure in slope materials that are homogeneous and where the conditions apply that were assumed in deriving the charts (see Section 9.3).

More comprehensive methods of analysis are presented in Section 9.6. These methods can be used, for example, where the material properties vary within the slope, or where part of the slide surface is at a soil/rock interface and the shape of the slide surface differs significantly from a simple circular arc.

This chapter primarily addresses the stability of slopes in two dimensions, and assumes that the slope can be modelled as a unit slice through an infinitely long slope, under plane-strain conditions; Section 9.6.5 discusses three-dimensional circular failure analysis.

Figure 9.1 (**See colour insert.**) Circular failure in highly weathered, granitic rock (Highway 1, near Devil's Slide, Pacifica, California).

9.2 CONDITIONS FOR CIRCULAR FAILURE AND METHODS OF ANALYSIS

In the previous chapters, it has been assumed that the failure of rock slopes is controlled by geological features such as bedding planes and joints that divide the rock into a discontinuous mass. Under these conditions, one or more of the discontinuities normally defines the slide surface. However, in the case of a closely fractured or highly weathered rock, a well-defined structural pattern no longer exists, and the slide surface is free to find the line of least resistance through the slope. Observations of slope failures in these materials suggest that this slide surface generally takes the form of a circle, and most stability theories are based upon this observation. Figure 9.1 shows a typical circular failure in a highly weathered rock slope above a highway.

The conditions under which circular failure will occur arise when the individual particles in a soil or rock mass are very small compared with the size of the slope. Hence, broken rock in a fill will tend to behave as a 'soil' and fail in a circular mode when the slope dimensions are substantially greater than the dimensions of the rock fragments. Similarly, soil consisting of sand, silt and smaller particle sizes will exhibit circular slide surfaces, even in slopes only a few metres in height. Highly altered and weathered rocks, as well as rock with closely spaced, randomly oriented discontinuities such as some rapidly cooled basalts, will also tend to fail in this manner. It is appropriate to design slopes in these materials on the assumption that a circular failure process will develop.

9.2.1 Shape of slide surface

The actual shape of the 'circular' slide surface is influenced by the geological conditions in the slope. For example, in a homogeneous weak or weathered rock mass, or a rock fill, the failure is likely to form as a shallow, large-radius surface extending from a tension crack close behind the crest to the toe of the slope (Figure 9.2a). This contrasts with failures in high-cohesion, low-friction materials such as clays where the surface may be deeper with a

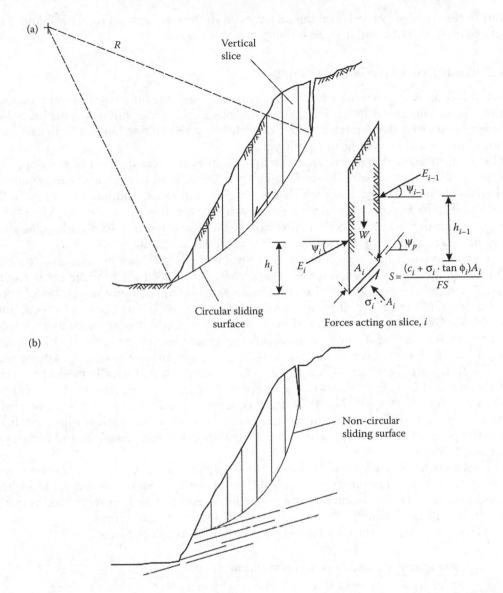

Figure 9.2 The shape of typical sliding surfaces: (a) large-radius circular surface in homogeneous, weak material, with the detail of forces on slice; (b) non-circular surface in weak, surficial material with stronger rock at base.

smaller radius that may exit beyond the toe of the slope. Figure 9.2b shows an example of conditions in which the shape of the slide surface is modified by the slope geology. Here, the circular surface in the upper, weathered rock is truncated by the shallow-dipping, stronger rock near the base. Stability analyses of both types of surface can be carried out using circular failure methods, although for the latter case it is necessary to use a procedure that allows the shape of the surface to be defined.

For each combination of slope parameters, a slide surface exists for which the factor of safety is a minimum – this is usually termed the 'critical surface'. The procedure to find the critical surface is to run a large number of analyses in which the centre coordinates and the radius of the circle are varied until the surface with the lowest factor of safety is found. This is an essential part of circular slope stability analysis.

9.2.2 Stability analysis procedure

The stability analysis of circular failure is carried out using the limit equilibrium procedure similar to that described in earlier chapters for plane and wedge failures. This procedure involves comparing the available shear strength along the sliding surface with the force required to maintain the slope in equilibrium.

The application of this procedure to circular failures involves division of the slope into a series of slices that are usually vertical, but may be inclined to coincide with prominent geological features. The base of each slice is inclined at angle ψ_p and has an area A. In the simplest case, the forces acting on the base of each slice are the shear resistance S due to the shear strength of the rock (cohesion c; friction angle ϕ), and forces E (dip angle ψ; height h above base) acting on the sides of the slice (see detail Figure 9.2a).

The analysis procedure is to consider equilibrium conditions slice by slice, and if a condition of equilibrium is satisfied for each slice, then it is also satisfied for the entire sliding mass. The number of equations of equilibrium available depends on the number of slices N, and the number of equilibrium conditions that are used. The number of equations available is $2N$ if only force equilibrium is satisfied, and $3N$ if both force and moment equilibrium are satisfied. If only force equilibrium is satisfied, the number of unknowns is $(3N - 1)$, while, if both force and moment equilibria are satisfied, the number of unknowns is $(5N - 2)$. Usually between 10 and 40 slices are required to realistically model the slope, and therefore, the number of unknowns exceeds the number of equations. The excess of unknowns over equations is $(N - 1)$ for force equilibrium analysis, and $(2N - 2)$ for analyses that satisfy all conditions of equilibrium. Thus, the analyses are statically indeterminate and assumptions are required to make up the imbalance between equations and unknowns (Duncan, 1996).

The various LEA procedures either make assumptions to balance the known and unknowns, or they do not satisfy all the conditions of equilibrium. For example, the Spencer method assumes that the inclination of the side forces is the same for every slice, while the Fellenius and Bishop methods do not satisfy all conditions of equilibrium.

The factor of safety of the circular failure based on LEA is defined as

$$FS = \frac{\text{Shear strength available to resist sliding } (c + \sigma \cdot \tan\phi)}{\text{Shear stress required for equilibrium on slip surface } (\tau_e)} \tag{9.1}$$

and rearranging this equation

$$\tau_e = \frac{c + \sigma \cdot \tan\phi}{FS} \tag{9.2}$$

The method of solution for the factor of safety is to use an iterative process in which an initial estimate is made for FS, and this is refined with each iteration.

The influence of various normal stress distributions upon the factor of safety of soil slopes has been examined by Frohlich (1955) who found that a lower bound for all factors of safety that satisfy statics is given by the assumption that the normal stress is concentrated at a single point on the slide surface. Similarly, the upper bound is obtained by assuming that the normal stress is concentrated at the two ends of the slide surface.

The unreal nature of these stress distributions is of no consequence since the object of the exercise, up to this point, is simply to determine the extremes between which the actual factor of safety of the slope must lie. In an example considered by Lambe and Whitman (1969), the upper and lower bounds for the factor of safety of a particular slope corresponded to 1.62 and 1.27, respectively. Analysis of the same problem by the Bishop simplified method of slices gives a factor of safety of 1.30, which suggests that the actual factor of safety may lie reasonably close to the lower-bound solution.

Further evidence that the lower-bound solution is also a meaningful practical solution is provided by an examination of the analysis that assumed the slide surface has the form of a logarithmic spiral (Spencer, 1969). In this case, the factor of safety is independent of the normal stress distribution, and the upper and lower-bounds coincide. Taylor (1937) compared the results from a number of logarithmic spiral analyses with results of lower-bound solutions* and found that the difference is negligible. Based on this comparison, Taylor concluded that the lower-bound solution provides a value of the factor of safety that is sufficiently accurate for the most practical problems involving simple circular failure of homogeneous slopes.

The basic principles of these methods of analyses are discussed in Section 9.6.

9.3 DERIVATION OF CIRCULAR FAILURE CHARTS

This section describes the use of a series of charts that can be used to rapidly determine the factor of safety of circular failures. These charts have been developed by running many thousands of circular analyses from which a number of dimensionless parameters were derived that relate the factor of safety to the material unit weight, friction angle and cohesion, and the slope height and face angle. It has been found that these charts give a reliable estimate for the factor of safety, provided that the conditions in the slope meet the assumptions used in developing the charts. In fact, the accuracy in calculating the factor of safety from the charts is usually greater than the accuracy in determining the shear strength of the rock mass.

Use of the stability charts presented in this chapter requires that the conditions in the slope meet the following assumptions:

1. The material forming the slope is homogeneous, with uniform shear-strength properties along the slide surface.
2. The shear-strength τ of the material is characterised by cohesion c and a friction angle ϕ, which are related by the equation $\tau = c + \sigma \cdot \tan \phi$ (see Section 1.4).
3. Failure occurs on a circular slide surface that passes through the toe of the slope.[†]
4. A vertical tension crack occurs in the upper surface or in the face of the slope.
5. The locations of the tension crack and of the slide surface are such that the factor of safety of the slope is a minimum for the slope geometry and ground water conditions considered.
6. Ground water conditions vary from a dry slope to a fully saturated slope under heavy recharge; these conditions are defined in Figure 9.4.

---

* The lower-bound solution discussed in this chapter is usually known as the friction circle method and was used by Taylor (1937) for the derivation of his stability charts.

† Terzaghi (1943), page 170, shows that the toe failure assumed for this analysis gives the lowest factor of safety provided that $\phi > 5°$. The $\phi = 0$ analysis involving failure below the toe of the slope through the base material has been discussed by Skempton and Hutchinson (1948) and by Bishop and Bjerrum (1960) and is applicable to failures that occur during or after the rapid excavation of a slope.

7. Circular failure charts are optimised for a rock mass density of 18.9 kN/m³ (120 lb/ ft³). Densities higher than this give high factors of safety, densities lower than this give low factors of safety. Detailed circular analysis may be required for slopes in which the material density is significantly different from 18.9 kN/m³ (120 lb/ft³).

The charts presented in this chapter correspond to the lower-bound solution for the factor of safety, obtained by assuming that the normal stress is concentrated on a single point on the slide surface. These charts differ from those published by Taylor in that they include the influence of a critical tension crack and of ground water.

9.3.1 Ground water flow assumptions

In order to calculate the forces due to water pressures acting on the slide surface and in the tension crack, it is necessary to assume a set of ground water flow patterns that coincide as closely as possible with conditions that are believed to exist in the field.

In the analysis of rock slope failures discussed in Chapters 7, 8 and 10, it is assumed that most of the water flow takes place in discontinuities in the rock and that the rock itself is practically impermeable (see Figure 6.8). In the case of slopes in soil or waste rock, the hydraulic conductivity of the mass of material is generally several orders of magnitude higher than that of intact rock and, hence, a general flow pattern will develop in the material behind the slope.

Figure 6.11a shows that, within the isotropic soil mass, equipotentials are approximately perpendicular to the phreatic surface. Consequently, the flow lines will be approximately parallel to the phreatic surface for the condition of steady-state drawdown. Figure 9.3 shows that this approximation has been used for the analysis of the water pressure distribution in a slope under conditions of normal drawdown. Note that the phreatic surface is assumed to coincide with the ground surface at a distance x, measured in multiples of the slope height, behind the toe of the slope. This may correspond to the position of a surface water source, or be the point where the phreatic surface is judged to intersect the ground surface.

The phreatic surface itself has been obtained, for the range of the slope angles and values of x considered, by solution of the equations proposed by Casagrande (1934), and discussed in the textbook by Taylor (1937). For the case of a saturated slope subjected to heavy surface recharge, the equipotentials and the associated flow lines used in the stability analysis are based upon the work of Han (1972). This work involved the use of an electrical resistance analogue method to study ground water flow patterns in slopes comprising isotropic materials.

Figure 9.4 shows five ground water conditions ranging from fully drained to saturated, based on the models shown in Figure 9.3. For conditions 2, 3 and 4, the position of the ground water is defined by the ratio x/H. These five ground water conditions are used in conjunction with the circular failure charts discussed in Section 9.4.

9.3.2 Production of circular failure charts

The circular failure charts presented in this chapter were produced by running a search routine to find the most critical combination of slide surface and tension crack for each of a wide range of slope geometries and ground water conditions. Provision was made for the tension crack to be located in either the upper surface, or in the face of the slope. Detailed checks were carried out in the region surrounding the toe of the slope where the curvature of the equipotentials results in local flow which differs from that illustrated in Figure 9.3a.

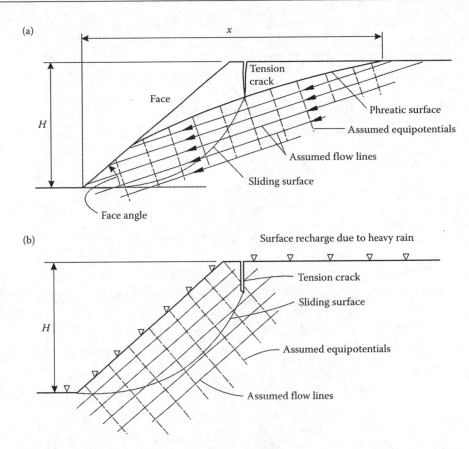

Figure 9.3 Definition of ground water flow patterns used in circular failure analysis of slopes in weak and closely fractured rock: (a) ground water flow pattern under steady state drawdown conditions where the phreatic surface coincides with the ground surface at a distance *x* behind the toe of the slope. The distance *x* is measured in multiples of the slope height *H*; (b) ground water flow pattern in a saturated slope subjected to surface recharge by heavy rain.

The charts are numbered 1–5 (Figures 9.6 to 9.10) to correspond with the ground water conditions defined in Figure 9.4.

9.3.3 Use of the circular failure charts

In order to use the charts to determine the factor of safety of a slope (height H, rock unit weight γ), the steps outlined below and shown in Figure 9.5 should be followed:

Step 1: Decide upon the ground water conditions which are believed to exist in the slope and choose the chart which is closest to these conditions, using Figure 9.4.

Step 2: Select rock strength parameters applicable to the material forming the slope (cohesion c and friction angle ϕ).

Step 3: Calculate the value of the dimensionless ratio ($c/(\gamma \cdot H \cdot \tan \phi)$) and find this value on the outer circular scale of the chart.

Step 4: Follow the radial line from the value found in step 3 to its intersection with the curve, which corresponds to the slope angle.

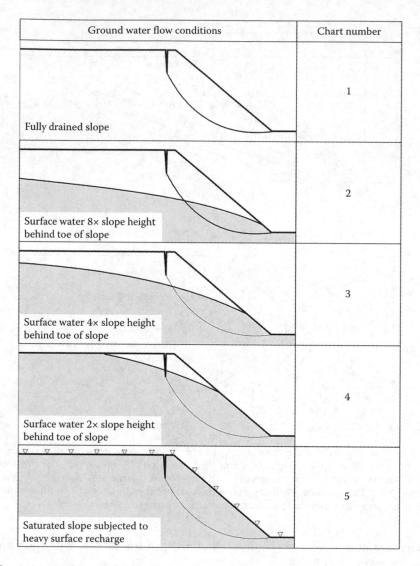

Ground water flow conditions	Chart number
Fully drained slope	1
Surface water 8× slope height behind toe of slope	2
Surface water 4× slope height behind toe of slope	3
Surface water 2× slope height behind toe of slope	4
Saturated slope subjected to heavy surface recharge	5

Figure 9.4 Ground water flow models used with circular failure analysis charts – Figures 9.6 to 9.10.

Step 5: Find the corresponding value of (tan ϕ/FS) or (c/($\gamma \cdot H \cdot FS$)), depending upon which is more convenient, and calculate the factor of safety.

Consider the following example:

A 15.2 m (50 ft) high cut with a face angle of 40° is to be excavated in overburden soil with a density of $\gamma = 15.7$ kN/m³ (100 lb/ft³), a cohesion of 38 kPa (5.5 psi) and a friction angle of 30°. Find the factor of safety of the slope, assuming that a surface water source is located 61 m (200 ft) behind the toe of the slope.

The ground water conditions indicate the use of chart number 3 (61/15.2 ≈ 4). The value of (c/($\gamma \cdot H \cdot \tan \phi$) = 0.28) and the corresponding value of (tan ϕ/FS), for a 40° slope angle, is 0.32. Hence, the factor of safety of the slope of 1.80.

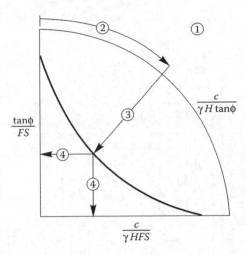

Figure 9.5 Sequence of steps involved in using circular failure charts to find the factor of safety of a slope.

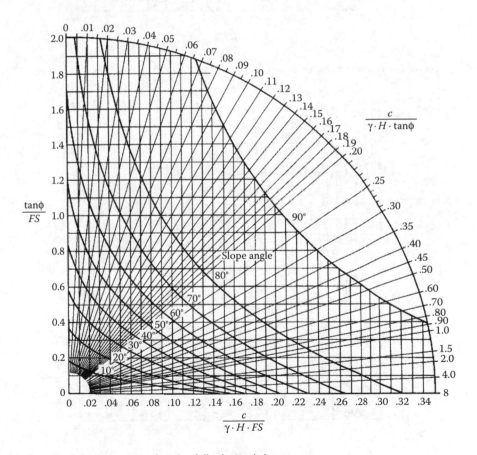

Figure 9.6 Circular failure chart number I – fully drained slope.

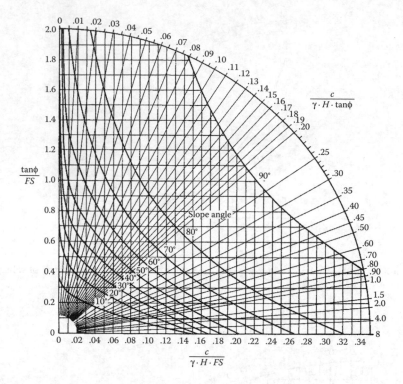

Figure 9.7 Circular failure chart number 2 – ground water condition 2 (Figure 9.4).

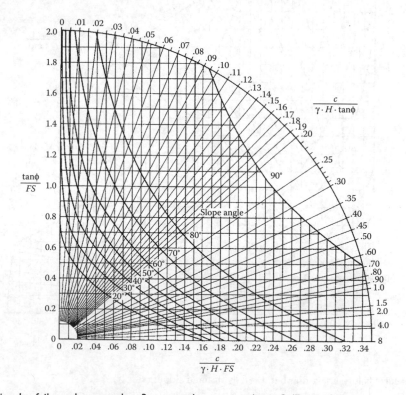

Figure 9.8 Circular failure chart number 3 – ground water condition 3 (Figure 9.4).

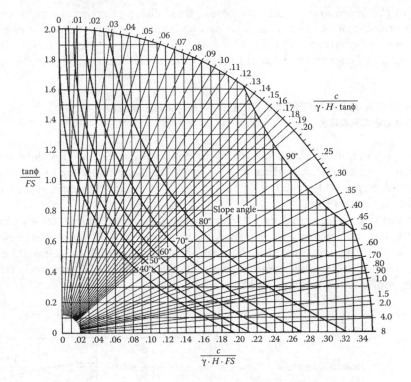

Figure 9.9 Circular failure chart number 4 – ground water condition 4 (Figure 9.4).

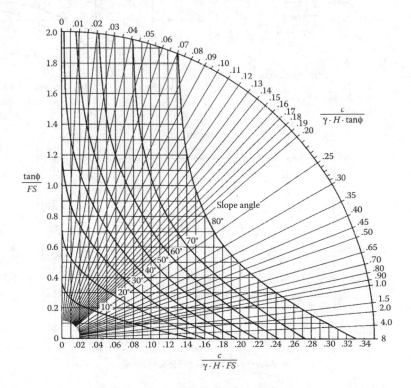

Figure 9.10 Circular failure chart number 5 – fully saturated slope.

Because of the speed and simplicity of using these charts, they are ideal for checking the sensitivity of the factor of safety of a slope to a wide range of conditions. For example, if the cohesion were to be halved to 20 kPa (2.9 psi) and the ground water pressure increased to that represented by chart number 2, the factor of safety drops to 1.28.

9.4 LOCATION OF CRITICAL SLIDE SURFACE AND TENSION CRACK

During the production of the circular failure charts presented in this chapter, the locations of both the critical slide surface and the critical tension crack for limiting equilibrium ($FS = 1$) were determined for each slope analysed. These locations are presented, in the form of charts, in Figures 9.11 and 9.12.

It was found that, once ground water is present in the slope, the locations of the critical circle and the tension crack are not particularly sensitive to the position of the phreatic surface and hence only one case, that for chart number 3, has been plotted. It will be noted that the location of the critical circle centre given in Figure 9.12 differs significantly from that for the drained slope plotted in Figure 9.11.

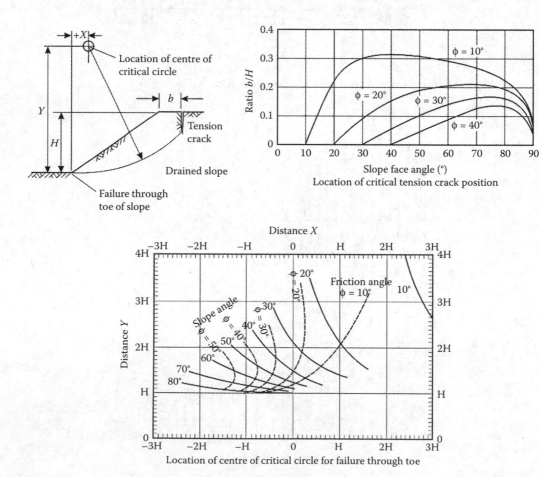

Figure 9.11 Location of critical sliding surface and critical tension crack for drained slopes.

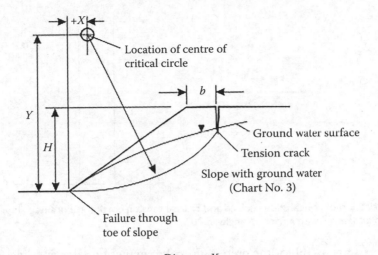

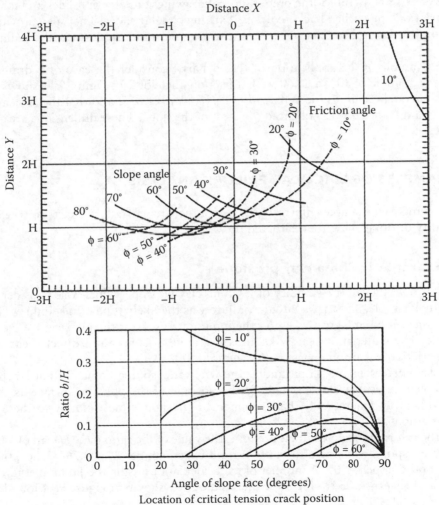

Figure 9.12 Location of critical sliding surface and critical tension crack for slopes with ground water present.

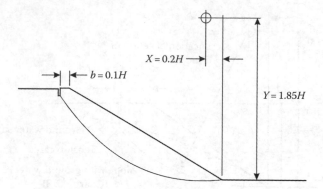

Figure 9.13 Location of critical slide surface and critical tension crack for a drained slope at an angle of 30° in a material with a friction angle of 20°.

These charts are useful for the preparation of drawings of potential slides and for estimating the friction angle when back-analysing existing circular slides. They also provide a start in locating the critical slide surface when carrying out more sophisticated circular failure analysis.

As an example of the application of these charts, consider the case of a drained slope having a face angle of 30° in a soil with a friction angle of 20°. Figure 9.11 shows that the critical slide circle centre is located at $X = 0.2H$ and $Y = 1.85H$ and that the critical tension crack is at a distance $b = 0.1H$ behind the crest of the slope. These dimensions are shown in Figure 9.13.

9.5 EXAMPLES OF CIRCULAR FAILURE ANALYSIS

The following two examples illustrate the use of the circular failure charts for the study of the stability of slope in highly weathered rock.

9.5.1 Example 1: China clay pit slope

Ley (1972) investigated the stability of a China clay pit slope, which was considered to be potentially unstable, and that a circular failure was the likely type of instability. The slope profile is illustrated in Figure 9.14 and the input data used for the analysis are included in this figure. The material, a heavily kaolinised granite, was tested in direct shear to determine the friction angle and cohesion (see Figure 5.16).

Two piezometers in the slope and a known water source some distance behind the slope enabled an estimate to be made of the position of the phreatic surface as shown in Figure 9.14. From Figure 9.14, chart number 2 corresponds most closely to these ground water conditions.

From the information given in Figure 9.14, the value of the ratio $(c/(\gamma \cdot H \cdot \tan \phi) = 0.0056)$, and the corresponding value of $(\tan \phi/FS)$ from chart number 2, is 0.76. Hence, the factor of safety of the slope is 1.01. A number of trial calculations using the Janbu method (Janbu, Bjerrum and Kjaernsli, 1956) for the critical slide circle shown in Figure 9.14 found a factor of safety of 1.03.

These factors of safety indicated that the stability of the slope was inadequate under the assumed conditions, and steps were taken to deal with the problem.

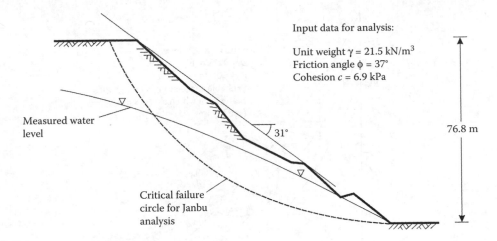

Input data for analysis:

Unit weight $\gamma = 21.5$ kN/m^3
Friction angle $\phi = 37°$
Cohesion $c = 6.9$ kPa

76.8 m

Measured water level

31°

Critical failure circle for Janbu analysis

Figure 9.14 Slope profile of China clay pit slope considered in Example 1.

9.5.2 Example 2: Highway slope

A highway plan called for a cut at an angle of 42°. The total height of the cut would be 61 m (200 ft) and it was required to check whether the cut would be stable. The slope was in weathered and altered material, and failure, if it occurred, would be a circular type. Insufficient time was available for ground water levels to be accurately established or for shear tests to be carried out. The stability analysis was carried out as follows:

For the condition of limiting equilibrium, $FS = 1$ and $(\tan \phi/FS = \tan \phi)$. By reversing the procedure outlined in Figure 9.5, a range of friction angles were used to find the values of the ratio $(c/(\gamma \cdot H \cdot \tan \phi))$ for a face angle of 42°. The value of the cohesion c that is mobilised at failure, for a given friction angle, can then be calculated. This analysis was carried out for dry slopes using Figure 9.6, chart number 1 (line B, Figure 9.15), and for saturated slopes using Figure 9.10, chart number 5 (line A, Figure 9.15). Figure 9.15 shows the range of friction angles and cohesions that would be mobilised at failure.

The shaded circle (D) included in Figure 9.15 indicates the range of shear strengths that were considered probable for the material under consideration, based upon the data presented in Figure 5.21. This figure shows that the available shear strength may not be adequate to maintain stability in this cut, particularly when the slope is saturated. Consequently, the face angle could be reduced, or ground water conditions investigated to establish actual ground water pressures and the feasibility of drainage.

The effect of reducing the slope angle can be checked very quickly by finding the value of the ratio $(c/(\gamma \cdot H \cdot \tan \phi))$ for a flatter slope of 30°, in the same way as it was found for the 42° slope. The dashed line (C) in Figure 9.15 indicates the shear strength which is mobilised in a dry slope with a face angle of 30°. Since the mobilised shear strength (line C) is less than the available shear strength (area D), the dry slope is likely to be stable.

9.6 DETAILED STABILITY ANALYSIS OF CIRCULAR FAILURES

The circular failure charts presented earlier in this chapter are based upon the assumption that the material forming the slope has uniform properties throughout the slope, and that

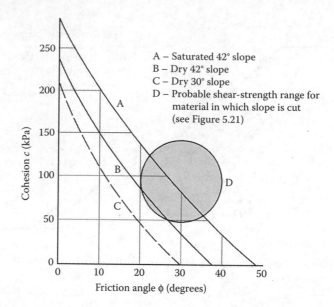

Figure 9.15 Comparison between shear strength mobilised and shear strength available for slope considered in Example 2.

failure occurs along a circular slide path passing through the toe of the slope. When these conditions are not satisfied, it is necessary to use one of the methods of slices published by Bishop (1955), Janbu (1954), Nonveiller (1965), Spencer (1967), Morgenstern and Price (1965) or Sarma (1979). This section describes in detail the simplified Bishop and Janbu methods of stability analysis for circular failure.

9.6.1 Bishop and Janbu methods of slices

The determination of the factor of safety by the Bishop simplified method of slices (1955) and the Janbu modified method of slices (1954) requires the slope and slide surface geometries, and the stability equations, as given in Figures 9.16 and 9.17, respectively. The Bishop method assumes a circular slide surface and that the side forces are horizontal; the analysis satisfies vertical forces and overall moment equilibrium. The Janbu method allows a slide surface of any shape, and assumes the side forces are horizontal and equal on all slices; the analysis satisfies vertical force equilibrium. As pointed out by Nonveiller (1965), the Janbu method gives reasonable factors of safety when applied to shallow slide surfaces that are typical in rock with an angle of friction in excess of 30° and for rock fill. However, the Janbu method may be seriously in error and should not be used for deep slide surfaces in materials with low-friction angles.

The procedures for using the Bishop and Janbu methods of slices are very similar and it is convenient to discuss them together.

Step 1: Slope and slide surface geometry – The geometry of the slope is defined by the actual or the designed profile as seen in a vertical section through the slope. In the case of a circular failure, the charts given in Figures 9.11 and 9.12 can be used to estimate the centre of the circle with the lowest factor of safety. In the Janbu analysis, the slide surface may be defined by known structural features or weak zones within the rock or soil mass, or it may be estimated in the same way as that for the Bishop analysis. In either case, the slide surface assumed for the first analysis may not give the lowest factor of safety, and a series of analyses are required with variations on this position to find the surface with the lowest factor of safety.

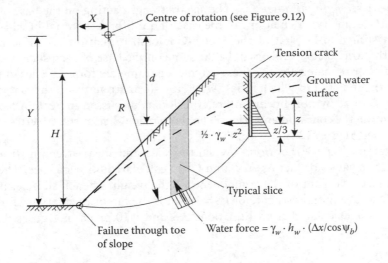

Centre of rotation (see Figure 9.12)

Tension crack

Ground water surface

Typical slice

Failure through toe of slope

Water force $= \gamma_w \cdot h_w \cdot (\Delta x / \cos \psi_b)$

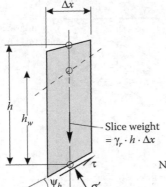

Slice weight $= \gamma_r \cdot h \cdot \Delta x$

Factor of safety:

$$FS = \frac{\Sigma(A/(1+B/FS))}{\Sigma(C)+Q} \qquad (9.3)$$

where

$$A = [c + (\gamma_r \cdot h - \gamma_w \cdot h_w) \cdot \tan\phi] \cdot (\Delta x / \cos\psi_b) \qquad (9.4)$$
$$B = \tan\psi_b \cdot \tan\phi \qquad (9.5)$$
$$C = \gamma_r \cdot h \cdot \Delta x \cdot \sin\psi_b \qquad (9.6)$$
$$Q = \tfrac{1}{2} \cdot \gamma_w \cdot z^2 \cdot (d/R) \qquad (9.7)$$

Note: Angle ψ_b is negative when sliding uphill

The following conditions must be satisfied for each slice:

1) $\sigma' = \dfrac{\gamma_r \cdot h - \gamma_w \cdot h_w - c\,(\tan\psi_b/FS)}{1+B/FS} \qquad (9.8)$

2) $\cos\psi_b\,(1+B/FS) > 0.2 \qquad (9.9)$

Figure 9.16 The Bishop simplified method of slices for the analysis of circular failure in slopes cut into materials in which failure is defined by the Mohr–Coulomb failure criterion.

Step 2: Slice parameters – The sliding mass assumed in step 1 is divided into a number of slices. Generally, a minimum of five slices should be used for simple cases. For complex slope profiles, or where the rock or soil mass comprises different materials, a larger number of slices may be required in order to define adequately the problem. The parameters that have to be defined for each slice are:

- Base angle ψ_b.
- The weight of each slice W is given by the product of the vertical height h, the unit weight γ_r of the rock or soil and the width of the slice Δx: $W = (h \cdot \gamma_r \cdot \Delta x)$.
- Uplift water force U on the base of each slice is given by the product of the height h_w to the phreatic surface, the unit weight γ_w of water and the width of the slice Δx, that is, $U = (h_w \cdot \gamma_w \cdot \Delta x)$.

Step 3: Shear-strength parameters – The shear strength acting on the base of each slice is required for the stability calculation. In the case of a uniform material in which the failure criterion is assumed to be that of the Mohr–Coulomb (Equation 1.1 in Section 1.4), the shear-strength parameters c and ϕ will be the same on the base of each slice. When the slope is cut in a number of materials, the shear-strength parameters for each slice must be chosen according to the material in which it lies. When the shear strengths of the materials forming the slope are defined by non-linear failure criterion as discussed in Section 5.5, it is necessary to determine the cohesion and friction angle for each slice at the effective normal stress for that slice (see Figure 5.23).

Step 4: Factor of safety iteration – When the slice and shear-strength parameters have been defined, the values of factors A, B and C are calculated for each slice. The water force Q is added to ΣC, the sum of the components of the weight of each slice acting parallel to the slide surface. An initial estimate of $FS = 1.00$ for the factor of safety is used, and a new factor of safety is calculated from Equations 9.3 and 9.10 given in Figures 9.16 and 9.17,

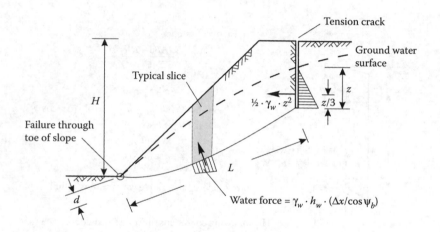

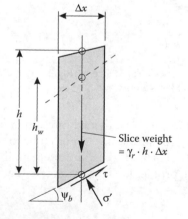

Factor of safety:

$$FS = \frac{f_o \cdot \Sigma \left(A/(1 + B/FS) \right)}{\Sigma (C) + Q} \qquad (9.10)$$

where

$$X = [c + (\gamma_r \cdot h - \gamma_w \cdot h_w) \tan\phi] \cdot (1 + \tan^2\psi_b) \, \Delta x \qquad (9.11)$$

$$Y = \tan\psi_b \cdot \tan\phi \qquad (9.12)$$

$$Z = \gamma_r \cdot h \cdot \Delta x \cdot \tan\psi_b \qquad (9.13)$$

$$Q = \tfrac{1}{2} \cdot \gamma_w \cdot z^2 \qquad (9.14)$$

Note: Angle ψ_b is negative when sliding uphill

Approximate correction factor f_o

$$f_o = 1 + K(d/L - 1.4(d/L))^2 \qquad (9.15)$$

for $c' = 0$; $K = 0.31$

$c' > 0$, $\phi' > 0$; $K = 0.50$

Figure 9.17 The Janbu modified method of slices for the analysis of non-circular failure in slopes cut into materials in which failure is defined by the Mohr–Coulomb failure criterion.

respectively. If the difference between the calculated and the assumed factors of safety is greater than 0.001, the calculated factor of safety is used as a second estimate of *FS* for a new factor of safety calculation. This process is repeated until the difference between successive factors of safety is less than 0.001. For both the Bishop and Janbu methods, approximately seven iteration cycles will be required to achieve this result for most slope and slide surface geometries.

Step 5: Conditions and corrections – For the Bishop simplified analysis, Figure 9.16 lists two conditions (Equations 9.8 and 9.9) that must be satisfied for each slice in the Bishop analysis. The first condition (Equation 9.8) ensures that the effective normal stress on the base of each slice is always positive. If this condition is not met for any slice, the inclusion of a tension crack into the analysis should be considered. If it is not possible to satisfy this condition by readjustment of the ground water conditions or the introduction of a tension crack, the analysis as presented in Figure 9.16 should be abandoned and a more elaborate form of analysis, to be described later, should be adopted.

Condition 2 (Equation 9.9) in Figure 9.16 was suggested by Whitman and Bailey (1967) and it ensures that the analysis is not invalidated by conditions that can sometimes occur near the toe of a slope in which a deep slide surface has been assumed. If this condition is not satisfied by all slices, the slice dimensions should be changed and, if this fails to resolve the problem, the analysis should be abandoned.

Figure 9.17 for the Janbu simplified method gives a correction factor f_o, which is used in calculating the factor of safety by means of the Janbu method. This factor allows for interslice forces resulting from the shape of the slide surface assumed in the Janbu analysis. The equation for f_o given in Figure 9.17 has been derived by Hoek and Bray (1981) from the curves published in Janbu (1954).

9.6.2 Use of non-linear failure criterion in Bishop stability analysis

When the material in which the slope is cut obeys the Hoek–Brown non-linear failure criterion discussed in Section 5.5, the Bishop simplified method of slices as outlined in Figure 9.18 can be used to calculate the factor of safety. The following procedure is used, once the slice parameters have been defined as described earlier for the Bishop and Janbu analyses:

1. Calculate the effective normal stress σ' acting on the base of each slice by means of the Fellenius equation (Equation 9.19 in Figure 9.18).
2. Using these values of σ', calculate $\tan \phi$ and c for each slice from Equations 5.25 and 5.26.
3. Substitute these values of $\tan \phi$ and c into the factor of safety equation in order to obtain the first estimate of the factor of safety.
4. Use this estimate of *FS* to calculate a new value of σ' on the base of each slice, using the Bishop equation (Equation 9.20 in Figure 9.18).
5. On the basis of these new values of σ', calculate new values for $\tan \phi$ and c.
6. Check that conditions defined by Equations 9.8 and 9.9 in Figure 9.16 are satisfied for each slice.
7. Calculate a new factor of safety for the new values of $\tan \phi$ and c.
8. If the difference between the first and second factors of safety is greater than 0.001, return to step 4 and repeat the analysis, using the second factor of safety as input. Repeat this procedure until the difference between successive factors of safety is less than 0.001.

Generally, about 10 iterations will be required to achieve the required accuracy in the calculated factor of safety.

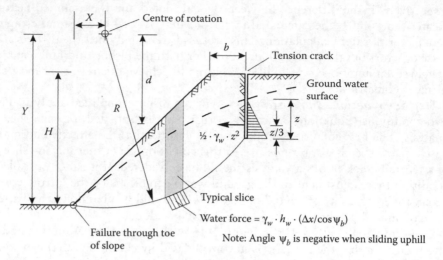

Factor of safety:

$$FS = \frac{\Sigma\,(c_i' + \sigma' \cdot \tan\phi_i')\,(\Delta x/\cos\psi_b)}{\Sigma(\gamma_r \cdot h \cdot \Delta x \cdot \sin\psi_b) + \frac{1}{2} \cdot \gamma_w \cdot z^2 \cdot d/R} \tag{9.16}$$

where

$$\sigma' = \gamma_r \cdot h \cdot \cos^2\psi_b - \gamma_w \cdot h_w \qquad \text{(Fellenius solution)} \tag{9.17}$$

and

$$\sigma' = \frac{\gamma_r \cdot h - \gamma_w \cdot h_w - (c_i' \cdot \tan\psi_b/FS)}{1 + \dfrac{\tan\phi_i' \cdot \tan\psi_b}{FS}} \qquad \text{(Bishop solution)} \tag{9.18}$$

$$\phi_i' = \sin^{-1}\left[\frac{6 \cdot a \cdot m_b(s + m_b \cdot \sigma_{3n}')^{a-1}}{2(1+a)(2+a) + 6 \cdot a \cdot m_b(s + m_b \cdot \sigma_{3n}')^{a-1}}\right] \tag{5.25}$$

$$c_i' = \frac{\sigma_{ci}[(1+2\cdot a)s + (1-a)m_b \cdot \sigma_{3n}'](\sigma + m_b \cdot \sigma_{3n}')^{a-1}}{(1+a)(2+a)\sqrt{1 + [6 \cdot a \cdot m_b(s + m_b \cdot s_{3n}')^{a-1}][(1+a)(2+a)]}} \tag{5.26}$$

where $\sigma_{3n} = \sigma_{3max}'/\sigma_{ci}$; see Equation 5.17 for
parameter a

The conditions which must be satisfied for each slice are:

1) $\sigma' > 0$, where σ' is calculated by Bishop's method

2) $\cos\psi_b\,[1 + (\tan\psi_b \cdot \tan\phi_i')/FS] > 0.2$

Figure 9.18 The Bishop simplified method of slices for the analysis of circular failure of a slope in material where strength is defined by non-linear criterion given in Section 5.5.

9.6.3 Example of the Bishop and Janbu methods of analysis

A slope is to be excavated in strong, blocky sandstone with closely spaced and persistent discontinuities. The slope will consist of three, 15 m (50 ft) high-bench faces with two 8 m (26 ft) wide benches, the primary function of which is to collect surface run-off and control erosion (Figure 9.19a). The bench faces will be at 75° to the horizontal, and the overall slope angle of he benched face is 60°, with the upper slope above the crest of the cut at an

(a)

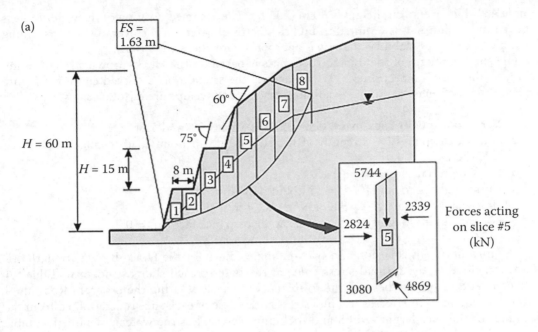

(b)

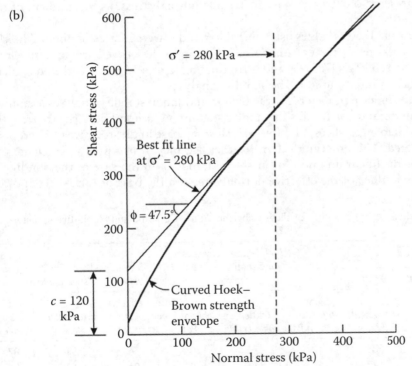

Figure 9.19 Circular slope stability analysis of benched slope in closely jointed sandstone; (a) section of slope showing water table, slice boundaries, tension crack and circular sliding surface for minimum factor of safety; (b) plot of the Hoek–Brown rock mass shear strength showing curved strength envelope and best fit line at a normal stress level of 280 kPa (40.6 psi).

angle 45°. The overall cut height is 60 m (200 ft). The assumed position of the water table is shown in the figure. It is required to find the factor of safety of the overall slope, assuming that a circular-type stability analysis is appropriate for these conditions.

The shear strength of the jointed rock mass is based on the Hoek–Brown strength criterion, as discussed in Section 5.5, which defines the strength as a curved envelope (Figure 9.19b); the rock mass parameters defining the curved envelope are as follows:

1. Very poor quality rock mass, GSI = 20 (see Tables 5.3 and 5.4)
2. Uniaxial compressive strength of intact rock (from point load testing) ≈ 150 MPa (21,800 psi)
3. Rock material constant, $m_i = 13$
4. Unit weight of rock mass, $\gamma_r = 25$ kN/m³ (165 lb/ft³)
5. Unit weight of water, $\gamma_w = 9.81$ kN/m³ (62.4 lb/ft³)
6. For careful blasting used in excavation, disturbance factor $D = 0.7$

The instantaneous cohesion and friction angle values for the Hoek–Brown strength criterion at the effective normal stress value at the base of each slide as shown in Table 9.1 are calculated from the best fit line to the curved envelope using the program RocData 4 (Rocscience Inc). The effective normal stress at the base of each slice is determined from the product of the slice height, width and rock unit weight, less the water pressure. The table shows that c increases and ϕ decreases with increasing normal stress, as the gradient of the best fit line decreases.

For the effective normal stress values listed in Table 9.1, the average stress is about 280 kPa (40.6 psi). At this stress level, the calculated best fit line to the curved strength envelope defines a cohesion of 120 kPa (17.4 psi) and a friction angle of 47.5°; this is the equivalent Mohr–Coulomb shear strength used in the stability analyses.

Table 9.1 shows the input parameters for the Bishop and Janbu stability analyses assuming a linear shear strength, and for the Bishop analysis assuming a non-linear shear strength. The slope is divided into eight slices, and for each slice the base angle, rock weight and pore pressure are calculated. The program SLIDE (Rocscience Inc) is then used to calculate the factors of safety for the linear and non-linear shear strengths. In the case of the non-linear strength analysis, the values of the effective normal stress on the base of each slice, and the

Table 9.1 Calculated shear-strength values of slices using the Mohr–Coulomb and Hoek–Brown failure criteria

Slice member	Slice parameters for all cases (Slice width = 7.03 m)			Mohr–Coulomb values for Bishop and Janbu analyses		Hoek–Brown non-linear failure criterion values for Bishop analysis		
	Angle of slice base (°)	Slice weight (kN)	Pore pressure $\gamma_w \cdot h_w$ (kPa)	Friction angle, ϕ (°)	Cohesion, c (kPa)	Base effective normal stress, σ_n (kPa)	Inst. friction angle, ϕ_i (°)	Inst. cohesion, c_i (kPa)
1	12	1585	10	47.5	120	174	51	87
2	18	2127	56	47.5	120	181	51	90
3	24	3875	94	47.5	120	325	46	133
4	30	4334	123	47.5	120	322	44	132
5	37	5744	140	47.5	120	411	45	157
6	45	5695	126	47.5	120	366	47	145
7	53	5203	92	47.5	120	287	47	122
8	64	3600	42	47.5	120	151	52	79

corresponding instantaneous friction angle and cohesion as shown in the last three columns of Table 9.1, are used in the stability analysis. The calculated factors of safety are:

Bishop simplified method of slices for Mohr–Coulomb shear strength	1.63
Janbu simplified method of slices for Mohr–Coulomb shear strength	1.49
Bishop simplified method of slices for non-linear shear strength	1.61

9.6.4 Circular failure stability analysis computer programs

The circular failure charts discussed in Section 9.3 provide a rapid means of carrying out stability analyses, but are limited to simple conditions as illustrated in the examples. More complex analyses can be carried out using the Bishop and Janbu methods discussed in Section 9.6.3, and the purpose of providing details on the procedures is to show the principles of the analyses.

Computer programs are available to carry out stability analyses of slopes where the circular failure charts are not applicable. Important features included in these programs, which allow them to be used for a wide range of conditions, are as follows:

1. Slope face can include benches and a variety of slope angles.
2. Boundaries between the materials can be positioned to define layers of varying thickness and inclination, or inclusions of any shape.
3. Shear strength of the materials can be defined in terms of the Mohr–Coulomb or Hoek–Brown criterion.
4. Ground water pressures can be defined on single or multiple water tables, or as specified pressure distributions.
5. External loads, in any direction within the plane of the slope cross-section, can be positioned at their correct location on the slope. Such loads can include bridge and building foundations and bolting forces.
6. Earthquake acceleration is applied as a horizontal force in order to carry out pseudo-static stability analysis.
7. The shape and position of the slide surface can be defined as a circular arc or straight line segments.
8. A search routine finds the slide surface with the minimum factor of safety.
9. Deterministic and probabilistic analysis methods calculate the factor of safety and probability of failure, respectively. The probabilistic analysis requires that the design parameters be defined as probability distributions rather than single values.
10. Error messages which identify negative stresses along the slide surface.
11. Drawing of slope showing slope geometry, material boundaries, ground water table(s) and slide surface(s).

One program that contains all these functions is the program XSTABL (Sharma, 1991). An example of the output (partial) produced by XSTABL is shown in Figure 9.20 for a benched slope excavated in sandstone, shale and siltstone above a proposed highway. For these conditions, the shape of the failure surfaces is influenced by the position and thickness of the beds of the weaker material (shale).

9.6.5 Three-dimensional circular failure analysis

Most programs used for LEA (SLOPE/W, XSTABL, SLIDE) analyse the stability of a unit width slice of the slope, which is a two-dimensional analysis that ignores any shear

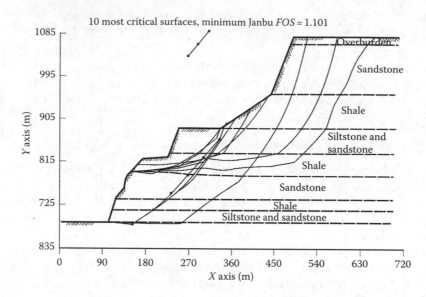

Figure 9.20 Two-dimensional stability analysis of a highway cut in sandstone/shale formation using XSTABL showing 10 lowest factor of safety slip surfaces.

stresses on the sides of the slice; this is the same principle that is used in the plane failure analysis described in Chapter 7. While two-dimensional procedures are a long-established and reliable method of analysis, there are circumstances where three-dimensional analysis will provide a much better representation of slide surface and slope geometry (Hungr, 1987).

One program that provides a three-dimensional analysis is TSLOPE (www.tagasoftcom), which divides the sliding mass into columns, rather than slices as used in the two-dimensional mode. Figure 9.21 shows an example of a model of an open-pit mine slope, with the mesh surface showing the mine benches, and the solid planes showing four faults that bound

Figure 9.21 Benched slope containing four faults modelled in three dimensional using program TSLOPE (Tagasoft, 2016).

an unstable rock mass. In the model, the failure surface is created by making a composite surface of the relevant fault segments.

TSLOPE also allows arbitrary sections to be drawn through the three-dimensional model so that conventional two-dimensional analyses can be carried out for any part of the slide to compare factors of safety for two-dimensional and three-dimensional models. The TSLOPE model can also be incorporated into open-pit mine-planning software packages that incorporate the geological structure.

9.6.6 Numerical slope stability analysis (strength reduction method)

This chapter has been concerned solely with the limit equilibrium method of analysis in which the factor of safety is defined by the ratio of the resisting to the displacing forces on the slide surface.

An alternative method of analysis is to examine the stresses and strains within the slope as a means of assessing stability conditions. If the slope is close to failure, then a zone of high strain will develop within the slope with a shape that will be approximately coincident with the circular slide surface. If the shear-strength properties of the rock are progressively reduced, a sudden increase in the movement along the shear zone will occur when the slope is on the point of failure. The approximate factor of safety of the slope can be calculated from the ratio of the actual shear strength of the rock, to the shear strength at which the sudden movement occurred. This method of stability analysis is termed strength reduction, and the strength ratio is the strength reduction factor (SRF); the method is described in more detail in Section 12.1.1.

9.7 SLOPE STABILITY ANALYSIS IN WEATHERED ROCK

Chapter 3 describes the occurrence, formation and properties of weathered rock. In summary, these materials occur in tropical climates where temperatures are high (>18°C) and heavy rainfall occurs frequently such that the rock is modified, *in situ*, by chemical and physical processes that reduce the strength of the rock mass. Weathering is a gradational process that can range from fresh rock at depth to residual soil at the surface, and is quantified in six grades as described in Table 3.1.

The weathering process is initiated by the flow of water through discontinuities, resulting in the progressive alteration of the rock at the surface of these fractures. As the weathering develops, more of the rock is altered to clay minerals, or is lost to solution, until the entire rock mass becomes a residual soil.

9.7.1 Slope design principles

Section 3.5 describes the principles of slope design for rock cuts that extend through residual soil and weathered rock into fresh rock, requiring a composite design with different stable slope angles in each material. For the weathered rock, the stable slope angle depends on both the strength of the weakened rock mass and the possible presence of remnant geologic structure. For weathered rock with no remnant structure, the design of excavated slopes is carried out using circular stability analysis as described in this chapter. The following section describes the application of circular stability analysis to determine the stable overall slope angles for two rock cuts that are entirely in weathered rock.

For geological conditions where the entire cut is in weathered rock, a multi-bench design is usually required, as shown in Figures 9.22 and 9.23, where the overall face angle

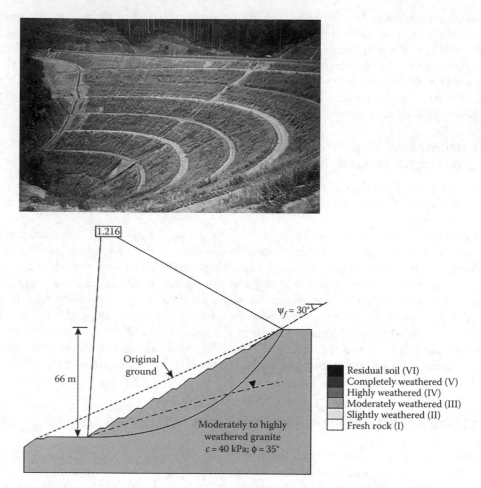

Figure 9.22 Circular stability analysis of 66 m (215 ft) high cut in moderately and highly weathered granite (Malaysia).

required for stability defines the combination of bench face height, bench width and inter-bench face angle.

Fundamental to the slope designs shown in Figures 9.22 and 9.23 is the incorporation of benches on the slope face to control surface water flows that would otherwise erode these low-strength materials (see Figure 1.2). Since weathered rock generally occurs in tropical climates where periods of intense rainfall are common, it is necessary to develop the benches as construction proceeds. A common vertical spacing between benches is 6–8 m (20–26 ft). Each bench incorporates a ditch lined with concrete or masonry that connects with a system of channels running down the face that convey the run-off water to a disposal area. Often the down-slope channels contain energy-dissipating concrete or masonry blocks to control the velocity of water flowing down the channels.

9.7.2 Slope designs in deep weathered rock

Figures 9.22 and 9.23 show two substantial excavations in weathered rock, where the overall face angle depends on the grade(s) of weathering and the corresponding rock mass strength(s), and the ground water pressures. Both these excavations were made in tropical

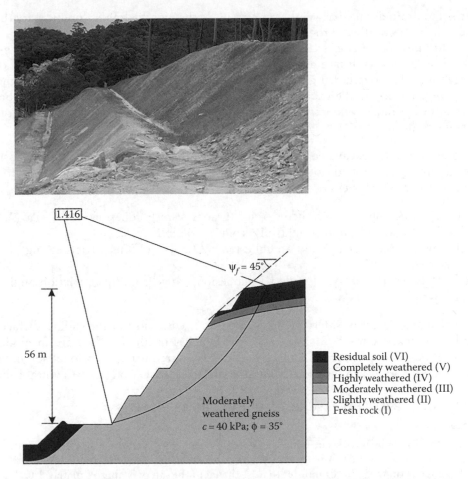

Figure 9.23 (**See colour insert.**) Circular stability analysis of 56 m (185 ft) high cut in gradational weathered gneissic rock (Sao Paulo, Brazil).

areas where extensive experience in the design of excavations in varying grades of weathered rock has been developed. This work had provided reliable information on the shear-strength properties of weathered rock masses. The stability analyses were carried out using the Bishop simplified method with the program Slide 5.0 (Rocscience Inc).

1. *Cut in weathered granite* (Figure 9.22) – 66 m (215 ft) high cut in weathered granite was made essentially parallel to the original ground surface, predominately in moderately to highly weathered rock (grades III/IV). The average shear strength of these materials that apply over the full height of the cut was (see Figure 5.25):

Cohesion = 40 kPa (5.8 psi), friction angle = 35°, unit weight = 20 kN/m³ (127 lb/ft³)

A circular stability analysis of this slope at an overall angle of 30°, with the water table at the location on the drawing, gave a factor of safety of 1.22. The drawing shows the critical circle for this slope, which is at a maximum depth of about 24 m (80 ft) below the slope surface. The stability of this slope is sensitive to the level of the water table that is likely to fluctuate significantly in response to periods of intense rainfall, and then to drain during dry weather as a result of the high-hydraulic conductivity that can occur in the weathered rock.

2. *Cut in weathered gneiss* (Figure 9.23) – 56 m (185 ft) high cut in weathered gneissic rock is at an overall face angle of 45° and was excavated through a sequence of progressively less weathered gneissic rocks. The ground water table was below the level of the cut because the excavation was made into a hillside above the valley bottom, and the hydraulic conductivity of the moderately weathered rock was sufficient for the slope to drain.

The stability of the excavated slope was studied using circular failure analysis assuming that structural geology had no influence on stability. The shear-strength values for each class of weathered rock were as follows:

Residual soil/completely weathered rock (grades VI/V): cohesion = 16.5 kPa (2.4 psi), friction angle = 25°, unit weight = 17 kN/m³ (108 lb/ft³)

Highly weathered rock (grade IV): cohesion = 22.5 kPa (3.3 psi), friction angle = 28.5°, unit weight = 19 kN/m³

Moderately weathered rock (grade III): cohesion = 40 kPa (5.8 psi), friction angle = 35°, unit weight = 20 kN/m³

The calculated factor of safety of this slope is 1.4, with most of the sliding surface lying in grade III rock. Figure 9.23 shows that the benches on the cut face are in moderately weathered rock where minor local instability occurred, requiring occasional soil nailing to maintain the design bench face angles and the drainage ditches on the horizontal benches.

EXAMPLE PROBLEM 9.1: CIRCULAR FAILURE ANALYSIS

Statement

A 22 m (72 ft) high rock cut with a face angle of 60° has been excavated in a massive, very weak volcanic tuff. A tension crack has opened behind the crest and it is likely that the slope is on the point of failure, that is, the factor of safety is approximately 1.0. The friction angle of the material is estimated to be 30°, its density is 25 kN/m³ (160 lb/ft³), and the position of the water table is shown on the sketch of the slope (Figure 9.24). The rock contains no continuous joints dipping out of the face, and the most likely type of failure mode is circular failure.

Required

1. Carry out a back-analysis of the failure to determine the limiting value of the cohesion when the factor of safety is 1.0.
2. Using the strength parameters calculated in (1), determine the factor of safety for a completely drained slope. Would drainage of the slope be a feasible method of stabilisation?
3. Using the ground water level shown in Figure 9.24 and the strength parameters calculated in (1), calculate the reduction in slope height, that is, amount of unloading of the slope crest required to increase the factor of safety to 1.3.
4. For the slope geometry and ground water level shown in Figure 9.24, find the coordinates of the centre of the critical circle and the position of the critical tension crack.

Solution

1. The ground water level shown in Figure 9.24 corresponds to ground water condition 3 in Figure 9.4, so circular failure chart number 3 (Figure 9.8) is used in the analysis. When $\phi = 30°$ and $FS = 1.0$, $\tan \phi/FS = 0.58$.

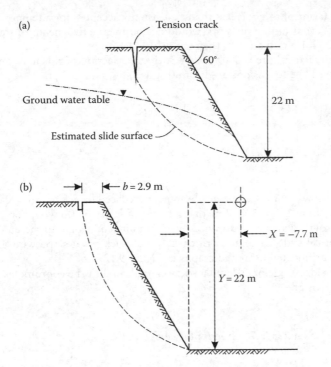

Figure 9.24 Slope geometry for Example Problem 9.1: (a) slope geometry for circular failure with ground water table corresponding to circular failure chart number 3; (b) position of critical slide surface and critical tension crack.

The intersection of this value for tan ϕ/FS and the curve for a slope angle of 60° gives

$$\frac{c}{\gamma \cdot H \cdot FS} = 0.086$$

$$\therefore c = 0.086 \cdot 25 \cdot 22 \cdot 1.0$$
$$= 47.3\,\text{kPa}$$

2. If the slope were completely drained, circular failure number chart 1 could be used for analysis.

$$\frac{c}{\gamma \cdot H \cdot \tan \phi} = \frac{47.3}{25 \cdot 22 \cdot \tan(30)}$$
$$= 0.15$$

The intersection of this inclined line with the curved line for a slope angle of 60° gives

$$\frac{\tan \phi}{FS} = 0.52$$

$$\therefore FS = \frac{\tan 30}{0.52}$$
$$= 1.11$$

This factor of safety is less than that usually accepted for a temporary slope (i.e. $FS \approx 1.2$), so draining the slope would not be an effective means of stabilisation.

3. When $FS = 1.3$ and $\phi = 30°$, then $\tan \phi / FS = 0.44$.

   On circular failure chart number 3, the intersection of this horizontal line with the curved line for a slope angle of 60° gives

$$\frac{c}{\gamma \cdot H \cdot FS} = 0.11$$
$$\therefore H = \frac{47.3}{25 \cdot 1.3 \cdot 0.11}$$
$$= 13.2 \, \text{m}$$

This shows that the slope height must be reduced by 8.8 m to increase the factor of safety from 1.0 to 1.3. Note that a factor of safety of 1.3 would only be achieved if the ground water level dropped by an amount equivalent to the unloading.

4. The critical slide circle and critical tension crack for a slope with ground water present are located using the graphs in Figure 9.12.

   For a slope angle of 60° and a friction angle of 30°, the coordinates of the centre of the circle are:

$X = -0.35 \cdot H$

$= -7.7 \, \text{m}$, that is, 7.7 m horizontally beyond the toe

$Y = H$

$= 22 \, \text{m}$, that is, 22 m above the toe

The location of the tension crack behind the crest is

$b/H = 0.13$

$b = 2.9 \, \text{m}$

This critical circle is shown in Figure 9.24b.

(a)

(b)

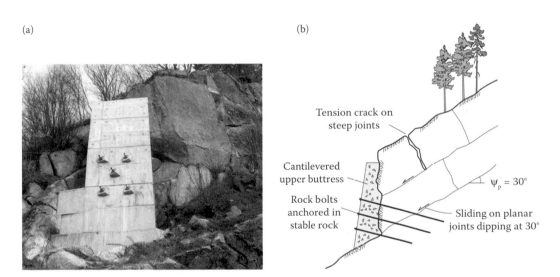

Tension crack on
steep joints

Cantilevered
upper buttress

Rock bolts
anchored in
stable rock

$\psi_p = 30°$

Sliding on planar
joints dipping at 30°

Figure 1.1 Rock slope in very strong granite in which upper blocks of rock can slide from right to left on planar joints dipping at about 30° (near Agassiz, British Columbia, Canada). (a) Concrete buttress and tie-back anchors; (b) slope geometry showing dimensions of sliding blocks, bolt lengths and buttress configuration (Gygax Engineering).

Figure 1.4 Cut face coincident with continuous, low-friction bedding planes in shale (Trans Canada Highway near Lake Louise, Alberta). (Image by A. J. Morris.)

Figure 2.1 Rock faces formed by persistent discontinuities: (a) plane failure formed by bedding planes in shale parallel to face with continuous lengths over the full height of the slope on Route 19 near Robbinsville, North Carolina; (b) wedge failure formed by two intersecting planes in sedimentary formation dipping out of the face on Route 60 near Phoenix, Arizona. (Image by C.T. Chen.)

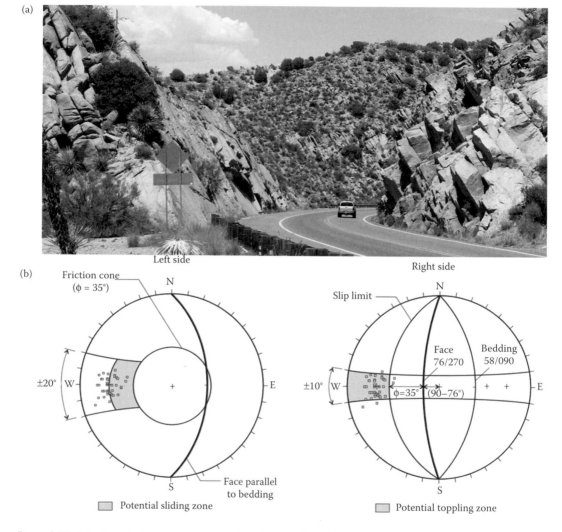

Figure 2.20 Relationship between structural geology and stability conditions on slope faces in through-cut: (a) photograph of through-cut showing two failure mechanisms – plane sliding on left side (west) and toppling on right side (east) on Route 60 near Globe, AZ (image by C. T. Chen); (b) stereographic plots showing kinematics analysis of left and right cut slopes.

Figure 3.7 Weathering sequence in granite showing irregular surface of fresh granite underlying weathered rock and residual soil (Sao Paulo, Brazil).

Figure 3.9 Columnar basalt flow overlying weathered soil that has undermined the basalt (Sea to Sky Highway, British Columbia, Canada).

(a)

(b)

Figure 3.15 Typical weathered limestone formation: (a) irregular ground surface and cavities of varying dimensions; (b) closeup of weathered rock showing network of fine cavities at Cottesloe Beach, Perth, Western Australia. (Image by I. McDougall.)

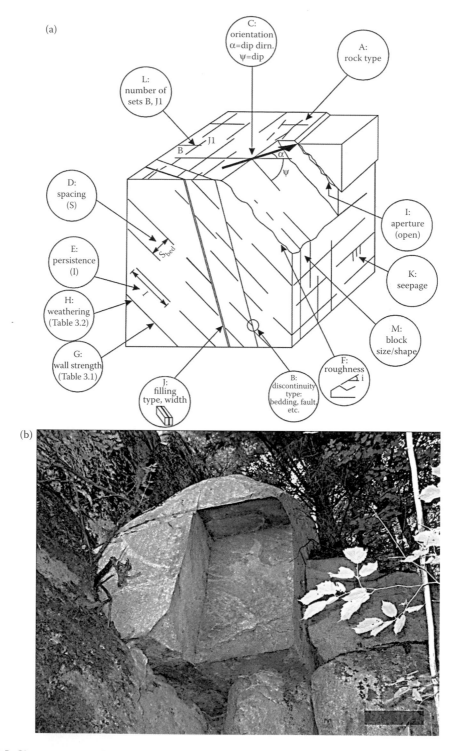

Figure 4.5 Characteristics of discontinuities in rock masses: (a) parameters describing the rock mass; letter ('A', etc.) refers to description parameters in text (Wyllie, 1999). (b) Photograph of blocky granite containing three joints orientated in orthogonal directions (near Hope, British Columbia).

Figure 7.1 Plane failure on smooth, persistent bedding planes in shale (Interstate 40, near Newport, Tennessee).

Figure 7.7 A tension crack behind a sliding rock mass in which significant horizontal displacement has occurred (above: Kooteney Lake, British Columbia).

Figure 8.1 Typical wedge failure involving sliding on two persistent joints with line of intersection of joints daylighting at toe of rock face, and an upper plane that formed a tension crack (strong, volcanic rock on Interstate 5, near Grants Pass, Oregon).

Figure 9.1 Circular failure in highly weathered, granitic rock (Highway 1, near Devil's Slide, Pacifica, California).

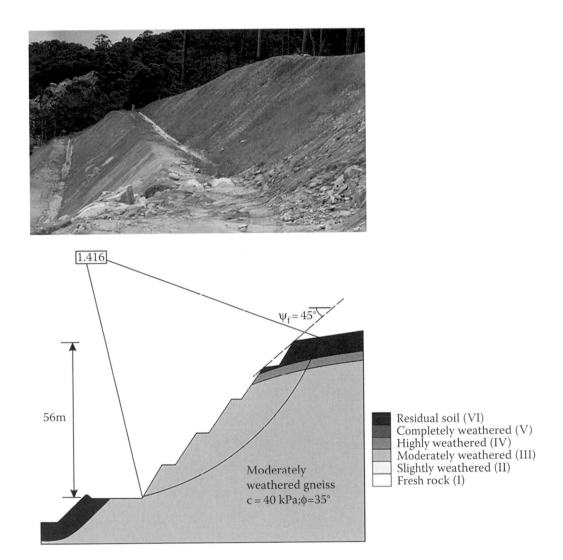

Figure 9.23 Circular stability analysis of 56 m (185 ft) high cut in gradational weathered gneissic rock (Sao Paulo, Brazil).

Figure 10.1 Toppling blocks in columnar basalt showing typical tension cracks with widths greater at the top and narrowing near the base (Sea to Sky Highway near Whistler, British Columbia, Canada).

Figure 13.10 Example of controlled blasting, with drill 'offset' at mid-height for rock cut on highway project (strong granite, Sea to Sky Highway, British Columbia, Canada).

Figure 14.17 Construction of reinforced concrete grillage to support cut slope in weathered rock (near Niigata, Japan).

Chapter 10

Toppling failure

10.1 INTRODUCTION

The failure modes discussed in the three previous chapters all relate to sliding of a rock or soil mass along an existing or induced sliding surface. This chapter discusses a different failure mode – that of toppling, which involves rotation of columns or blocks of rock about a fixed base. Similarly to plane and wedge failure, the stability analysis of toppling failures involves first carrying out a kinematic analysis of the structural geology to identify potential toppling conditions, and then if this condition exists, performing a stability analysis specific to toppling failures. Figure 10.1 shows typical toppling blocks in columnar basalt.

One of the earliest references to toppling failures is by Muller (1968) who suggested that block rotation or toppling may have been a contributory factor in the failure of the north face of the Vajont slide (Figure 10.2). Hofmann (1972) carried out a number of model studies under Muller's direction to investigate block rotation. Similar model studies were carried out by Ashby (1971), Soto (1974) and Whyte (1973), while Cundall (1971), Byrne (1974) and Hammett (1974) incorporated rotational failure modes into the computer analysis of rock mass behaviour. Figure 10.3 shows a computer model of a toppling failure in which the solid blocks are fixed and the open blocks are free to move. When the fixed blocks at the face are removed, the tallest columns of blocks topple because their centre of gravity lies outside the base. The model illustrates a typical feature of toppling failures in which the tension cracks are wider at the top than at the base. This condition, which can best be observed when looking along the strike, is useful in the field identification of topples.

Papers concerning field studies of toppling failures include de Freitas and Watters (1973) who discuss slopes in Britain, and Wyllie (1980) who demonstrates stabilisation measures for toppling failures related to railway operations.

Most of the discussion that follows in this chapter is based on a paper by Goodman and Bray (1976) in which a formal mathematical solution to a simple toppling problem is shown. This solution, which is reproduced here, represents a basis for designing rock slopes in which toppling is present, and has been further developed into a more general design tool (Zanbak, 1983; Adhikary, Dyskin, Jewell et al., 1997; Bobet, 1999; Sagaseta, Sánchez and Cañizal, 2001).

10.2 TYPE OF TOPPLING FAILURES

Goodman and Bray (1976) have described a number of different types of toppling failures that may be encountered in the field, and each is discussed briefly in the following pages. The importance of distinguishing between types of toppling is that two distinct

Figure 10.1 **(See colour insert.)** Toppling blocks in columnar basalt showing typical tension cracks with widths greater at the top and narrowing near the base (Sea to Sky Highway near Whistler, British Columbia, Canada).

Figure 10.2 Suggested toppling mechanism of the north face of Vajont slide (Müller, 1968).

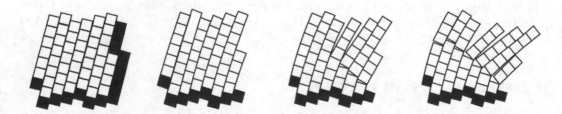

Figure 10.3 Computer-generated model of toppling failure; solid blocks are fixed in space while open blocks are free to move (Cundall, 1971).

methods of stability analysis for toppling failures can occur as described in the following pages – block and flexural toppling – and it is necessary to use the appropriate analysis in design.

10.2.1 Block toppling

As illustrated in Figure 10.4a, block toppling occurs when, in a strong rock, individual columns are formed by a set of discontinuities dipping steeply into the face, and a second set of widely spaced orthogonal joints defines the column height. The short columns forming the toe of the slope are pushed forward by the loads from the longer overturning columns behind, and this sliding of the toe allows further toppling to develop higher up on the slope. The base of the failure generally consists of a stepped surface rising from one cross-joint to the next. Typical geological conditions in which this type of failure may occur are bedded sandstone and columnar basalt in which orthogonal jointing is well developed.

10.2.2 Flexural toppling

The process of flexural toppling is illustrated in Figure 10.4b that shows continuous columns of rock, separated by well-developed, steeply dipping discontinuities, breaking in flexure as they bend forward. Typical geological conditions in which this type of failure may occur are thinly bedded shale and slate in which orthogonal jointing is not well developed. Generally, the basal plane of a flexural topple is not as well defined as a block topple.

Sliding, excavation or erosion of the toe of the slope allows the toppling process to start and it retrogresses back into the rock mass with the formation of deep tension cracks that

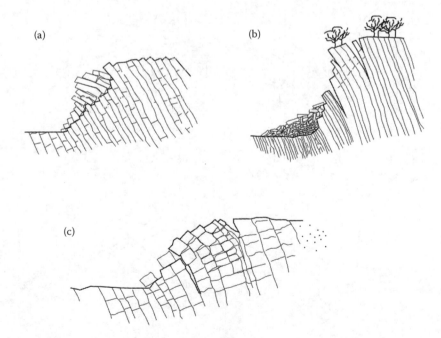

Figure 10.4 Common classes of toppling failures: (a) block toppling of columns of rock containing widely spaced orthogonal joints; (b) flexural toppling of slabs of rock dipping steeply into face; (c) block flexure toppling characterised by pseudo-continuous flexure of long columns through accumulated motions along numerous cross-joints (Goodman and Bray, 1976).

become narrower with depth. The lower portion of the slope is covered with disordered fallen blocks and it is sometimes difficult to recognise a toppling failure from the bottom of the slope. Detailed examination of toppling slopes shows that the outward movement of each cantilevered column produces an interlayer slip and a portion of the upper surface of each plane is exposed in a series of back-facing, or obsequent scarps, such as those illustrated in Figure 10.4a.

10.2.3 Block–flexure toppling

As illustrated in Figure 10.4c, block–flexure toppling is characterised by pseudo-continuous flexure along long columns that are divided by numerous cross-joints. Instead of the flexural failure of continuous columns resulting in flexural toppling, toppling of columns in this case results from accumulated displacements on the cross-joints. Because of the large number of small movements in this type of topple, fewer tension cracks develop under these conditions than in flexural toppling, and fewer edge-to-face contacts and voids than in block toppling.

10.2.4 Secondary toppling modes

Figure 10.5 illustrates a number of possible secondary toppling mechanisms suggested by Goodman and Bray (1976). In general, these failures are initiated by undercutting of the toe of the slope, either by natural agencies such as scour or weathering, or by human activities. In all cases, the primary failure mode involves sliding or physical breakdown of the rock,

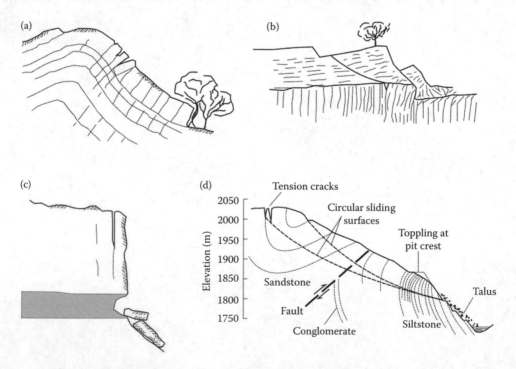

Figure 10.5 Secondary toppling modes: (a) Toppling at head of slide. (b) Toppling at toe of slide with shear movement of upper slope (Goodman and Bray, 1976). (c) Toppling of columns in strong upper material due to weathering of underlying weak material. (d) Toppling at pit crest resulting in circular failure of upper slope (Wyllie and Munn, 1979).

and toppling is induced in the upper part of the slope as a result of this primary failure (Figure 10.5a and b).

Figure 10.5c illustrates a common occurrence of toppling failure in horizontally bedded sandstone and shale formations. The shale is usually significantly weaker and more susceptible to weathering than the sandstone, while the sandstone often contains vertical stress relief joints. As the shale weathers, support for the sandstone is undermined and columns of sandstone, with their dimensions defined by the spacing of the vertical joints, topple from the face. At some locations, the overhangs can be as wide as 5 m (16 ft), and failures of substantial volumes of rock can occur with little warning.

An example of a combined toppling and circular slide is shown in the failure of a pit slope in a coal mine where the beds at the crest of the pit dipped at 70° into the face, and their strike was parallel to the face (Figure 10.5d). Mining of the pit slope at an angle of 50° initiated a toppling failure at the crest of the pit where movement monitoring showed that the columns of sandstone initially moved upwards and towards the pit. This movement resulted in a circular failure that extended to a height of 230 m (750 ft) above the base of the topple. Detailed monitoring of the slope showed that a total movement of about 30 m (100 ft) occurred on the slope above the pit, resulting in cracks opening in the crest of the mountain that were several metres wide and up to 9 m (30 ft) deep. As described in Section 15.7.1, continuous movement monitoring was used to allow mining to proceed under the moving slope, and finally the slope was stabilised by back-filling the pit (Wyllie and Munn, 1979).

A further example of the toppling mechanism is illustrated in Figure 10.6 (Sjöberg, 2000). In open pit mines where the depth of the slope progressively increases, minor toppling movement may eventually develop into a substantial failure. Careful monitoring of

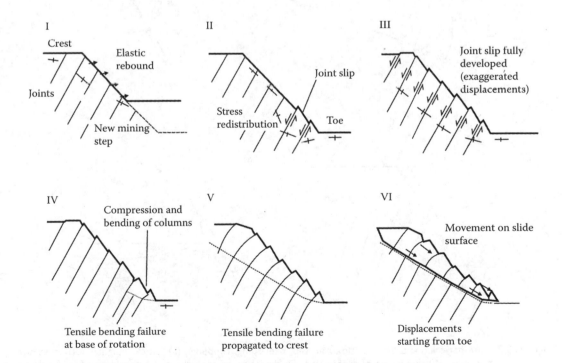

Figure 10.6 Failure stages for large-scale toppling failure in a slope (Sjöberg, 2000).

the movement, and recognition of the toppling mechanism, can be used to anticipate when hazardous conditions are developing.

10.3 KINEMATICS OF BLOCK TOPPLING FAILURE

The potential for toppling can be assessed from two kinematic tests described in this section. These tests examine first the shape of the block, and second the relationship between the dip of the planes forming the slabs and the face angle. It is emphasised that these two tests are useful for identifying potential toppling conditions, but the tests cannot be used alone as a method of stability analysis.

10.3.1 Block shape test

The basic mechanics of the stability of a block on a plane are illustrated in Figure 10.7a (see also Figure 1.11). This diagram shows the conditions that differentiate stable, sliding or toppling blocks with height y and width Δx on a plane dipping at an angle ψ_p. If the

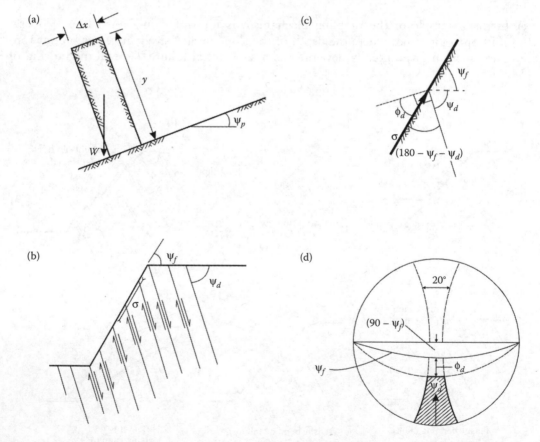

Figure 10.7 Kinematic conditions for flexural slip preceding toppling: (a) block height/width test for toppling; (b) directions of stress and slip direction in rock slope; (c) conditions for interlayer slip; (d) kinematic test defined on lower hemisphere stereographic projection.

friction angle between the base of the block and the plane is ϕ_p, then the block will be stable against sliding when the dip of the base plane is less than the friction angle, that is, when

$$\psi_p < \phi_p - \text{Stable} \tag{10.1}$$

but will topple when the centre of gravity of the block lies outside the base, that is, when

$$\frac{\Delta x}{y} < \tan \psi_p - \text{Topple} \tag{10.2}$$

For example, for a 3 m (10 ft) wide block on a base plane dipping at 10°, toppling will occur if the height exceeds 17 m (56 ft).

10.3.2 Interlayer slip test

A requirement for toppling to occur in the mechanisms shown in Figures 10.4 and 10.6 is shear displacement on the face-to-face contacts on the front and back faces of the blocks. Sliding on these faces will occur if the following conditions are met (Figure 10.7b). The state of stress close to the slope face is uniaxial with the direction of the normal stress σ aligned parallel to the slope face. When the layers slip past each other, σ must be inclined at an angle ϕ_d with the normal to the layers, where ϕ_d is the friction angle of the sides of the blocks. If ψ_f is the dip of slope face and ψ_d is the dip of the planes forming the sides of the blocks, then the condition for interlayer slip is given by (Figure 10.7c)

$$(180 - \psi_f - \psi_d) \geq (90 - \phi_d) \tag{10.3}$$

or

$$\psi_d \geq (90 - \psi_f) + \phi_d \tag{10.4}$$

10.3.3 Block alignment test

The other kinematic condition for toppling is that the strike of the planes forming the blocks is approximately parallel to the slope face so that each layer is free to topple with little constraint from adjacent layers. Observations of topples in the field show that instability is possible where the dip direction of the planes forming the sides of the blocks α_d is within about 10° of the dip direction of the slope face α_f, or

$$\left| (\alpha_f - \alpha_d) \right| < 10° \tag{10.5}$$

The two conditions defining kinematic stability of topples given by Equations 10.4 and 10.5 can be depicted on the stereonet (Figure 10.7d). On the stereonet, toppling is possible for planes for which the poles lie within the shaded area, provided also that the base friction properties and shape of the blocks meet the conditions given by Equations 10.1 and 10.2, respectively.

10.4 LIMIT EQUILIBRIUM ANALYSIS OF TOPPLING ON A STEPPED BASE

The method of toppling analysis described in this section utilises the same principles of limiting equilibrium that have been used throughout this book. While this method of analysis is limited to a few simple cases of toppling failure, it provides a basic understanding of the factors which are important in toppling, and allows stabilisation options to be evaluated. The stability analysis involves an iterative process in which the dimensions of all the blocks and the forces acting on them are calculated, and then the stability of each is examined, starting at the uppermost block. Each block will be stable, toppling or sliding, and the overall slope is considered unstable if the lowermost block is either sliding or toppling. A basic requirement of this analysis is that the friction angle on the base of each block is greater than the dip angle of the base so that sliding on the base plane does not occur in the absence of any external force acting on the block (see Equation 10.1).

The limit equilibrium method of analysis is ideally suited to incorporating external forces acting on the slope to simulate a wide variety of actual conditions that may exist in the field. For example, if the lower block or blocks are unstable, then tensioned anchors with a specified tensile strength and plunge can be installed in these blocks to prevent movement. Also, ground motion due to earthquakes can be simulated by a pseudo-static force acting on each block (see Section 11.6), water forces can act on the base and sides of each block, and loads produced by bridge foundations can be added to any specified block (Wyllie, 1999).

As an alternative to the detailed analysis described in this section, Zanbak (1983) developed a series of design charts that can be used to identify unstable toppling slopes and estimate the support force required for limiting equilibrium.

10.4.1 Angle of base plane (ψ_b)

The base of the toppling blocks is a stepped surface with an overall dip of ψ_b (Figure 10.8). Note that no explicit means of determining a value of the parameter ψ_b has been found. However, it is necessary to use an appropriate value for ψ_b in the analysis because this has a significant effect on the stability of the slope. That is, as the base angle becomes flatter, the lengths of the blocks increase and the tendency of the taller blocks to topple increases, resulting in decreased stability of the slope. If the base angle is coincident with the base of the blocks (i.e. $\psi_b = \psi_p$), then the geometry of toppling requires dilatancy δ of the blocks along the base plane and shearing on the faces of the blocks (Figure 10.9). However, if the base is stepped (i.e. $\psi_b > \psi_p$), then each block can topple without dilatancy, provided the displacement occurs on the face-to-face contacts (Figure 10.8). It is expected that more energy is required to dilate the rock mass than to develop shear along existing discontinuities, and so a stepped base is more likely than a planar base. The examination of base friction, centrifugal and numerical models (Goodman and Bray, 1976; Pritchard and Savigny, 1990, 1991; Adhikary, Dyskin, Jewell et al., 1997) shows that base planes tend to be stepped, and the approximate dip angle is in the range

$$\psi_b \approx (\psi_p + 10°) \text{ to } (\psi_p + 30°) \tag{10.6}$$

It is considered that an appropriate stability analysis procedure for situations where the value of ψ_b is unknown is to carry out a sensitivity analysis within the range given by Equation 10.6 and find the value of ψ_b that gives the least stable condition.

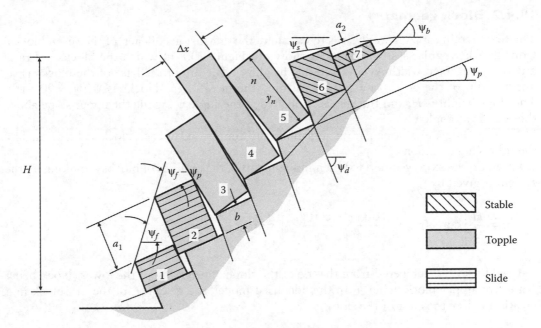

Figure 10.8 Model for limiting equilibrium analysis of toppling on a stepped base (Goodman and Bray, 1976).

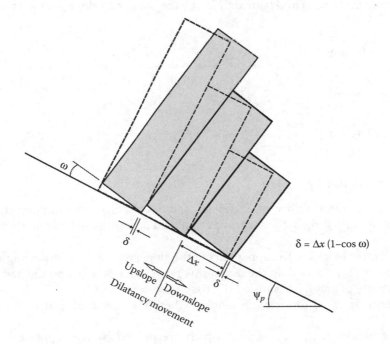

Figure 10.9 Dilatancy of toppling blocks with base plane coincident with normal to dip of blocks (Zanbak, 1983).

10.4.2 Block geometry

The first step in toppling analysis is to calculate the dimensions of each of the seven blocks. Consider the regular system of blocks shown in Figure 10.8 in which the blocks are rectangular with fixed width Δx and variable height y_n. The dip of the base of the blocks is ψ_p and the dip of the orthogonal planes forming the faces of the blocks is $\psi_d (\psi_d = 90 - \psi_p)$. The slope height is H, and the face is excavated at the angle ψ_f, while the upper slope above the crest is at angle ψ_s.

These geologic and slope parameters allow the dimensions of each block to be defined by the following equations.

Based on the slope geometry shown in Figure 10.8, the number of blocks n making up the system is given by

$$n = \frac{H}{\Delta x}\left[\text{cosec}(\psi_b) + \left(\frac{\cot(\psi_b) - \cot(\psi_f)}{\sin(\psi_b - \psi_f)} \right) \cdot \sin(\psi_s) \right] \qquad (10.7)$$

The blocks are numbered from the toe of the slope upwards, with the lowest block being 1 and the upper block being n. In this idealised model, the height y_n of the nth block in a position below the crest of the slope is

$$y_n = n(a_1 - b) \qquad (10.8)$$

while above the crest

$$y_n = y_{n-1} - a_2 - b \qquad (10.9)$$

The three constants a_1, a_2 and b are defined by the block and slope geometry and are given by

$$a_1 = \Delta x \cdot \tan(\psi_f - \psi_p) \qquad (10.10)$$

$$a_2 = \Delta x \cdot \tan(\psi_p - \psi_s) \qquad (10.11)$$

$$b = \Delta x \cdot \tan(\psi_b - \psi_p) \qquad (10.12)$$

10.4.3 Block stability

Figure 10.8 shows the stability of a system of blocks subject to toppling, in which it is possible to distinguish three separate groups of blocks according to their mode of behaviour:

1. A set of stable blocks in the upper part of the slope, where the friction angle of the base of the blocks is greater than the dip of this plane (i.e. $\phi_p > \psi_p$), and the height is limited so the centre of gravity lies inside the base ($\Delta x/y > \tan \psi_p$).
2. An intermediate set of toppling blocks, where the centre of gravity lies outside the base.
3. A set of blocks in the toe region, which are pushed by the toppling blocks above. Depending on the slope and block geometries, the toe blocks may be stable, toppling or sliding.

Figure 10.10 demonstrates the terms used to define the dimensions of the blocks, and the positions and directions of all the forces acting on the blocks during both toppling and sliding. Figure 10.10a shows a typical block (n) with the normal and shear forces developed on the base (R_n, S_n), and on the interfaces with adjacent blocks (P_n, Q_n, P_{n-1}, Q_{n-1}). When the block is one of the toppling set, the points of application of all forces are known, as shown in Figure 10.10b. The points of application of the normal forces P_n are M_n and L_n on the upper and lower faces, respectively, of the block, and are given by the following relationships.

If the nth block is below the slope crest:

$$M_n = y_n \tag{10.13}$$

$$L_n = y_n - a_1 \tag{10.14}$$

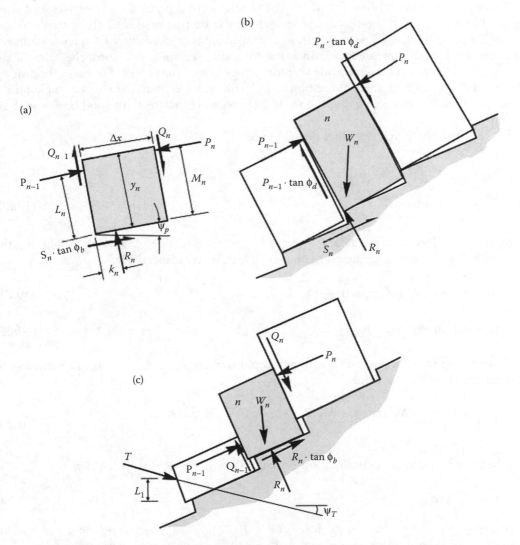

Figure 10.10 Limiting equilibrium conditions for toppling and sliding of nth block: (a) forces acting on nth block; (b) toppling of nth block; (c) sliding of nth block (Goodman and Bray, 1976).

If the nth block is the crest block:

$$M_n = y_n - a_2 \qquad\qquad (10.15)$$

$$L_n = y_n - a_1 \qquad\qquad (10.16)$$

If the nth block is above the slope crest:

$$M_n = y_n - a_2 \qquad\qquad (10.17)$$

$$L_n = y_n \qquad\qquad (10.18)$$

For an irregular array of blocks, y_n, L_n and M_n can be determined graphically.

When sliding and toppling occur, frictional forces are generated on the bases and sides of the blocks. In many geological environments, the friction angles on these two surfaces are likely to be different. For example, in a steeply dipping sedimentary sequence comprising sandstone beds separated by thin seams of shale, the shale will form the sides of the blocks, while joints in the sandstone will form the bases of the blocks. For these conditions, the friction angle of the sides of the blocks (ϕ_d) will be lower than the friction angle on the bases (ϕ_p). These two friction angles can be incorporated into the limit equilibrium analysis as follows:

For limiting friction on the sides of the block:

$$Q_n = P_n \cdot \tan\phi_d \qquad\qquad (10.19)$$

$$Q_{n-1} = P_{n-1} \cdot \tan\phi_d \qquad\qquad (10.20)$$

By resolving forces perpendicular and parallel to the base of a block with weight W_n, the normal and shear forces acting on the base of block n are, respectively,

$$R_n = W_n \cdot \cos\psi_p + (P_n - P_{n-1}) \cdot \tan\phi_d \qquad\qquad (10.21)$$

$$S_n = W_n \cdot \sin\psi_p + (P_n - P_{n-1}) \qquad\qquad (10.22)$$

Considering rotational equilibrium, it is found that the force P_{n-1} that is just sufficient to prevent toppling has the value

$$P_{n-1,t} = \frac{P_n(M_n - \Delta x \cdot \tan\phi_d) + (W_n/2) \cdot (y_n \cdot \sin\psi_p - \Delta x \cdot \cos\psi_p)}{L_n} \qquad\qquad (10.23)$$

When the block under consideration is one of the sliding set (Figure 10.10c),

$$S_n = R_n \cdot \tan\phi_p \qquad\qquad (10.24)$$

However, the magnitudes of the forces Q_{n-1}, P_{n-1} and R_n applied to the sides and base of the block, and their points of application L_n and K_n, are unknown. Although the problem is indeterminate, the force P_{n-1} required to prevent sliding of block n can be determined if it is assumed that $[Q_{n-1} = (\tan \phi_d \cdot P_{n-1})]$. Then the shear force just sufficient to prevent sliding has the value

$$P_{n-1,s} = P_n - \frac{W_n(\cos \psi_p \cdot \tan \phi_p - \sin \psi_p)}{(1 - \tan \phi_p \cdot \tan \phi_d)} \tag{10.25}$$

10.4.4 Calculation procedure for toppling stability of a system of blocks

The calculation procedure for examining the toppling stability of a slope comprising a system of blocks dipping steeply into the face is as follows:

1. The dimensions of each block and the number of blocks are defined using Equations 10.7 to 10.12.
2. Values for the friction angles on the sides and base of the blocks (ϕ_d and ϕ_p) are assigned based on laboratory testing, or inspection. The friction angle on the base should be greater than the dip of the base to prevent sliding (i.e. $\phi_p > \psi_p$).
3. Starting with the top block, Equation 10.2 is used to identify if toppling will occur, that is, when ($\Delta x/y < \tan \psi_p$). For the upper toppling block, Equations 10.23 and 10.25 are used to calculate the lateral forces required to prevent toppling and sliding, respectively.
4. Let n_1 be the uppermost block of the toppling set.
5. Starting with block n_1, determine the lateral forces $P_{n-1,t}$ required to prevent toppling, and $P_{n-1,s}$ to prevent sliding. If $P_{n-1,t} > P_{n-1,s}$, the block is on the point of toppling and P_{n-1} is set equal to $P_{n-1,t}$, or if $P_{n-1,s} > P_{n-1,t}$, the block is on the point of sliding and P_{n-1} is set equal to $P_{n-1,s}$.

    In addition, a check is made that a normal force R acts on the base of the block, and that sliding does not occur on the base, that is,

    $$R_n > 0; \quad \text{and} \quad (|S_n| > R_n \cdot \tan \phi_p)$$

6. The next lower block ($n_1 - 1$) and all the lower blocks are treated in succession using the same procedure. It may be found that a relatively short block in the lower part of the slope that does not satisfy Equation 10.2 for toppling may still topple if the moment applied by the thrust force on the upper face is great enough to satisfy the condition stated in (5) above. If the condition $P_{n-1,t} > P_{n-1,s}$ is met for all blocks, then toppling extends down to block 1 and sliding does not occur.
7. Eventually, a block may be reached for which $P_{n-1,s} > P_{n-1,t}$. This establishes block n_2, and for this and all lower blocks, the critical state is one of sliding. The stability of the sliding blocks is checked using Equation 10.24, with the block being unstable if ($S_n = R_n \cdot \tan \phi_b$). If block 1 is stable against both sliding and toppling (i.e. the resultant force acting on the last block, $P_0 < 0$), then the overall slope is considered to be stable. If block 1 either topples or slides (i.e. $P_0 > 0$), then the overall slope is considered to be unstable.

10.4.5 Cable force required to stabilise a slope

If the calculation process described in Section 10.4.4 shows that block 1 is unstable, then a tensioned cable can be installed through this block and anchored in stable rock beneath the zone of toppling to prevent movement. The design parameters for anchoring are the bolt tension, the plunge of the anchor and its position on block 1 (Figure 10.10c).

Suppose that an anchor is installed at a plunge angle ψ_T through block 1 at a distance L_1 above its base. The anchor tension required to prevent toppling of block 1 is

$$T_t = \frac{W_1 / 2(y_1 \cdot \sin\psi_p - \Delta x \cdot \cos\psi_p) + P_1(y_1 - \Delta x \cdot \tan\phi_d)}{L_1 \cdot \cos(\psi_p + \psi_T)} \tag{10.26}$$

while the required anchor tension to prevent sliding of block 1 is

$$T_s = \frac{P_1(1 - \tan\phi_p \cdot \tan\phi_d) - W_1(\tan\phi_p \cdot \cos\psi_p - \sin\psi_p)}{\tan\phi_p \cdot \sin(\psi_p + \psi_T) + \cos(\psi_p + \psi_T)} \tag{10.27}$$

When the force T is applied to block 1, the normal and shear force on the base of the block are, respectively,

$$R_1 = P_1 \cdot \tan\phi_d + T \cdot \sin(\psi_p + \psi_T) + W_1 \cdot \cos\psi_p \tag{10.28}$$

$$S_1 = P_1 - T \cdot \cos(\psi_p + \psi_T) + W_1 \cdot \sin\psi_p \tag{10.29}$$

The stability analysis for a slope with a tensioned anchor in block 1 is identical to that described in Section 10.4.3, apart from the calculations relating to block 1. The required tension is the greater of T_t and T_s defined by Equations 10.26 and 10.27.

10.4.6 Factor of safety for limiting equilibrium analysis of toppling failures

For both reinforced and unreinforced slopes, the factor of safety can be calculated by finding the friction angle for limiting equilibrium. The procedure is to first carry out the limiting equilibrium stability analysis as described in Section 10.4.4 using the estimated values for the friction angles. If the lowest block ($n = 1$) is unstable, then one or both of the friction angles are increased by increments until the value of the normal force acting in the lowest block (P_0) is very small (in Figure 10.10c); if block ($n - 1$) is the lowest block, then $P_{n-1} = P_0$. Conversely, if block $n = 1$ is stable, then the friction angles are reduced until P_0 is very small. These values of the friction angles are those required for limiting equilibrium.

The limiting equilibrium friction angles are termed the required friction angles, while the actual friction angles of the block surfaces are termed the available friction angles. The factor of safety for toppling can be defined by dividing the tangent of the friction angle believed to apply to the rock layers ($\tan\phi_{available}$), by the tangent of the friction angle required for equilibrium ($\tan\phi_{required}$).

$$FS = \frac{\tan\phi_{available}}{\tan\phi_{required}} \tag{10.30}$$

The actual factor of safety of a toppling slope depends on the details of the geometry of the toppling blocks. Figure 10.8 shows that once a column overturns by a small amount, edge-to-face contacts occur between the blocks, and the friction required to prevent further rotation increases. Hence, a slope just at limiting equilibrium is metastable. However, rotation equal to $2 \cdot (\psi_b - \psi_p)$ will convert the edge-to-face contacts along the sides of the columns into continuous face contacts and the friction angle required to prevent further rotation will drop sharply, possibly even below that required for initial equilibrium. The choice of factor of safety, therefore, depends on whether or not deformation can be tolerated.

The restoration of continuous face-to-face contact of toppled columns of rock is probably an important arrest mechanism in large-scale toppling failures. In many cases in the field, large surface displacements and tension crack formation can be observed and yet the volumes of rock that fall from the face are small.

10.4.7 Example of limit equilibrium analysis of toppling

The following is an example of the application of the Goodman and Bray limit equilibrium analysis to calculate the factor of safety and required bolting force of the toppling failure illustrated in Figure 10.11a.

A rock face 92.5 m (303 ft) high (H) is cut at an angle of 56.6° (ψ_f) in a layered rock mass dipping at 60° into the face ($\psi_d = 60°$); the width of each block is 10 m (32.8 ft) (Δx). The angle of the slope above the crest of the cut is 4° (ψ_s), and the base of the blocks is stepped 1 m at every block (atn $(1/10)$) = 5.7°, and $\psi_b = (5.7 + \psi_p) = 35.7°$. Based on this geometry, 16 blocks are formed between the toe and crest of the slope (Equation 10.7); block 10 is at the crest. Using Equations 10.10 to 10.12, the constants are $a_1 = 5.0$ m, $a_2 = 5.2$ m and $b = 1.0$ m (16.4 ft, 17.1 ft and 3.3 ft). These constants are used to calculate the height y_n of each block, and the width-to-height ratio $(\Delta x/y_n)$ as shown in the table in Figure 10.11b.

The friction angles on the faces and bases of the blocks are equal and have a value of 38.15° $(\phi_{available})$. The unit weight of the rock is 25 kN/m³ (160 lb/ft³). It is assumed that the slope is dry, and that no external forces act.

The stability analysis is started by examining the toppling/sliding mode of each block, beginning at the crest. Since the friction angle on the base of the blocks is 38.15° and the dip of the base is 30°, the upper blocks are stable against sliding. Equation 10.2 is then used to assess the toppling mode. Since (tan ψ_p = tan 30 = 0.58), blocks 16, 15 and 14 are stable, because for each, the ratio $(\Delta x/y_n)$ is greater than 0.58. That is, these three blocks are short and their centre of gravity lies inside the base.

For block 13, the ratio $(\Delta x/y_n)$ has the value 0.45, which is less than 0.58 and the block topples. Therefore, P_{13} is equal to 0 and P_{12} calculated as the greater of $P_{12,t}$ and $P_{12,s}$ given by Equations 10.23 and 10.25, respectively. This calculation procedure is used to examine the stability of each block in turn progressing down the slope. As shown in the table of forces in Figure 10.11b, $P_{n-1,t}$ is the larger of the two forces until a value of $n = 3$, whereupon $P_{n-1,s}$ is larger. Thus, blocks 4–13 constitute the potential toppling zone, and blocks 1–3 constitute a sliding zone.

The factor of safety of this slope can be found by increasing the friction angle until the base blocks are just stable. It is found that the required friction angle for limit equilibrium conditions is 39°, so the factor of safety as given by Equation 10.30 is

$$FS = \frac{\tan 38.15}{\tan 39} = 0.97$$

The analysis also shows that the required tension in an anchor installed horizontally in block 1 to just stabilise the toe blocks is 500 kN/m (34 kips/ft) length of slope. This compares with the maximum value of P (in block 5) equal to 4837 kN/m (331 kips/ft).

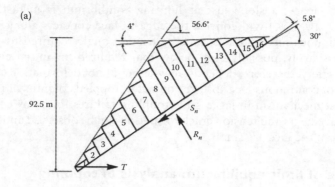

(b)

n	y_n	$\Delta x/y_n$	M_n	L_n	$P_{n.t}$	$P_{n.s}$	P_n	R_n	S_n	S_n/R_n	Mode
16	4.0	2.5			0	0	0	866	500	0.577	
15	10.0	1.0			0	0	0	2155	1250	0.577	STABLE
14	16.0	0.6			0	0	0	3463	2000	0.577	
13	22.0	0.5	17	22	0	0	0	4533.4	2457.5	0.542	
12	28.0	0.4	23	28	292.5	−2588.7	292.5	5643.3	2966.8	0.526	T
11	34.0	0.3	29	34	825.7	−3003.2	825.7	6787.6	3520.0	0.519	O
10	40.0	0.3	35	35	1556.0	−3175.0	1556.0	7662.1	3729.3	0.487	P
9	36.0	0.3	36	31	2826.7	−3150.8	2826.7	6933.8	3404.6	0.491	P
8	32.0	0.3	32	27	3922.1	−1409.4	3922.1	6399.8	3327.3	0.520	L
7	28.0	0.4	28	23	4594.8	156.8	4594.8	5872.0	3257.8	0.555	I
6	24.0	0.4	24	19	4837.0	1300.1	4837.0	5352.9	3199.5	0.598	N
5	20.0	0.5	20	15	4637.5	2013.0	4637.5	4848.1	3159.4	0.652	G
4	16.0	0.6	16	11	3978.1	2284.1	3978.1	4369.4	3152.5	0.722	
3	12.0	0.8	12	7	2825.6	2095.4	2825.6	3707.3	2912.1	0.7855	
2	8.0	1.3	8	3	1103.1	1413.5	1413.5	2471.4	1941.3	0.7855	SLIDING
1	4.0	2.4	4	–	−1485.1	472.2	472.2	1237.1	971.8	0.7855	

(c)

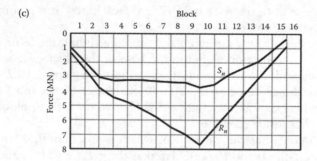

Figure 10.11 Limited equilibrium analysis of a toppling slope: (a) slope geometry; (b) table listing block dimensions, calculated forces and stability mode; (c) distribution of normal (R) and shear (S) forces on base of blocks (Goodman and Bray, 1976).

If $\tan \phi$ is reduced to 0.650, it will be found that blocks 1–4 in the toe region will slide while blocks 5–13 will topple. The tension in an anchor installed horizontally through block 1, required to restore equilibrium, is found to be 2013 kN/m (138 kips/ft) of slope crest. This is not a large number, demonstrating that support of the 'keystone' is remarkably effective in increasing stability. Conversely, removing or weakening the keystone of a toppling slope that is near failure can have serious consequences.

When the distribution of P forces has been defined in the toppling region, the forces R_n and S_n on the base of the blocks can be calculated using Equations 10.21 and 10.22. Assuming ($Q_{n-1} = P_{n-1} \cdot \tan \phi_s$), the forces R_n and S_n can also be calculated for the sliding

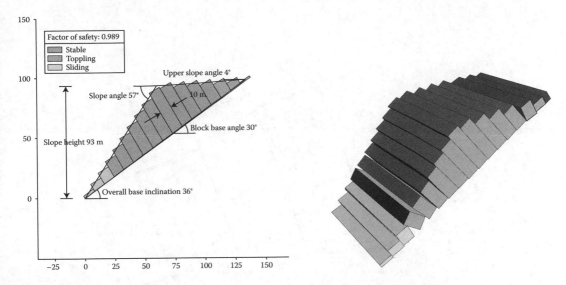

Figure 10.12 Results of analysis of toppling failure shown in Figure 10.11 using RocTopple (RocScience).

region. Figure 10.11c shows the distribution of these forces throughout the slope. The conditions defined by $R_n > 0$ and $|S_n| < R_n \cdot \tan \phi_p$ are satisfied everywhere.

Figure 10.12 shows the results of an analysis of the toppling failure detailed in Figure 10.11 using the program TOPPLE. The diagrams identify the stable, toppling and sliding blocks, and the toppling displacement in the lower part of the slope.

10.4.8 Application of external forces to toppling slopes

In circumstances where external forces act on the slope, it is necessary to investigate their effect on stability. Examples of external forces include water forces acting on the sides and bases of the blocks, earthquake ground motion simulated as a horizontal force acting on each block (see Section 11.6), and point loads produced by bridge piers located on a specific block(s). Another external force that has been considered in the worked example in Section 10.4.7 are rock anchors that are secured in stable ground beneath the toppling mass and then tensioned against the face.

A feature of limit equilibrium analysis is that any number of forces can be added to the analysis, provided that their magnitude, direction and point of application are known. Figure 10.13 shows a portion of a toppling slope in which the water table is sloping. The forces acting on block n include the force Q inclined at an angle ψ_Q below the horizontal, and three water forces V_1, V_2 and V_3, as well as the forces P_n and P_{n-1} produced by the blocks above and below. By resolving all these forces normal and parallel to the base of the blocks, it is possible to modify Equations 10.23 and 10.25 as follows. Considering rotational equilibrium, the force $P_{n-1,t}$ that is just sufficient to prevent toppling of block n has the value:

$$
\begin{aligned}
P_{n-1,t} = \{ &P_n(M_n - \Delta x \cdot \tan \phi_d) + W_n/2\, (y_n \cdot \sin \psi_p - \Delta x \cdot \cos \psi_p) \\
&+ V_1 \cdot y_w/3 + \gamma_w \cdot \Delta x^2/6 \cdot \cos \psi_p (z_w + 2 y_w) - V_3 \cdot z_w/3 \\
&+ Q[-\sin(\psi_Q - \psi_p) \cdot \Delta x/2 + \cos(\psi_Q - \psi_p) \cdot y_n]\} L_n^{-1}
\end{aligned}
$$

$$(10.31)$$

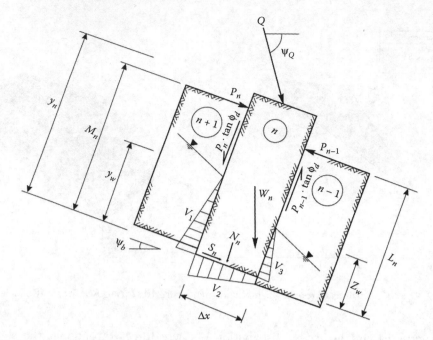

Figure 10.13 Toppling block with external forces.

Assuming that the blocks are in a state of limiting equilibrium, the force just sufficient to prevent sliding of block n has the value

$$
\begin{aligned}
P_{n-1,s} = P_n + \{&-W(\cos\psi_p \cdot \tan\phi_p - \sin\psi_p) \\
&+ V_1 - V_2 \cdot \tan\phi_p - V_3 \\
&+ Q \cdot [-\sin(\psi_Q - \psi_p) \cdot \tan\phi_p + \cos(\psi_Q - \psi_p)]\} \cdot (1 - \tan\phi_p \cdot \tan\phi_d)^{-1}
\end{aligned}
\tag{10.32}
$$

where

$$
\begin{aligned}
V_1 &= \frac{1}{2}\gamma_w \cdot \cos\psi_p \cdot y_w^2; \\
V_2 &= \frac{1}{2}\gamma_w \cdot \cos\psi_p (y_w + z_w) \cdot \Delta x \\
V_3 &= \frac{1}{2}\gamma_w \cdot \cos\psi_p \cdot z_w^2
\end{aligned}
\tag{10.33}
$$

The limit equilibrium stability analysis then proceeds as before using the modified versions of the equations for $P_{n-1,t}$ and $P_{n-1,s}$.

10.5 STABILITY ANALYSIS OF FLEXURAL TOPPLING

Figure 10.4b shows a typical flexural toppling failure in which the slabs of rock flex maintain fact-to-face contact. The mechanism of flexural toppling is different from the block toppling mechanism described in Section 10.4. Therefore, it is not appropriate to use limit equilibrium stability analysis for the design of flexural toppling slopes. Techniques that have been used to study the stability of flexural toppling include base friction models (Goodman, 1976), centrifuges (Adhikary, Dyskin, Jewell et al., 1997) and numerical modelling (Pritchard and Savigny, 1990, 1991). All these models show the common features of this failure mechanism, including interlayer shearing, obsequent scarps at the crest, opening of tension cracks that decrease in width with depth, and a limiting dip angle ψ_b for the base of toppling. As discussed in Section 10.4.1, dip angle ψ_b is steeper than the plane ψ_p (normal to the dip of the slabs) by about 10–30°.

The centrifuge modelling by Adhikary et al. has been used to develop a series of design charts that relate stability to the slope face angle, the dip of the blocks into the face and the ratio of the slope height to the width of the slabs. Another input parameter is the tensile strength of the rock, because bending of the slabs induces tensile cracking in their upper face. The design charts provide information, for example, on the allowable face angle for specific geological conditions and slope height.

Alternatively, computer simulations of block movement provide a means of studying a wide range of geometric and material properties. One of the most suitable computer programs for these simulations is UDEC (Universal Distinct Element Code), which has been developed by Itasca Consulting Group in Minnesota (Itasca Consulting Group Inc., 2000). Figure 10.14 shows the results of an analysis carried out on an open pit mine slope (Pritchard and Savigny, 1990). The main features of UDEC analysis for studying toppling slopes are:

1. Incorporates a number of materials each with differing strength properties.
2. Recognises the existence of contacts or interfaces between discrete bodies, such as slabs of rocks formed by discontinuities dipping steeply into the slope face.

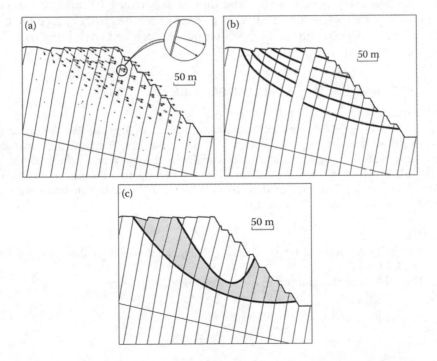

Figure 10.14 UDEC model of toppling pit slopes: (a) pure flexural toppling deformation with grid point velocity vectors; (b) contours of horizontal displacement for toppling slopes; (c) area of failed nodes due to flexure (Pritchard and Savigny, 1990).

3. Calculates the motion along contacts by assigning a finite normal stiffness along the discontinuities that separate the columns of rock. The normal stiffness of a discontinuity is defined as the normal closure that occurs on the application of a normal stress and can be measured from a direct shear test (see Figure 5.17).
4. Assumes deformable blocks that undergo bending and tensile failure.
5. Allows finite displacements and rotations of the toppling blocks, including complete detachment, and recognises new contacts automatically as the calculation progresses.
6. Uses an explicit 'time'-marching scheme to solve the equation of motion directly. This allows modelling of progressive failure, or the amount of creep exhibited by a series of toppling blocks for a chosen slope condition, such as excavation at the toe of the slope. Note that the time step in the analysis is not actual time but a simulation of progressive movement.
7. Allows the user to investigate different stabilisation measures, such as installing rock bolts or installing drain holes, to determine which scenario has the most effect on block movements.

Because of the large number of input parameters that are used in UDEC and the power of the analysis, the most reliable results are obtained if the model can be calibrated against an existing toppling failure in geological conditions similar to those in the design slope. The ideal situation is in mining operations where the development of the topple can be simulated by UDEC as the pit is deepened and movement is monitored. This allows the model to be progressively updated with new data.

Chapter 12 discusses numerical modelling of slopes in more detail.

The application of kinematic stability tests and reinforcement design for a flexural toppling failure is described by Davies and Smith (1993). The toppling occurred in siltstones in which the beds were very closely spaced and dipped at between 90° and 70° into the face. Excavation for a bridge abutment resulted in a series of tension cracks along the crest, and stabilisation of the slope required the installation of tensioned rock bolts to enforce the slope, and excavation to reduce the slope angle.

EXAMPLE PROBLEM 10.1: TOPPLING FAILURE ANALYSIS

Statement

Consider a 6-m-high slope with an overhanging face at an angle of 75°. A fault at the toe of the slope, and dipping at an angle of 15° out of the face, is weathering and undercutting the face. A tension crack, which is wider at the top than at the bottom, has developed 1.8 m behind the crest of the slope, indicating that the face is marginally stable (Figure 10.15). The friction angle ϕ of the fault is 20° and the cohesion c is 25 kPa (3.6 psi). The slope is dry.

Required

1. Calculate the factor of safety of the block against sliding if the density of the rock is 23.5 kN/m³.
2. Is the block stable against topping as defined by the relation

$$\frac{\Delta x}{y} > \tan \psi_p - \text{stable}$$

3. How much more undercutting of the fault must occur before toppling failure takes place?
4. What stabilisation measures would be appropriate for this slope?

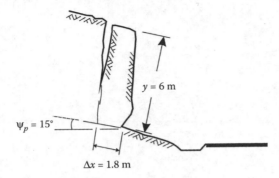

Figure 10.15 Toppling block illustrating Example Problem 10.1.

Solution

1. The factor of safety against sliding is determined by the methods described in Chapter 7; the equation for a dry slope is

$$FS = \frac{c \cdot A + W \cdot \cos \psi_p \cdot \tan \phi_p}{W \cdot \sin \psi_p}$$

$$= \frac{25 \cdot 1.8 + 254 \cdot \cos 15 \cdot \tan 20}{254 \cdot \sin 15}$$

$$\approx 2.0$$

where
   A = base area of block
      = 1.8 m²/m
   W = weight of block/m
      = 23.5 · 1.8 · 6
      = 254 kN/m

2. From the dimensions given in Figure 10.15, the following values are obtained to test stability conditions:
   $\Delta x/y = 1.8/6 = 0.3$

   $\tan 15° = 0.27$

   The block is stable against toppling because 0.3 > 0.27.

3. If weathering results in the width of the base of the block being reduced by a further 0.2 m, toppling is likely to occur because $\Delta x/y = (1.8-0.2)/6 \approx 0.27$.

4. Stabilisation measures which could be used on this slope include:
   a. Prevention of ground water infiltration to limit build-up of water pressure both in the tension crack and on the fault at the toe
   b. Application of reinforced shotcrete to the fault to prevent further weathering
   c. Trim blasting to reduce the slope angle and the dimension y, if this can be achieved without destabilising the block.

Chapter 11

Seismic stability analysis of rock slopes

11.1 INTRODUCTION

In seismically active areas of the world, the design of rock slopes should take into account the effects on the stability of earthquake-induced ground motions. This chapter describes the influence of ground motions on stability, and discusses slope design procedures that incorporate seismic acceleration. These design procedures are generally applicable to rock and soil slopes and both are covered in this chapter, with discussion, where appropriate, on how rock and soil slopes differ.

Much of the information on seismic slope stability has been obtained from NCHRP Report 611 – Seismic Analysis and Design of Retaining Walls, Buried Structures, Slopes and Embankments (TRB, 2008a) and related documents TRB (2008b) and FHWA (1998a).

Numerous rock falls and landslides have been induced by seismic ground motions (Youd, 1978; Van Velsor and Walkinshaw, 1992; Harp and Nobel, 1993; Ling and Cheng, 1997). For example, in 1980, a magnitude 6.0–6.1 earthquake at Mammoth Lakes, California, dislodged a 21.4 ton boulder that bounced and rolled a horizontal distance of 421 m (1380 ft) from its source, and in 1983 in a magnitude 4.2 earthquake in Idaho, a 20.5 ton boulder travelled a distance of about 95 m (310 ft) (Kobayashi, Harp and Kagawa, 1990). On a larger scale, the magnitude 7.9 Denali, Alaska, event in 2002 induced numerous landslides in strong rock with volumes of millions of cubic metres (Jibson, Harp, Schulz et al., 2004). Also, the 2011 earthquakes in Christchurch, New Zealand, triggered thousands of boulder-size rock falls, causing damage to hundreds of homes (Massey et al., 2014). Figure 11.1 shows an earthquake-induced landslide in Japan.

11.1.1 Relative abundance of landslide types

Study of earthquake-induced landslides where widespread instability occurred, shows that rock falls and rock slides are the most common type of events. The following is a list of the relative abundance of earthquake-induced landslides (Keefer, 1984):

- Very abundant (>100,000 occurrences in 40 earthquakes) – rock falls and slides, and disrupted soil slides
- Abundant (10,000–100,000 occurrences in 40 earthquakes) – soil spreads/slumps
- Moderately common (1000–10,000 occurrences in 40 earthquakes) – soil falls, rapid soil flows, rock slumps
- Uncommon (100–1000 occurrences in 40 earthquakes) – subaqueous slides, slow earth flows, rock block slides, rock avalanches

Figure 11.1 Rock fall induced by magnitude 6.4 earthquake on Niijima Island in Japan contained by an MSE wall (Protec Engineering, 2002).

It is expected that the relatively high frequency of rock slides is related to the significant reduction in shear strength that can occur on rock discontinuities following shear displacement of a few centimetres (inches).

11.1.2 Susceptibility of slopes to ground motions

Studies on the number and distribution of landslides and rock falls near earthquakes have shown that the concentrations of landslides can be as high as 50 events per square kilometre. These data have been used to assess the geological and topographical conditions for which the landslide and rock fall hazard is high (Keefer, 1992; Harp and Wilson, 1995, Harp and Jibson, 2002, Harp et al., 2003). It has also been found that the following five slope parameters have the greatest influence on stability during earthquakes:

- *Slope angle* – rock falls and slides rarely occur on slopes with angles less than about 25°
- *Weathering* – highly weathered rock comprising core stones in a soil matrix, and residual soil are more likely to fail than fresh rock
- *Induration* – poorly indurated rock in which the particles are weakly bonded is more likely to fail than stronger, well-indurated rock
- *Discontinuity characteristics* – rock containing closely spaced, open discontinuities is more susceptible to failure than massive rock in which the discontinuities are closed and healed
- *Water* – slopes in which the water table is high, or where recent rainfall has occurred, are susceptible to failure

The relationship between these five conditions and the slope failure hazard is illustrated in the decision tree in Figure 11.2. It is also of interest that the hazard is high for preexisting

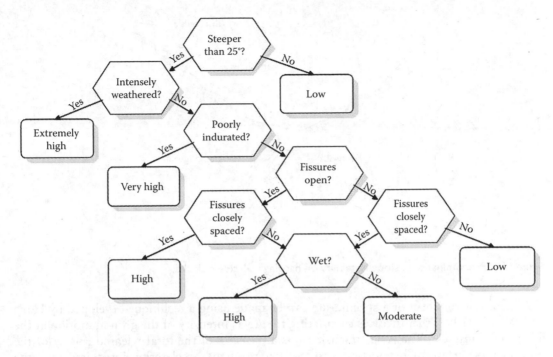

Figure 11.2 Decision tree for susceptibility of rock slopes to earthquake-induced failure (Keefer, 1992).

landslides with slopes flatter than 25°, and for slopes with local relief greater than about 2000 m (6500 ft), probably because seismic shaking is amplified by the topography (Harp and Jibson, 2002). Furthermore, a rock fall hazard may exist at high altitude and cold climates where freeze–thaw action loosens the surficial rock.

The decision tree shown in Figure 11.2 can be used, for example, as a screening tool in assessing rock fall and slide hazards along transportation and pipeline corridors.

The conditions that influence slope stability shown in Figure 11.2 all relate to the properties of the material forming the slope. Another factor that has been found to influence seismic slope stability is the orientation of the fault that was the source of the earthquake. That is, as shown in Figure 11.3, damage is found to occur predominately on the slopes above the hanging wall side of the fault where attenuation relationships show that peak ground acceleration (PGA) values are higher than on the footwall side (Abrahamson, 2000).

11.1.3 Slope failure thresholds related to Arias intensity

In order to quantify the relationship between earthquake magnitude and landslide occurrence, a detailed inventory of landslides induced by the 1994 magnitude 6.7 Northridge earthquake in Los Angeles identified about 11,000 landslides in an area of approximately 10,000 km² (Jibson and Harp, 1995). These slides occurred primarily in the Santa Susana Mountains where the slopes comprise Late Miocene through Pleistocene clastic sediments that have little or no cementation, and that have been folded and uplifted by rapid tectonic deformation (Jibson, Harp and Michael, 1998). While most of these slides were shallow with volumes of a few cubic metres, slides occurred with volumes of hundreds of thousands of cubic metres, as well as over one million cubic metres.

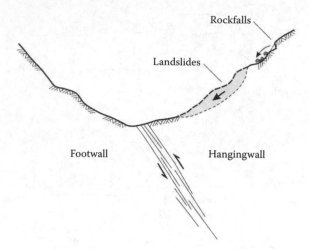

Figure 11.3 Predominance of slope instability on hanging wall side of faults.

A hazard assessment of a specific site can be made using a technique developed by Harp and Wilson (1995) that involves calculating the Arias intensity of the ground motion at the location of interest. The Arias intensity I_a is a measure of the total released energy of the ground motion and is proportional to the square of the acceleration integrated over the entire time history (units of m/s). The Arias intensity is discussed further in Section 11.4.2.

The occurrence of rock falls and slides has been correlated with the Arias intensity of the recorded strong-motion records for the 1987 Whittier and 1987 Superstition earthquakes in California. The study showed that the Arias intensity threshold for slides in Miocene and Pliocene (Tertiary) deposits was in the range of 0.08–0.6 m/s, and the threshold for Precambrian and Mesozoic rocks with open discontinuities was in the range of 0.01–0.07 m/s. It was also confirmed that rock falls and slope failures were rare if the discontinuities were tight, as shown in Figure 11.2.

Detailed slide inventories have also been used to find the maximum epicentral distance to the landslide and the aerial density of slides. For the Northridge, Whittier and Superstition events discussed above, the maximum epicentral distance of landsides was about 60 km (37 miles) (affecting an area of 1500 km²), which is about the average in relation to historical worldwide earthquakes of this magnitude (Keefer, 1984). However, the 2011 magnitude 5.8 Mineral earthquake in Virginia, USA, was unusual in that it induced landslides at distances up to 245 km (150 miles) from the epicentre, over an area of 33,400 km² (Jibson, 2013).

The historical records show that in mountainous terrain the average area *A,* affected by landslides as a function of earthquake magnitude *M,* is given by

$$\log A = M - 3.46 \pm 0.47 \tag{11.1}$$

11.1.4 Landslide dams

The information presented in the previous sections demonstrates the common occurrence of landslides with volumes of hundreds of thousands, or millions, of cubic metres that are triggered by earthquakes. Where these occur in river valleys, they can block the river and create landslide dams that can be several hundred metres high and impound lakes with volumes of millions of cubic metres of water (Costa and Schuster, 1988). Landslide dams can be formed, of course, in both seismic and non-seismic conditions. Records from 128 case studies show that 51% were caused by rainfall and snowmelt, about 39% were the result of earthquakes, and volcanic eruptions were the cause of 8% of the dams.

A minority of landslide dams, and the lakes that they create, are permanent resulting in a sudden change in the topographic, geologic and hydrologic conditions of the area. An example of a permanent dam is Spirit Lake with a volume of 259 million cubic metres (340 million cubic yards) that was formed in 1980 by the landslide created by the eruption of Mt. St. Helens in Washington State, USA. Another permanent dam was created by the Thistle Slide in 1983 in Utah, USA (see Section 1.1.2). These events can result in large-scale loss of life and damage to homes, farms and transportation networks where they are buried in the slide or inundated by the lake. The loss of life can be particularly severe when the dam fills rapidly such that the evacuation time is limited. For permanent dams, the hydraulic conductivity is high enough that the river inflow equals the leakage through the landslide debris, and this occurs without significant scour or erosion of the debris.

The more common behaviour of landslide dams is that they fail soon after formation, usually by overtopping as the lake fills, and then headward erosion of the landslide material. Records of 73 case studies show that 27% of dams failed within a day, 41% failed within a week and 85% failed within a year. These records show the great danger created by landslide dams is the severe downstream flooding that can occur, possibly with little warning when the dam fails.

Immediate response to a landslide dam event may be to evacuate all persons downstream of the dam that are considered to be within the flood zone. The evacuation order would remain in effect until the situation has stabilised, or until longer-term remedial measures can be implemented. Remedial work for landslide dams usually involves cutting a spillway to prevent overtopping. For example, a drainage channel was cut through the Sheep Mountain landslide dam in Montana that was formed as a result of a magnitude 7.3 earthquake in 1959. Also, a drainage tunnel was driven through the rock abutment of the Thistle Slide in Utah to drain the lake formed by the 1983 landslide on the Spanish Fork River. One of the considerations in the selection of a method to drain the lake is control of the discharge rate to limit downstream flooding.

Figure 11.4 shows a temporary dam formed by rock fall with a volume of about 30,000 cubic metres that was breached within a day due to the erosion of the slide debris by the river flow.

11.2 GROUND MOTIONS: PUBLISHED CODES AND STANDARDS

The design of slopes that may be subjected to seismic ground motions requires quantitative information on the magnitude of the motion (Abrahamson, 2000). This information may be either the peak ground acceleration (PGA) for pseudo-static stability analysis (see Section 11.6), or the acceleration–time history of the motions for Newmark displacement analysis (see Section 11.7). This section describes procedures for obtaining the relevant ground motion parameters.

For geotechnical design, it is usual to use the horizontal component of the peak ground acceleration (PHGA) since it is this motion that causes most damage. However, the damage potential of ground motions may also depend on the duration of strong shaking, frequency, energy content of the motion and the peak vertical ground acceleration (PVGA), velocity and displacement.

While the PGA is the parameter that is used in geotechnical design, earthquakes are usually quantified in terms of their magnitude M, which is a measure of the energy released by

Figure 11.4 Breach of landslide dam caused by rock fall (Seymour River, Vancouver, Canada).

the earthquake. Of the various magnitude scales, the moment magnitude M_w, which is a measure of the kinetic energy released, is a widely used standard, and it corresponds to the Richter scale M_L for magnitudes up to 6.

11.2.1 Sources of information

The usual first step in obtaining design information on ground motions is to consult seismic hazard maps that show zones of equal peak acceleration levels. The building codes of countries with seismic areas publish maps in which the country is divided into zones showing, for example, the PGA (as a fraction of gravity acceleration), as well as the horizontal and vertical components of the motions. It may also be possible to obtain values for the peak horizontal ground velocity (PHGV) and the peak horizontal ground displacement (PHGD), and the spectral acceleration (see Section 11.2.3). One of the values of using the zoned seismic hazard maps is that it encourages uniform design practices throughout each zone.

Figure 11.5 shows the time histories of the acceleration, velocity and displacement for the magnitude 2015 Kathmandu earthquake. The plots show the relationship between the relatively high-frequency ground acceleration and the lower-frequency velocity and displacement, with time. The displacement plot shows the PGD at 150 cm (4.9 ft) that is indicative of the widespread and extensive damage caused by this earthquake.

Ground motion parameters are also available on the internet. For example, in the United States, it is possible to find acceleration levels for postal zip codes (www.earthquake.usgs.gov/hazards), and in Canada the Canadian Geologic Survey provides design PGA values based on the site geographic co-ordinates (www.EarthquakesCanada.ca). Advantages of the internet-based information are that they can be updated as new information becomes available, and they allow a finer relationship between acceleration and location that avoids the sudden changes in design acceleration that occurs at zone boundaries on the hazard zone maps. Furthermore, the data are provided in probabilistic terms in which the PGA is related to the return period, or probability of exceedance per annum (see Section 11.2.2).

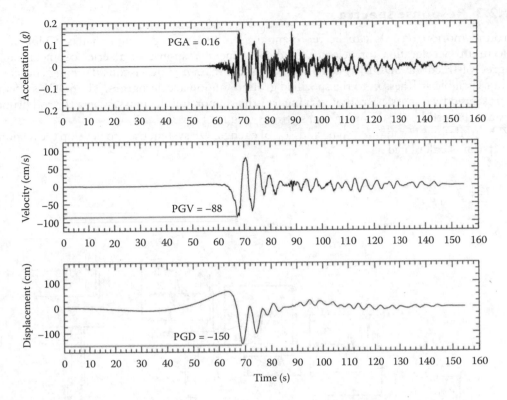

Figure 11.5 Time histories of ground motion acceleration, velocity and displacement for 2015 magnitude 7.8 Kathmandu earthquake, channel HNE (east-west) (PEER - Pacific Earthquake Engineering Reserach Center, 2015).

11.2.2 Earthquake probability: Return periods

Ground motion parameters can be reported in terms of their probability of occurrence, defined as the return period, or recurrence interval, which is the probability that an event will occur within the time defined by the return period. That is, for a 100-year return period, earthquakes can occur on average every 100 years, but in any one 100-year period, no events, or several events, may occur. Values of the return period are determined by studying the historical record of earthquakes and calculating the average recurrence interval of each event magnitude.

The return period (R) can also be expressed as the probability of exceedance p, in a stated number of years t, such as 5% exceedance in 50 years ($R = 1000$ years)

$$R = \left(\left(\frac{p}{100} \right) \cdot \left(\frac{1}{t} \right) \right)^{-1} \tag{11.2}$$

Since small earthquakes are more common than large earthquakes, return periods also vary according to magnitude. For example, for the same source, the PGA is $0.026 \cdot g$ for $R = 100$ years ($p = 40\%$ in $t = 50$ years), and the PGA is $0.091 \cdot g$ for $R = 1000$ years ($p = 5\%$ in $t = 50$ years).

11.2.3 Response spectra

Ground motion records can be represented as a response spectrum that is a plot of the maximum acceleration (or velocity, or displacement) response of a series of linear, single degree-of-freedom (SDOF) systems with the same mass m, and viscous damping coefficient c, but variable stiffness k, to the specified ground motion accelerogram. The SDOF system is represented by a massless rod of varying lengths supporting the mass, where the rod properties are defined by the damping and stiffness (Figure 11.6).

The undamped fundamental period, T_0, of each SDOF system used to develop the response spectrum is calculated as

$$T_0 = 2\pi\sqrt{\frac{m}{k_n}}$$

(11.3)

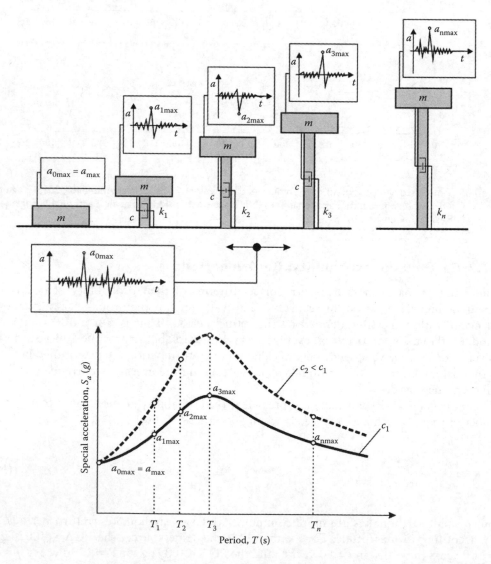

Figure 11.6 Schematic representation of acceleration response spectra showing relationship between ground motion (a_0) and system accelerations (a_1 to a_n) (Federal Highway Administration, 1998a).

The damping of an SDOF system is represented by the viscous damping coefficient c, commonly referred to as the spectral damping; in geotechnical engineering, the value of c is usually 5%.

For a design earthquake in western Canada, spectral accelerations $S_{(T)}$ for a range of periods T at 5% damping and for two probabilities of occurrence are as follows:

2% probability of exceedance in 50 years (0.000404 per annum, or $R = 2475$ years):

$$S_{(T=0.2)} = 0.267 \cdot g \quad S_{(T=0.5)} = 0.156g \quad S_{(T=1.0)} = 0.075g \quad S_{(T=2.0)} = 0.041 \cdot g \quad PGA = 0.134 \cdot g$$

10% probability of exceedance in 50 years (0.0021 per annum, or $R = 476$ years):

$$S_{(T=0.2)} = 0.117 \cdot g \quad S_{(T=0.5)} = 0.068 \cdot g \quad S_{(T=1.0)} = 0.032 \cdot g \quad S_{(T=2.0)} = 0.018 \cdot g \quad PGA = 0.065 \cdot g$$

Plots of these $S_{(T)}$ values normalised to PGA (i.e. $S_{(T)}$/PGA) against period T have the standard form of spectrum response plot shown in Figure 11.6.

Acceleration response spectra, which provide quantitative information on both the intensity and frequency content of the acceleration–time history, are mainly used in building design. Response spectra are rarely used in geotechnical design, except as an aid in the selection of time histories for input to site response and deformation analyses (see Section 11.7).

11.3 SEISMIC HAZARD ANALYSIS

For projects where seismic ground motions are critical to performance, or where a particular seismic hazard exists such as a potentially active fault, it may be necessary to carry out a seismic hazard analysis, in addition to obtaining information on ground motion parameters from published sources as described in the previous section.

The process by which the design ground motion parameters are established is termed 'seismic hazard analysis', which involves the following three steps (Federal Highway Administration, 1998a; Glass, 2000):

1. Identification of seismic sources capable of producing strong ground motions at the site
2. Evaluation of seismic potential for each fault that is a capable source of earthquakes
3. Evaluation of the intensity of design ground motions at the site where the hazard analysis is being conducted

Implementation of these three steps involves the activities described in the following sections.

11.3.1 Seismic sources

Earthquakes are the result of fault movement, so identification of seismic sources includes establishing the types of faults and their geographic location, depth, size and orientation. Figure 11.7 shows the terms used to define distances used in earthquake engineering.

R_E = epicentral distance
R_R = closest distance to the fault plane (rupture distance)
R_S = seismogenic distance
R_H = hypocentral distance
R_{JB} = closest horizontal distance to the vertical projection of the fault plane (Joyner and Boore distance)

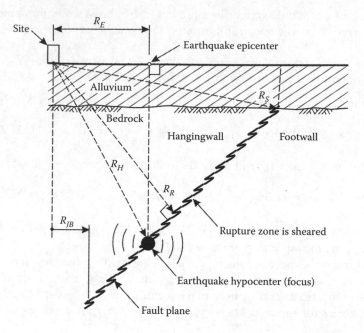

Figure 11.7 Definition of distances used in earthquake engineering (Federal Highway Administration, 1998a).

Information on faulting is usually available from publications such as geological maps and reports prepared by government geological survey groups and universities, and any previous projects that have been undertaken in the area. Also, identification of faults can be made from the study of aerial photographs, geological mapping, geophysical surveys and trenching. On aerial photographs, such features as fault scarplets, rifts, fault slide ridges, shutter ridges, fault saddles and offsets in such features as fences and railway lines may identify active faults (Cluff, Hansen, Taylor et al., 1972). In addition, records of seismic monitoring stations provide information on the location and magnitude of recent earthquakes that can be correlated to fault activity.

11.3.2 Seismic potential

Movement of faults within the Holocene epoch (approximately the last 11,000 years) is generally regarded as the criterion for establishing that the fault is active (United States Environmental Protection Agency [USEPA], 1993). Although the occurrence interval of major earthquakes may be greater than 11,000 years, and not all faults rupture to the surface, lack of evidence that movement has occurred in the Holocene is generally sufficient evidence to dismiss the potential for ground surface rupture. Most Holocene fault activity in North America has occurred west of the Rocky Mountains, and may be identified by detailed mapping, followed by trenching, geophysics or drilling. In regions where no surface expression of fault rupture exists, seismic source characterisation depends primarily on microseismic studies and the historic record of felt earthquakes.

11.3.3 Ground motion intensity

Once the seismic sources capable of generating strong ground motions at a site have been identified and characterised, the intensity of the ground motions can be evaluated either from published codes and standards, or from ground motion characterisation as discussed in Section 11.4.

11.4 GROUND MOTION CHARACTERISATION

Where a seismic hazard analysis of a site has identified a fault(s) with the potential for generating strong ground motions, characterisation of these motions can be carried out to develop site-specific acceleration–time histories for design. The value of seismic analysis, in contrast to the use of published codes as discussed in Section 11.2, is the ability to incorporate the latest developments in local seismicity and site characterisation.

The development of ground motion parameters' histories is described below.

11.4.1 Attenuation relationships

Attenuation equations are used to calculate the ground motion parameters of interest at a given site produced by an earthquake of a given magnitude and site-to-source distance. That is, the equations quantify the attenuation and modification to the ground motions that occur between the source and the site as the result of the energy propagating from the source, and being absorbed by the rock through which it passes. Attenuation relationships may be developed from either statistical regression analysis of previous earthquakes, or from theoretical models of the propagation of strong ground motions.

Site characteristics that are important in developing attenuation equations and matching time histories include earthquake magnitude, source mechanism (fault movement type, subduction zone), focal depth, site-to-source distance, site geology, PGA, frequency content, duration and energy content. Since these conditions are highly variable in seismic areas, the attenuation equations are site-specific and may be regularly updated as new, relevant earthquake time histories become available.

The following are two examples of attenuation equations to illustrate both the form of the equations, and the parameters that define the attenuation relationships.

- *Taiwan* – regression analysis of time histories from over 6000 earthquakes recorded by 650 stations across the island:

$$\ln Y = a \cdot \ln(X + h) + b \cdot X + c \cdot M_w + d \pm \sigma \tag{11.4}$$

  where Y is the ground motion parameter, X the hypocentral distance, M_w the moment magnitude, a the geometrical spreading coefficient, b the inelastic attenuation coefficient, c the magnitude coefficient, d a constant, h the close-in distance saturation coefficient and σ the standard deviation (Liu and Tsai, 2005).

- *Subduction zones* – regression analysis of time histories from 168 subduction earthquakes, with moment magnitude M_w greater than 5, recorded on the west coast of north, central and south America, and in Japan and the Solomon Islands:

$$\ln(PGA) = C_1 + C_2 \cdot M_w + C_3 \cdot \ln\left[R_H + e^{C_4 - (C_2/C_3)M_w}\right] + C_5 \cdot M \cdot Z_t + C_9 \cdot H + C_{10} \cdot Z + \eta + \epsilon \tag{11.5}$$

  where R_H is the hypocentral distance, H the focal depth and the other terms are regression coefficients (Youngs, Chiou, Silva et al., 1997).

11.4.2 Earthquake energy: Arias intensity

The energy content of the acceleration–time history provides a means of characterising strong ground motions, and can be quantified by the Arias intensity I_a, which includes both the acceleration peaks and the duration of ground motions. As discussed in Section 11.7, the Arias intensity can be correlated with slope movement calculated according to Newmark displacement analysis, and guidelines have been developed on the thresholds of movement that correspond to slope stability conditions.

The Arias intensity is used in two types of engineering applications. First, slope instability can be correlated with the level of energy release defined by the Arias intensity, and to define thresholds of slope instability related to the type of slope material. Examples of slope instability, based on records of earthquakes in California, gave the following Arias intensity thresholds:

- Slopes in Precambrian and Mesozoic rocks where the discontinuities were open, Arias intensity threshold in the range of 0.01–0.07 m/s
- Slopes in Miocene and Pliocene (Tertiary) deposits, Arias intensity threshold in the range of 0.08–0.6 m/s

The second engineering application for Arias intensity is in Newmark seismic deformation analysis, as the deformation potential is directly proportional to the energy content; Newmark deformation analysis is discussed in Section 11.7.

The Arias intensity is proportional to the square of the acceleration integrated over the entire acceleration–time history, defined as

$$I_a = \frac{\pi}{2} g \int_0^{T_d} [a(t)]^2 \, dt \quad \text{(units : m/s)} \tag{11.6}$$

where $a(t)$ is a single component acceleration–time series for a strong-motion record with total duration T_d (seconds), t is the time in seconds and g the acceleration of gravity.

As an alternative to integrating ground motion records, an analysis of 1208 records from 76 earthquakes around the world, having magnitudes ranging from 4.7 to 7.6, has been carried out to find an empirical correlation between acceleration–time histories and Arias intensity (Travasarou, Bray and Abrahamson, 2003). Factors that were incorporated into the analysis were the site conditions (i.e. fresh and weathered rock, deep and shallow soil deposits), fault type (i.e. normal, strike slip, reverse-oblique, reverse) as well as the moment magnitude M_w, and rupture distance R_R (km) (see Figure 11.7). The reported Arias intensity is the average of the two perpendicular horizontal components of the recorded strong ground motions; this component is most commonly used in slope design.

The predictive equation for the Arias intensity I_a, for $4.7 \le M_W \le 7.6$ and $0.1 \le R_R \le 250$ km, is

$$\ln(I_a) = c_1 + c_2(M_w - 6) + c_3 : \ln\left(\frac{M_w}{6}\right) + c_4 \cdot \ln\left(\sqrt{R_R^2 + h^2}\right) + S_C(s_{11} + s_{12}(M_w - 6))$$
$$+ S_D(s_{21} + s_{22}(M_w - 6)) + f_1 \cdot F_N + f_2 \cdot F_R \tag{11.7}$$

where h is a fictitious hypocentral distance determined by the regression analysis.

The terms S_C and S_D are indicator variables for the site condition (see Table 11.1):

$S_C = S_D = 0$ – Miocene and older strata: limestone, igneous and metamorphic rock (site class B)

$S_C = 1$, $S_D = 0$ – Pliocene, Pleistocene strata: conglomerates, pyroclastic, lateritic terraces (site class C)

$S_C = 0$, $S_D = 1$ – late Pleistocene, Holocene strata: fluvial terraces, stiff clays, sandy soils (site class D)

The terms F_N and F_R are indicator variables for the fault type:

$F_N = F_R = 0$ – strike slip faults
$F_N = 1$, $F_R = 0$ – normal faults
$F_N = 0$, $F_R = 1$ – reverse, or reverse-oblique faults

The correlation coefficients for Equation 11.7 are as follows:

$c_1 = 2.800$; $c_2 = -1.981$; $c_3 = 20.720$; $c_4 = -1.703$; $h = 8.78$
$s_{11} = 0.454$; $s_{12} = 0.101$; $s_{21} = 0.479$; $s_{22} = 0.334$; $f_1 = -0.166$; $f_2 = 0.512$

Figure 11.8 is an acceleration–time history for the 1989 M_w 6.9 Loma Prieta earthquake with the corresponding Arias intensity plot showing the evolution of energy release and how almost all the energy is released during the 8 s of the highest acceleration levels; this is termed a Husid plot (Husid, 1969).

11.4.3 Earthquake duration

The earthquake duration may be related to cumulative effects such as permanent seismic deformation of slopes, liquefaction of soil and loss of shear strength of the slope materials. In rock slopes, small displacements may cause the shear strength to diminish significantly from peak to residual values, which would be detrimental to stability.

Earthquake durations can be defined on an Arias intensity plot as the time interval between 5% and 95% of the total intensity on a Husid plot, as shown in Figure 11.8. Alternatively, an empirical relationship between duration T (seconds) and earthquake magnitude M is given by

$$\log T = 0.432 \cdot M - 1.83 \tag{11.8}$$

11.5 SEISMIC STABILITY ANALYSIS OF ROCK SLOPES

The methods of slope stability analysis described in Chapters 7 to 10, namely for plane, wedge, circular and toppling failures, are limit equilibrium methods in which the ratio of the resisting to displacing forces on a delineated slip surface(s) defines the factor of safety of the slope. This analysis method to find the static factor of safety can be readily adapted to incorporate ground motion accelerations that assess the influence of earthquakes on the stability of slopes. Also, seismic analysis can be applied to both rock and soil slopes with the incorporation of site appropriate factors.

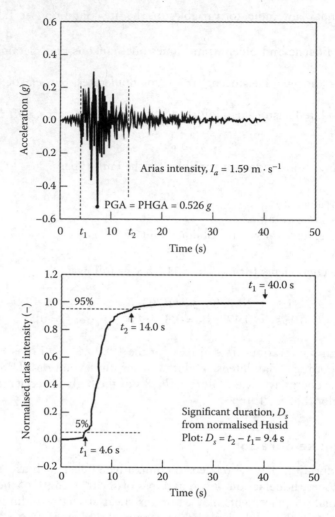

Figure 11.8 Acceleration time history for 1989, magnitude 6.9 Loma Prieta earthquake showing normalised Arias intensity plot, and the duration of strong shaking (Lee and Green, 2008).

Methods developed to assess the stability or displacement of slopes during earthquakes fall into three categories – pseudo-static analysis, permanent displacement analysis (Newmark) and stress deformation analysis as described below:

- Pseudo-static analysis is a modification of widely used limit equilibrium analysis and is the simplest analytical method (see Section 11.6).
- Newmark displacement analysis is a more realistic method of assessing seismic effects and procedures are available to calculate slope displacement for a range of site conditions (see Sections 11.7).
- Stress deformation analysis involves the application of highly complex models that are usually only used for critical structures, such as earth dams where a high density of site data is available (Jibson, 2011). This method of analysis is not discussed in this book.

11.6 PSEUDO-STATIC STABILITY ANALYSIS

The limit equilibrium method of determining the factor of safety of a sliding block as described in Chapter 7 on plane failure (Section 7.3) can be modified to incorporate the effect on stability of seismic ground motions, as described in this section.

11.6.1 Limit equilibrium analysis

Figure 11.9a shows a rock slope with face angle ψ_f and a potentially unstable block on a sliding plane dipping at ψ_p out of the face and a vertical tension crack at distance b behind the crest. The weight of the block is W, the sliding surface has area A and shear strength defined by cohesion c, and friction angle, φ.

The pseudo-static limit equilibrium method for seismic slope design involves simulating the ground motions as a static horizontal force F_H acting in a direction out of the face.

The magnitude of force F_H is

$$F_H = W \cdot k \tag{11.9}$$

where k is the seismic coefficient (units: g). The value of the seismic coefficient can take into account the magnitude of the ground motions, the slope materials and the height of the slope (see Equation 11.11).

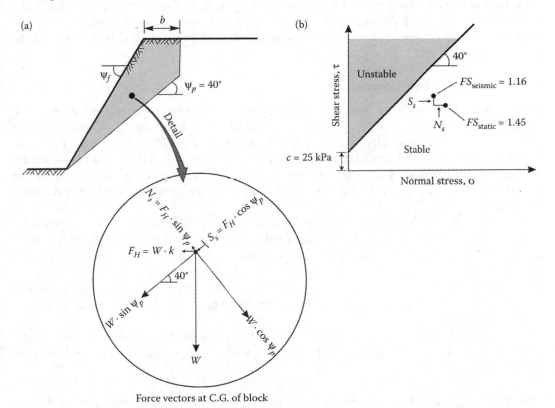

Figure 11.9 Seismic stability analysis of plane failure. (a) Slope model showing magnitude of horizontal seismic force F_H and normal and shear components; (b) Mohr diagram showing influence of force F_H on factor of safety.

By limit equilibrium analysis, the pseudo-static factor of safety FS of the slope incorporating the seismic coefficient k (assuming the slope is drained, $U = V = 0$) is

$$FS = \frac{\text{Resisting force}}{\text{Driving force}}$$

$$FS = \frac{c \cdot A + (W(\cos \psi_p - k \cdot \sin \psi_p)) \cdot \tan \phi}{W(\sin \psi_p + k \cdot \cos \psi_p)}$$

(11.10)

Figure 11.9b shows that the effect on stability of the horizontal seismic force F_H is to diminish the normal force, N_S ($-k \cdot \sin \psi_p$), which reduces the resisting force, and to increase the driving force, S_S ($+k \cdot \cos \psi_p$), such that the factor of safety is reduced from FS_{static} to FS_{seismic}. The usual guideline for pseudo-static slope design is that, to limit displacement, the static factor of safety should be a minimum of 1.5, and the seismic factor of safety a minimum of 1.1.

The advantage of pseudo-static stability analysis is that it is easy to use, and this function is included in many commercial slope stability analysis programmes. However, the disadvantages of the analysis method are that it is first an unrealistic model of the effects of ground motions on slope stability. Second, pseudo-static analysis provides conservative results since the assumption is made that the actual transient ground motions, with reversing directions and a duration of a few seconds, are replaced by a constant force acting, in the most adverse direction, over the entire life of the slope. Therefore, as discussed below, it is important to use values of the seismic coefficient that are appropriate for the site conditions.

11.6.2 Seismic coefficient

Most of the early work on pseudo-static analysis was for the design of soil slopes, and for earth dams where up to 1 m (3 ft) of displacement can be tolerated provided that no loss of shear strength occurred during cyclic loading (Seed, 1979; Pyke, 1999). For these conditions, it was recommended that k should have a value between 0.1 and 0.15. However, for most civil slopes, such as bridge abutments and fills on transportation systems, where tolerable movement is probably not more than 50 mm, higher values for k are often required for design.

Prior to 1991, guidelines for the value of k were not linked directly to the earthquake magnitude and other site conditions. Improved procedures have now been developed that calculate k as a function of allowable displacement, earthquake magnitude and spectral acceleration, the basis for which is calibration related to allowable displacement (California Department of Conservation, 2008; Bray and Travasarou, 2009). These documents recommend k values of ($0.4 \cdot g$) to ($0.75 \cdot g$) to limit displacements to 50 mm (2 in.), depending on the magnitude M.

In conducting pseudo-static stability analysis, the value of the seismic coefficient k can be modified to account for details of the site conditions by the application of the following two site factors:

- *Site coefficient* F_{PGA} – used to modify k to account for site conditions including characteristics of the rock or soil and the magnitude of the ground motions (see Section 11.6.3)
- *Wave scattering factor* α – proportional to the slope height and site characteristics (see Section 11.6.4)

These two parameters are incorporated into pseudo-static analysis according to the following relationship:

$$k = \left(\frac{PGA \cdot F_{PGA} \cdot \alpha}{2} \right) \qquad (11.11)$$

where k is the seismic coefficient used in Equation 11.9 and PGA is the peak horizontal ground acceleration (PGHA) level obtained from published seismic information for the site. The factor of (1/2) is applied to this equation to account for the conservatism of the pseudo-static method in which the transient ground motions acting for a very brief time are applied as a permanent horizontal force acting out of the face.

Values for k discussed in the previous paragraphs are generally applicable to slopes in soil, and in weathered rock, where the rock mass contains no distinct sliding surface and so no significant loss of shear strength occurs with displacement. For slopes in soft soil, k may have values greater than (PGA/2).

With respect to seismic design of rock slopes, F_{PGA}, which is referenced to soft rock, has a value of less than 1, while α is generally equal to 1 as discussed in the following sections. Therefore, Equation 11.11 is usually equal to (PGA/2) for rock slopes. However, two conditions may exist for which it is advisable to use a k value greater than (PGA/2). First, where the slope contains a distinct sliding surface for which a significant decrease in shear strength may occur with limited displacement, sliding planes on which the strength would be sensitive to movement include smooth, planar joints or bedding planes with no infilling. For example, Figure 5.17 shows the results of a direct shear test where the peak shear strength occurs at a shear displacement of only 4 mm (0.15 in.). The second condition that may require a k value greater than (PGA/2) is where the rock slope is a topographic high point and amplification of the ground motions may be expected (Sepulveda, Murphy, Jibson et al., 2005).

No definitive guidelines have been drawn up for values of k to use in design. Therefore, in critical situations, it may also be advisable to check the sensitivity of the factor of safety to the value of k, and to calculate possible seismic deformations using the Newmark analysis as discussed in Section 11.7. In general, slopes in brittle rock have little capacity for deformation and the factor of safety is probably the better measure of slope stability than the magnitude of the deformation.

11.6.3 Site condition spectral acceleration coefficient F_{PGA}

The published ground motion spectra are referenced to soft rock ground conditions, where the average shear wave velocity within the upper 30 m (100 ft) of the geologic profile ranges from 760 to 1510 m/s (2500–5000 ft/s), which is designated as Site Class B (Table 11.1). Actual accelerations at the surface will depend on amplification or attenuation of the reference ground motions by the surficial materials. That is, ground motions will be amplified as the soil becomes less dense, and will be attenuated by hard rock. These effects can be quantified by multiplying the reference PGA value by the site coefficient F_{PGA} that is a function of both the soil profile (site class) and the magnitude of the PGA (Table 11.1).

11.6.4 Height-dependent seismic coefficient, α

The pseudo-static stability analysis may take into account the slope height, and the rigidity of the material forming the slope. That is, the value of the PGA is reduced in soil and highly weathered rock to account for the wave scattering, or non-coherence that takes place within the slope, the effect of which is to diminish the level of ground shaking in the slope compared to the reference input time history (Federal Highway Administration, 2011).

Table 11.1 Values of F_{PGA} at zero-period on acceleration spectrum

| Site class | Soil profile name | Average properties in upper 30 m (100 ft) | | Peak ground acceleration coefficient (F_{PGA}) | | | | |
		Shear wave velocity (m/s)	Standard penetration resistance (blows/ft)	PGA < 0.1	PGA = 0.2	PGA = 0.3	PGA = 0.4	PGA > 0.5
A	Hard rock	$v_s > 1500$	N/A	0.8	0.8	0.8	0.8	0.8
B	Rock	$760 \leq v_s \leq 1510$	N/A	I	I	I	I	I
C	Weathered rock, very dense soil	$360 \leq v_s \leq 760$	$N > 50$	1.2	1.2	1.1	I	I
D	Stiff soil	$180 \leq v_s \leq 360$	$15 \leq N \leq 50$	1.6	1.4	1.2	1.1	I
E	Soft soil	$v_s \leq 180$	$N < 15$	2.5	1.7	1.2	0.9	0.9

Source: Adapted from American Association of State Highway and Transportation Officials (AASHTO) 2012. *LRFD Bridge Design Specifications, Customary US Units.* AASHTO, Washington, DC, Contract No.: LRFDUS-6.

With respect to slope height, the reduction in k to account for wave scattering is not applied for slope heights less than about 8 m (25 ft), and the maximum reduction factor is applied to slope heights of about 30 m (100 ft).

For slope heights between 6 and 30 m (20–100 ft), the slope height reduction factor α is calculated as follows:

$$\alpha = (1 + 0.01 \cdot H \cdot [(0.5 \cdot \beta) - 1]) \tag{11.12}$$

where H is the slope height in feet, and

$$\beta = F_v \cdot \frac{S_1}{PGA} \tag{11.13}$$

F_v is the spectral acceleration coefficient related to the site class as defined in Table 11.2, and the ratio (S_1/PGA) is the normalised spectral acceleration at a period of 1 s (see Figure 11.12, for example).

Table 11.2 Site classes and spectral acceleration coefficients, F_v

| Site class | Soil profile name | Spectral acceleration coefficient F_v at period 1 s (S_1) | | | | |
		$S_1 < 0.1$	$S_1 = 0.2$	$S_1 = 0.3$	$S_1 = 0.4$	$S_1 > 0.5$
A	Hard rock	0.8	0.8	0.8	0.8	0.8
B	Rock	I	I	I	I	I
C	Weathered rock, very dense soil	1.7	1.6	1.5	1.4	1.3
D	Stiff soil	2.4	2	1.8	1.6	1.5
E	Soft soil	3.5	3.2	2.8	2.4	2.4

Source: Adapted from Federal Highway Administration (FHWA). 2011. *LRFD seismic analysis and design of transportation geotechnical features and structural foundations.* NHI Course No. 130094. Reference Manual. J. E. Kavazanjian et al., eds., U.S. Department of Transportation, Washington, DC, Report No.: FHWA-NHI-11-032.

For slopes in strong rock, which can be considered as a rigid mass compared to soil, the wave motions are equal (or coherent) throughout the slope and no height reduction factor is applied ($\alpha = 1$).

11.6.5 Vertical ground motions

Study of the effect of the vertical component of seismic ground motions on the factor of safety has shown that incorporating the vertical component will not change the factor of safety by more than about 10%, provided that ($k_V < k_H$) (Federal Highway Administration, 1998a). Furthermore, the vertical acceleration would only be applied when the vertical and horizontal components are exactly in phase. Based on these results, it is usually acceptable to ignore the vertical component of seismic ground motions.

11.6.6 Slope design procedure

As a summary of the slope design procedures discussed in this section, the following are the suggested steps for a pseudo-static slope stability analysis:

1. Conduct the static limit equilibrium slope stability analysis to check whether the static factor of safety is adequate for the project, usually at least 1.5.
2. Establish the PGHA for the site for the applicable return period using sources of information discussed in Section 11.2.
3. Determine the site class corresponding to geology in the upper 30 m of the site (100 ft) (Table 11.1).
4. Determine value for site factor F_{PGA} corresponding to site class and value of PGHA (Table 11.1).
5. From spectral acceleration plot, find the value of spectral acceleration S_1 at period $T = 1$ s; for normalised spectra (e.g. Figure 11.12), S_1 = (normalised acceleration PGA).
6. Determine the value for spectral acceleration coefficient F_v corresponding to site class and value of S_1 (Table 11.2).
7. Calculate the height reduction factor α from Equation 11.12.
8. Calculate the overall seismic coefficient as $k = (PGA \cdot F_{PGA} \cdot \alpha \, / 2)$; use this value of k to calculate pseudo-static factor of safety in Equation 11.10. For slope displacement limited to 25–50 mm, the value of k should be not less than 0.5 of PGHA.
9. Conduct the pseudo-static stability analysis to check whether the factor of safety is at least 1.1. If this factor of safety cannot be met, the slope angle could be reduced, or the amount of support (rock bolts, for example) increased.

EXAMPLE OF PSEUDO-STATIC STABILITY ANALYSIS

The rock slope (planar model) is shown in Figure 11.9a; the stability analysis is conducted as follows:

Slope parameters

Height $H = 15.24$ m (50 ft); distance of tension crack behind crest $b = 9.1$ m (30 ft), face angle $\psi_f = 76°$; sliding plane angle $\psi_p = 40°$; shear strength of sliding plane: $c = 0$, $\varphi = 45°$; geology – moderately weathered rock, site class C (Table 11.1), drained ($U = V = 0$), $\gamma_r = 26$ kN/m³.

Seismic calculations

Assume site in WUS with earthquake magnitude 7 with PGA $= 0.25 \cdot g$. From spectral acceleration plot for rock slopes (Figure 11.12a), normalised spectral acceleration at a period of 1 s ($T = 1$ s) is 0.9, and $S_1 = (0.9 \cdot PGA) = (0.9 \cdot 0.25) = 0.225 \cdot g$.

From Table 11.1 for site class C and PGA = 0.25·g, site acceleration coefficient $F_{PGA} = 1.2$.

From Table 11.2 for site class C and $S_1 = 0.225 \cdot g$, spectral acceleration coefficient $F_v = 1.6$.

In Equation 11.13, parameter $\beta = (F_v \cdot S_1/PGA) = (1.6 \cdot 0.9) = 1.44$, and for 50 ft slope height (H), reduction factor $\alpha = 0.86$.

Seismic coefficient $k = (PGA \cdot F_{PGA} \cdot \alpha)/2 = (0.25 \cdot 1.2 \cdot 0.86)/2 = 0.258/2 = 0.13 \cdot g \approx PGA/2$.

Stability calculations

Weight of the sliding block of rock is 774 kN/m

For a PGA of 0.25·g, and a k value of 0.13·g, the horizontal force $F_H = 774 \cdot 0.13 = 100$ kN

For these slope parameters, and $c = U = V = 0$

$FS_{static} = 1.19$ using Equation 7.4

$FS_{seismic} = 0.92$ using Equation 11.10

11.7 NEWMARK DISPLACEMENT ANALYSIS

When a rock slope is subject to seismic shaking, failure does not necessarily occur when the dynamic transient stress reaches the shear strength of the rock. Furthermore, if the factor of safety on a potential sliding surface drops below 1.0 for a limited time(s) during the ground motion, it does not necessarily imply a serious problem. What really matters is the magnitude of permanent displacement caused at the times that the factor of safety is less than 1.0 (Lin and Whitman, 1986; Jibson, 1993). The permanent displacement of rock and soil slopes as the result of earthquake ground motions can be calculated using a method developed by Newmark (1965). This is a more realistic method of analysing the seismic effects on rock slopes than the pseudo-static method of analysis (see Section 11.6).

11.7.1 Principle of slope displacement due to ground motions

The principle of Newmark's method is illustrated in Figure 11.10 in which it is assumed that the potential sliding block is a rigid body on a yielding base (ground). Displacement of the block occurs when the ground is subjected to a uniform horizontal acceleration pulse of magnitude ($k \cdot g$) and duration t_0. The velocity of the ground is a function of time t and is designated $v_g(t)$, and its velocity at time t is v_b. Assuming a frictional contact between the block and the ground, the velocity of the block will be v_b, and the relative velocity between the block and the ground will be u where

$$u = v_g - v_b \tag{11.14}$$

The resistance to motion is accounted for by the inertia of the block. The maximum force that can be used to accelerate the block is the shearing resistance on the base of the block, which has a friction angle of ϕ degrees. This limiting force is proportional to the weight of the block (W) and is of magnitude ($W \cdot \tan \phi$), corresponding to a yield acceleration ($k_y = g \cdot \tan \phi$), as shown by the dashed line on the acceleration plot (Figure 11.10b). The shaded area shows that the ground acceleration pulse exceeds the acceleration of the block, resulting in slippage.

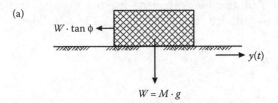

(a)

$W \cdot \tan \phi$

$W = M \cdot g$

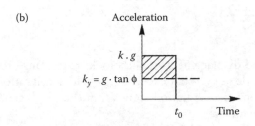

(b)

Acceleration

$k \cdot g$

$k_y = g \cdot \tan \phi$

t_0 Time

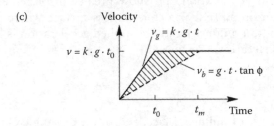

(c)

Velocity

$v_g = k \cdot g \cdot t$

$v = k \cdot g \cdot t_0$

$v_b = g \cdot t \cdot \tan \phi$

t_0 t_m Time

Figure 11.10 Displacement of rigid block on rigid base: (a) block on moving base; (b) acceleration plot; (c) velocity plot (Newmark, 1965).

Figure 11.10c shows the velocities as a function of time for both the ground and the block accelerating forces. The maximum velocity for the ground accelerating force has a magnitude v, which remains constant after an elapsed time of t_0. The magnitude of the ground velocity v_g is given by

$$v_g = (k \cdot g) \cdot t_0 \tag{11.15}$$

while the velocity of the block v_b is

$$v_b = g \cdot t \cdot \tan \phi \tag{11.16}$$

After time t_m the two velocities are equal (v) and the block comes to rest with respect to the base, that is, the relative velocity, $u = 0$. The value of t_m is calculated by equating the ground velocity to the block velocity to give the following expression for the time t_m

$$t_m = \frac{v}{g \cdot \tan \phi} \tag{11.17}$$

The displacement δ_m of the block relative to the ground at time t_m is obtained by computing the area of the shaded region in Figure 7.13c as follows:

$$
\begin{aligned}
\delta_m &= \frac{1}{2} v \cdot t_m - \frac{1}{2} v \cdot t_0 \\
&= \frac{v^2}{2 \cdot g \cdot \tan\phi} - \frac{v^2}{2 \cdot a \cdot g} \\
&= \frac{v^2}{2 \cdot g \cdot \tan\phi} \cdot \left(1 - \frac{\tan\phi}{k}\right)
\end{aligned}
\tag{11.18}
$$

Equation 11.18 gives the displacement of the block in response to a single acceleration pulse (duration t_0, magnitude $(k \cdot g)$) that exceeds the yield acceleration $(g \cdot \tan\phi)$, assuming infinite ground displacement. The equation shows that the displacement is proportional to the square of the ground velocity.

While Equation 11.18 applies to a block on a horizontal plane, a block on a sloping plane will slip at a lower yield acceleration and show greater displacement, depending on the direction of the acceleration pulse. For a cohesionless surface where the factor of safety of the block FS is equal to $(\tan\phi / \tan\psi_p)$ and the applied acceleration is horizontal, Newmark shows that the yield acceleration k_y is given by

$$
k_y = (FS - 1) \cdot \sin\psi_p
\tag{11.19}
$$

where ϕ is the friction angle of sliding surface, and ψ_p the dip angle of this surface. Note that for $\psi_p = 0$, $k_y = g \cdot \tan\phi$. Also Equation 11.19 shows that for a block on a sloping surface, the yield acceleration is higher when the acceleration pulse is in the downdip direction compared to the pulse in the updip direction (Figure 11.11).

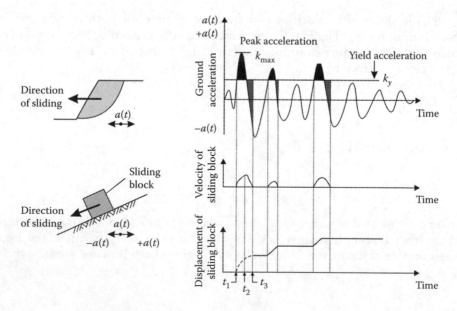

Figure 11.11 Integration of ground motion accelerogram to determine downslope movement (Federal Highway Administration, 1998a).

The displacement δ_m of a block on an inclined plane can be calculated by combining Equations 11.18 and 11.19 as follows:

$$\delta_m = \frac{(k \cdot g \cdot t)^2}{2 \cdot g \cdot k_y} \cdot \left(1 - \frac{k_y}{k}\right) \qquad\qquad (11.20)$$

In an actual earthquake, the first pulse would be followed by a number of pulses of varying magnitude, some positive and some negative, which will produce a series of displacement pulses. This method of displacement analysis can be applied to the case of a transient sinusoidal acceleration $(a(t)g)$ illustrated in Figure 11.11. If, during a period of the acceleration pulse, the shear stress on the sliding surface exceeds the shear strength, displacement will take place. Displacement will take place more readily in a down slope direction; this is illustrated in Figure 11.11, where the shaded areas are the portion of each pulse in which movement takes place.

Integration of the yield portions of the acceleration pulses gives the velocity of the block. It will start to move at time t_1 when the yield acceleration is exceeded, and the velocity will increase up to time t_2 when the acceleration drops below the yield acceleration. The velocity drops to zero at time t_3 as the acceleration direction begins to change from down slope to up slope. Integration of the velocity pulses gives the displacement of the block, with the duration of each displacement pulse being $(t_3 - t_1)$.

The simple displacement models shown in Figures 11.10 and 11.11 have since been developed to more accurately model displacement due to actual earthquake motions. With respect to slope stability, Jibson developed procedures for estimating the probability of landslide occurrence as a function of Newmark displacement based on observations of landslides caused by the 1994 Northridge earthquake in California (Jibson, Harp and Michael, 1998). Also, correlations have been developed between Newmark displacement and Arias intensity and the yield acceleration of the ground motions.

Some of the assumptions of the Newmark analysis are that the sliding mass is a rigid-plastic body that does not deform internally and deforms plastically at constant stress along a discrete basal surface when the yield acceleration is exceeded. Other assumptions are that the yield acceleration is not strain-dependent and thus remains constant throughout the analysis, and that no upslope displacement is permitted.

11.7.2 Thresholds of allowable slope movement

Newmark displacement analysis is useful for design where guidelines are available on the relationship between slope stability and the calculated displacement. While the Newmark method of analysis is highly idealised and the calculated displacements should be considered order-of-magnitude estimates of actual field behaviour, California Division of Mines and Geology (1997) has developed the guidelines on likely slope behaviour as listed in Table 11.3.

When applying these displacement criteria in rock slope design, consideration should be given to the amount of displacement that will have to occur before the residual shear strength is reached. For example, if the sliding surface is a persistent discontinuity surface containing a weak infilling, 20–150 mm (0.8–6 in.) of movement may be sufficient for the strength to be reduced to the residual value; movements of this magnitude may also be the threshold for shallow slides. In contrast, a weathered rock or a soil may undergo up to a metre of displacement with little reduction in shear strength. Section 11.1.2 showed that rock falls and slides are the most abundant types of slope instability, compared to soil slides, which is indicative of the sensitivity of rock slopes to failure in response to limited slope movement.

Table 11.3 Guidelines for relationship between slope displacement and likely slope instability

Calculated displacement	Slope stability condition
0–150 mm (0–6 in.)	Unlikely to correspond to serious landslide movement and damage to structures
150–1000 mm (6–40 in.)	Slope deformations may be sufficient to cause serious ground cracking or enough strength loss to result in continuing post-seismic failure
>1000 mm (>40 in.)	Damaging landslide movement and slopes should be considered unstable

Source: Adapted from California Division of Mines and Geology (CDMG). 1997. *Guildelines for Evaluating and Mitigating Seismic Hazards in California.* Califiornia Div. of Mines and Geology. Special Report 117. www/consrv.ca.gov/dmg/ pubs/sp/117/

11.7.3 Spectral acceleration characteristics

Methods of reporting earthquake ground motions include time histories as shown in Figure 11.5, and acceleration (or velocity or displacement) response spectra plots that provide quantitative information on both the intensity and frequency content of the acceleration–time history (Figure 11.12). The plots in Figure 11.12 show the average spectra for (a) rock and (b) soil sites for the western United States (WUS) and central and eastern United States (CEUS), for magnitude 5–8 earthquakes. The spectra are, of course, not universally applicable due to the wide variety of site conditions, but do provide a guideline on possible acceleration values to use in design. On the plots, the acceleration is normalised to PGA, that is, the acceleration at each frequency is divided by the maximum acceleration (PGA). The plots show that rock and soil sites are generally similar except for the ground motions in rock in CEUS, where high accelerations occur at relatively high frequencies compared to the WUS.

The following sections demonstrate how the spectral acceleration plots are used in the calculation of Newmark displacements.

11.7.4 Calculation of Newmark sliding block displacement

Slope stability analysis using the Newmark displacement method involves calculating slope displacement by the methods described in this section, and then comparing this number with the guideline values for thresholds of allowable slope movement listed in Table 11.3. That is, displacement in excess of 150 mm (6 in.) may result in damage to structures.

Calculation of Newmark displacement involves the use of equations developed from correlation of slope movement with strong-motion time acceleration records as demonstrated in the examples described in this section. The two methods of calculating the displacement are to use either the ratio between the yield acceleration and the maximum acceleration, or to use the Arias intensity of the ground motions. Selection of the appropriate analysis method would depend on the information available on the required input parameters.

Yield acceleration/maximum acceleration displacement calculation – correlations have been made of slope movement for ground motion records in the United States (Transportation Research Board, 2008a). The calculations showed that different correlation equations apply to the WUS and the CEUS, and for rock and soil sites.

Equation 11.21 provides the mean displacement d, for WUS, rock and soil sites, and for CEUS, soil sites

$$\log(d) = -1.51 - 0.74 \cdot \log\left(\frac{k_y}{k_{max}}\right) + 3.27 \cdot \log\left(1 - \frac{k_y}{k_{max}}\right) - 0.8 \cdot \log(k_{max}) + 1.58 \cdot \log \text{PGV}$$

(11.21)

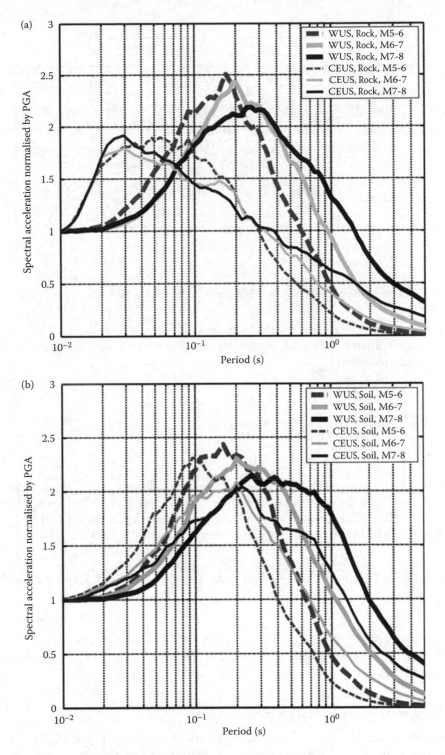

Figure 11.12 Average normalised spectral acceleration records. (a) Rock records; (b) soil records, for WUS and CEUS (Transportation Research Board, 2008a)

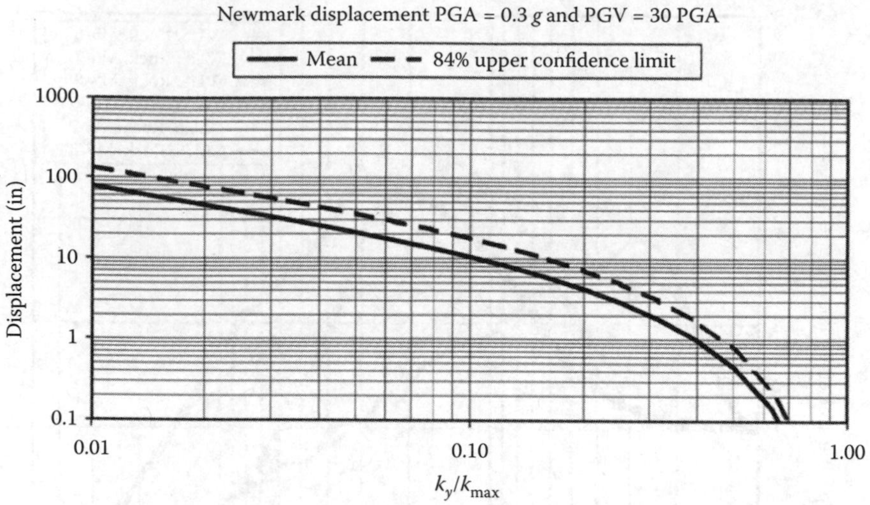

Figure 11.13 Mean Newmark displacement and 84% confidence level for PGA = 0.3 · g and PGV = 30 · PGA.

where k_y is the yield acceleration for the slope as defined by Equation 11.19, and k_{max} is the maximum acceleration making allowance for wave scattering slope height effect, α (see Equation 11.12), such that

$$k_{max} = \alpha \cdot F_{PGA} \cdot PGA \tag{11.22}$$

The peak ground velocity, PGV, is defined by

$$PGV = 55 \cdot F_v \cdot S_1 \; (in./s) \tag{11.23}$$

where F_v is a site factor related to the soil or rock conditions in the slope as defined in Table 11.2, and S_1 is the spectral acceleration at a period of 1 s that is found from the spectral acceleration plots for rock and soil such as that shown in Figure 11.12.

Equation 11.24 provides the mean displacement d, for CEUS, rock sites

$$\log(d) = -1.31 - 0.93 \cdot \log\left(\frac{k_y}{k_{max}}\right) + 4.52 \cdot \log\left(1 - \frac{k_y}{k_{max}}\right) - 0.46 \cdot \log(k_{max}) + 1.12 \cdot \log PGV \tag{11.24}$$

The units applicable to Equations 11.21 and 11.24 are: d – in.; PGV – in./s.

Figure 11.13 shows an example of a plot of acceleration ratio (k_y/k_{max}) against displacement d based on Equation 11.21.

11.7.5 Newmark slope design procedure

As a summary of the slope design procedures discussed in this section, the following are the suggested steps for calculating the Newmark displacement analysis of a slope:

1. Conduct the static limit equilibrium slope stability analysis to check whether the static factor of safety is adequate for the project, usually at least 1.5.

2. Establish the PGHA for the site for the applicable return period.
3. Determine the site class corresponding to site geology (Table 11.1).
4. Determine the value for site factor F_{PGA} corresponding to site class and value of PGA (Table 11.1).
5. From spectral acceleration plot, find the value of spectral acceleration S_1 at a period of 1 s; for normalised spectra (e.g. Figure 11.12), S_1 = (normalised acceleration) · (PGA).
6. Determine value for spectral acceleration coefficient F_v corresponding to site class and value of S_1 (Table 11.1).
7. Calculate the height reduction factor α from Equation 11.12.
8. Modify the PGA to account for site geology and wave scattering slope height effects: $k_{max} = (\alpha \cdot F_{PGA} \cdot PGA)$.
9. Calculate the yield acceleration for the slope from Equation 11.19 using the static factor of safety established in step 1 above.
10. Establish the acceleration ratio (k_y/k_{max}).
11. Calculate the peak ground velocity PGV from Equation 11.23.
12. Calculate the slope displacement from Equation 11.21 or 11.24, to suit site conditions.
13. Compare the calculated slope displacement with guidelines on relationship between displacement and stability listed in Table 11.3.

EXAMPLE OF NEWMARK DISPLACEMENT ANALYSIS

The rock slope (planar model) is shown in Figure 11.9a; the displacement analysis is conducted as follows:

Slope parameters

Height $H = 15$ m (50 ft); distance of tension crack behind crest $b = 9$ m (30 ft), face angle $\psi_f = 76°$; sliding plane angle $\psi_f = 40°$; shear strength of sliding plane: $c = 0$, $\varphi = 45°$; geology – moderately weathered rock, site class C (Table 11.1).

Seismic calculations

Assume site in WUS with earthquake magnitude 7 with PGA = 0.25·g. From spectral acceleration plot for rock slopes (Figure 11.12a), normalised spectral acceleration = 0.9, and $S_1 = (0.9 \cdot PGA) = (0.9 \cdot 0.25) = 0.225 \cdot g$.

From Table 11.1 for site class C and PGA = 0.25·g, site acceleration coefficient $F_{PGA} = 1.2$.

From Table 11.2 for site class C and $S_1 = 0.23·g$, spectral acceleration coefficient $F_v = 1.6$.

In Equation 11.13, parameter $\beta = (F_v \cdot S_1/PGA) = (1.6 \cdot 0.9) = 1.44$, and for 50 ft slope height (H), reduction factor $\alpha = 0.86$.

Calculate the yield acceleration using Equation 11.19: $k_y = [((\tan 45)/(\tan 40) - 1) \cdot (\sin 40)] = 0.12·g$.

Calculate the PGV using Equation 11.23: PGV = 55 · 1.6 · 0.225 = 20 in./s.

Seismic coefficient $k_{max} = (PGA \cdot F_{PGA} \cdot \alpha) = (0.25 \cdot 1.2 \cdot 0.86) = 0.258·g$.

Establish the acceleration ratio: $k_y/k_{max} = (0.12/0.247) = 0.47$.

Displacement calculations

Calculate the displacement using Equation 11.21: $d = 0.31$ in. (7.9 mm).

If the friction angle is reduced from 45° to 42°, then $k_y = 0.05·g$, $k_y/k_{max} = (0.05/0.26) = 0.19$ and $d = 4.7$ in. (120 mm).

These two displacement calculations show the effect of the reduced friction angle (i.e. reduced shear resistance) is to increase the slope displacement.

These calculated displacements are consistent with the values defined by the plot of k_y/k_{max} versus displacement in Figure 11.13. Comparison of these displacements with the guidelines on slope stability listed in Table 11.3 shows that damage to structures is unlikely.

11.7.6 Arias intensity displacement calculation

Section 11.4.2 discusses how the Arias intensity, I_a, can be developed from the acceleration–time history of the strong-motion record, and how the intensity is an expression of the energy content of the earthquake. The Arias intensity of an earthquake can be found from integration of the acceleration–time history (Equation 11.6), or from empirical correlations where I_a is a function of moment magnitude M_w, rupture distance R and site factors related to the geology and fault type (Equation 11.7).

The Arias intensity of design ground motions can be used to calculate the Newmark displacement of a slope. A rigorous analysis of the displacement would require double integration of the portions of strong-motion accelerations that exceed yield acceleration, an activity that is both impractical and inappropriate. In order to facilitate the Newmark analysis, Jibson (2007) developed a simplified method wherein an empirical regression equation is used to estimate Newmark displacement as a function of Arias intensity I_a, and yield acceleration k_y. The most recent form of this equation, based on the analysis of 875 single component strong-motion records from 30 earthquakes, is

$$\log d = 2.401 \cdot \log I_a - 3.481 \cdot \log k_y - 3.23 \pm 0.656 \tag{11.25}$$

The units in this equation are as follows: displacement d in centimetres; Arias intensity I_a in m/s; yield acceleration k_y in g's.

Equation 11.25 can be readily applied by the use of the programme SLAMMER (Jibson, Rathje, Jibson et al., 2013) that can be downloaded from the USGS website. SLAMMER incorporates a total of 11 regression equations that use various combinations of yield acceleration, peak acceleration and velocity, Arias intensity and magnitude to calculate displacement. In addition, analyses can be carried for rigid blocks, blocks that deform internally (decoupled analysis) and analyses that take into account the effect of plastic sliding displacement of the ground motions (coupled analysis). Another function of SLAMMER is to calculate the probability of failure as a function of the displacement based on the following relationship:

$$p(f) = 0.335 \cdot (1 - \exp(-0.048 \cdot d^{1.565})) \tag{11.26}$$

This equation was developed from analysis of about 11,000 landslides that were triggered by the 1994 magnitude 6.7 Northridge earthquake, and therefore is mainly applicable to shallow slides in Southern California.

As an example of the application of Arias intensity to calculate slope movement, consider the slope shown in Figure 11.9, where the yield acceleration is $0.12 \cdot g$. If this slope were located close to the seismic station where the ground motion record of the Loma Prieta earthquake shown in Figure 11.8 ($I_a = 1.59$ m/s) was recorded, the calculated displacement for a rigid block is 29 mm (1.1 in.), and the corresponding probability of failure is 0.075.

Chapter 12

Numerical analysis

Loren Lorig * *and Douglas Stead*[†]

12.1 INTRODUCTION

The previous chapters discussed limit equilibrium methods of slope stability analysis for rock bounded by specified slide planes, where limit equilibrium methods are used to estimate the stability of a slope in terms of the factor of safety. However, limit equilibrium methods assume rigid/perfectly plastic material behaviour and are therefore incapable of computing displacements and/or deformation. By contrast, numerical models (sometimes called deformation models or displacement methods) are capable of computing displacements and deformation and can use a variety of material models to simulate rock slope behaviours. Numerical methods are more recent developments than limit equilibrium methods and their use has increased significantly since 2005, in part due to increased computer speed, but also due to more general acceptance by practitioners.

Typical applications for numerical analyses in civil engineering are the study of landslides, where, for example, the slope contains a variety of geological formations or structures, where the topography cannot be modelled by a simple face and upper slope, and the mechanism of instability is more complex than a simple planar slide.

Numerical models are computer programs that attempt to represent the mechanical response of a rock mass subjected to a set of initial conditions such as *in situ* stresses and water levels, boundary conditions and induced changes such as slope excavation. With numerical models, a 'full' solution of the coupled stress/displacement, equilibrium and constitutive equations is made. Given a set of properties, the system is found either to be stable or unstable. If an equilibrium result is obtained, the resultant stresses and displacements at any point in the rock mass can be compared with measured values. If an unstable/collapse result is obtained, the predicted mode of failure is demonstrated. By performing a series of simulations with varying properties, the factor of safety can be found corresponding to the point of instability.

Numerical models divide the rock mass into zones. Each zone is assigned a material model and properties. The material models are idealised stress/strain relations that describe how the material behaves. The simplest model is a linear elastic model, which uses the elastic properties (Young's modulus and Poisson's ratio) of the material. Elastic–plastic models use strength parameters to limit the shear stress that a zone may sustain.[‡] The zones may be connected together, termed a continuum model, or separated by discontinuities, termed a discontinuum model. Discontinuum models allow slip and separation at explicitly located surfaces within the model.

---

* Irasca Consulting Corporation, Minneapolis, Minnesota.
† Simon Fraser University, British Columbia, Canada.
‡ In numerical analysis, the terms 'elements' and 'zones' are used interchangeably. However, the term 'element' is used more commonly in finite element analysis, and the term 'zone' in finite difference analysis.

Numerical models tend to be general purpose in nature – that is, they are capable of solving a wide variety of problems. While it is often desirable to have a general-purpose tool available, it requires that each problem be constructed individually. The zones must be arranged by the user to fit the limits of the geomechanical units and/or the slope geometry. Hence, numerical models often require more time to set up and run than special-purpose tools such as limit equilibrium methods.

The reasons why numerical models are used for slope stability studies are as follows:

- Numerical models can be extrapolated confidently outside their databases in comparison to empirical methods in which the failure mode is explicitly defined.
- Numerical analysis can incorporate key geologic features such as faults and ground water providing more realistic approximations of the behaviour of real slopes than analytic models. In comparison, non-numerical analysis methods such as analytic, physical or limit equilibrium may be unsuitable for some sites or tend to oversimplify the conditions, possibly leading to overly conservative solutions.
- Numerical analysis can help to explain observed physical behaviour.
- Numerical analysis can evaluate multiple possibilities of geological models, failure modes and design options.

12.1.1 Factor of safety determination

For slopes, the factor of safety is often defined as the ratio of the actual shear strength to the minimum shear strength required to prevent failure. A logical way to compute the factor of safety with a finite element or finite difference program is to reduce the shear strength until collapse occurs. The factor of safety is the ratio of the rock's actual strength to the reduced shear strength at failure. This shear-strength reduction technique was used first with finite elements by Zienkiewicz et al. (1975) to compute the safety factor of a slope composed of multiple materials.

To perform slope stability analysis with the shear-strength reduction method (SRM), simulations are run for a series of increasing trial factors of safety (f). Actual shear strength-properties, cohesion (c) and friction angle (ϕ), are reduced for each trial according to the following equations:

$$c_{trial} = \left(\frac{1}{f}\right) c \tag{12.1}$$

$$\phi_{trial} = \arctan\left(\frac{1}{f}\right) \tan \phi \tag{12.2}$$

If multiple materials and/or joints are present, the reduction is made simultaneously for all materials. The trial factor of safety is increased gradually until the slope fails. At failure, the factor of safety equals the trial factor of safety (i.e. $f = FS$). Dawson et al. (1999) show that the shear-strength reduction factors of safety are generally within a few percent of limit analysis solutions when an associated flow rule, in which the friction angle and dilation angle are equal, is used.

12.1.2 Comparison of numerical and limit equilibrium analysis methods

The shear-strength reduction technique has two main advantages over limit equilibrium slope stability analyses. First, the critical slide surface is found automatically, and it is not

Table 12.1 Comparison of numerical and limit equilibrium analysis methods

Analysis result	Numerical solution	Limit equilibrium
Equilibrium	Satisfied everywhere	Satisfied only for specific objects, such as slices
Stresses	Computed everywhere using field equations	Computed approximately on certain surfaces
Deformation	Part of the solution	Not considered
Failure	Yield condition satisfied everywhere; slide surfaces develop 'automatically' as conditions dictate	Failure allowed only on certain pre-defined surfaces; no check on yield condition elsewhere
Kinematics	The 'mechanisms' that develop satisfy kinematic constraints	A single kinematic condition is specified according to the particular geologic conditions

necessary to specify the shape of the slide surface (e.g. circular, log spiral, piecewise linear) in advance. In general, the failure surface geometry for slopes is more complex than simple circles or segmented surfaces. Second, numerical methods automatically satisfy translational and rotational equilibrium, whereas not all limit equilibrium methods do satisfy equilibrium. Consequently, the shear-strength reduction technique usually will determine a factor of safety equal to or slightly less than limit equilibrium methods for similar failure modes.

A summary of the differences between a numerical solution and the limit equilibrium method is shown in Table 12.1.

12.1.3 Typical results

Morgenstern (1992) emphasised that a major role of the factor of safety is to provide an empirical basis to ensure that the slope will function in a serviceable manner. That is, if the factor of safety is chosen correctly, one knows by experience that the deformations will not be excessive. In many slope problems, it is not necessary to know the magnitude of the deformations.

All numerical models can produce displacement and stresses within the modelled region. For this reason, numerical models are sometimes referred to as stress models or deformation models. Numerical models produce displacement rates (i.e. velocities or rates of slope movement) only if the constitutive models include time-dependent (e.g. creep) behaviour. However, velocities can be estimated by computing displacements resulting from different slope conditions that typically represent excavation and/or water pressure changes occurring over the time period during which the changes occurred. In other words, 'time' is related to the excavation sequence and/or depressurisation.

Narendranathan et al. (2013) compiled rates of slope movement and factors of safety from various case histories and these are summarised in Table 12.2. The case histories used to establish this relation are mostly from open pits in strong rock. Given that weak rocks have deformation moduli much less than the moduli of strong rock, often by an order of magnitude, a slope in weak rock could undergo greater deformations than a slope in hard rock at the same level of stability. Hence, it would appear that correlations given in Table 12.2 should be used with extreme caution in weak rocks.

One advantage of numerical analysis is the ability to provide safety factor contours that show the regions of a slope most susceptible to failure. The SRM can be applied to produce multiple potential failure surfaces in one simulation by monitoring failure in terms of the development of unstable regions (defined by high-gridpoint velocities) as the strength of the material is incrementally reduced. Examples from two- and three-dimensional models are shown in Figures 12.1 and 12.2, respectively.

Table 12.2 Rates of slope movement and factors of safety for pit slopes in strong rock suggested by Narendranathan et al. (2013)

Displacement rate (mm/month)	Factor of safety
<1.5	>1.5
1.5 to 3	1.2 to <1.5
3 to 7.5	1.1 to <1.2
>7.5	<1.1

Source: Based on the work of Call, R. D. 1992. Slope stability. In: *SME Mining Engineering Handbook*. Society of Mining Engineers of AIME, Englewood, CO, Vol. 1, pp. 881–96; Sullivan, T. D. 1993. Understanding pit slope movements. In: *Proceedings of the Geotechnical Instrumentation and Monitoring in Open Pit and Underground Mining*, T. Szwedzicki, ed., A. A. Balkema, Rotterdam, pp. 435–45; and Zavodni, Z. M. 2000. Time-dependent movements of open-pit slopes. In: *Proceedings of the Slope Stability in Surface Mining*, W. A. Hustrulid, M. K. McCarter and D. J. A. Van Zyl, eds., Society of Mining, Metallurgy and Exploration, Littleton, CO, pp. 81–7.

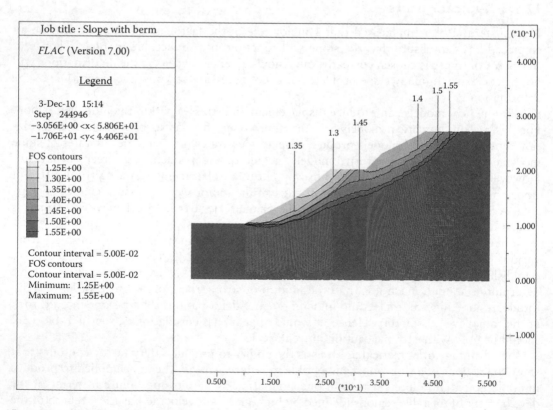

Figure 12.1 Safety factor contours for slope with mid-height berm.

FS	
	<1.0
	1.0–1.1
	1.1–1.2
	1.2–1.3
	≥1.3

Figure 12.2 Factor of safety contours in a three-dimensional model.

12.2 OVERVIEW OF NUMERICAL MODELS

All rock slopes contain discontinuities, and representation of these discontinuities in numerical models differs depending on the type of model, as discussed below.

12.2.1 Discontinuum and continuum models

The basic types of models are: discontinuum models and continuum models. Discontinuities in discontinuum models are represented explicitly – that is, the discontinuities have a specific orientation and location. Discontinuities in continuum models are represented implicitly, with the intention that the behaviour of the continuum model is substantially equivalent to the real jointed rock mass being represented.

Discontinuum codes start with a method designed specifically to model faults and joints (discontinua) and treat continuum behaviour as a special case. These codes are generally referred to as discrete element codes. Two widely used discrete element codes for slope stability studies are *UDEC* (Itasca, 2014b) and its three-dimensional equivalent, *3DEC* (Itasca, 2013). A discrete element code embodies an efficient algorithm for detecting and classifying contacts. It will maintain a data structure and memory allocation scheme that can handle hundreds or thousands of discontinuities. The discontinuities divide the problem domain into blocks. Blocks within discrete element codes may be rigid or deformable, with continuum behaviour being assumed within deformable blocks.

Selection of the structural geometry that defines the shape and extent of these blocks is a crucial step in discontinuum analyses. Typically, only a small percentage of the faults and joints can be included in order to create models of reasonable size for practical analysis. The structural geometry data must be filtered to select only the faults and joints most critical to the mechanical response. This is done by identifying the structures that are most susceptible to slip and/or separation for the prescribed loading condition. This may involve determining whether sufficient kinematic freedom is provided, especially in the case of toppling, and calibrating the analysis by comparing observed behaviour to model response. A termination criterion, stipulating whether the structure terminates in the rock mass or against other faults or joints, is also required. This criterion is fundamental to providing the rock mass with strength derived from rock bridges and other natural rock mass features that are not considered when all the structures in the rock mass have infinite persistence.

Continuum models assume that material is continuous throughout the slope, that is discontinuities are absent. Finite element and finite difference continuum codes that are widely used by slope design practitioners include RS^2 (Rocscience), *FLAC* (Itasca, 2016) and *FLAC3D* (Itasca, 2012). All continuum numerical models divide the rock mass into elements. Each element is assigned a material model and material properties. The material models are stress–strain relations that describe how the material behaves during loading or unloading. Continuum models typically allow for both isotropic and anisotropic material behaviour. Major structural features such as faults are represented as zero-thickness interfaces between major regions of continuum behaviour, or as specific zones with finite thickness.

12.2.2 Analysis methods

Finite element programs are probably more familiar, but the finite difference method is perhaps the oldest numerical technique used to solve sets of differential equations. Both finite element and finite difference methods produce a set of algebraic equations to solve. While the methods used to derive the equations are different, the resulting equations are the same. Finite difference programs generally use an 'explicit' time-marching scheme to solve the equations, whereas finite element methods usually solve systems of equations in matrix form.

Although a static solution to a problem is usually of interest, the dynamic equations of motion are typically included in the formulation of finite difference programs. One reason for doing this is to ensure that the numerical scheme is stable when the physical system being modelled is unstable. With non-linear materials, the possibility of physical instability always exists – for example, the failure of a slope. In real life, some of the strain energy is converted to kinetic energy. Explicit finite difference programs model this process directly, because inertial terms are included. In contrast, programs that do not include inertial terms must use some numerical procedure to treat physical instabilities. Even if the procedure is successful at preventing numerical instability, the *path* taken may not be realistic. The consequence of including the full law of motion in finite difference programs is that the user must have some physical feel for what is happening. Explicit finite difference programs are not black boxes that will 'give the solution'. The behaviour of the numerical system must be interpreted.

FLAC and *UDEC* are two-dimensional finite difference programs developed specifically for geomechanical analysis. These codes can simulate varying loading and water conditions, and have several pre-defined material models for representing rock mass continuum behaviour. Both codes are unique in their ability to handle highly non-linear and unstable problems. The three-dimensional equivalents of these codes are *FLAC3D* (Fast Lagrangian Analysis of Continua in Three Dimensions [Itasca, 2012]) and *3DEC* (Itasca, 2013).

12.2.3 Rock mass material models

It is impossible to model all discontinuities in a large slope, although it may be possible to model the discontinuities for a limited number of benches. Therefore, in large slopes, much of the rock mass must be represented by an equivalent continuum in which the effect of the discontinuities is to reduce the intact rock elastic properties and strength to those of the rock mass. This is true whether or not a discontinuum model is used. As mentioned in the introduction to this chapter, numerical models divide the rock mass into zones. Each zone is assigned a material model and material properties. The material models are stress/strain relations that describe how the material behaves. The simplest model is a linear elastic model that uses only the elastic properties (Young's modulus and Poisson's ratio) of the material.

12.2.4 Rock mass material models: Isotropic materials

The strength of isotropic rock masses may be defined by either the Hoek–Brown or strain-weakening criteria as discussed below.

Mohr–Coulomb – linear elastic–perfectly plastic stress–strain relation is the most commonly used rock mass material models. These models typically use the Mohr–Coulomb strength parameters (cohesion and friction angle) to limit the shear stress that a zone may sustain. The tensile strength is limited by the specified tensile strength, which in many analyses is taken to be 10% of the rock mass cohesion. Using this model, the rock mass behaves in an isotropic manner.

Hoek–Brown – the most common failure criterion for rock masses is the Hoek–Brown failure criterion (see Section 5.5). The Hoek–Brown failure criterion is an empirical relation that characterises the stress conditions that lead to failure in intact rock and rock masses. It has been used successfully in design approaches that use limit equilibrium solutions. It also has been used indirectly in numerical models by finding equivalent Mohr–Coulomb shear-strength parameters according to best-fit lines for specific confining stresses, to the curved rock mass strength envelope defined by the Hoek–Brown failure criterion (see Figure 5.23). The best-fit Mohr–Coulomb parameters are then used in traditional Mohr–Coulomb-type constitutive relations and the parameters may or may not be updated during analyses. The procedure is awkward and time-consuming. Currently, most numerical models include a constitutive relation based on a Hoek–Brown failure criterion. However, it is important to note that the Hoek–Brown failure criterion is generally not applicable to weak rocks, that is, $GSI < 30$, or rock subject to brittle response, that is, $GSI > 80$ (Geological Strength Index, see Section 5.5).

Strain weakening – gradual accumulation of strain on principal structures within the rock mass usually results from excavation, and 'time' is related to the excavation sequence. In order to study the progressive failure effects due to excavation, one must either introduce characteristics of the post-peak or post-failure behaviour of the rock mass into a strain-weakening model or introduce similar characteristics into the explicit discontinuities.

In contrast to elastic–perfectly plastic behaviour, cohesion loss (and tensile strength loss) results in residual strengths that are less than the peak (i.e. yield) strength. These materials are typically simulated using some form of strain-weakening material model as shown in Figure 12.3. The stress–strain curve is linear to the point of yield; in that range, the total

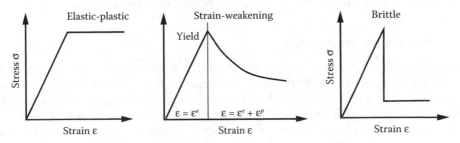

Figure 12.3 Typical behaviour models used in numerical analyses.

strain (ε) is elastic only (i.e. $\varepsilon = \varepsilon^e$). After yield, the total strain (ε) is composed of elastic (ε^e) and plastic parts (ε^p) (i.e. $\varepsilon = \varepsilon^e + \varepsilon^p$). In the general strain-weakening model, the user defines the cohesion, friction, dilation and tensile strength variance as a function of the plastic portion, ε^p, of the total strain. In brittle materials, such as cemented sediments, including leached rocks, cohesion and tensile loss is instantaneous once the peak strength is reached (see Figure 12.3).

In practice, two difficulties are associated with strain-weakening rock mass models. The first is estimating the post-peak strength and the strain over which the strength reduces. Few empirical guidelines exist for estimating the required parameters. This means that the properties must be estimated through calibration. The second difficulty is that, for a simulation in which the response depends on shear localisation and in which material weakening is used, the results will depend on the zone sizes. However, it is quite straightforward to compensate for this form of mesh dependence.

Strain-weakening models require special attention when computing safety factors. If a strain-weakening constitutive model is used, the weakening logic should be turned off during the shear-strength reduction process or the factor of safety will be underestimated. When the slope is excavated, some zones will have exceeded their peak strength, and some amount of weakening will have taken place. During the strength reduction process, these zones should be considered as a new material with lower strength, but no further weakening should be allowed due to the plastic strains associated with the gradual reduction of strength.

12.2.5 Rock mass material models: Anisotropic

Most rock masses are anisotropic to some degree. As noted by Sainsbury et al. (2016), the strength and deformation behaviour of a rock mass is governed strongly: first by the 'intact' strength of rock and, second by weakness surfaces such as joints, bedding planes, foliation and other discontinuities. Anisotropic rock mass strength and deformation behaviour results when a majority of the discontinuities are aligned in a preferred direction such as the shale shown in Figure 12.4.

For anisotropic rock masses, discontinuum analysis techniques provide the most rigorous assessment of the strength and deformation behaviour. In this case, the joint fabric and intact rock are explicitly simulated. However, owing to current computational limitations, it is not possible to explicitly simulate the detailed joint fabric that controls anisotropy in large-scale pit slopes, particularly in three dimensions. As a result, ubiquitous joint constitutive models are commonly used. The ubiquitous joint model is an anisotropic plasticity model that includes implicit weak planes of specific orientation embedded in a Mohr–Coulomb solid. Yield may occur in either the solid or along the weak plane, or both, depending on the stress state, the orientation of the weak plane and the material properties of the solid and weak plane. Anisotropy is generally less pronounced in weak rocks as the fabric strength is often close to the matrix/rock mass strength, except possibly in bedded rocks.

12.2.6 Joint material models

The material model used most commonly to represent joints is a linear elastic–perfectly plastic model. The limiting shear strength is defined by the usual Mohr–Coulomb parameters of friction angle and cohesion. A peak and residual shear-strength relation can also be specified for the joints. The residual strength is used after the joint has failed in shear at the peak strength. The elastic behaviour of the joints is specified by joint normal and shear stiffnesses, which may be linear or piecewise linear.

Figure 12.4 Anisotropic rock mass – shale (from Sainsbury et al., 2008).

12.3 MODELLING CONSIDERATIONS

Numerical models that predict deformations and/or stability to an acceptable degree of accuracy might typically include:

1. Representative stress–strain relations, including behaviour from peak to residual shear strengths
2. Anisotropy
3. Variable pore pressure distributions
4. Non-homogeneity arising from a variation of material properties with depth, layering and/or discontinuities
5. Influence of initial *in situ* stress
6. Staging/benching excavation sequence

All of these factors have an influence on the deformations of a slope, and their determination for a deformation analysis suitable for accurately forecasting slope displacement often requires calibration with unstable or deforming slopes. Consequently, the use of numerical models for deformation analyses is not always routine, but can be used to predict slope stability. Conditions that can affect deformation and/or stability in numerical models are discussed below.

12.3.1 Two-dimensional versus three-dimensional modelling

The first step in creating a model is to decide whether to perform two-dimensional or three-dimensional analyses. Until recently (2016), three-dimensional analyses were uncommon, but advances in personal computers now permit three-dimensional analyses to be performed routinely. Strictly speaking, three-dimensional analyses are recommended/required if

1. The direction of the principal geologic structure does not strike within about 20° of the strike of the slope
2. The axis of material anisotropy does not strike within about 20° of the slope
3. The directions of principal stresses are not parallel or not perpendicular to the slope
4. The distribution of geomechanical units varies along the strike of the slope
5. The slope geometry in plan cannot be represented by two-dimensional analysis, which assumes axisymmetric or plane strain

Despite the foregoing, many design analyses for slopes assume a two-dimensional geometry comprising a unit slice through an infinitely long slope, under plane strain conditions. In other words, the radii of both the toe and crest are assumed to be infinite. This is not the condition encountered in reality – particularly in open pit mining where the radii of curvature can have an important effect on safe slope angles. The convex nature of any 'nose' geometries in the slope may lead to stability issues due to a lack of lateral restraint. Conversely, concave slopes are typically more stable than plain strain slopes due to the lateral restraint provided by material on either side of a potential failure in a concave slope. Lorig and Varona (2004) showed that concave slopes could have safety factors as much as 50% higher than two-dimensional plane strain slopes; Fredlund (2014) found similar results. Consequently, the trend is towards three-dimensional analyses in order to more realistically assess the stability of complex geometries, especially in weaker materials.

12.3.2 Boundary location

Boundaries can be real or artificial. Real boundaries in slope stability problems correspond to the natural or excavated ground surface that is usually stress-free. Artificial boundaries do not exist in reality. Problems in geomechanics, including slope stability problems, usually require that the infinite extent of a real problem domain be truncated artificially to include only the immediate area of interest. Figure 12.5 shows

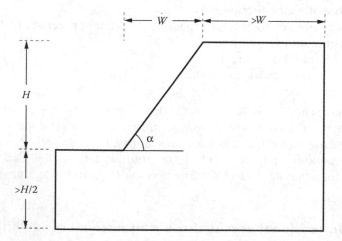

Figure 12.5 Typical recommendations for the locations of artificial far-field boundaries in slope stability analyses.

typical recommendations for locations of the artificial far-field boundaries in mechanical analysis of slope stability problems. Artificial boundaries can be prescribed displacement or prescribed stress. Prescribed displacement boundaries inhibit displacement in the vertical or horizontal direction, or both, and represent the condition at the base of the model and the toe of the slope. Displacement at the base of the model is always fixed in both the vertical and horizontal directions to inhibit rotation of the model and sliding along the base.

Where the slope model incorporates ground water flow analysis, the model boundaries are usually greater than those for a drained slope model to allow for the complete development of the flow net.

Two assumptions are made regarding the displacement boundaries near the toe of any slope. One is that the displacements near the toe are inhibited only in the horizontal direction. This is the mechanically correct condition for a problem that is perfectly symmetric with respect to the plane or axis representing the toe boundary. Strictly speaking, this condition only occurs in slopes of infinite length, which are modelled in two dimensions and assume plane strain, or in slopes that are axially symmetric, in which the pit is a perfect cone. In reality, these conditions are rarely satisfied. Therefore, some models are extended laterally to avoid the need to specify any boundary condition at the toe of the slope. It is important to note that difficulties with the boundary condition near the slope toe are usually a result of the two-dimensional assumptions. This difficulty seldom exists in three-dimensional models.

The far-field boundary locations and conditions must be specified in any numerical model for slope stability analyses. The general notion is to select the far-field location so that it does not significantly influence the results. If this criterion is met, it is not important whether the boundary is prescribed displacement or prescribed stress. In most slope stability studies, a prescribed displacement boundary is used. In some cases, a prescribed stress boundary has been used without significantly differing from the results of a prescribed displacement boundary.

The magnitude of the horizontal stress for the prescribed stress boundary must match the assumptions regarding the initial stress state for the model to be in equilibrium. However, following any change in the model, such as an excavation increment, the prescribed stress boundary causes the far-field boundary to displace towards the excavation while maintaining its original stress value. For this reason, a prescribed stress boundary is also referred to as a following stress or constant stress boundary, because the stress does not change and it follows the displacement of the boundary. Following stresses usually occur where slopes are cut into areas where the topography rises behind the slope. Even where slopes are excavated into an inclined topography, the stresses would flow around the excavation to some extent, depending on the effective width of the excavation perpendicular to the downhill topographic direction.

The effects of boundary conditions on analysis results can be summarised as follows:

- A fixed boundary causes both stresses and displacements to be underestimated, whereas a stress boundary does the opposite.
- The two types of boundary conditions bracket the true solution; conducting tests with smaller models and then averaging the results may lead to a reasonable estimate of the true solution.

One final point to be kept in mind is that for three-dimensional excavations such as quarries or open pits, the stresses are free to flow beneath and around the sides of the excavation.

It is therefore likely that, unless faults of very low strength exist parallel to the analysis plane, a constant stress or following stress boundary will overpredict the stresses acting horizontally.

12.3.3 Initial conditions

Initial conditions are those that existed prior to excavation. The initial conditions of importance at a site are the *in situ* stress field and the ground water conditions. The role of stresses has traditionally been ignored in slope analyses, possibly for the following reasons:

- Limit equilibrium analyses, which are widely used for stability analyses, cannot include the effect of stresses. Nevertheless, they are thought to provide reasonable estimates of stability in many cases, particularly where geological structure is absent.
- Most stability analyses have been performed for soils, where the range of possible *in situ* stresses is more limited than for rocks. Furthermore, many soil analyses have been performed for constructed embankments such as dams, where only gravity-induced stresses are applied.
- Most slope failures are gravity driven and the effects of *in situ* stress are thought to be minimal.
- *In situ* stresses are not routinely measured for slopes and their effects are largely unknown.

One advantage of stress analysis programs such as numerical models is their ability to include pre-excavation initial stress states in the stability analyses and to evaluate their importance. Initial conditions become more important as brittle failure is introduced, because initial stress conditions play a role in progressive rock failure. In general, it is impossible to say what effect the initial stress state will have on a particular problem, because behaviour depends on factors such as the orientation of major structures, rock mass strength and water conditions. Hoek et al. (2009) provide an overview of common effects of *in situ* stress on slope behaviour. Additional observations on the effects of *in situ* stress on stability include the following:

- The larger the initial horizontal stresses, the larger the horizontal elastic displacements, although this is not much help as elastic displacements are not particularly important in slope stability studies.
- Initial horizontal stresses in the plane of analysis that are less than the vertical stresses tend to slightly decrease stability and reduce the depth of significant shearing with respect to a hydrostatic stress state. This observation may seem counterintuitive, as smaller horizontal stresses would be expected to increase stability. The explanation lies in the fact that the lower horizontal stresses actually slightly decrease normal stress on potential shearing surfaces and/or joints within the slope. This observation was confirmed in a *UDEC* analysis of a slope in Peru (Carvalho et al., 2002), where *in situ* horizontal stresses lower than the vertical stress led to deeper levels of joint shearing in toppling structures compared to cases involving horizontal stresses equal to or greater than the vertical stress.
- It is important to note that the regional topography may limit the possible stress states, particularly at elevations above regional valley floors. Three-dimensional models have been very useful in addressing some regional stress issues.

12.3.4 Discretisation

To adequately capture stress and strain gradients within the slope, it is necessary to use relatively fine discretisations near the face. By experience, it has been found that at least 30 lower-order elements (elements with constant or uniform stress) are required over the slope height of interest. Finite element programs using higher-order elements may require fewer zones than the constant strain/constant stress elements common in finite difference codes. If the zone size is uncertain, the best criterion is usually 'convergence': showing that mesh refinement will not significantly change the outcome.

12.3.5 Incorporating water pressure

Pore pressures can play a significant role in slope stability as discussed in Chapter 6 and in Beale and Read (2013). In actively excavated slopes, the slope geometry and hydraulic stresses are constantly evolving, so it is usually necessary to determine the pore pressures acting within a slope by performing transient simulations. For slopes that have existed for some time, steady-state solutions are normally acceptable.

The most common modelling approach for slope stability problems involving ground water currently relies on a two-step process: first, determining the transient or steady-state pore pressure distributions or phreatic surfaces at a particular point in time using flow codes, and second, the pore pressure distributions or phreatic surfaces are mapped into mechanical codes for the determination of slope stability. Most practitioners apply 100% of the predicted water pressures to all components (i.e. rock mass, joints and faults) of the rock. Terzaghi's effective stress principle is used to account for water pressure within the mechanical codes. Typically, little to no 'coupling' or interaction occurs between the two steps.

The most rigorous method of incorporating ground water into the slope model is to perform a complete flow analysis, and use the resultant pore pressures in the stability analyses. A less rigorous, but more common, method is to specify a water table, and the resulting pore pressures are given by the product of the vertical depth below the water table, the water density and gravity. In this sense, the water table approach is equivalent to specifying a piezometric surface. Both methods use similar phreatic surfaces. However, the water table method underpredicts actual pore pressure concentrations near the toe of a slope and slightly overpredicts the pore pressure behind the toe by ignoring the inclination of equipotential lines (Figure 12.6).

Seepage forces must also be considered in the analysis. The hydraulic gradient is the difference in water pressure that exists between the two points at the same elevation and results from seepage forces (or drag) as water moves through a porous medium. Flow analysis automatically accounts for seepage forces.

To evaluate the error resulting from specifying a water table without doing a flow analysis, two identical problems were run. In one case, a flow analysis was performed to determine the pore pressures. In the second case, the pressures were determined using only a piezometric surface that was assumed to be the phreatic surface taken from the flow analysis. The right-hand boundary was extended to allow the far-field phreatic surface to coincide with the ground surface at a horizontal distance of 2 km behind the toe. Hydraulic conductivity within the model was assumed to be homogeneous and isotropic. The error caused by specifying the water table can be seen in Figure 12.6. The largest errors, of up to 45%, are found just below the toe, while errors in pore pressure values behind the slope are generally less than 5%. The errors near the phreatic surface are insignificant, as they result from the relatively small pore pressures just below the phreatic surface where small errors in small values result in large relative errors.

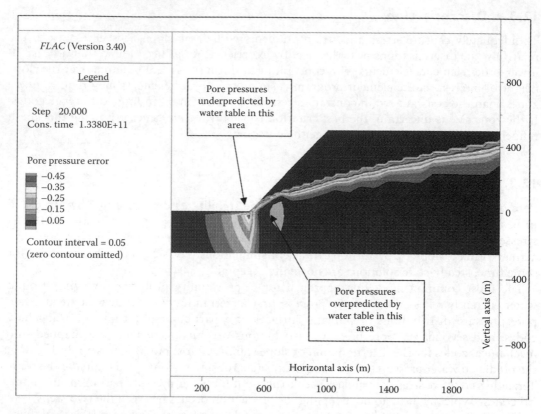

Figure 12.6 Error in pore pressure distribution caused by specifying water table compared to performing a flow analysis.

For a phreatic surface at the ground surface at a distance of 2 km, a factor of safety of 1.1 is predicted using circular failure chart number 3 (refer to Figures 9.4 and 9.8). The factor of safety determined by *FLAC* was approximately 1.15 for both cases. The *FLAC* analyses give similar safety factors because the distribution of pore pressures in the area behind the slope where failure occurs is very similar for the two cases. The conclusion drawn here is that the difference is insignificant in predicted stability between a complete flow analysis and simply specifying a piezometric surface. However, it is not clear if this conclusion can be extrapolated to other cases involving, for example, anisotropic flow.

12.4 SIMPLE STABILITY ANALYSIS

This section compares the results of limit equilibrium stability analyses with the same models analysed by numerical methods.

12.4.1 Rock mass failure

Numerical analysis of slopes involving purely rock mass failure is studied most efficiently using continuum codes such as RS²/PHASE2, FLAC or FLAC3D. As mentioned in Section 12.2.1, discontinuities are not considered explicitly in continuum models; rather, they are assumed to be smeared throughout the rock mass. Assuming that rock mass shear-strength properties can be estimated reasonably, the analysis is straightforward. The process

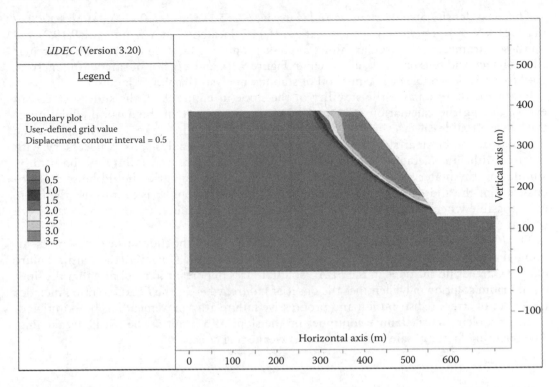

Figure 12.7 Rock mass failure mode for slope determined with *UDEC*.

for initially estimating the rock mass properties is often based on empirical relations as described, for example, by Hoek et al. (2002). These initial properties are then modified, as necessary, through the calibration process.

Failure modes involve mainly shearing through the rock mass. For homogeneous slopes where the slide surface is often approximately circular, the failure surface initiates at the toe and becomes nearly vertical near the ground surface. The failure mode for the parameters listed above is shown in Figure 12.7; the calculated safety factor is 1.64.

The following is a comparison of a slope stability analysis carried out using limit equilibrium circular failure analysis (the Bishop method) and numerical stability analysis. In Section 9.6.3, the stability of a benched slope is strong, but closely fractured sandstone, including a water table and tension crack, is described (see Figure 9.19). The rock mass is classified as a Hoek–Brown material with the following strength parameters:

$m_i = 13$
$GSI = 20$
$\sigma_c = 150$ MPa
Disturbance factor, $D = 0.7$

The tensile strength is estimated to be 0.009 MPa. For Bishop's analysis method, the Mohr–Coulomb strength is estimated by fitting a straight line to the curved Hoek–Brown failure envelope at the normal stress level estimated from the slope geometry. Using this procedure, friction angle and cohesion were (see Table 9.1)

$\phi = 47°$
$c = 0.12$ MPa (17.4 psi)

The mass density of the rock mass and water were 2550 kg/m³ (25.0 kN/m³) and 1000 kg/m³ (9.81 kN/m³), respectively. The phreatic surface is located as shown in Figure 9.19. Based on these parameters, the circular failure analysis produces a location for the critical circular slide surface and tension crack, as shown in Figure 9.19, and a factor of safety ranging from 1.49 to 1.63, depending on the method of stability analysis (Bishop or Janbu).

Using *FLAC* to analyse the stability of the slope in Figure 9.19, the slide surface can evolve during the calculation in a way that is representative of the natural evolution of the physical slide surface in the slope. It is not necessary to make an estimate for the location of the circular slide surface when beginning an analysis, as is necessary with limit-equilibrium methods. *FLAC* will find the slide surface and failure mechanism by simulating the material behaviour directly. A reasonably fine grid should be selected to ensure that the slide surface will be well defined as it develops. It is best to use the finest grid possible when studying problems involving localised failure. In this case, a zone size of 2 m was used.

The *FLAC* analysis showed a factor of safety of 1.26, with the slide surface closely resembling that produced from the Bishop solution (Figure 12.8). However, the tensile failure extends farther up the slope in the *FLAC* solution. It is important to recognise that the limit equilibrium solution only identifies the onset of failure, whereas the *FLAC* solution includes the effect of stress redistribution and progressive failure after movement has been initiated. In this problem, tensile failure continues up the slope as a result of the tensile weakening. The resulting factor of safety allows for this weakening effect.

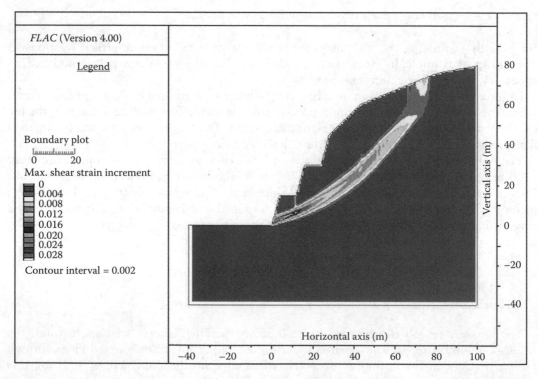

Figure 12.8 Failure mode (circular) and tension crack location determined with *FLAC* for slope in closely fractured sandstone slope (refer to Figure 9.19).

12.4.2 Plane failure

Failure modes that involve rigid blocks sliding on planar joints that daylight in the slope face are most efficiently solved using the analytical methods. For comparison purposes, a *UDEC* analysis is performed for blocks dipping at 35° out of the slope. The joints are assumed to have a cohesion of 100 kPa (14.5 psi) and a friction angle of 40°. The resulting safety factor is 1.32, which agrees with the analytic value given by Equation 7.4 in Chapter 7, assuming that no tension crack forms. The plane failure mode in the *UDEC* analysis is shown in Figure 12.9.

If a tension crack does form, then the factor of safety is slightly reduced. Deformable blocks with elastic–plastic behaviour are required to form tension cracks within the *UDEC* analysis. When deformable zones are used, the resultant safety factor is 1.27, similar to the value of 1.3 given by the analytic solution. The difference may be that the analytic solution assumes a vertical tension crack, whereas the *UDEC* analysis indicates that the tension crack curves where it meets the sliding plane (see Figure 12.10).

Similar analyses can be performed for slopes containing non-daylighting failure planes. In this case, failure involves sliding on discontinuities in the upper part of the slope, and shearing through the rock mass in the lower part of the slope, as shown in Figure 12.11. Here, the cohesionless sliding planes dip at 70° and are spaced 20 m apart. The resultant factor of safety is about 1.5.

This mode of slope failure which is also termed 'dip slope' stability is discussed in Section 7.4, where a zone of crushed rock is developed in the transition zone between the upper, active wedge and the lower, passive wedge that daylights in the face.

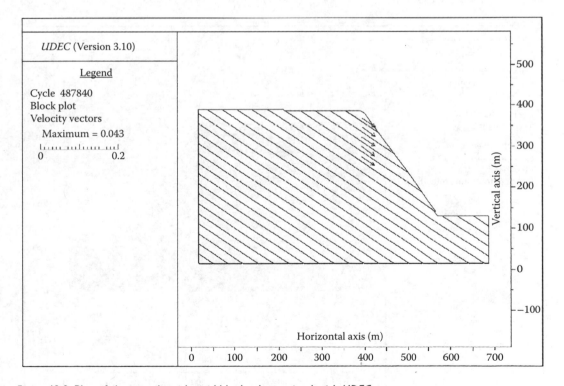

Figure 12.9 Plane failure mode with rigid blocks determined with *UDEC*.

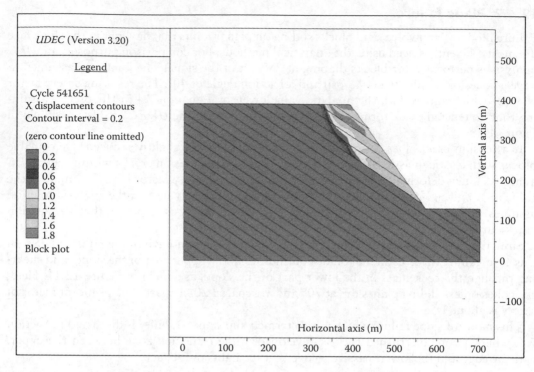

Figure 12.10 Plane failure mode with deformable blocks determined with *UDEC*.

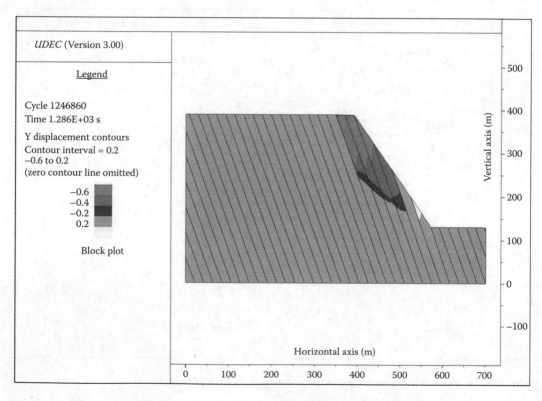

Figure 12.11 Non-daylighting plane failure mode (or dip slope) determined with *UDEC*.

12.4.3 Wedge failure

Analyses involving wedge failures are similar to those involving plane failures, except that the analyses must be performed in three dimensions. As with plane failure, sliding analysis of daylighting rigid blocks is best solved using analytic methods, as described in Chapter 8. Analyses involving formation of tension cracks and/or non-daylighting wedges require numerical analysis. Candidate codes include *FLAC3D* and *3DEC*. The plasticity formulation in *FLAC3D* uses a mixed discretisation technique and presently provides a better solution than *3DEC* in cases where rock mass failure dominates. On the other hand, setting up problems involving more than one sliding plane in *FLAC3D* is more difficult and time consuming than similar problems in *3DEC*. The reader is referred to Hungr and Amann (2011) for comparison of *3DEC* with limit equilibrium studies of rock slope wedges.

12.4.4 Toppling failure

Toppling failure modes involve rotation and thus usually are difficult to solve using limit equilibrium methods. As the name implies, block toppling involves free rotation of individual blocks (Figure 10.3), whereas flexural toppling involves bending of rock columns or plates (see Figure 10.4).

As discussed in Chapter 10, block toppling occurs where narrow slabs are formed by joints dipping steeply into the face, combined with flatter cross-joints. The cross-joints provide release surfaces for rotation of the blocks. In the most common form of block toppling, the blocks, driven by self-weight, rotate forward out of the slope. However, backward or reverse toppling can also occur when joints parallel to the slope face and flatter cross-joints are particularly weak. In cases of both forward and backward toppling, stability depends on the location of the centre of gravity of the blocks relative to their base (see Figure 1.11).

Figure 12.12 shows the results of an analysis involving forward block toppling. The steep joint set dips at 70° with a spacing of 20 m (66 ft), while the cross-joints are orthogonal and spaced at 30 m (100 ft). The resultant safety factor is 1.13. Figure 12.13 shows the result of an analysis involving backward block toppling. In this case, the face-parallel joints are spaced at 10 m (33 ft), and the horizontal joints are spaced at 40 m (130 ft). The factor of safety for this failure mode is 1.7.

Flexural toppling occurs when one dominant, closely spaced, set of joints dips steeply into the face, with insufficient cross-jointing to permit free rotation of blocks. The columns bend out of the slope like cantilever beams. Figure 12.14 shows the results of analysis with joints spaced at 20 m. The factor of safety is 1.3, with the safety factor being reduced as the joint spacing decreases because narrower blocks have a greater tendency to topple than wide blocks (see Figure 1.11). Problems involving flexural toppling require finer zoning than problems involving block toppling. Because flexural toppling involves high stress gradients across any rock column, it is necessary to provide sufficient zones to accurately represent the stress gradients due to bending. In the modelling of centrifuge tests reported by Adhikary and Guo (2002), *UDEC* modelling required four zones across each column, resulting in a model with nearly 20,000 three-noded triangular zones.

12.4.5 Complex kinematics

Many landslides involve complex kinematics with a combination of multiple failure modes such as two-dimensional/three-dimensional sliding, toppling, rock fall and intact rock fracture. The importance of variations in the orientation of the failure surface, both along dip and dip direction, may have implications for the observed deformation and failure mechanism

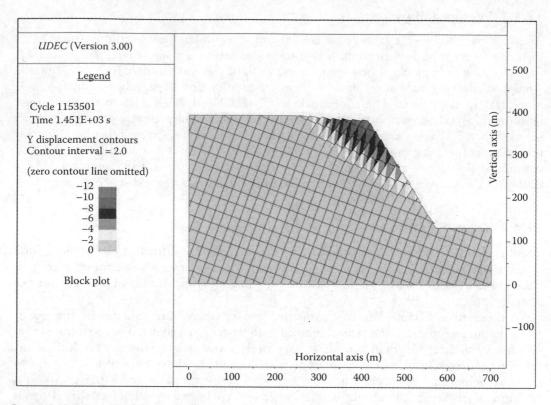

Figure 12.12 Forward block toppling failure mode determined with UDEC.

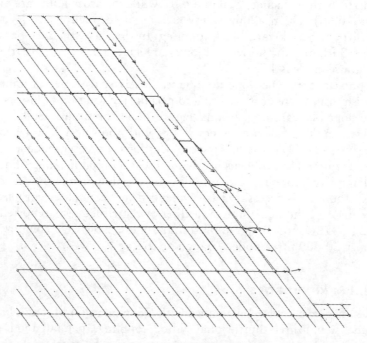

Figure 12.13 Reverse (backward) block toppling failure mode determined with UDEC; arrows show movement vectors.

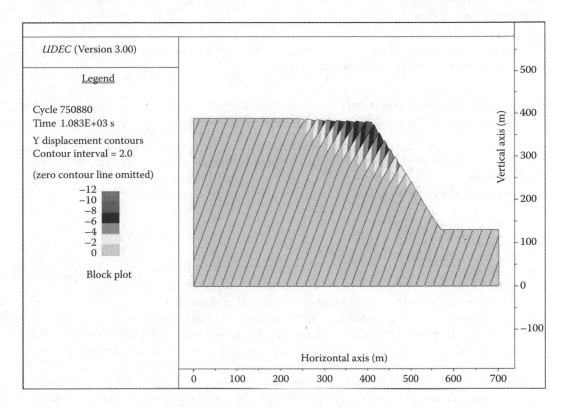

Figure 12.14 Flexural toppling failure mode determined with *UDEC*.

(Stead and Eberhardt, 2013). The nature of the kinematic release mechanisms of a land-slide, both the flanks and main scarp (see Figure 1.7), must be considered in detail before deciding on a representative numerical modelling approach. The decision as to whether a two-dimensional or three-dimensional analysis is undertaken can be subject to consider-able model uncertainty if the kinematic controls on movement are not fully understood. A toolbox approach is often valuable using simple methods to understand three-dimensional controls before moving on to more sophisticated numerical approaches.

Brideau (2010) provides an illustrative account of the kinematic controls on major land-slides using both detailed engineering geological mapping and subsequently a combined block-theory-numerical modelling approach. A detailed investigation of the factors influenc-ing planar and toppling failure of rock slopes is undertaken in two-dimensional and three-dimensional models using the *UDEC/3DEC* codes; the importance of joint persistence is also considered. Havaej (2015) illustrate the use of simple limit equilibrium pentahedral wedge analyses incorporating a basal sliding surface (*Swedge* – Rocscience), prior to more complex three-dimensional distinct element/discrete fracture models applied to the investi-gation of rock slope stability. It is essential to examine the geomorphological evidence of slope movements and, most importantly, slope monitoring records wherever available to allow the degree of complexity of failure mechanisms to be ascertained before developing slope models. Modelling of the Aknes slope in Norway provides a useful reference on the integration of engineering geological mapping, rock testing and slope monitoring into the development of a three-dimensional time-dependent continuum model that incorporated a rock bridge degradation approach (Grøneng et al., 2010).

12.5 SPECIAL TOPICS

This section describes a number of techniques that can be used in numerical analysis to model specific sites features such as reinforcement.

12.5.1 Stabilisation methods

Slope stabilisation options typically include rock reinforcement, support buttresses, ground water depressurisation and re-sloping, all of which can be simulated in a straightforward manner within numerical models. Perhaps the most challenging simulation is reinforcement, which is often used to stabilise civil slopes, and occasionally critical mine slopes. The different types of reinforcement typically represented in numerical models are

- Fully grouted rock bolts
- Cable bolts or tendons
- End-anchored rock bolts
- Vertical piles

In assessing the support provided by rock reinforcement, two components of restraint should be considered. First, reinforcement provides local restraint where it crosses discontinuities. Second, the intact rock is restrained due to inelastic deformation in the failed region surrounding an excavation. Such situations arise in modelling inelastic deformations associated with failed rock and/or reinforcement systems such as cable bolts, in which the cement or resin grout bonding agent may fail in shear over some length of the reinforcement. Cable elements allow the modelling of a shearing resistance along their length, as provided by the shear resistance generated by the bond between the grout and either the cable or the rock. The cable is assumed to be divided into a number of segments of length L, with nodal points located at each segment end. The mass of each segment is lumped at the nodal points, as shown in Figure 12.15. Shearing resistance is represented by spring/slider connections between the structural nodes and the rock in which the nodes are located.

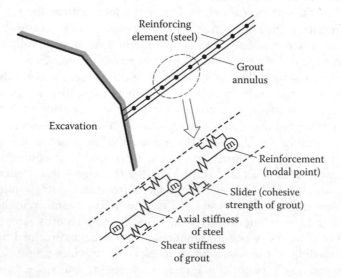

Figure 12.15 Conceptual mechanical representation of fully bonded reinforcement, accounting for shear behaviour of the grout annulus.

End-anchored rock bolts are the simplest to model. They simply supply axial restraint to the portions of the model in which they are anchored where the axial stiffness, K, is given by

$$K = \frac{AE}{L} \qquad (12.3)$$

where A is the cross-sectional area of the bolt, E the modulus of the steel and L is the distance between the anchoring point and the slope face.

12.5.2 Input data from remote sensing

Remote sensing using techniques such as LiDAR or digital photogrammetry has proved extremely useful in characterising rock slope geometry, failure surface geometry and discontinuity characteristics in addition to rock mass quality and seepage. Remote sensing can also provide important model constraint through monitoring of slope movements. A LiDAR or photogrammetry point cloud representing the rock slope or landslide geometry can be processed to form a digital terrain model, or DTM, using a wide variety of photogrammetry and LiDAR point cloud processing software (Figure 12.16). Havaej (2015) describes a workflow consisting of simplification of DTMs using public domain software and subsequent pre-processing using *Rhino* (McNeel and Associates, 2016) and *Kubrix* (Itasca, 2015) to allow import of DTMs into two-dimensional and three-dimensional geomechanical models.

The DTM allows accurate representation of the landslide geometry and for failed slopes, the rupture surface, and flank and main scarp release surfaces. The remote-sensing DTM can also be used to measure discontinuity dip and dip direction to identify discontinuity sets and characterise the geometry of discontinuity sets (spacing, intensity and trace length/continuity). Discontinuity roughness can also be measured using both LiDAR and photogrammetry.

Measurement of discontinuity orientation, intensity, continuity and termination nature can be used to stochastically generate discrete fracture networks (DFNs) that can be imported into (or generated within) geomechanical models. Both photogrammetry and LiDAR have been used to undertake rock mass and seepage characterisation for rock slopes. Recent important developments in remote sensing, including structure from motion photogrammetry, drone-based sensors, full-waveform LiDAR and thermal and hyperspectral imaging, may have significant potential for improved slope characterisation and data for input into numerical models.

12.5.3 Brittle fracture modelling

Conventional continuum and discontinuum models do not allow the brittle fracture associated with rock slope failures to be accurately simulated. It is now commonly accepted that many slope failures involve brittle fracture from the micro to the macro-scale, and the importance of rock bridge failure between non-persistent joints. To allow simulation of brittle fracture in rock slopes, four main methods have been utilised as discussed below:

1. *Voronoi polygonisation* – the Voronoi polygonisation method can be used to simulate brittle fracture in both continuum (RS^2, Rocscience Inc.) and discontinuum models (*UDEC/3DEC*). The Voronoi contacts are given both intact rock and rock mass properties, and normal/shear stiffness that may fail in either shear or tension, resulting in fracturing that changes the kinematic freedom of the slope. Cracks forming along the Voronoi contacts allow varying magnitudes of displacements depending upon the shape of the Voronoi (polygonal or triangular). The Voronoi element itself can be rigid or deformable. It is important to calibrate the Voronoi DEM models using procedures described in Kazerani and Zhao (2010) and Gao and Stead (2014).

(a)

(b)

Figure 12.16 Example of photogrammetric model of rock slope: (a) ground based LiDAR scan of multi-bench quarry face; (b) model of bench developed from LiDAR point cloud showing grey discs defining three joint sets labelled "floors", "grains" and "shortahs" (after Havaej et al., 2016a).

Alzo'ubi (2009) presents a *UDEC* Voronoi damage model that is used to successfully model both natural and engineered slopes, including the incorporation of joint sets and rock support in the left abutment of the Revelstoke Dam in Canada. Wolter et al. (2015) describe the preliminary use of *UDEC* Voronoi polygonisation to simulate brittle fracture associated with the earthquake that triggered the 1959 Madison Canyon landslide in the United States, while Havaej et al. (2014a) used *UDEC* Voronoi to model bedding parallel translational slope failures. Tuckey (2012) and Vivas Becerra (2014) used *UDEC* Voronoi to investigate damage associated with active–passive biplanar rock slope failures including the effect of pore water pressures on brittle fracturing, and Gao and Stead (2014) show the application of Voronoi in the three-dimensional modelling of brittle fracture within the *3DEC* code.

2. *Particle flow codes (PFC)* – particle flow codes defined by *PFC2D/PFC3D* (Itasca, 2014, 2014a) and *YADE* (Scholtes and Donze, 2012) have been used to model rock slope failures in which intact rock breakage during slope failure is simulated by the failure of bonds between circular or spherical particles (Lisjak and Grasselli, 2011).

Wang et al. (2003) described the application of *PFC2D* to the modelling of footwall coal mine slopes ('dip slopes'), while Poisel and Preh (2008) successfully used *PFC2D/3D* to model initiation and runout of several large landslides. Lorig et al. (2009) describe the use of *PFC2D* incorporating a fracture network to model a large open pit toppling instability at the Chuquicamata mine in Chile, and Scholtes and Donze (2012) describe the use of the YADE particle flow code to simulate rock bridge failure in rock slopes. The application of particle flow codes to large rock slopes, particularly in three dimensions, has been limited due to the computing processing requirements and the need for model calibration.

A successful approach discussed by Mas Ivars et al. (2011) is the use of a synthetic rock mass model in which a DFN is incorporated within *PFC* models of the intact rock to provide an SRM model. Simulation of varying scales of rock mass samples under compressive and tensile loading conditions is undertaken to investigate scale effects on the strength of the rock mass. Based on this SRM test data, rock slope analyses have been undertaken using a continuum strain weakening ubiquitous joint approach in *FLAC3D* (Sainsbury et al., 2008, 2016); this method has significant potential for large-scale slope problems.

3. *Lattice-spring method* – a recently (2016) introduced method of modelling landslides is the lattice-spring method in which the particles of a *PFC* model are replaced by nodes, and the bond–bond contacts are represented by springs, where the intact rock fracture in a rock slope is represented by the breakage of the springs.

Cundall and Damjanac (2009) describe the application and verification of the *slope model* code (Itasca, 2015), a lattice-spring-based method. *Slope model* is a three-dimensional code that can incorporate DFNs and can simulate stress-induced brittle fracture within a rock slope. This code is also able to simulate brittle fracture and hydro-mechanical coupling. Havaej et al. (2014a,b) show the application of *slope model* and the lattice-spring method to investigate the brittle fracture associated with the 1963 Vajont rock slide, non-daylighting/toe-breakout wedge failures and footwall slope failures. Tuckey (2012), Tuckey et al. (2013) and Vivas Becerra et al. (2015) show the successful application of *slope model* simulations to active–passive wedge slope failures.

4. *Finite discrete element (FDEM)* – the fourth method of simulating brittle fracture in rock slopes is based on the FDEM approach (Munjiza et al., 1995). Eberhardt et al. (2004) show the application of the FDEM code, *ELFEN* (Rockfield, 2016), to the simulation of brittle fracture associated with the 1991 Randa rock slide in Switzerland, while Havaej et al. (2014a) present the application of *ELFEN* in modelling of footwall/bedding parallel slope failure mechanisms. Vyazmensky et al. (2010), Elmo et al. (2011) and Hamdi et al. (2014) show the successful use of the *ELFEN* code incorporating DFNs, FDEM-DFN, in the brittle fracture modelling of rock, while Lisjak and Grasselli (2011) present the application of the FDEM approach to modelling of brittle fracture in rock slopes using the *YGEO* and *IRAZU* codes (Geomechanica Inc., 2015).

12.5.4 Short-term versus long-term stability

Most geomechanical models and slope analyses are focussed on the short-term stability of a slope. However, the importance of geological and geomorphic processes in engineering geology and the need to consider landform evolution have been emphasised by Griffiths et al. (2012). Kemeny (2003) provides an insightful discontinuum *UDEC* analysis showing the importance of time-dependent degradation on the stability of rock slopes over periods of hundreds to thousands of years, and Grøneng et al. (2010), using a time-dependent constitutive criterion based on landslide slope monitoring data, considered the importance of rock bridge failure on the stability of the Aknes slide in Norway.

Numerous researchers have emphasised that the stress-driven failure of rock bridges in rock slopes may have an important role in the long-term stability of slopes. Tang et al. (2015) have described the application of damage-based geomechanical models constrained against microseismic records to demonstrate the importance of rock slope damage on the performance of the left-bank rock slope of the Jingping I hydropower station. The long-term rock slope stability may be related to numerous processes resulting in strength degradation. For example, Alzo'ubi (2009) emphasised the importance of weathering on the reduced tensile strength of the rock in modelling of the Revelstoke Dam in Canada abutment slope using a *UDEC* Voronoi damage model.

The importance of stress cycling and fatigue in rock slopes has recently been modelled by several workers, including the effects of thermal variations (Gischig, 2011), seasonal ground water table movements (Smithyman, 2007) and repeated earthquake activity (Wolter et al., 2015). These factors have all been suggested as conditioning factors that may result in the reduction of the long-term stability of rock slopes.

12.5.5 Runout prediction

This section discusses the runout distance of landslides based on both empirical studies and numerical modelling.

Empirical (Fahrböschung) – the runout of a landslide is often a critical factor in assessing the associated risk. Based on studies of the literature on landslide runout, empirical approaches have been proposed recognising the so-called 'fahrböschung' angle that links the volume of the landslide and the length of runout. Figure 12.17 shows the definition of the fahrböschung angle and a graph of source volume versus runout from Whittall (2015). An extensive volume of literature exists of landslide runout and the reader is encouraged to refer to work by Corominas (1996), McDougall et al. (2012), Whittall (2015) and McKinnon (2010). Whittall (2015) provides a current summary of landslide runout literature and a discussion of landslide runout terminology, including the fahrböschung angle (and relationships with slope angle, failure volume and potential energy), the travel angle, runout length, excessive travel distance and inundation area. The reader is referred to Hungr et al. (1984), Hungr (2005) and Rickemann (2005), who discuss empirical approaches and important concepts such as material entrainment, volume-balance approaches, yield rate (volume of material eroded per metre of channel length) and erosion depth.

Numerical modelling of landslide runout – an area of considerable current research involves the numerical modelling of landslide runout. Models have been based on particle flow code approaches (Poisel and Preh, 2008), as well as continuum finite element and rheological approaches. McDougall et al. (2012) provide an excellent summary and discussion of landslide runout modelling approaches. Common models used in the simulation of runout include *DANW* and *DAN3D* (Hungr, 1995; McDougall and Hungr, 2004). *DANW* enables modelling of runout on two-dimensional geometrical sections but is considered a one-dimensional approach as it uses a depth-averaging approach. *DAN3D*, based on a smooth-particle, hydrodynamic approach, allows simulation of runout with digital elevation models as input; it also, however, uses a depth-averaging approach and is therefore a two-dimensional model. Runout models may allow the simulation of runout length, velocity, division of flow around topographic obstacles and entrainment of material over which the debris passes. The rheological models that are used to simulate runout include simple

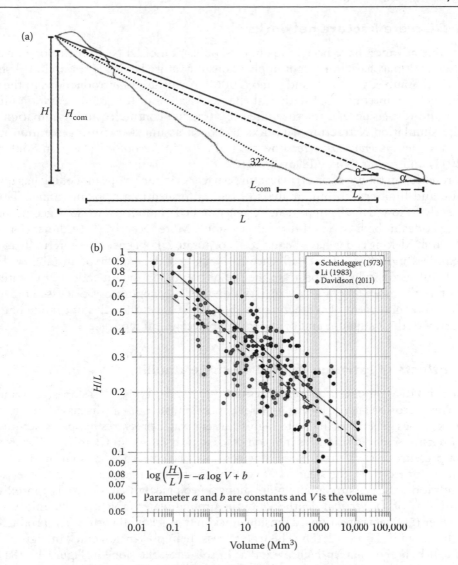

Figure 12.17 (a) Terminology used in runout modelling showing the fahrböschung angle α, the travel angle θ, and idealised runout angle, 32°, where H is the maximum vertical height, H_{com} is the vertical height of the centre of mass, L is the maximum horizontal length, L_{com} is the runout related to the centre of mass and L_e is the exceeded length travelled beyond that expected of simple sliding block (modified after McKinnon, 2010). (b) Empirical graph showing H/L versus volume of recorded landslides (modified after Whittall, 2015).

frictional-based criterion, Bingham flow and turbulent Voellmy-based approaches incorporating a turbulence coefficient; different rheological models may be used for different reaches or sections of the runout path as considered appropriate.

The applications of runout modelling for major landslides, including the Frank Slide, Turtle Mountain Alberta and the Zymoetz River rock avalanche in Canada, are described by Hungr and Evans (1996), Hungr (2014) and McDougall et al. (2006), respectively. Recent research has considered the use of remote-sensing characterisation, geomechanical models and rheological models to allow a total slope approach to the study of landslide initiation and runout simulation. Strouth and Eberhardt (2009) describe the investigation of the Newhalem rock avalanche in Washington, USA, using a linked approach involving *3DEC* and *DAN3D*.

12.5.6 Discrete fracture networks

Considerable advances have been made in the application of DFNs in rock slope investigations; this work may have important implications for future landslide research. Dershowitz et al. (2004), Staub et al. (2002) and Elmo (2006) provide useful introductions to the use of DFNs in rock engineering. The statistical characterisation of fracture orientation (dip, dip direction and dispersion, K), fracture intensity, fracture continuity and termination mode allow the simulation of fracture networks based on assumed fracture generation models using codes such as *FracMan* (Dershowitz et al., 2004), *Resoblock* (Merrien-Soukatchoff et al., 2011) and *Fracas* (Xu and Dowd, 2010).

Early research in the use of discrete fracture networks in rock slope modelling has involved kinematic and limit equilibrium analysis of blocks formed in a slope or tunnel. This was subsequently followed by the importing of generated DFN models into slope geomechanical modelling codes in both two and three dimensions. More recently, three-dimensional geomechanical modelling codes have begun to incorporate DFN generators internally to allow more seamless modelling of the influence of block geometry on slope instability. Havaej et al. (2016b) successfully demonstrate the use of the discontinuum code *3DEC* with DFN generation in the modelling of a slate quarry slope. DFN generators, as discussed previously, are an integral component of synthetic rock mass approaches to rock slope modelling using a wide array of discontinuum and continuum geomechanical models.

12.5.7 Effects of extreme rainfall on slope stability

Extreme rainfall events are common causes of slope instability. For soil slopes, extreme rainfall can cause erosion and/or loss of apparent cohesion (suction) as discussed in Chapter 3, while rock slopes generally fail due to high-transient water pressures in open fractures, particularly tension-induced fractures (tension cracks) as discussed in Chapter 6. For example, transient pressures equivalent to 40 m of water have been measured at some mines.

Consider, for example, the steady-state condition shown in Figure 12.18a. Under steady-state conditions, the pressure at any point along a geological discontinuity is approximated by the product of the vertical depth below the ground water table and the unit weight of water. Under transient conditions, a tension crack may quickly fill with water, producing the condition shown in Figure 12.18b. The water pressure in the tension crack in Figure 12.18b is significantly higher than in Figure 12.18a. In this case, the slope in Figure 12.18a had a safety factor of 1.18, whereas the slope in Figure 12.18b had a safety factor of 1.01.

12.5.8 Geostatistics

The importance of considering the probability of failure is now recognised, and most limit equilibrium and an increasing number of numerical models allow deterministic, sensitivity analysis and probabilistic approaches to design. The use of geostatistics in rock slope analysis has been the subject of comparatively limited research. Geostatistics has been defined in the literature as the study of phenomena that vary in space and/or time (Deutsch, 2002). Geostatistical analysis requires the accurate characterisation of the underlying spatial structures of data using techniques such as semivariograms, covariograms and correlograms, combined with methods including kriging and/or sequential Gaussian simulation (Clark, 1979; Isaaks and Srivastava, 1989; Srivastava and Parker, 1989; Mayer, 2015).

Jefferies et al. (2008) provide a clear demonstration of the need to consider a geostatistical approach to slope design wherein the spatial variation of the rock mass strength is considered. Using a *FLAC* model of a simple slope, they showed that this approach highlights the

(a)

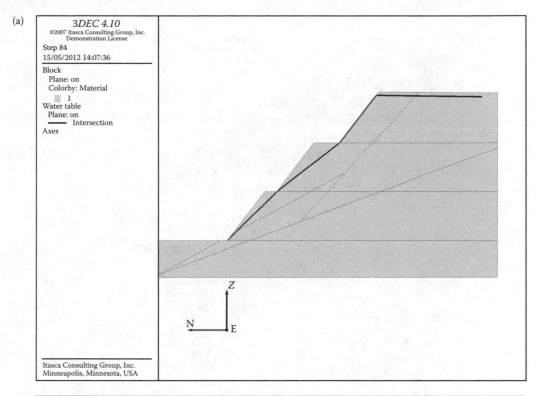

(b)

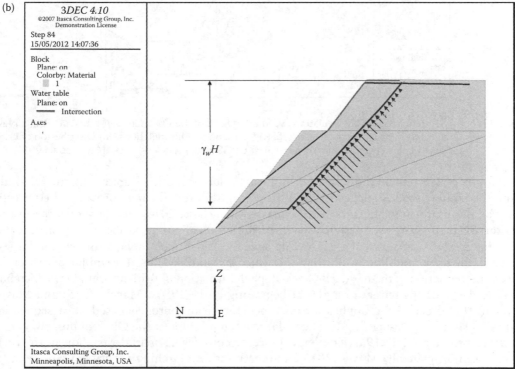

Figure 12.18 Slope models incorporating ground water pressures: (a) steady-state situation with a ground water table; (b) transient water pressure distribution in tension crack.

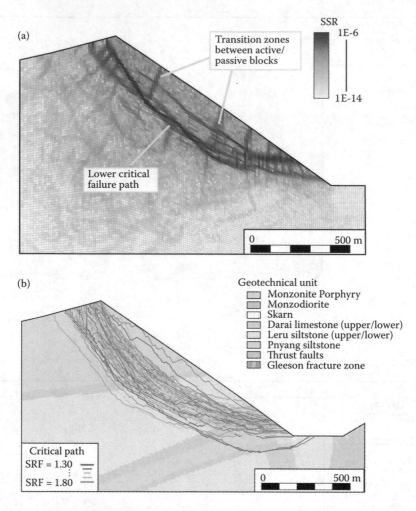

Figure 12.19 (a) Sequential Gaussian simulation (SGS) using *FLAC* of the Ok Tedi open-pit slope, Papua New Guinea, showing shear strain rates (SSR) from one of the simulations. (b) Distribution of critical paths behind the pit wall from stochastic SGS *FLAC* modelling (after Mayer et al., 2014).

limitations of conventional analysis using mean value strengths. Limited additional study using the similar models suggests that 30th percentile strengths may provide a better estimate of the uniform strength of a rock mass. Kalenchuk (2010) clearly shows the use of geostatistical approaches in deriving the three-dimensional failure surface and ground water table for a *3DEC* analysis of the Downie slide in British Columbia, Canada, the largest active prehistoric landslide in North America. Clayton (2014) used a similar approach in the characterisation of the Mitchell Creek slope deformation in British Columbia, which has been active since the retreat of a glacier beginning in the 1950s. Mayer (2015) and Mayer et al. (2014) describe the application of a geostatistical approach based on a sequential Gaussian simulation using *FLAC2D* of a large open pit slope at the Ok Tedi mine in Papua New Guinea (Figure 12.19a). The importance of considering spatial variation in the rock mass is demonstrated by Mayer (2015), and methods for searching for critical failure paths in the SGG *FLAC* model are illustrated in Figure 12.19b. In critical slopes with sufficient data, the importance of spatial variation in mechanical properties should be considered along with the use of geostatistical-driven geomechanical models.

Chapter 13

Blasting

13.1 INTRODUCTION

Excavation of rock slopes usually involves blasting, and it is appropriate that the subject receive attention in this book on rock slope engineering. The fragmentation of rock by explosives is a major subject in its own right and the fundamentals are dealt with in a number of excellent textbooks and handbooks (Langefors and Kihlstrom, 1973; Hemphill, 1981; C.I.L, 1983; Atlas Powder Company, 1987; FHWA, 1991; Persson, Holmburg and Lee, 1993; Oriard, 2002; ISEE, 2015).

13.1.1 Blasting operations

The responsibility for the design and implementation of blasting operations varies with the type of project. Mines and large quarries will usually have a department responsible for all aspects of blasting, possibly with some assistance from technical representatives of the explosive supplier. This approach is suited to circumstances where blasting is carried out frequently, and good experience of local conditions has been developed. This situation allows the blast designs to be adjusted to accommodate such factors as variations in geology and optimisation of equipment operations. The primary requirement of blasting in open-pit mines is to produce a muck pile that is fragmented to suit operation of the excavating and hauling equipment. Also, times for equipment moves before and after the blast are minimised if flyrock is controlled. A further requirement of blasting is to control damage to the rock and minimise slope instability when excavating final faces. Production blasting methods are discussed in Section 13.3.

On civil projects, in contrast to mining operations, blasting is usually the responsibility of the contractor, with the principle duty of the owner's representative being to check that the desired results are produced. That is, the work is carried out under a performance-type contract in which the required results are specified, but the methods of achieving these are left to the contractor. This situation requires that the owner understands blasting methods so that the procedures and results can be reviewed and, if necessary, modifications discussed with the contractor. The owner should also ensure that accurate records are maintained of each blast as required by most legislation governing blasting. The records are also useful in relating the results obtained to the methods used, and for cost–control purposes.

Rock excavation for civil construction often requires the formation of cut faces that will be stable for many years, and as steep as possible to minimise excavation volume and land use. While these two requirements are contradictory, the stability of cuts will be enhanced, and the maximum safe slope angle increased, by using a blasting method that does the least possible damage to the rock behind the final face. Section 13.4 describes methods of minimising blasting damage to the face that are included in the general term 'controlled blasting'.

When blasting in urban or industrial areas, precautions are required to control damage to residences and other sensitive structures. Section 13.5 describes methods of controlling structural damage due to blast vibrations, as well as minimising hazards of flyrock, air blast and noise.

13.1.2 Mechanism of rock fracturing by explosives

The mechanism by which rock is fractured by explosives is fundamental to the design of blasting patterns, whether for production or controlled blasting. It also relates to the damage to surrounding structures and disturbance to people living in the vicinity. The following is a description of this mechanism (Hagan, 1992. Queensland, Australia. Personal communication; Konya and Walter, 1991; Persson, Holmburg and Lee, 1993; Oriard, 2002).

When an explosive is detonated, the solid is converted within a few thousandths of a second into a high temperature and high-pressure gas. When confined in a blast hole, this very rapid reaction produces pressures, that can reach 18,000 atmospheres, to be exerted against the blast hole wall. This energy is transmitted into the surrounding rock mass in the form of a compressive strain wave that travels at a velocity of 2000–6000 m/s.

As the strain wave enters the rock surrounding the blast hole, the material for a distance of one to two charge radii (in hard rock, more in soft rock) is crushed by compression (Figure 13.1a). As the compressive wave front expands, the stress level quickly decays below the dynamic compressive strength of the rock, and beyond this pulverised zone the rock is subjected to intense radial compression that causes tangential tensile stresses to develop. Where these stresses exceed the dynamic tensile strength of the rock, radial fractures form. The extent of these fractures, which depends on the energy available in the explosive and the strength of the rock, can equal 40–50 times the blast hole diameter. As the compressive wave passes through the rock, concentric shells of rock undergo radial expansion resulting in tangential relief-of-load fractures in the immediate vicinity of the blast hole.

These concentric fractures follow cylindrical surfaces, and are subsequently created nearer and nearer to the free face. When the compressive wave reaches a free face, it is reflected as a tensile strain wave. If the reflected tensile wave is sufficiently strong, 'spalling' occurs progressively from any effective free face back towards the blast hole. This causes unloading of the rock mass, producing an extension of previously formed radial cracks (Figure 13.1b). Rock is much weaker in tension than compression, so the reflected strain wave is particularly effective in fracturing rock.

The process of fracture formation due to strain wave energy typically occurs throughout the burden within 1 or 2 ms after detonation, whereas the buildup of explosive gases takes in the order of 10 ms. As the rock is unloaded due to radial expansion and reflection of the compressive wave, it is now possible for the expanding gases to wedge open the strain wave-generated cracks and begin to expel the rock mass (Figure 13.1c). This stage is characterised by the formation of a dome around the blast hole. As wedging action takes place due to the heaving and pushing effect of the expanding gases, considerably more fracturing occurs due to shear failure as the rock mass is expelled in the direction of the free face. In highly fissured rocks, fragmentation and muck pile looseness are caused mostly by expanding gases.

The fragmentation achieved by the process described in the previous paragraphs is highly dependent upon the confinement of the explosive, the contact of explosives with the rock in the blast holes (termed coupling ratio: explosive diameter/hole diameter), the burden distance and the detonation sequencing of the blast holes. That is, if confinement of the charge by stemming (gravel in the top of the hole) is inadequate, energy will be lost up the blast holes, and inadequate explosive/rock coupling results in poor transmission of strain energy to the rock mass. Also, excessive burdens result in choking and poor movement of the

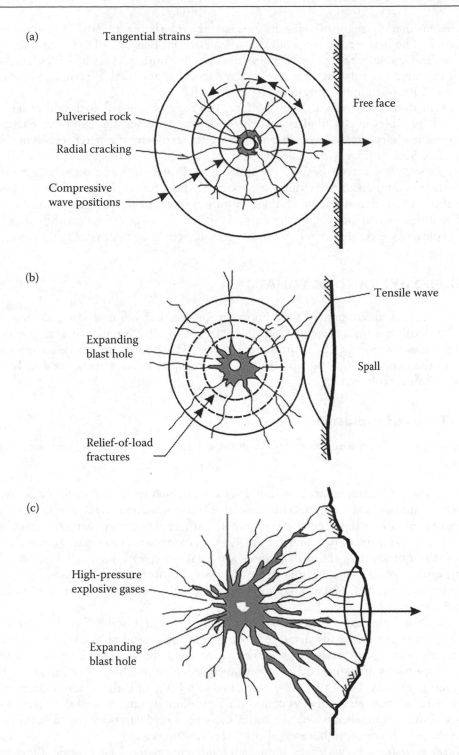

Figure 13.1 Mechanism of rock breakage by explosives: (a) crushing of rock around the blast hole and development of radial fractures; (b) reflection of shock wave in tension from the free face; (c) extension of radial cracks to free face and displacement of fractured rock mass.

rock, whereas inadequate burden results in waste of explosive energy and excessive throw of blasted rock. The best results are produced when effective delaying of individual blast holes ensures utilisation of free faces by reducing the effective burden. This provides freedom for the rock to move towards the free face and reduces damage to the surrounding rock (see Section 13.3, Production blasting).

To prevent damage to rock behind the face, the zone of crushed rock and radial cracking around the holes in the final row is controlled by reducing the explosive energy, and decoupling the charge in the hole, that is, having the explosive diameter less than the hole diameter (see Section 13.4, Controlled blasting).

As the shock wave travels beyond the limit of rock breakage and into the surrounding rock, it sets up vibrations both within the rock and at the ground surface. Structures located close to the blast, and through which these vibration waves pass, may be damaged by twisting and rocking motion induced by the ground motion. Damage can be controlled by reducing the explosive weight detonated per delay (see Section 13.5, Blast damage control).

13.2 EXPLOSIVES AND DETONATORS

The following is a discussion of the important factors that influence the selection of the appropriate explosive for a project. A wide variety of explosive products is available that are marketed under trade names by their manufacturers. These names give little information on the type of the explosive, or its properties, so it is necessary to consult the product literature to find a suitable explosive(s).

13.2.1 Types of explosives

The following is a brief discussion on the common types of explosives, and their primary ingredients:

- *Dynamite* – dynamite contains liquid sensitisers such as nitroglycerin or nitroglycol, sodium nitrate and carbonaceous absorbents. In ammonia dynamites, some of the liquid sensitiser is replaced with ammonium nitrate. Dynamites are very sensitive and have fair water resistance, and are available as cardboard packaged cartridges.
- *Gelatin dynamite* – gelatin dynamites are gelatinised with nitrocellulose, which is a thickening agent that gives the explosive water resistance and prevents fluid leakage from the cartridges. Gelatin dynamites have a high-detonation velocity and excellent water resistance.
- *Emulsions* – emulsions are intimate mixtures of two immiscible liquids called phases, with one phase uniformly dispersed throughout the second phase, with liquid separation prevented with active ingredients called emulsifiers. Thus, emulsion explosives are dispersions of water oxidiser solutions in a fuel oil medium. The oil or fuel phase, which is usually diesel fuel (FO), surrounds and coats all the oxidiser droplets. The water or oxidiser phase always contains ammonium nitrate, as well as other salts such as sodium nitrate. Emulsions are available as packaged cartridges or in bulk; they are detonator-sensitive and have excellent water resistance.
- *Water gels/slurries* – water gel comprises oxidising agents, such as ammonium nitrate, and fuels, such as aluminium, coal or ethyl glycol, dissolved or dispersed in a continuous water phase. The entire system is thickened and made water-resistant by the addition of gellants and cross-linking agents. Because water gels contain substantial amount of water, they are less sensitive than dynamites.

- *Bulk explosives* – these explosives are a combination of a dry ammonium nitrate oxidiser (AN) in the form of prills with a diameter of 1–2 mm, and diesel oil fuel (FO). The optimum mixture for ANFO is 94% AN and 6% FO in order to achieve maximum strength and velocity of detonation (VOD). ANFO cannot be detonated with a detonator and must be initiated with a relatively high-strength booster (see Section 13.2.7). ANFO has very poor water resistance, but water proof ANFO is available. ANFO is only available as a bulk explosive and not as cartridges, and the minimum hole diameter in which it can be used is 50 mm (2 in.).

Table 13.1 provides information on typical properties of the main classes of explosives. Each class of explosive has a set of characteristics suitable to specific applications and conditions such as the size of the blast hole, the presence of water and the need to control flyrock and noise. For example, ANFO and bulk emulsions are commonly used for large-scale blasts in open pits and quarries, while watergels and dynamites are used in smaller construction projects. Also, as shown in Figure 13.3, it is necessary, when using ANFO as the main explosive, to use a higher strength explosive in the lower end of the hole ('toe load'). The function of the toe load is to both ensure complete detonation of the ANFO, and to break the rock in the floor of the bench where it is most highly confined.

13.2.2 Explosive strength

The strength of an explosive is a measure of the work done by a certain weight or volume of explosive. This strength can be expressed in absolute units, or as a ratio relative to a standard explosive. Often the bulk strength of explosives is related to the strength of ANFO that is assigned an arbitrary bulk strength of 100. ANFO is the term used for the most widely used explosive comprising ammonium nitrate prills (0.5 mm [0.02 in.] diameter spheres) and 6% fuel oil.

One measure of the strength of an explosive is its VOD; the higher the velocity the greater the shattering effect. However, explosive strength, density and degree of confinement are also factors that should be considered in selecting an explosive for a specific purpose. Table 13.1 lists the specific gravity (SG), VOD, absolute bulk strength and water resistance of different classes of explosives.

Explosive strength is defined by weight and bulk strengths. Weight strengths are useful when comparing blast designs in which explosives of different strengths are used, and when comparing the cost of explosives because explosives are sold by weight. The bulk (or volume) strength (ABS) in calories per cubic centimetre is related to the weight strength by the SG, and this value is important in calculating the volume of blast hole required to contain

Table 13.1 Typical properties of explosives products

Explosive type	Density (g/cm³)	V.O.D. (m/s)	Absolute bulk strength, ABS (cal/cm³)	Water resistance
Dynamites	1.3–1.45	4300–6000	1240–1510	Good to excellent
Wall control dynamites	0.95–1.40	2700–3000	1230–1440	Good
ANFO	0.84	3600	700	None
Emulsions	1.15–1.27	4700–5500	890–935	Excellent
Bulk emulsions	1.25	4200–5600	815–880	Excellent
Boosters	1.65	7300	Pressure = 220 kbars	Excellent

Source: Dyno Nobel. 2013. Product technical information. Dyno Nobel, http://www.dynonobel.com/.

a given amount of explosive energy. A higher-bulk strength requires less blast hole capacity to contain a required charge.

13.2.3 Detonation sensitivity

The sensitivity of an explosive is a characteristic that determines the method by which a charge is detonated, the minimum diameter of the charge and the safety with which the explosive can be handled. Highly sensitive explosives will detonate when used in smaller diameter charges, and can be detonated with a relatively low-strength detonator. As the sensitivity of the explosive is decreased, the diameter of the charge and the energy of the booster/detonator must be increased (see Section 13.2.7, Detonators).

13.2.4 Water resistance

Blast holes drilled in rock below the water table will often be filled with water, in which case it will be necessary to use an explosive that is resistant to water, keeping in mind that for large blasts the explosive may remain in the ground for several hours before the blast is detonated.

13.2.5 Loading method

The method of loading explosives in blast holes is related to the scale of the excavation operation and the access for equipment. In large quarries and open-pit mines where the blast hole diameters can be as great as 450 mm (18 in.) and bulk explosives such as ANFO and emulsions are used, it is usual that they are loaded on site with trucks that can mix the ingredients to create custom explosives. In contrast, for tunnels and blasting in steep terrain where the hole diameter is about 50–100 mm (2–4 in.), cartridge explosives are used that can be transported to the face in a pickup truck or by hand, and allow the explosive load to be precisely adjusted to the particular requirements of the project.

13.2.6 Explosive cost

The least expensive explosive is usually ANFO, partially because it can be transported as a non-explosive product and only becomes a high explosive when the diesel fuel is added on site. Other bulk explosives such as emulsions are also less expensive than specialised dynamite explosives that are produced in smaller quantities, and are packaged in cartridges.

13.2.7 Detonators

Modern explosives can be safely dropped, punctured and burnt with little risk of an accidental explosion, and it is necessary to use detonators to actually initiate a blast. As discussed in Section 13.2.3, various types of explosives have different 'sensitivities' that relate to the method that is required to detonate the explosive. Since a fundamental component of blasting is the hole detonation sequence, the detonators incorporate timing elements that allow the initiation sequence of the holes to be controlled (see Section 13.3.8).

Sensitive explosives such as dynamites can be detonated by a detonator that comprises, in an aluminium tube about 10 mm (0.5 in.) diameter and 100 mm long, an element that burns at a known rate to determine the delay timing, and a high-explosive element (e.g. PETN or pentaerythritol tetranitrate) and that initiates the blast. The detonator itself is initiated by either an electrical current or by non-electric shock tube (NONEL).

The delay interval for detonators is 25 ms for short period (SP) delays that are commonly used for surface blasting, and 200 ms for long period (LP) delays that are more commonly used for tunnelling. A deficiency of these detonators is that the error in actual firing time compared to the rated firing time may be as high as 15%, and such errors may result in holes firing out of sequence, especially if the delay interval is short. As discussed in Section 13.3.8, blast hole detonation sequence is an important component of blast design, and out of sequence detonation is detrimental to good blasting results.

Precise detonation timing can be achieved by the use of electronic delay detonators in which the chemical delay element is replaced with an integrated circuit chip. Experience with electronic detonation shows that their use improves fragmentation and reduces overbreak, ground vibration, airblast and flyrock. A further advantage of electronic detonators is the ability to programme the delay interval in the field to suit site conditions (McKinstry, Floyd and Bartley, 2002; Watson, 2002). At present (2016), electronic detonators cost as much as three times that of conventional detonators and this is holding back their wide use.

For low sensitivity explosives such as ANFO and water gels, it is necessary to initiate the blast with boosters that are cast cylinders, with diameters compatible with the drill hole diameter, comprising PETN and trinitrotoluene (TNT). The boosters are embedded in the explosive and are themselves initiated with a detonator, either electric, NONEL or electronic.

13.3 PRODUCTION BLASTING

The basic economics of rock excavation using explosives are shown in Figure 13.2 (Harries and Merce, 1975). The production of a well-fragmented and loosely packed muck pile that

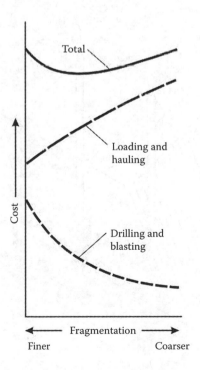

Figure 13.2 Effect of fragmentation on the cost of drilling, blasting, loading and hauling.

has not been scattered around the excavation area facilitates loading and hauling operations. This condition is at the minimum total cost point on the graph. However, close to the final face, drilling and blasting costs will increase because more closely spaced and lightly loaded holes will be required to limit damage to the rock behind the face. Conversely, the production of riprap will involve the use of holes with a spacing greater than the largest block size required.

In order to achieve the optimum results of blasting under all conditions, a thorough understanding of the following parameters is required:

1. Type, weight, distribution of explosive
2. Nature of the rock
3. Bench height
4. Blast hole diameter
5. Burden
6. Spacing
7. Sub-drill depth
8. Stemming
9. Initiation sequence for detonation of explosives
10. Powder factor

Factors 3–8 are illustrated in Figure 13.3, and this section describes procedures for calculating the parameters involved in designing production blasts. This discussion is based on work by Dr. C. J. Konya (Konya and Walter, 1991). It should be noted that the equations provided in the following sections defining blasting parameters are guidelines only and will need to be modified, as necessary, to suit local site conditions.

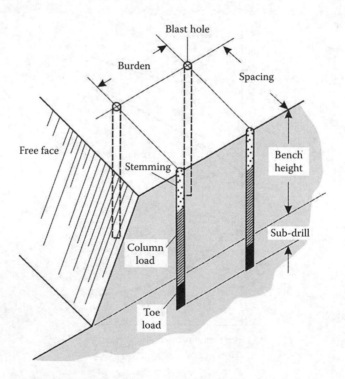

Figure 13.3 Definition of bench blasting terms.

13.3.1 Bench height

Bench heights are usually determined by the geometry of the site, with single benches being used where the excavation depth is up to about 8 m (26 ft). On large construction projects and in open-pit mines and quarries, multi-benches operations are conducted. For these operations, the selection of the optimum bench height to maximise the overall cost efficiency of drilling and blasting requires the correct combination of drilling and loading equipment. Furthermore, regulations may limit the bench height, in relation to the maximum reach of the excavating equipment (bench height is usually 1.5 times the equipment vertical reach), to minimise the risk of damage or injury in the event of collapse of the face. The following are some of the factors that should be considered in the selection of the bench height:

1. Optimum blast hole diameter increases with bench height. In general, an increase in blast hole diameter results in a decrease in drilling costs.
2. For vertical blast holes and sloping bench face, the front row toe burden may become excessive as the bench height increases. Where small diameter blast holes are drilled in high benches, blast holes may need to be angled, at least in the front row.
3. Drilling accuracy becomes more critical in higher benches. Where precise alignment of holes on the final face is required, the maximum bench height is normally limited to 8 or 9 m (26–30 ft).

13.3.2 Burden

The influence of the effective burden (the distance between rows of holes and the nearest free face) on fragmentation is related to the mechanism of rock fracture described in Section 13.2. The blast is most efficient when the shock wave is reflected in tension from a free face so that the rock is broken and displaced to form a well-fragmented muck pile with little scattering of rock around the site. This efficiency depends to a large extent on having the correct burden. Too small a burden will allow the radial cracks to extend to the free face resulting in venting of the explosion gases with consequent loss of efficiency and the generation of flyrock and air blast problems. Too large a burden, where the shock wave is not reflected from the free face, will choke the blast resulting in poor fragmentation and a general loss of efficiency.

The relationship between the bench height H and the burden B can be expressed in terms of the 'stiffness ratio', H/B. If this ratio is low such that the burden is about equal to the bench height, then the blast will be highly confined resulting in severe backbreak, airblast, flyrock and vibration. In contrast, if the $H/B > 4$, little confinement of the explosive gases occurs and they will be vented at the free face resulting in airblast and flyrock. It is found that a stiffness ratio of 3–4 produces good results, or the burden B in relation to the bench height H, is

$$B = 0.33 \cdot H \text{ to } 0.25 \cdot H \tag{13.1}$$

The burden distance calculated from Equation 13.1 depends not only upon the blast hole pattern, but also upon the sequence of firing. As illustrated in Figure 13.4a, a square blast hole pattern which is fired row by row starting at the face gives an effective burden equal to the spacing between successive rows parallel to the face. On the other hand, an identical pattern of blast holes fired *en echelon* results in different burdens and spacings, and a spacing and burden ratio greater than 1 (Figure 13.4b).

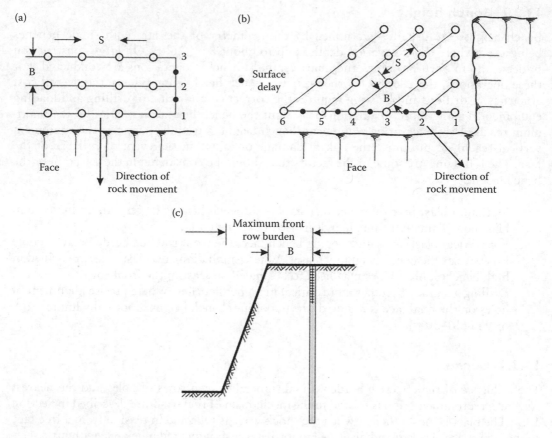

Figure 13.4 Definition of blast hole spacing (S) and burden (B): (a) burden and spacing for a square blasting pattern; (b) burden and spacing for an *en echelon* blasting pattern; (c) effect of face angle on front row burden.

An important factor in designing a blast is the choice of the front row burden. Once this row has been detonated and effectively broken, new free faces are created for the next row, until the last row is fired. If vertical blast holes are used and the bench face is inclined, the front row burden will increase with depth (Figure 13.4c). Allowance can be made for this variation by using a higher energy bottom load in the front row of holes. Alternatively, the blast hole can be inclined to give a more uniform burden. When the free face is uneven, the use of 'easer' holes to reduce the burden to acceptable limits is advisable.

For multiple row blasts, the rows towards the back of the blast are progressively more confined, and consequently the degree of fragmentation may diminish. This effect can be corrected for by decreasing the burden of the third and subsequent rows by a factor of 0.9.

13.3.3 Blast hole diameter

Blast hole diameters on construction projects may range from about 40 mm (1.5 in.) for hand held-drills, to 100 mm (4 in.) for 'air-trac' drills and 200 mm (8 in.) for 'track drills'. All these drills are rotary percussive and are powered with compressed air. In open-pit mines, electric powered rotary drills are commonly used with which holes up to 600 mm (24 in.) are drilled with roller tri-cone bits (Australian Drilling Industry, 1996).

Persson (1975) shows that the cost of drilling and blasting decreases as the hole size increases. This is because the hole volume increases with the square of hole diameter so that the same volume of explosive can be loaded into fewer holes. This cost saving is offset by the greater shattering of the rock that is produced by the more highly concentrated explosive which can result in less stable slopes, and larger rock fragments that the excavating equipment may not be able to handle.

Once the burden B (units: m) has been set to provide an appropriate stiffness ratio (Equation 13.1), the diameter of the explosive d_{ex} (units: mm) can be determined as follows:

$$d_{ex} \approx \frac{B \cdot 1000}{\left(\dfrac{24\gamma_{ex}}{\gamma_r} + 18 \right)} \tag{13.2}$$

where γ_{ex} and rock γ_r are the unit weights of the explosive and rock, respectively.

Equation 13.2 can be used where the energy of the explosive and its unit weight are correlated. However, emulsion slurry explosives have a range of energies with a near constant unit weight, and for these conditions, it is appropriate to calculate the burden using the relative bulk strength (RBS) compared to ANFO (RBS = 100). The blast hole diameter taking into account the RBS of the explosive is given by

$$d_{ex} \approx \frac{B \cdot 1000}{8 \left(\dfrac{RBS}{\gamma_r} \right)^{0.33}} \tag{13.3}$$

13.3.4 Nature of the rock

The nature and degree of heterogeneity of the rock mass is important in blast design. That is, discontinuities such as joints, bedding planes, faults and soft seams can allow the explosive's energy to be wastefully dissipated rather than to break the rock. In some cases, the discontinuities can dominate the fracture pattern produced by the explosive, and the influence of the structural geology overshadows that of the rock's mechanical and physical properties. Best fragmentation is usually obtained, where the face is parallel to the major discontinuity set.

Heave energy of the explosive is important in highly fissured rock. The explosion gases jet into, wedge open, and extend the pre-existing cracks. Therefore, the overall degree of fragmentation tends to be controlled by the structural geology. For example, closely spaced joints and bedding planes result in increased fragmentation because few new fracture surfaces need to be created in the blast. For these conditions, longer stemming columns and correspondingly lower powder factors or energy factors can be used, and low-density, low-velocity explosives (e.g. ANFO) are preferred to higher velocity explosives that cause excessive fragmentation immediately around the charge. Satisfactory results are achieved when the charges provide sufficient heave energy to displace the rock into a loose, readily excavated muck pile, without scattering the rock around the site.

The effects of structural geology on blast design can be quantified by applying two correction factors to the calculated burden distance. These factors relate to both the orientation of persistent discontinuities relative to the face k_ψ, and fracture characteristics k_s, as shown in Tables 13.2 and 13.3, respectively. The correction factors are applied to the calculated burden distance as follows:

$$B' = k_\psi \cdot k_s \cdot B \tag{13.4}$$

Table 13.2 Correction factors for dip of structure

Orientation of structure	Correction factor, k_ψ
Structure dipping steeply out of face	1.18
Structure dipping steeply into face	0.95
Other orientation of structure	1.00

Table 13.3 Correction factors for discontinuity characteristics

Structural characteristics	Correction factor, k_s
Closely jointed, closely spaced, weakly cemented seams	1.30
Thin, well-cemented layers with tight joints	1.10
Massive, intact rock	0.95

13.3.5 Sub-drill depth

Sub-drilling, or drilling to a depth below the required grade, is necessary in order to break the rock at bench level. Poor fragmentation at this level will form a series of hard 'toes' and an irregular bench floor resulting in high-operating costs for loading and hauling equipment. However, excessive sub-drilling can result in unnecessary drilling and explosive costs.

Breakage of the rock usually projects from the base of the toe load in the form of an inverted cone with sides inclined at 15–25° to the horizontal, depending upon the strength and structure of the rock (Figure 13.5). In multi-row blasting, the breakage cones link up to give a reasonably even transition from broken to undamaged rock. Experience has shown that a sub-drill depth of 0.2–0.5 times the burden is usually adequate for effective digging to grade. Where benches are to be formed on the final face, it is advisable to eliminate the sub-drilling in the back rows to help maintain stability of the next bench crest below the blast level.

13.3.6 Stemming

Figure 13.3 shows that the upper part of blast holes contains 'stemming' (usually angular gravel) and the explosive does not extend to the top of the hole. The function of the

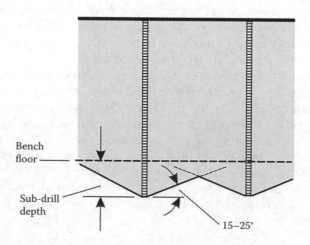

Bench floor

Sub-drill depth

15–25°

Figure 13.5 Rock breakage at the bottom of blast holes with the use of sub-drilling.

stemming is to contain the explosive gases and direct the explosive energy into the rock mass. Stemming material comprising well-graded, angular gravel is more effective than drill cuttings that are more easily ejected from the hole. The optimum size of the stemming material increases with the diameter of the blast hole, and the average size of the stemming particles should be about 0.05 times the blast hole diameter.

The effects of stemming length on blast results are similar to those of burden distance discussed previously. That is, a short stemming length will allow the explosive gases to vent, generating flyrock and airblast problems and reducing the effectiveness of the blast, while too great a stemming length will give poor fragmentation of the rock above the column load.

The common stemming length is about 0.7 times the burden, which is adequate to keep material from ejecting prematurely from the hole. If unacceptably large blocks are obtained from the top of the bench, even when the minimum stemming column consistent with fly-rock and airblast problems is used, fragmentation can be improved by locating a small 'pocket' charge centrally within the stemming (Hagan, 1975).

13.3.7 Hole spacing

When cracks are opened parallel to the free face as a result of the reflected tensile strain wave, gas pressure entering these cracks exerts an outward force which fragments the rock and heaves it on to the muck pile. Obviously, the lateral extent to which the gas can penetrate is limited by the size of the crack and the volume of gas available, and a stage will be reached when the force generated is no longer large enough to fragment and move the rock. If the effect of a single blast hole is reinforced by holes on either side, the total force acting on the strip of burden material will be evened out and uniform fragmentation of this rock will result.

Figure 13.6 illustrates drill hole patterns with a variety of burden/spacing ratios. While the square pattern is the easiest to layout, in some conditions better fragmentation is obtained

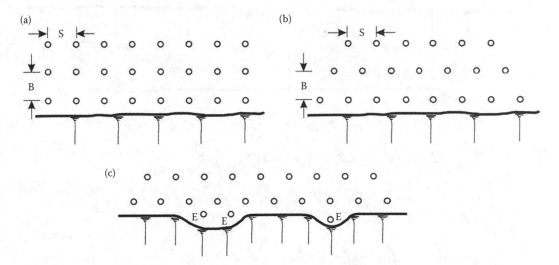

Figure 13.6 Typical blast hole patterns used in production blasting: (a) square pattern with burden/spacing ration 1:1; (b) staggered pattern with burden/spacing ratio 1:1:15; (c) easer holes (E) to assist movement of front row burden.

when the spacing is greater than the burden. For a series of delayed holes, the spacing S can be calculated from the following two equations:

For a stiffness ratio H/B between 1 and 4:

$$S = \frac{(H + 7 \cdot B)}{8} \tag{13.5}$$

and for a stiffness ratio H/B greater than 4:

$$S = 1.4 \cdot B \tag{13.6}$$

where H is the bench height and B is the burden.

13.3.8 Hole detonation sequence

A typical blast for a construction project or open-pit mine may contain as many as 100 blast holes, which in total contain several thousand kilograms of explosives. Simultaneous detonation of this quantity of explosives would not only produce very poorly fragmented rock, but also would damage the rock in the walls of the excavation and create large vibrations in nearby structures. In order to overcome this situation, the blast is broken down into a number of sequential detonations by delays. When the front row is detonated and moves away from the rock mass to create a new free face, it is important that time be allowed for this new face to be established before the next row is detonated. Figure 13.7 shows examples of detonation

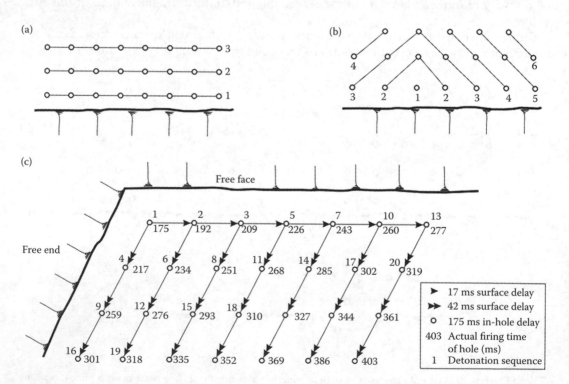

Figure 13.7 Typical detonation sequence: (a) square 'row-by-row' detonation sequence; (b) square 'V' detonation sequence; (c) hole-by-hole detonation using both surface and in-hole non-electric delays (W. Forsyth).

sequences. Row-to-row blasts are parallel to the free face, with the row closest to the face being detonated first in the sequence (Figure 13.7a). The V cut shown in Figure 13.7b, where the rows are inclined to the face, is used both to open a new free face, and when blasting in strongly jointed rock where near vertical joints strike across the bench at an angle to the face.

In many cases, blasting results can be improved by introducing hole-by-hole firing, where every blast hole is initiated in sequence at a unique time (Figure 13.7c). Where appropriate delays are selected, hole-by-hole initiation exploits the positive benefits of blast hole interaction while avoiding most of the negative effects. This can lead to finer fragmentation, looser muck piles, less overbreak, lower ground vibration and better control over the position and profile of the final muck pile.

Hole-by-hole firing can be achieved by using in-hole detonators. However, the available range of commercial in-hole delays is limited and, hence hole-by-hole initiation is usually achieved by using a surface delay system to control blast hole sequencing. If long inter-row or inter-hole delays are required, a combination of surface and in-hole delays will avoid downline cutoffs caused by ground movement during the blast. The hole-to-hole detonation sequence shown in Figure 13.7c is produced by the use of an identical in-hole delay of 175 ms in each hole, together with surface delays of 17 ms along the free face, and 42 ms delays between the rows. The diagram shows the actual firing times of each hole using the delay arrangement as follows:

Hole #1: 175 ms; hole #2: 175 + 17 = 192 ms; hole #3: 175 + 17 + 17 = 209 ms; hole #4: 175 + 42 = 217 ms; hole #5: 175 + 17 + 17 + 17 = 226 ms; hole #6: 175 + 17 + 42 = 234 ms, etc.

The required delay interval is related to the distance between holes by the following two relationships:

For row-to-row detonation:

$$\text{Time delay between rows (units: ms)} \approx (10 \text{ to } 13) \cdot (\text{burden, units: m}) \qquad (13.7)$$

For example, for a 5 m burden, the delay between rows is 50–65 ms.

For hole-to-hole detonation:

$$\text{Time delay between holes (units: ms)} \approx (\text{delay constant}) \cdot (\text{hole spacing, units: m}) \qquad (13.8)$$

Values for the delay constant depend on the rock type as shown in Table 13.4. For example, when blasting granite with a hole-to-hole spacing of 6 m (20 ft), the delay interval would be about 30 ms.

These two relationships for calculating delay times take into account both the time required to displace the broken rock and establish a new free face, and the natural scatter that occurs in the actual firing times of delays (see Section 13.2.7, Detonators).

Table 13.4 Delay constant–rock type relationship for hole-to-hole delay

Rock type	Delay constant (ms/m)
Sand, loam, marl, coal	6–7
Soft limestone, shale	5–6
Compact limestone and marble, granite, basalt, quartzite, gneiss, gabbro	4–6
Diabase, diabase porphyrites, compact gneiss and micaschist, magnetite	3–4

13.3.9 Fragmentation

The objective of selecting the appropriate combination of design parameters, as discussed on the preceding pages, is to achieve a desired result for the blast. These objectives may include a specified size range of rock fragments, or a final face with minimal blast damage, or noise and ground vibration levels within specified limits.

The basic design parameter for production blasting, where the objective is to produce a certain degree of fragmentation, is the 'powder factor' – the weight of explosive required to break a unit volume of rock, for example, kg/m^3 or lb/yd^3. The powder factor is derived from the procedures discussed in this section to calculate the blast hole diameter and depth, the hole pattern and the amount and type of explosive in each hole. That is, for an explosive diameter d_{ex} and unit weight γ_{ex}, the weight of explosive and the volume of rock are given by

Weight of explosive/hole, $W_{ex} = (2 \cdot \pi \cdot d_{ex}) \cdot (\gamma_{ex}) \cdot$ (bench height – stemming length + sub-drill depth)

and

Volume of rock per hole, $V =$ (bench height) (burden) $\cdot$ (spacing)

$$\text{Powder factor} = \frac{W_{ex}}{V} \tag{13.9}$$

The effect of the powder factor on fragmentation is shown in Figure 13.8, where the average fragment size is related to the powder factor for a range of burdens. These graphs, which are based on theoretical equations developed by Langefors and Kihlstrom (1973) and Persson, Holmburg and Lee (1993) together with extensive field testing, can be used to check on the likely results for a given blast design and assess the possible effects of design modifications.

13.3.10 Evaluation of a blast

Once the dust has settled and the fumes have dispersed after a blast, an inspection of the area should be carried out. The main features of a satisfactory blast are as follows (Figure 13.9):

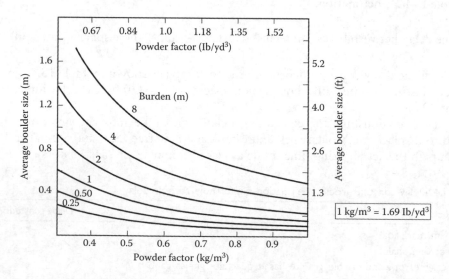

Figure 13.8 Relationship between average boulder size, powder factor and burden in bench blasting.

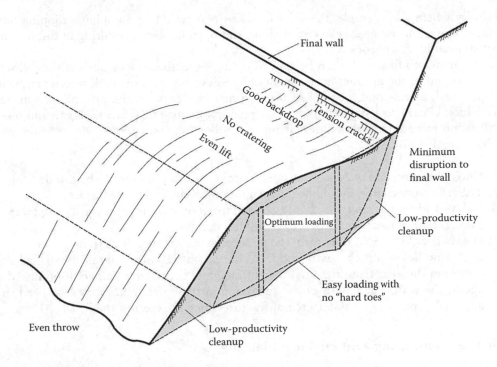

Figure 13.9 Features of a satisfactory production blast.

- The front row should have moved out evenly but not too far – excessive throw is unnecessary and expensive to clean up. The heights of most benches are designed for efficient loader operation and low muck piles, due to excessive front row movement, results in low-loader productivity.
- The main charge should have lifted evenly and cratering should, at worst, be an occasional occurrence. Flat or wrinkled areas are indicative of misfires or poor delaying.
- The back of the blast should be characterised by a drop, indicating a good forward movement of the free face. Tension cracks should be visible in front of the final excavation lines, although excessive cracking behind the final excavation line represents damage to the slopes and waste of explosive.

The quality of blast has a significant effect on components of the rock excavation cost such as secondary drilling and blasting of oversize boulders, loading rate, the condition of the haul roads and loader and truck maintenance. For example, oversized fragments, hard toes, tight areas and low muck piles (caused by excessive throw) have the most significant detrimental effect on the excavation rates. Therefore, careful evaluation of the blast to determine how improvements could be made to the design is usually worthwhile.

13.4 CONTROLLED BLASTING TO IMPROVE STABILITY

Slope instability is often related to blast damage to the rock behind the face. Blast-induced instability is usually surficial, extending possibly 1–3 m (3–10 ft) behind the face for construction scale blasts, that can result in rock falls occurring over time as water and ice open cracks and loosen blocks. It is also possible the blast damage can cause larger-scale instability where, for example, the slope contains persistent bedding planes dipping out of the face. The explosive gases can travel along and open the beds, resulting in displacement of substantial blocks of rock.

Blast damage to final walls can be limited by implementing one or both of the following procedures. First, the production blast should be designed to limit rock fracturing behind the final wall, and second, controlled blasting methods such as line drilling, pre-shearing and cushion blasting can be used to produce precisely defined final faces (Hagan and Bulow, 2000). With respect to production blasting, the following precautions help to avoid excessive backbreak:

1. Choke blasting into excessive burden or broken muck piles should be avoided.
2. The front row charge should be adequate to move the front row burden.
3. Delays and timing intervals should be used for good movement towards free faces and the creation of new free faces for the following rows.
4. Delays should be used to control the maximum instantaneous charge.
5. Back row holes, together with reduced charge 'buffer holes', should be drilled at an optimum distance from the final face to facilitate excavation and yet minimise damage to the wall. The length of the stemming in the buffer holes may also need to be adjusted depending on the degree of fracture along the crest of the bench.

13.4.1 Pre-shearing and cushion blasting

On permanent slopes for many civil projects, even small slope failures are not acceptable, and the use of controlled blasting to limit damage to the final wall is often required; an example of controlled blasting is shown in Figure 13.10. The principle behind these methods is that closely spaced, parallel holes drilled on the final face are loaded with a light explosive charge that has a diameter smaller than that of the hole. An air gap around the

Figure 13.10 **(See colour insert.)** Example of controlled blasting, with drill 'offset' at mid-height for rock cut on highway project (strong granite, Sea to Sky Highway, British Columbia, Canada).

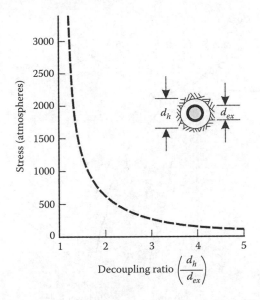

Figure 13.11 Stress level for decoupled blast holes (Konya and Walter, 1991).

explosive provides a cushion that significantly diminishes the shock wave that is transmitted to the rock. If the decoupling ratio, that is, the ratio of hole diameter to the explosive diameter, is greater than about 2, then the pressure of the shock wave will be about 10%–20% of that produced by an explosive tightly packed in the hole (Figure 13.11). This pressure is not sufficient to produce crushing of the rock around the hole (see Figure 13.1), but radial fractures form preferentially between the holes to create a clean break on the plane of the holes. It is found that it is not necessary to fire the holes on the final line on the same delay, and that the same results are obtained if every hole is fired on a separate delay (Oriard, 2002).

The following is a discussion on various methods of controlled blasting, and their advantages and disadvantages. The three basic methods of controlled blasting are *line drilling, cushion or trim blasting* and *pre-shear blasting*. Line drilling involves drilling holes precisely along the required break line at a spacing of two to four holes diameters, and then leaving a number of unloaded holes between the loaded holes. Line drilling is only used where very precise wall control is needed, such as corners in excavations. Cushion and pre-shear blasting are the most commonly used methods, with the main difference between the two being that in cushion blasting the final row holes are detonated last in the sequence, while in pre-shearing the final line holes are detonated first in the sequence. The following are some of the factors that should be considered in selecting either cushion or pre-shear blasting:

1. *Burden* – pre-shearing should only be used where the burden is adequate to contain the explosive energy that is concentrated along the shear line. For narrow burdens ('sliver' cuts), cushion blasting should be used because the energy produced by the pre-shear holes may be sufficient to displace the entire mass of rock between the shear line and the face before the other holes have detonated. For the blast shown in Figure 13.12, the relief holes would be detonated first to create a free face to which the shear line could break, with the holes detonated one by one along the face. As a guideline for the use of pre-shearing, the burden should at least be equal to the bench height, and preferably greater (Oriard, 2002).

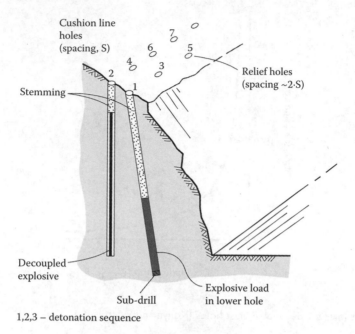

Cushion line
holes
(spacing, S)

Relief holes
(spacing ~2·S)

Stemming

Decoupled
explosive

Sub-drill

Explosive load
in lower hole

1,2,3 – detonation sequence

Figure 13.12 Blast hole layout and detonation sequence for cushion blast to excavate narrow 'sliver' cut.

2. *Discontinuities* – in closely jointed/sheared rock, the highly confined gases generated along the shear line in pre-shearing may vent into the cracks resulting in damage to the rock. In cushion blasting, the rock is somewhat less confined, and more stable slopes may be produced.

3. *Vibration* – the shock wave generated by highly confined pre-shearing may be greater than that of cushion blasting. As noted earlier in this section, it is not necessary to detonate the shear line holes on a single delay, so employing hole-by-hole delays can limit ground motion when vibrations must be closely controlled.

A general comment on the design of controlled blasts is that it is often necessary to carry out a number of trial blasts at the start of a project, and when the rock conditions change, to determine the optimum hole layout and explosive charge. This requires flexibility on the part of the contractor, and the use of end-product rather than method specifications.

The cost savings achieved by controlled blasting cannot be measured directly. However, it is generally accepted that, for permanent slopes on many civil projects, these savings are greater than the extra cost of drilling closely spaced, carefully aligned holes and loading them with special charges. The savings are the result of being able to excavate steeper slopes, and reduce excavation volume and land take. It is also found that less time is spent scaling loose rock from the face after the blast, and the resulting face is more stable and requires less maintenance during its operational life. From an aesthetic point of view, steep cuts have smaller exposed areas than flat slopes, although the traces of the final line drill holes on the face may be considered objectionable.

13.4.2 Drilling

The maximum depth that can be successfully excavated by cushion or pre-shear blasting depends on the accuracy of the hole alignment. Deviations of more than about 150 mm

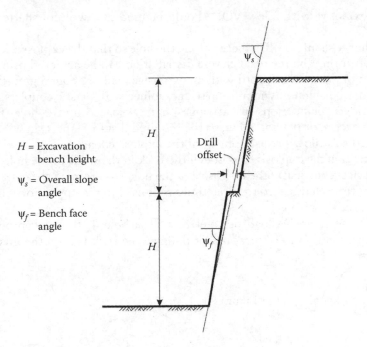

Figure 13.13 Layout of final line blast holes on multi-bench excavation taking into account drill offsets.

(6 in.) from the plane of the holes generally give poor results due to the uneven distribution of explosive on the shear plane, and the creation of an irregular face. While holes up to 27 m (90 ft) deep have been successfully cushion blasted, the depths of final line holes are usually limited to about 8–10 m (25–33 ft). Drill hole alignment can be enhanced with the use of high-rigidity drill steel, button bits and limiting the down-hole pressure on the bit.

The penetration rates of the drill should also be considered when determining the depth to be cushion blasted. If, for example, the penetration rate beyond a given depth becomes excessively slow, it may be more economical to carry out the excavation in a series of benches in order to keep penetration rates and drilling costs at acceptable levels. Also, when laying out final line drill holes in a benched excavation, allowance should be made for a minimum 0.3 m (1 ft) offset per bench since it is not possible to position the drill flush to the wall of the upper bench. This situation results in the overall slope of a multi-bench excavation being flatter than the bench face angle, and this difference should be allowed for in laying out the final line holes (Figures 13.10 and 13.13).

In unconsolidated sedimentary formations where it is difficult to hold a smooth wall, or when blasting around curved areas or in corners, unloaded guide holes between cushion holes may be used (Wyllie, 1999). Generally, small diameter guide holes are employed to reduce drilling costs. Where only the top of the formation is weathered, the guide holes need to be drilled only to that depth and not the full depth of the cushion holes.

13.4.3 Explosive load

The required explosive load in the final line holes can be determined by the following process:

First, the charge diameter should be smaller than the hole so that the decoupling ratio is at least 2 (see Figure 13.11).

Second, an explosive with a low VOD should be used that will not shatter the rock (see Table 13.1).

Third, the charge should be distributed along the hole so that the explosive load on the face is reasonably uniform. The required charge distribution can be achieved using either a continuous column of a detonating cord with an explosive load of about 5 g/m (Figure 13.14a), or a string of small diameter dynamite cartridges joined with plastic couplers. Alternatively, dynamite cartridges (or portions of cartridges) can be spaced up the hole (Figure 13.14b) using wooden spacers, or placing them, at the required distance apart, in 'C'-shaped plastic tubes; these methods allow precise control of the explosive load.

It is preferable that the explosive not touch the rock in the walls of the hole, which can be achieved using sleeves to centre the explosive in the hole. To promote shearing at the bottom of the hole, a bottom charge two to three times that used in the upper portion of the hole is generally employed.

The approximate explosive load per metre of blast hole l_{ex} that will produce sufficient pressure to cause splitting to occur, but not damage the rock behind the final face, can be approximated by

$$l_{ex} = \frac{(d_h)^2}{12,100} - \text{kg/m}, \quad \text{for } d_h \text{ in mm}$$
(13.10)

or

$$l_{ex} = \frac{(d_h)^2}{28} - \text{lb/ft}, \quad \text{for } d_h \text{ in inches}$$
(13.11)

where d_h is the hole diameter.

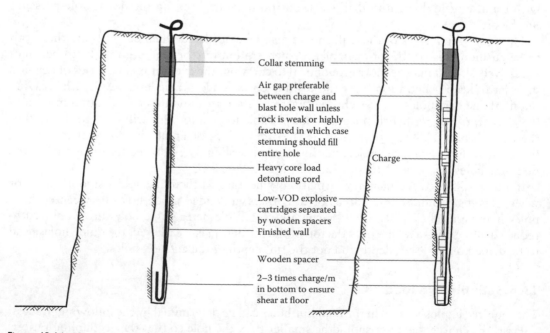

Collar stemming

Air gap preferable
between charge and
blast hole wall unless
rock is weak or highly
fractured in which case
stemming should fill
entire hole

Charge

Heavy core load
detonating cord

Low-VOD explosive
cartridges separated
by wooden spacers

Finished wall

Wooden spacer

2–3 times charge/m
in bottom to ensure
shear at floor

Figure 13.14 Alternative explosive placements for controlled blasting (adapted from ISEE, 2014).

13.4.4 Stemming

Similar to production holes, the upper part of final line holes is unloaded and stemmed with angular gravel to contain the explosive gases in the hole. For hole diameters up to about 100 mm, the stemming length is in the range of 0.6–1.0 m (2–3 ft), and varies with the formation being shot. In homogeneous formations, only the top of the hole is stemmed so that air gap around the charge serves as a protective 'cushion'. However, when stemming is not used around the charges, the explosive gases can find weak zones in the formation and tend to vent before the desired shear between holes is obtained. Similarly, the gases may find areas of weakness back into the finished wall and produce overbreak. Therefore, in weak and highly fractured, or faulted rock, complete stemming between and around individual charges may be desirable.

13.4.5 Spacing and burden

The spacing between blast holes on the final line differs slightly between pre-shearing and cushion blasting, as given by the following relationships:

Pre-shearing: Spacing $\approx$ (10 to 12) $\cdot$ (hole diameter)   (13.12)

Cushion blasting: Spacing $\approx$ (16) $\cdot$ (hole diameter)   (13.13)

For pre-shearing, the burden is effectively infinite, but as discussed in Section 13.4.1, it should at least equal the bench height. For cushion blasting, the burden should be larger than the spacing so that the fractures preferentially link between holes along the final wall and do not extend into the burden. The burden is given by

Cushion blasting: Burden ≥ 1.3 $\cdot$ (hole spacing)   (13.14)

13.4.6 Hole detonation sequence

As discussed in Section 13.4.1, the main difference between pre-shearing and cushion blasting is the detonation sequence. In pre-shearing, the final row holes are detonated first in the sequence, and in cushion blasting, the final row holes are detonated last in the sequence. Furthermore, satisfactory results are obtained by detonating each hole along the final line on separate delays, and it is not necessary to use a single delay for the full length of the final wall blast.

Figure 13.15 shows a possible blast hole layout and detonation sequence for pre-shearing of a multi-bench excavation. This excavation is suitable for pre-shearing because the burden is greater than the bench height and premature displacement of the burden is not possible. The blast hole layout comprising the final line holes, the buffer holes and production holes follows the guidelines on burden and spacing distances provided in this chapter (Equations 13.1 to 13.14), with allowance for the geometry of the bench. An essential component of the blast is the 'buffer' holes that are loaded with about 50% of the load, and a spacing of about 75%, of the production holes. The function of the buffer holes is to break the rock in front of the shear line, but not to damage the rock behind the final face. The vertical buffer holes should be positioned so that they are at least 1 m (3 ft) from the final line holes.

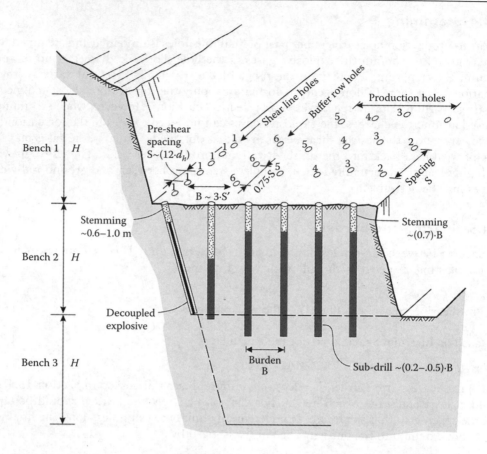

Figure 13.15 Blast hole layout and explosive loading for a multi-bench excavation using pre-shear blasting on the final face.

For a pre-shear blast, the shear line holes are detonated first in the sequence, followed by the production and buffer holes. However, it is possible that this could also be a cushion blast in which the final line holes are detonated last in the sequence.

13.5 BLAST DAMAGE AND ITS CONTROL

Blasting in urban areas must often be controlled to minimise the risk of damage to structures, and disturbance to people living and working in the vicinity. Three types of damage that can be caused by blasting include (Figure 13.16):

1. *Ground vibrations* – structural or cosmetic damage resulting from ground vibration induced by the shockwave spreading out from the blast area (see Section 13.6).
2. *Flyrock* – impact damage by rock ejected from the blast (see Section 13.7).
3. *Airblast and noise* – damage due to overpressure generated in the atmosphere (see Section 13.8).

The basic mechanism of blasting that causes damage to nearby structures is the excess energy of the shock wave, generated by the explosive detonation, as it spreads out from the

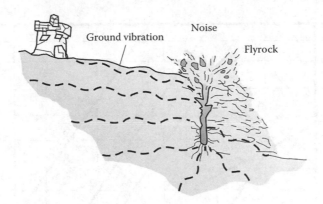

Figure 13.16 Causes of blast damage.

immediate blast area. As the shock wave spreads out, its energy diminishes due both to the energy consumed in breaking and deforming the rock, and because it occupies a progressively larger volume of rock with time. At distances where the energy is insufficient to break the rock, vibrations will be generated that can be large enough to damage structures. Furthermore, if the explosive gases are not adequately confined by the burden, the shock wave released at the face can generate both flyrock that is ejected from the blast, and considerable noise.

When designing a blast in an urban area, one must take into consideration not only the potential for damage in the vicinity of the blast, but also possible disturbance to people at a considerable distance from the site. Disturbance to people living outside the potential damage zone can give rise to complaints, and possibly spurious damage claims. This section provides information on allowable vibration and air blast levels, and discusses methods of preventing both damage to structures, and disturbance to people.

On most urban construction projects, blasting is only one of the many possible sources of vibration. Figure 13.17 shows how the peak article velocity produced by various types of construction equipment compares with the vibrations produced by detonation of 0.45 kg (1 lb) of dynamite. This chart represents approximate values, and actual vibration levels will vary from site to site. Detonation of an explosive charge produces a short duration, transient vibration compared to other sources, such as heavy machinery which produce steady-state (or pseudo-steady) vibrations. Generally, the steady-state vibrations are more disturbing to people than transient vibrations, even if the latter have a higher peak particle velocity (PPV). Also, vibrations produced by the steady-state sources can damage structures close to the source, so in some circumstances the effects of non-blasting vibrations should be considered in potential damage evaluation.

13.6 DAMAGE FROM GROUND VIBRATION

The following is a discussion of the properties of ground vibrations, and how these properties relate to damage to structures that are strained by these motions.

13.6.1 Ground motions

When an explosive charge is detonated near a free surface, the elastic response of the rock to the shock wave is the generation of two body waves and one surface wave. The faster

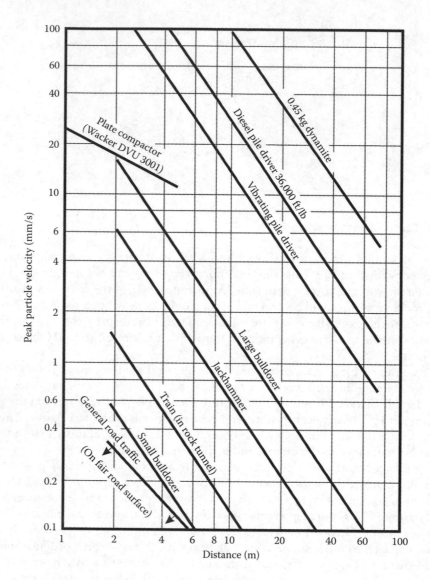

Figure 13.17 Peak particle velocities produced by construction machinery compared to explosive charge (adapted from Wiss, 1981).

of the two body waves propagated within the rock is a compressive, or *P*, wave, while the slower type is a shear, or *S*, wave. The surface wave, or *R* wave is slower than either the *P* or *S* wave; it is named after Raleigh who proved its existence. In terms of vibration damage, the *R* wave is the most important since it propagates along the ground surface, and because its amplitude decays more slowly with distance travelled than the *P* or *S* waves.

The wide variations in geometric and geological conditions on typical blasting sites preclude the solution of ground vibration magnitudes by means of elastodynamic equations. Therefore, the most reliable predictions are given by empirical relationships developed from observations of actual blasts.

Of the three most easily measured properties of the stress waves, that is, acceleration, velocity and displacement, it is generally agreed that velocity can be most readily correlated

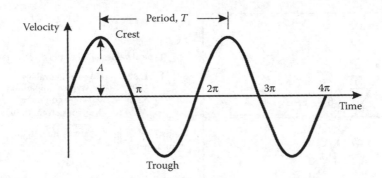

Figure 13.18 Sinusoidal wave motion for typical ground vibration.

with damage to structures. The stress wave has three components – vertical, longitudinal and transverse, and it is necessary to measure all three components and use the greatest, termed the PPV, to assess damage potential. The difference between the particle velocity of a stress wave, and the velocity of propagation of the wave can be explained as follows. As the wave passes through the ground, each particle of the rock and soil undergoes an elliptical motion, and it is the velocity of this oscillating motion (possibly up to 0.5 m/s) that is measured in assessing blast damage. In contrast, the velocity of propagation of the wave is in the range 300–6000 m/s, but this has no direct bearing on damage.

Ground motion can be described as a sinusoidal wave in which the displacement of the particle δ with time t is given by (Figure 13.18)

$$\delta = A \cdot \sin(\omega \cdot t) \tag{13.15}$$

where A is the amplitude of the wave and ω is the angular velocity. The magnitude of the angular velocity is given by

$$\omega = 2 \cdot \pi \cdot f - 2 \cdot \pi \cdot \frac{1}{T} \tag{13.16}$$

where f is the frequency (vibrations per second, or Hertz) and T is the period (time for one complete cycle). The wavelength L of the vibration is the distance from crest to crest of one full cycle and is related to the period T and the velocity of propagation V by

$$L = V \cdot T \tag{13.17}$$

If the particle displacement, δ, has been measured, then the velocity v and acceleration a can be calculated by differentiation as follows:

$$v = \frac{d\delta}{dt} = A \cdot \omega \cdot \cos(\omega \cdot t)$$

and

$$a = \frac{dv}{dt} = -A \cdot \omega^2 \cdot \sin(\omega \cdot t) \tag{13.18}$$

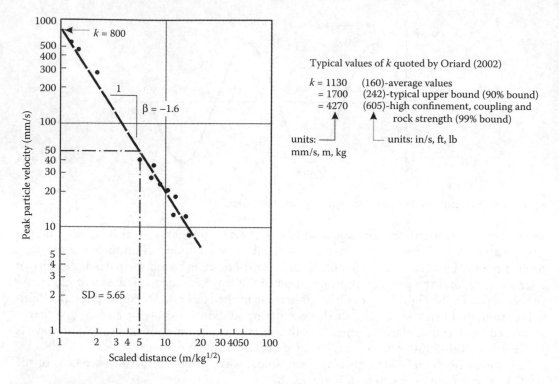

Figure 13.19 Hypothetical plot of measured particle velocity versus scaled distance from blast to determine attenuation parameters k and β.

The most reliable relationship between blast geometrics and ground vibration is that relating particle velocity to *scaled distance*. The scaled distance, SD, is defined by the function ($SD = R/W^{1/2}$), where R (m, ft) is the radial distance from the point of detonation and W (kg, lb) is the mass of explosive detonated per delay. Field tests have established that the maximum particle velocity, V (mm/s) (or PPV), is related to the scaled distance by following attenuation relationship (Oriard, 1971):

$$V = k\left(\frac{R}{\sqrt{W}}\right)^{\beta} \qquad (13.19)$$

where k and β are parameters that have to be determined by measurements on each blasting site. Equation 13.19 plots as a straight line on log–log paper, where the value of k is given by the PPV intercept at a scaled distance of unity, and β is given by the slope of the line. An example of such a plot is given in Figure 13.19.

13.6.2 Ground motion measurements

In order to obtain data for the preparation of a plot such as that in Figure 13.19, it is necessary to make vibration measurements with a suitable monitoring instrument. Table 13.5 shows typical specifications for a seismograph suitable for measuring blast vibrations, as well as vibrations produced by a wide range of construction equipment such as pile drivers. Of importance for monitoring non-blasting applications is the useful frequency range (linear

Table 13.5 Instantel DS-677 technical specifications

Ranges	All channels are autoranging Seismic (three channels): (0–254 mm/s)[a] Air pressure (one channel): 88–148 dB peak (200 kPa, 29 psi)
Trigger levels	Seismic: 0.13–254 mm/s Air pressure: 100–148 dB
Resolution	Seismic: 0.13 mm/s Air pressure: 1 dB (0.02 Pa)
Frequency response	All channels 2–250 Hz, ±3 dB (independent of record time)
Recording times	Up to 9000 s in 30 s increments
Acceleration	Computed up to 30 g with a resolution of 0.013 g (0.01 displayed)
Displacement	Computed to 38 mm metric resolution 0.000127 mm, 0.001 displayed

[a] Measurement units (imperial or metric) are user selectable in the field. Air pressure values are rms weighted.

response) of the equipment; for the Instantel DS677 this range is 1–315 Hz, which means that vibrations with frequencies outside this range will not be detected. Some of the features of vibration monitoring instruments are measurement of both ground vibration – velocity, acceleration, displacement and frequency in vertical, longitudinal and transverse axes, and air pressure. In addition, the instruments incorporate a trigger that starts recording when a pre-set vibration level or air pressure is detected. The results are digitally recorded so the data can be downloaded to a computer for storage and further analysis.

Measurement of ground vibrations involves installing geophones on the ground close to the structure of interest, or on the structure itself. It is important that the geophone be properly coupled to the ground or structure. The International Society of Rock Mechanics (ISRM, 1991) suggests that the method of mounting is a function of the particle acceleration of the wave train being monitored. When vertical accelerations are less than 0.2 g, the geophone may be placed on a horizontal surface without being anchored in place. When the maximum particle accelerations are between 0.2 and 1.0 g, the geophones should be completely buried when monitoring in soil, or firmly attached to rock, asphalt or concrete. Appropriate methods of attaching a geophone to these surfaces include double-sided tape, epoxy or quick setting cement. If these methods are unsatisfactory or if maximum particle accelerations exceed 1.0 g, then only cement or bolts are appropriate to secure the geophone to a hard surface. Note that weighting the geophone with a sandbag, for example, is not effective because the bag will move by the same amount as the geophone. In all cases, the geophone should be level and aligned with the longitudinal axis along the radial line to the blast.

13.6.3 Vibration damage threshold

Much work has been carried out to estimate threshold vibration levels for damage to many different types of structures (Siskind, Stachura and Raddiffe, 1976; Siskind, Staff, Kopp et al., 1980; Stagg, Siskind, Stevens et al., 1984). These PPV levels, which are listed in Table 13.6, have been established from many observations in the field and are now used with confidence as a basis for design of blasts where ground vibrations must be controlled.

The damage criterion most frequently used in urban areas is a PPV of 50 mm/s (2 in./s), which is the level below which the risk of damage to most residential structures is very slight. However, for poorly constructed buildings, and old buildings of historic interest, allowable

Table 13.6 PPV threshold damage levels

Velocity (mm/s)	Effect/damage
3–5	Vibrations perceptible to humans
10	Approximate limit for poorly constructed, and historic buildings
33–50	Vibrations objectionable to humans
50	Limit below which risk of damage to structures is very slight (<5%)
125	Minor damage, cracking of plaster, serious complaints
230	Cracks in concrete blocks

vibration levels may be as low as 10 mm/s. The required limit should be determined from an inspection of the structural condition of the building.

Figure 13.20 shows the relationship between the distances of a structure from the blast, the explosive weight detonated per delay and the expected PPV at the structure. These graphs have been drawn up using the relationship given in Equation 13.19. Plots such as that in Figure 13.20 can be used as a guideline in the assessment of blast damage for residential structures. If the blast design and this graph indicate that the vibration level is close to the damage threshold, then it is usually wise to carry out trial blasts and measure the vibration levels produced. These data can be used to produce a graph similar to that shown in Figure 13.19, from which the values of the parameters k and β for the site can be determined. The parameters can be used to draw up a blast design based on the calculated allowable charge

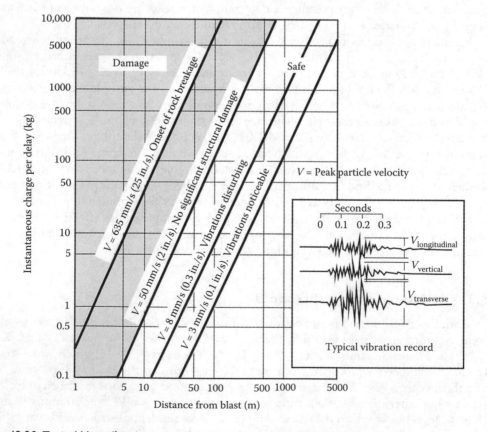

Figure 13.20 Typical blast vibration control diagram for residential structures.

weight per delay. It has been found that the values of k and β will vary considerably from site to site, so vibration measurements are useful in critical situations unless conservative restrictions are placed on allowable explosive weights per delay.

Figure 13.20 shows that 100 kg (220 lb) of explosive detonated per delay may cause minor cracking of plaster in houses at distances less than 100 m (300 ft) from the blast, while the vibrations will be noticeable at distances up to about 850 m (2800 ft). Halving of the mass of explosive detonated per delay to 50 kg (110 lb) will reduce these distances to 70 m (230 ft) and 600 m (2000 ft), respectively. Thus, the use of delays to limit the weight of explosive detonated per delay is the most effective method of controlling both damage, and reactions of the public to blasting operations.

13.6.4 Effect of vibration frequency on ground vibrations

The frequency of the vibration, in addition to particle velocity, is also of importance in assessing damage potential. If the principle frequency, that is, the frequency of greatest amplitude pulse, is approximately equal to the natural frequency of the structure, then the risk of damage is greater than if the principle and natural frequencies are significantly different (Dowding, 1985). The natural frequency of two-storey residential buildings is in the range of 5–20 Hz, and the natural frequency diminishes with increasing height of the structure. The principal frequency of a blast will vary with such factors as the type of blast, the distance between the blast and the structure, and the material through which the ground vibrations travel. Typical construction blasts produce vibrations with principal frequencies in the range of about 50–200 Hz. It is found that large quarry and mine blasts produce vibrations with lower-principal frequencies than construction blasts, and that principal frequencies decrease with increasing distance from the blast due to frequency attenuation, that is, the high frequency, relatively low-energy vibrations are attenuated more than low-frequency, high-energy vibrations.

In comparison to ground vibrations induced by blasting, earthquake ground motions typically have lower frequencies (0.1–5 Hz) because of the greater distance from the source. The higher amplitude accounts for the greater damage usually caused by earthquakes (see Chapter 11).

Figure 13.21 shows the relationship between PPV, vibration frequency and approximate damage thresholds for residential structures. This diagram demonstrates that with increasing frequency, the allowable PPV also increases.

13.6.5 Effect of geology on ground vibrations

It has been found that blast vibrations are modified by the presence of soil overburden at the measurement location. In general, vibrations measured on overburden have a lower frequency and higher amplitude than those measured on rock at the same distance from the blast. A consequence of this is that if the particle velocity is approximately the same at two locations, the lower frequency of the vibrations in overburden will make the blast vibrations more readily felt by humans.

13.6.6 Vibrations in uncured concrete

On some construction projects, blasting operations may need to be carried out close to uncured concrete. Under these circumstances, explosive charge weights per delay should be designed to keep ground vibrations within limits that are determined by the age of the concrete, the distance of the concrete from the blast and the type of structure (Oriard and Coulson, 1980; Oriard, 2002). Table 13.7 shows an approximate relationship between

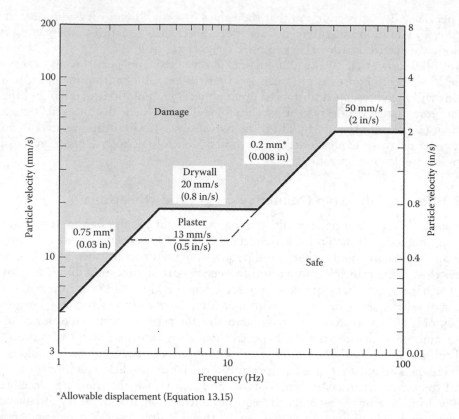

*Allowable displacement (Equation 13.15)

Figure 13.21 Allowable vibration levels for residential structures due to blasting (Siskind, Staff, Kopp et al., 1980).

allowable PPV levels and the concrete age for mass concrete. At ages less than 4 h, the concrete has not yet set and ground vibration levels are permissible, with gradually increasing levels allowed as the concrete sets.

Table 13.7 also shows that vibration levels must be reduced with increasing distance, as indicated by the distance factor (DF). Concrete can withstand higher-vibration levels at higher frequencies, because at low frequencies greater deflections will be induced in the structure. This is of particular concern for structural walls of freshly poured concrete. Vibration frequencies decrease as the distance from the blast increases because attenuation of frequency occurs with distance. The result of this frequency attenuation is that, at equal curing times, higher-vibration levels are permitted at closer distances.

In critical conditions, it is recommended that vibration measurements and strength tests be conducted to confirm the performance of the concrete and the relationships given in Table 13.7.

Values of the DF are obtained from the following relationship:

Distance factor, DF	Distance from blast, m (ft)
1.0	0–15 (0–50)
0.8	15–46 (50–150)
0.7	46–76 (150–250)
0.6	>76 (>250)

Table 13.7 Relationship between age of concrete and allowable PPV

Concrete age from batching	Allowable PPV, mm/s (in./s)
0–4 h	100 (4) × DF
4 h to 1 day	150 (6) × DF
1–3 days	225 (9) × DF
3–7 days	300 (12) × DF
7–10 days	375 (5) × DF
More than 10 days	500 (20) × DF

Source: Oriard, L. L. 2002. *Explosives Engineering, Construction Vibrations and Geotechnology.* International Society of Explosives Engineers, Cleveland, OH, 680pp.

13.6.7 Electronic equipment and machinery

Some types of electronic/electrical equipment are sensitive to vibrations. Studies have been done on computer disk drives, telecommunication equipment such as relays and fibre optic cables as well as electrical equipment such as older model power transformers that have mercury cutout switches. Vibrations from construction activities such as pile driving, as well as blasting can interfere with the operation of this equipment. Guidelines from manufacturers on allowable vibration levels are usually quoted in gravity units (g) showing the earthquake ground motions that they can withstand. In some circumstances, it may be necessary to conduct carefully documented test blasts to evaluate the vibration tolerance of the equipment.

13.6.8 Human response to blast vibrations

Humans are very sensitive to vibrations and can feel the effects of a blast well outside the potential damage zone. Figure 13.22 shows the relationship between PPV, frequency and the possible human response to vibrations. This indicates that low-frequency vibrations are more readily felt than high-frequency vibrations. The frequency of blast vibrations is usually in the range of 50–200 Hz.

13.6.9 Control of vibrations

The magnitude of blast vibrations at a particular location is dependent upon the distance from the blast and the charge weight per delay. Of the blast design factors discussed in Section 13.4, the most important regarding damage control is the use of delays and the correct detonation sequence so that each hole, or row of holes, breaks towards at least one free face.

To control vibration levels at a particular distance from the blast, it is necessary to limit the explosive charge detonated per delay according to the relationship given in Equation 13.19. Calculation of the allowable charge weight per delay, from either trial blasts or design charts, will determine how many holes can be detonated on a single delay. If the charge weight in a single hole is more than that allowable, then either shorter holes must be drilled, or a decked charge could be used. In a decked charge, the explosive load is separated into two parts with stemming material, and each charge is detonated on a separate delay. The minimum delay interval between charges or holes in order to limit the risk of constructive interference of vibrations produced by each charge is about 15–20 ms.

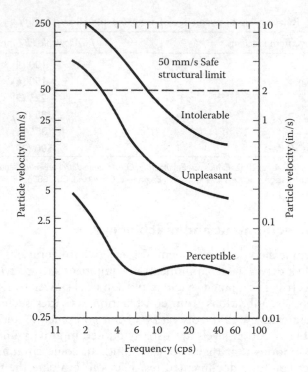

Figure 13.22 Human response to blast vibrations related to particle velocity and frequency (adapted from Wiss, 1981).

13.6.10 Pre-blast surveys

At locations where damage to structures due to blast vibrations is possible, it is usual to carry out pre-blast surveys on all structures within the potential damage zone. These surveys should record, with photographs and/or videos where appropriate, all pre-existing cracks, other structural damage, and settlement problems. The Office of Surface Mining (OSM) (2001) has drawn up a standardised method of recording structural damage which ensures that the survey is systematic and thorough (Figure 13.23). In addition to the building survey, a public relations programme informing people of the blasting operations, and carefully documented vibration measurements will usually minimise complaints and spurious damage claims.

13.7 CONTROL OF FLYROCK

Flyrock is the uncontrolled ejection of rock fragments from the blast, which can be a very hazardous condition because of its unpredictability. Common causes of flyrock generation are shown in Figure 13.24. For example, when the front row burden is inadequate or when the stemming column is too short to contain the explosive gases, a crater is formed and rock ejected from the crater. This figure also shows that flyrock can be caused by poor drill hole alignment, and by geologic conditions that allow venting of the explosive gases along the discontinuities in the rock mass.

In practice, the complete control of flyrock is difficult, even if the blast is designed with the recommended stemming and burden dimensions. Therefore, in areas where damage to

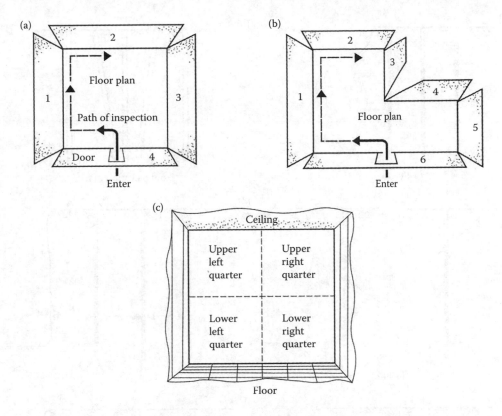

Figure 13.23 Method of making building damage surveys: (a) wall identification procedures; (b) multi-wall identification procedure; (c) detailed wall identification procedure (Office of Surface Mining [OSM], 2001).

structures is possible, blasting mats should be used to control flyrock. Blasting mats consist of rubber tires or strips of conveyor belting chained together. When the mat could be displaced by the explosive energy, they should be weighted with soil, or anchored to the bedrock.

13.8 CONTROL OF AIR BLAST AND NOISE

These two problems are taken together because they both arise from the same cause. Air blast, which occurs close to the blast itself, can cause structural damage such as the breaking of windows. Noise, into which the air blast degenerates with distance from the blast, can cause discomfort and give rise to complaints from those living close to the blasting operations.

Factors contributing to the development of air blast and noise include overcharged blast holes, poor stemming, uncovered detonating cord, venting of explosive gases along cracks in the rock extending to the face, and the use of inadequate burdens giving rise to cratering. In fact, many of the hole layout and charging conditions shown in Figure 13.24 can result in airblast in addition to flyrock.

The propagation of the pressure wave depends upon atmospheric conditions including temperature, wind and the pressure–altitude relationship. Cloud cover can also cause

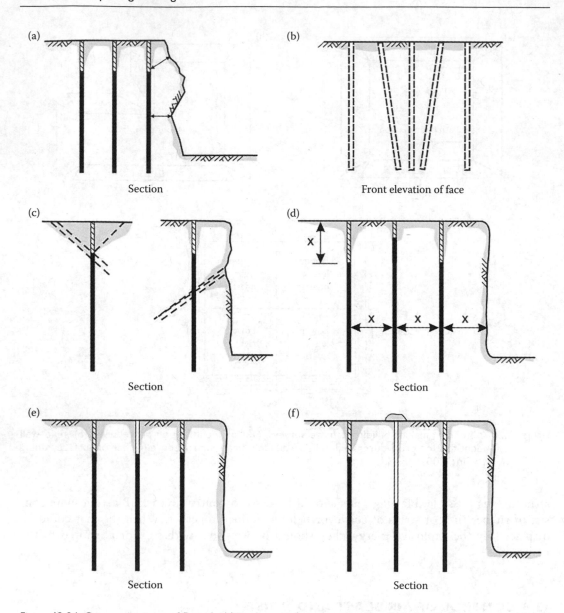

Figure 13.24 Common causes of flyrock: (a) inadequate front row burden; (b) hole misalignment resulting in concentration of explosives; (c) weak seams vent gas to rock faces; (d) holes loaded close to bench surface; (e) some holes with no stemming; (f) blocked holes loaded with fixed weight of explosives or number of cartridges (after C.I.L., 1983).

reflection of the pressure wave back to ground level at some distance from the blast. Figure 13.25 gives an indication of how the propagation of the shockwave is affected by the variation of temperature with altitude. This shows that air blast problems can be most severe during temperature inversions.

Figure 13.26 gives a useful guide to the response of structures and humans to sound pressure level (Ladegaard-Pedersen and Dally, 1975). The maximum safe air blast levels recommended by the U.S. Bureau of Mines is 136 dB for pressure measurements made using

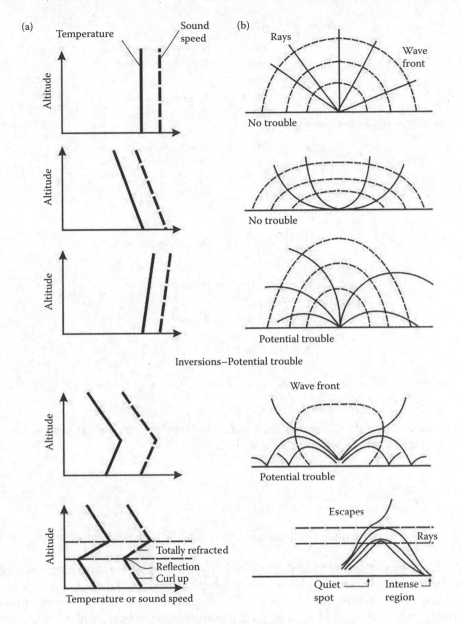

Figure 13.25 Effect of atmospheric conditions on air blast: (a) variation of temperature and sound speed with altitude; (b) corresponding pattern of sound rays and wave front (Baker, 1973).

a linear peak weighting network. The relationship between decibels and pressure P (kPa) is given by the following equation:

$$dB = 20 \cdot \log_{10} \left(\frac{P}{P_0} \right) \tag{13.20}$$

where P_0 is the overpressure of the lowest sound that can be heard, about 2×10^{-5} kPa (3×10^{-6} psi).

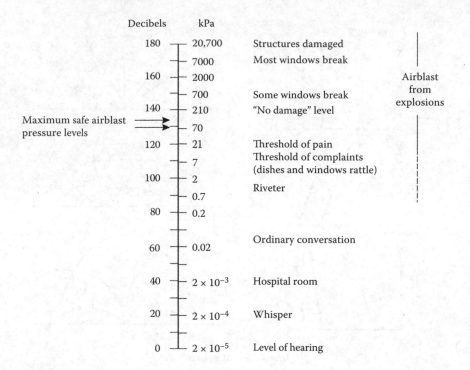

Figure 13.26 Human and structural response to sound pressure level (Ladegaard-Pedersen and Dally, 1975).

The decrease of sound pressure with distance can be predicted by means of cube root scaling. The scaling factor with distance K_R is given by

$$K_R = \frac{R}{\sqrt[3]{W}} \qquad (13.21)$$

where R is the radial distance from the explosion and W is the weight of charge detonated per delay.

Figure 13.27 gives the results in the original English units of pressure measurements carried out by the U.S. Bureau of Mines in a number of quarries. The burden, B, was varied and the length of stemming was 2.6 ft/in. (0.31 m stemming per cm) diameter of borehole. For example, if a 1000 lb (454 kg) charge is detonated with a burden of 10 ft (3.1 m), then the overpressure at a distance of 500 ft (152 m) is found as follows:

$$\frac{R}{\sqrt[3]{W}} = \frac{500}{\sqrt[3]{1000}} = 50$$

and

$$\frac{B}{\sqrt[3]{W}} = \frac{10}{\sqrt[3]{1000}} = 1$$

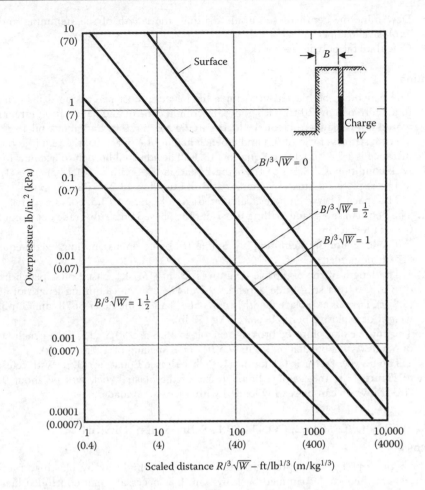

Figure 13.27 Overpressures as a function of scaled distance for bench blasting (Ladegaard-Pedersen and Dally, 1975).

From Figure 13.27, overpressure equals about 0.015 psi (0.1 kPa), or 74 db.

These calculations are related to the air blast produced by the explosive charge itself. However, a significant component of the air blast is produced by detonating cord and a reduction in noise can be achieved by covering the detonating cord with sand. Alternatively, a detonating system such as Nonel can be used, which comprises a fine plastic tube that propagates the shock wave with little noise.

EXAMPLE PROBLEM 13.1: BLAST DESIGN

Statement

A 6 m high rock bench is to be excavated by blasting. The rock contains thin, well-cemented layers that dip steeply out of the face, and the SG of the rock is 2.6. The available explosive has an SG of 1.3. The broken rock will be excavated by a front-end loader with a maximum vertical reach of 5 m.

Required

- Determine a suitable blast hole pattern, that is, the burden, spacing, and the diameter of the explosive.

- Determine the depth of sub-grade drilling, the length of the stemming and the explosive load per hole.
- Calculate the powder factor (kg/m^3).

Solution

1. The bench can be blasted in a single lift because most percussion drills can drill to a depth of 6 m (20 ft) with good directional control and penetration rate, and it would not be dangerous for the loader to dig a 6 m (20 ft) high muck pile.

   For a stiffness ratio of 0.3 and a bench height of 6 m (20 ft), the required burden distance is 1.8 m (6 ft) (see Equation 13.1). The adjusted burden to account for the rock condition is 2.3 m (7.5 ft), using Equation 13.4 with $k_\psi = 1.18$ and $k_s = 1.1$.

   The explosive diameter, calculated from Equation 13.2, is 78 mm (3 in.).

   The spacing between holes, calculated from Equation 13.5, is 2.8 m (9 ft).

2. The depth of sub-grade drilling is 0.7 m (2 ft), based on a sub-grade depth equal to 0.3 of the burden.

   The length of the stemming is 1.6 m (5 ft), based on a stemming length equal to 0.7 of the burden.

   The length of the explosive column is 5.1 m (16.7 ft), based on a bench height of 6 m (20 ft), a sub-grade depth of 0.7 m (2 ft) and a stemming length of 1.6 m (5.2 ft). For this explosive length, an SG of 1.3 and a diameter of 78 mm (3 in.), the weight of explosive per hole is 31.7 kg (70 lb).

3. The volume of rock to be broken per hole is 38.6 m^3 (50.5 yd^3), for a bench height of 6 m (20 ft), a burden of 2.3 m (7.5 ft) and a spacing of 2.8 m (9.2 ft).

   The powder factor is 0.8 kg/m^3 (1.35 lb/yd^3) (see Equation 13.9). With reference to Figure 13.8, the average boulder size of the blasted rock will be about 0.4 m (1.3 ft), which can be readily loaded with a front-end loader.

EXAMPLE PROBLEM 13.2: CONTROLLED BLASTING DESIGN

Statement

A 6 m high slope has been previously excavated for a highway by blasting, and it is required that the face be trimmed back by 3 m. It is necessary that controlled blasting be used for this trimming operation so that minimal overbreak occurs and the new face is stable.

Required

- Determine the most appropriate type of controlled blasting for this operation.
- If the explosive available has a diameter of 25 mm (1 in.), calculate the blast hole diameter, explosive charge (kg/m of hole) and hole spacing.
- If the explosive has an SG of 1.1, determine how the explosive can be distributed in the hole to achieve the required charge.
- Assess whether is it necessary for a second row of blast holes to be drilled between the present face and the final line. If a second row is necessary, determine how far this row should be from the final line, and what detonation sequence should be used.

Solution

1. Cushion blasting should be used to remove the 3 m (10 ft) thick slice of rock. In highway construction, it is rarely necessary to use line drilling because it is expensive and is only used where very high-quality slopes are required. Pre-shearing would be used only where the burden is at least equal to the bench height.

2. For an explosive diameter of 25 mm (1 in.), a 50 mm (2 in.) diameter hole would produce a decoupling ratio of 2. The required explosive charge per metre of blast hole is 0.21 kg/m (0.14 lb/ft), from Equation 13.10.

   The spacing between holes would be 0.8 m (2.6 ft), from Equation 13.13.

3. The weight of explosive for a 25 mm (1 in.) diameter cartridge with a unit weight of 1.1 is 0.54 kg/m (0.36 lb/ft). To produce an explosive charge of 0.21 kg/m (0.14 lb/ft), a possible distribution of this explosive up the hole would have cartridge lengths of 200 mm (8 in.), with 300 mm (12 in.) long spacers between each cartridge.

4. From Equation 13.14, the required burden is at least 1 m (3 ft). Since the total burden on the cushion blast is 3 m (10 ft), it would be necessary to drill a row or holes about 1.5 m (5 ft) behind the face. The denotation sequence would be to fire the front row at least 20 ms before the final row, based on Equation 13.7.

EXAMPLE PROBLEM 13.3: BLAST DAMAGE CONTROL

Statement

An historic building is located 140 m (460 ft) from the blast site, and a hospital is located at a distance of 1 km (1100 yd).

Required

1. Determine the maximum allowable charge weight per delay to minimise the risk of damage to the historic building.
2. Determine whether the ground vibrations would be disturbing to patients in the hospital.

Solution

1. Allowable charge weights per delay are determined from damage threshold vibration levels for different structures listed in Table 13.6 and from Equation 13.19.

   If $k = 1600$ and $\beta = -1.5$ and the threshold for damage to the historic building is 10 mm/s (0.04 in./s), then at $R = 140$ m (460 ft), the allowable instantaneous charge, W, is 23 kg/delay (51 lb/delay).

Figure 13.22 shows that vibrations are objectionable to humans when the PPV exceeds about 33–50 mm/s (1.3–2 in./s). For a charge weight per delay of 23 kg (51 lb), Equation 13.19 shows that the vibration levels at the hospital would be about 0.5 mm/s (0.02 in./s), so they are unlikely to be perceptible by the patients (perceptible threshold is about 3–5 mm/s) (0.1–0.2 in./s).

Chapter 14

Stabilisation of rock slopes

14.1 INTRODUCTION

In mountainous terrain, the operation of highways and railways, power generation and transmission facilities and the safety of residential and commercial developments often requires stable slopes and control of rock falls. This applies to both excavated and natural slopes. In contrast, open-pit mines and large quarries tolerate a certain degree of slope instability unless this is a hazard to the miners or causes a significant loss of production. For example, minor failures of benches usually have little effect on mine operations unless the fall lands on a haul road and results in tire or equipment damage. In the event of a large-scale slope failure in an open-pit mine, often the only economic and feasible stabilisation measure is drainage, which may involve long horizontal drains, pumped wells or drainage tunnels (see Section 14.4.7). A possible method of managing large-scale slope instability is to monitor the movement so that operations can continue beneath the moving slope. Procedures for monitoring movement and interpreting the results are discussed in Chapter 15.

This chapter is concerned with civil slopes because the high cost of failures means that stabilisation programmes are often economically justified. For example, on highways, even minor falls can cause damage to vehicles, injury or death to drivers and passengers and possibly discharge of toxic substances where transport vehicles are damaged. Also, substantial slope failures on transportation systems can severally disrupt traffic, usually resulting in both direct and indirect economic losses. For railroads and toll highways, closures result in a direct loss of revenue. Figure 14.1 shows a rock slide that occurred from a height of about 300 m and closed both the road and an adjacent railway.

While the cost of a slide, such as that shown in Figure 14.1, is substantial, the cost of even a single vehicle accident can be significant. For example, costs may be incurred for hospitalisation of the driver and passengers, for repair to the vehicle, and in some cases for legal charges and compensation payments. In addition, costs for stabilisation of the slope will involve both engineering and contracting charges, usually carried out at premium rates because of the emergency nature of the work.

Many transportation systems were constructed over a century ago in the case of railroads, and decades ago in the case of many highways. At that time, the blasting techniques that were often used in construction caused significant damage to the rock. Furthermore, since the time of construction, deterioration of stability conditions is likely due to weathering of the rock, loosening of the surficial blocks by ice and water and the growth of tree roots. All these effects can result in ongoing instability that may justify remediation programmes.

For urban developments in mountainous terrain, hazards such as rolling boulders and landslides that can threaten or destroy structures. The most effective protection against these conditions is initial hazard mapping, and then zoning unsafe areas to preclude development where mitigation cannot be carried out.

Figure 14.1 Highway closure caused by rock fall in very strong, blocky granite from a height of about 300 m.

In order to minimise the cost of falls, rock slope stabilisation programmes are often a preferred alternative to relocation or abandonment of the facility. Such programmes involve a number of interrelated issues including geotechnical and environmental engineering, safety, construction methods and costs, and contracting procedures. Methods for the design of stabilisation measures for rock slopes are described in this chapter, and in other references such as Brawner and Wyllie (1975); Fookes and Sweeney (1976); Wyllie (1991); Schuster (1992); FHWA (1993, 1998b); Transportation Research Board (TRB) (1996) and Morris and Wood (1999). These procedures have been used extensively since the 1970s and can, therefore, be used with confidence for a wide range of geological conditions. However, as described in this chapter, it is essential that appropriate method(s) be used for the particular conditions at each site.

The first part of this chapter discusses the cause of rock falls, and the planning and management of stabilisation programmes. For transportation systems with a large number of rock slopes, these programmes usually involve making an inventory of rock fall hazards and stability conditions, maintaining these records in a database and linked GIS map, and preparing a prioritised stabilisation schedule. The remainder of the chapter discusses stabilisation measures, categorised according to rock reinforcement, rock removal and rock fall protection.

14.2 CAUSES OF ROCK FALLS

The State of California made a comprehensive study of rock falls that occurred on the state highway system to assess both the causes of the rock falls, and the effectiveness of the various remedial measures that have been implemented (McCauley, Works and Naramore, 1985). With the diverse topography and climate within California, their records provide a useful guideline on the stability conditions of rock slopes and the causes of falls. Table 14.1 shows the results of a study of a total of 308 rock falls on California highways in which 14 different causes of instability were identified.

Of the 14 causes of rock falls, six are directly related to water, namely rain, freeze–thaw, snowmelt, channelled runoff, differential erosion and springs and seeps. Also, one cause that

Table 14.1 Causes of rock falls on highways in California

Cause of rock fall	Percentage of falls
Rain	30
Freeze–thaw	21
Fractured rock	12
Wind	12
Snowmelt	8
Channelled runoff	7
Adverse planar fracture	5
Burrowing animals	2
Differential erosion	1
Tree roots	0.6
Springs or seeps	0.6
Wild animals	0.3
Truck vibrations	0.3
Soil decomposition	0.3

is indirectly related to rainfall is the growth of tree roots in cracks that can open fractures and loosen blocks of rock on the face. These seven causes of rock falls together account for 68% of the total falls. These statistics are confirmed by the authors' experience in the analysis of rock fall records over a 25-year period on a major railroad in western Canada in which approximately 70% of the events occurred during the winter. The weather conditions during the winter included heavy rainfall, prolonged periods of freezing temperatures and daily freeze–thaw cycles in the fall and spring. The results of a similar study carried out by Peckover (1975) are shown in Figure 14.2. They clearly show that the majority of rock falls occurred between October and March, the wettest and coldest time of the year in western Canada. Also, in this geographical area, studies have been conducted on the relationship between the frequency and volume of rock slides (Hungr, Evans and Hazzard, 1999). The study showed that for falls with volumes less than 1 m^3, as many as 50 falls occurred per year, whereas for falls with volumes of 10,000 m^3, occurred every 10–50 years approximately.

The other major cause of rock falls in the California study is the particular geologic conditions at each site, namely fractured rock, adverse planar fractures (fractures dipping out of slope face), and soil decomposition. These three causes represented 17% of the falls, and the total rock falls caused by water and geologic factors accounted for 85% of the falls. These statistics demonstrate that water and geology are the most important factors influencing rock slope stability.

It appears that the study in California was carried out during a time when no significant earthquakes occurred, because these events frequently trigger rock falls, and cause displacement and failure of landslides. Chapter 11 discusses the influence of seismic ground motions on slope stability.

14.3 ROCK SLOPE STABILISATION PROGRAMMES

Transportation systems in mountainous terrain may require excavations of hundreds of rock slopes with a variety of rock fall hazards that can be a significant cost to the operator. Under these circumstances, a long-term slope stabilisation programme is often justified; this section describes the steps involved in implementing such a programme.

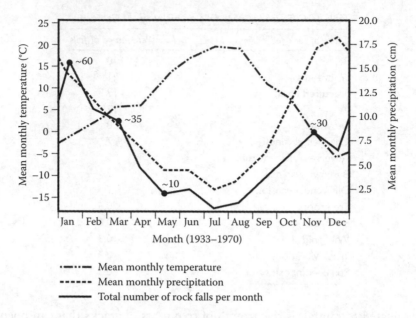

Figure 14.2 Correlation of number of rock falls with temperature and precipitation on railway lines in Fraser Canyon, British Columbia (Peckover, 1975).

14.3.1 Planning stabilisation programmes

When implementing a programme to stabilise a large number of slopes, the best use of available funds is often made by setting up a systematic programme that identifies and rates the most hazardous sites. Annual stabilisation work can then be scheduled, with the most hazardous sites having the highest priority. Figure 14.3 shows an example of how such a programme may be structured.

The objective of the programme shown in Figure 14.3 is to be proactive in identifying and stabilising slopes before rock falls and accidents occur. This requires a careful examination of each site to identify the potential hazard, and estimate the likely benefit of the stabilisation work. In contrast, a reactive programme places the emphasis in areas where rock falls and accidents have already occurred, and where the hazard may then be diminished.

An effective proactive approach to stabilisation requires a consistent, long-term programme under the direction of a team experienced in both the engineering and construction aspects of this work. Another important component of this work is to keep accurate records, with photographs, of slope conditions, rock falls and stabilisation work. This information will document the location of hazardous areas and determine the long-term effectiveness of the programme in reducing the incidence of rock falls. These records can be most conveniently handled using database programs that readily allow updating and retrieval of records.

14.3.2 Rock slope inventory systems

An important component of a rock fall stabilisation programme is to make an inventory of the rock slopes on the transportation system comprising both the location and a description of the hazard and risk of each slope. The hazard and risk can be quantified by assigning scores to the slope characteristics that influence stability, from which it is possible to

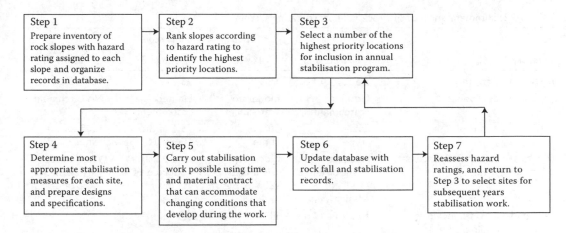

Figure 14.3 Rock slope stabilisation programme for transportation systems.

determine a total stability score for each rock slope. The total score can then be used to rank the slopes, identify the highest risk locations and plan a stabilisation programme.

Early work on slope inventories was carried out by Brawner and Wyllie (1975) and Wyllie (1987), which was adapted by Pierson, Davis and Van Vickle (1990) into a process for the rational management of rock slopes on highways that has been named the rock fall hazard-rating system (RHRS). The first step in this process is to make an inventory of the stability conditions of each slope so that they can be ranked according to their rock fall hazard (Steps 1 and 2 in Figure 14.3).

The rock fall areas identified in the inventory are ranked by scoring the categories as shown in Table 14.2. These nine categories represent the significant elements of a rock slope that contribute to the overall hazard. The four columns correspond to logical breaks in the hazard represented by each category. The criteria scores increase exponentially from 3 to 81 points, and represent a continuum of points from 1 to 100. These exponential scores allow for a rapid increase in score for poor stability conditions that clearly identifies the more hazardous sites. Using a continuum of points allows flexibility in evaluating the relative impact of conditions that are variable by nature. Some categories require a subjective evaluation while others can be directly measured and then scored.

Once scores have been assigned to each category, a total score for each slope can be calculated by one of two methods. First, the nine scores can be summed to give a maximum possible total of 729 (9 times 81). A second method is to add the scores separately for *hazard factors* that relate to slope conditions, and *risk factors* that relate to the rock fall path and traffic conditions; the two scores are then multiplied to obtain the total score for the site. In Table 14.2, hazard and risk factors are as follows:

Hazard factors:
   a. Slope height – maximum height of rock fall sources
   f. Geology – stability controlled by either structural geology or differential weathering
   g. Block size – controlled by joint spacing and rock strength
   h. Climate – effects of both precipitation and freezing temperatures
   i. Rock fall history – record of past events
Risk factors:
   b. Ditch effectiveness – width and depth of ditch related to slope height and angle defines efficacy of the ditch to contain falls

Table 14.2 Summary sheet of the rock fall hazard-rating system

Category		Rating criteria and score			
		Points 3	Points 9	Points 27	Points 81
(a) Slope height (m)		7.5	15	23	30
(b) Ditch effectiveness		Good catchment	Moderate catchment	Limited catchment	No catchment
(c) AVR (percent of time)		25% of the time	50% of the time	75% of the time	100% of the time
(d) Percentage of DSD (percent of design value)		Adequate sight distance, 100% of design value	Moderate sight distance, 80% of design value	Limited sight distance, 60% of design value	Very limited sight distance 40% of design value
(e) Roadway width including paved shoulders (m)		13.5	11	8.5	6
(f) Geologic character					
Case 1	Structural condition	Discontinuous joints, favourable orientation	Discontinuous joints, random orientation	Continuous joints, adverse orientation	Continuous joints, adverse orientation
	Rock friction	Rough, irregular	Undulating	Planar	Clay infilling, or slickensided
Case 2	Structural condition	Few differential erosion features	Occasional erosion features	Many erosion features	Major erosion features
	Difference in erosion rates	Small difference	Moderate difference	Large difference	Extreme difference
(g) Block size (m)		0.3	0.6	1.0	1.2
Quantity of rock fall event (m³)		3	6	9	12
(h) Climate and presence of water on slope		Low-to-moderate precipitation; no freezing periods, no water on slope	Moderate precipitation, or short-freezing periods, or intermittent water on slope	High precipitation or long-freezing periods, or continual water on slope	High precipitation and long-freezing periods, or continual water on slope and long-freezing periods
(i) Rock fall history		Few falls	Occasional falls	Many falls	Constant falls

Source: Pierson, L., Davis, S. A. and Van Vickle, R. 1990. The rock fall hazard rating system, implementation manual. Technical Report #FHWA-OR-EG-90-01, Washington, DC; Wyllie, D. C. 1987. Rock slope inventory system. *Proceedings of the Federal Highway Administration Rock Fall Mitigation Seminar,* FHWA, Region 10, Portland, Oregon.

   c. Average vehicle risk (*AVR*) – percent of time that vehicles are below the slope
   d. Sight distance – ability of drivers to see rock fall on road and take evasive action
   e. Roadway width – space available to contain falls and for drivers to take evasive action

The advantage of using the multiplied score is that, for example, the influence of improving the ditch to reduce the ditch effectiveness score to 3 compared to an original score of 27 will significantly reduce the total score, and the relative priority level, for the site. In comparison, this change in the added total will be less significant.

14.3.3 Hazard-rating criteria

The following is a brief description of each of the nine categories that is used to rate the rock fall hazard along a highway (see Table 14.2):

a. *Slope height* – vertical height of the slope measured from the highest point from which rock fall is expected. If falls originate from the natural slope above the cut, the cut height plus the additional slope height (vertical dimension) is used.

b. *Ditch effectiveness* – effectiveness of a ditch is measured by its ability to prevent falling rock from reaching the travelled way. In estimating the ditch effectiveness, factors to consider are: (1) slope height and angle; (2) ditch width, depth and shape; (3) anticipated block size and quantity of rock fall and (4) effect on rock fall trajectories of slope irregularities (launching features). A launching feature can negate the benefits expected from a fallout area. The ditch effectiveness can be assessed both by comparing the dimensions with those recommended in the ditch design chart shown in Figure 14.23 (Ritchie, 1963), and from information provided by maintenance personnel.

c. *Average vehicle risk (AVR)* – represents the percentage of time a vehicle will be present in the rock fall section. The percentage is obtained from the average daily traffic count (vehicles per day, *ADT*), the length of the rock fall hazard area (*L*, km) and the posted speed limit (*S*, km/h) by the following relationship:

$$AVR = \frac{ADT}{24 \text{ (h per d)}} \cdot \frac{L}{S} \cdot 100\% \tag{14.1}$$

For example, if the length of the slope is 0.2 km in an area where the posted speed is 90 kph and the average daily traffic count is 7000 vehicles per day, the *AVR* is 65% and the corresponding hazard score is 18. A rating of 100% means that on an average at least one vehicle can be expected to be under the slope at all times, and the hazard score is 81.

d. *Percentage of decision sight distance (DSD)* – DSD is used to determine the length of roadway (in m) needed to make a complex or instantaneous decision. The *DSD* is critical when obstacles on the road are difficult to perceive, or when unexpected or unusual manoeuvres are required. Sight distance is the shortest distance along a roadway that an object is continuously visible to the driver. Throughout a rock fall section, the sight distance can change appreciably. Horizontal and vertical highway curves, together with obstructions such as rock outcrops and roadside vegetation, can severely limit the available sight distance.

The relationship between *DSD* and the posted speed limit used in the inventory system is based on the U.S. highway design criteria, modified as shown in Table 14.3 (AASHTO, 1984).

The actual sight distance is related to the *DSD* by Equation 14.2 as follows:

$$DSD \, (\%) = \frac{\text{Actual sight distance}}{\text{Decision sight distance}} \times 100\% \tag{14.2}$$

Table 14.3 DSD to avoid obstacles

Posted speed limit, kph (mph)	Decision sight distance, m (ft)
48 (30)	137 (450)
64 (40)	183 (600)
80 (50)	229 (750)
97 (60)	305 (1000)
113 (70)	335 (1100)

For example, if the actual sight distance is restricted by the road curvature to 120 m in a zone with a posted speed limit of 80 kph, then the *DSD* is 52%. Based on the charts provided in the RHRS manual, the score for this condition is 42.

e. *Roadway width* – dimension that represents the available manoeuvring room to avoid a rock fall, measured perpendicular to the highway centreline from edge of pavement to edge of pavement. The minimum width is measured when the roadway width is not constant.

f. *Geologic character* – geologic conditions that cause rock fall generally fit into two cases. *Case 1* is for slopes where joints, bedding planes or other discontinuities are the dominant structural features of a rock slope. *Case 2* is for slopes where differential weathering forming overhangs or over-steepened slopes is the dominant condition that controls rock fall. The case which best fits the slope should be used when doing the evaluation. If both situations are present, both are scored but only the worst case (highest score) is used in the rating.

Geologic character – Case 1 Structural condition – adverse discontinuities are those with orientations that promote plane, wedge or toppling failures.

Rock friction – friction on a discontinuity is governed by the characteristics of the rock material and any infilling, as well as the surface roughness (see Section 5.2).

Geologic character – Case 2 Structural condition – differential erosion/weathering or over-steepening is the dominant condition that leads to rock fall. Erosion/weathering features include over-steepened slopes, unsupported rock units or exposed resistant rocks.

Difference in weathering rates – different rates of erosion/weathering within a slope directly relate to the potential for a future rock fall event. The score should reflect how quickly weathering is occurring; the size of rocks, blocks or units being exposed; the frequency of falls; and the amount of material released during a rock fall.

g. *Block size or volume of rock fall per event* – type of event most likely to occur, related to the spacings and continuous lengths of the discontinuity sets. The score should also take into account any tendency of the blocks of rock to break up as they fall down the slope.

h. *Climate and presence of water on slope* – water and freeze–thaw cycles contribute to both the weathering and movement of rock materials. If water is known to flow continually or intermittently on the slope, it is rated accordingly. This rating could be based on the relative precipitation over the region in which the ratings are being made, and incorporate the influence of freeze–thaw cycles.

i. *Rock fall history* – historical information is an important check on the potential for future rock falls. Development of a database of rock falls allows more accurate conclusions to be made of the rock fall potential.

14.3.4 Database analysis of slope inventory

It is common practice to enter the results of the slope inventory into a computer database. The database can be used both to analyse the data contained in the inventory, and to facilitate updating the inventory with new information on rock falls and construction work. The following are some examples of database analysis:

- Rank the slopes in order of increasing point score to identify the most hazardous sites.
- Correlate the rock fall frequency with such factors as weather conditions, rock type and slope location.
- Assess the severity of rock falls from analysis of delay hours or road closures caused by falls.
- Assess the effectiveness of stabilisation work from change in annual number of rock falls.

14.3.5 Selection of high-priority sites

It has been found that when managing ongoing stabilisation programmes with durations of possibly several decades and involving many hundreds of rock slopes, it is valuable to establish a rational process for selecting the high-priority sites. This becomes necessary because slope stability conditions deteriorate over time, and may not be possible to re-rate every slope, every year, according to the scoring system shown in Table 14.2. However, it may be possible to inspect all the higher-priority slopes on an annual basis to assess stability, and from this assessment to determine whether stabilisation is required, and within what time frame. This involves assigning each slope an 'Inspection Rating' and a corresponding 'Required Action' (see Table 14.4). The filled boxes in Table 14.4 indicate allowable actions for each of the ratings. For example, for an 'Urgent' slope the permissible actions are to 'Llimit Service' and work at the site within 1 month, or to carry out a 'Follow-up' inspection to assess stability conditions in more detail. However, for either an 'Urgent' or 'Priority' slope, it is not permissible to assign 'No Action' to the site.

The following are examples of criteria that could be used to assign inspection ratings, using a combination of measurement, the rating scores given in Table 14.2 and subjective observations of stability:

1. *Urgent* – obvious recent movement or rock falls, kinematically feasible block with dimensions large enough to be a hazard; weather conditions are detrimental to stability. Failure possible within next few months.
2. *Priority* – likely movement since last inspection, of block large enough to be a hazard; failure possible within next 2 years approximately.

Table 14.4 Inspection ratings and corresponding actions

		Required action				
		Limit service; work within 1 month	Work in current year	Follow-up inspection	Work in 1–2 years	No action
Inspection	Urgent	×		×		
Rating	Priority		×	×	×	
	Observe					×
	Okay					×

3. *Observe* – possible recent movement, but no imminent instability. Check specific stability conditions in the next inspection.
4. *OK* – no evidence of slope movement.

It has been found that the two primary benefits in assigning a required action for every slope are as follows. First, this forces the inspector to make a decision on the need for mitigation and its urgency, and second, it automatically draws up a list of work sites for the current year and the next 2 years. This list becomes the basic planning tool for the ongoing programme.

14.3.6 Selection of stabilisation measures

This section provides guidelines on selecting the method, or methods, of stabilisation that are most appropriate for the topographical, geological and operational conditions at the site. Methods of slope stabilisation slope stabilisation can be categorized as follows:

1. Reinforcement
2. Rock removal
3. Protection

Figure 14.4 includes 16 of the more common stabilisation measures divided into these categories. The following are examples of the factors that will influence the selection of appropriate stabilisation methods. Where the slope is steep and the toe is close to the highway or railway, no space will be available to excavate a catch ditch or construct a

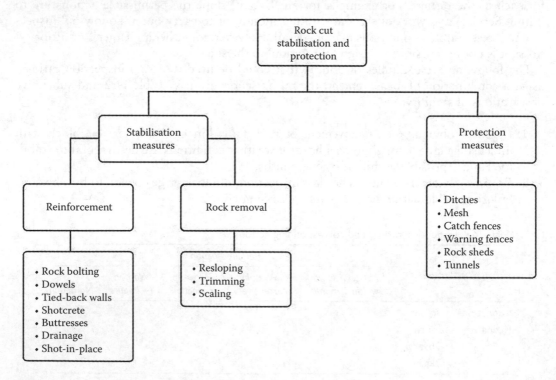

Figure 14.4 Categories or rock slope stabilisation measures.

barrier. Therefore, alternative stabilisation measures may be to remove loose rock, secure it in place with bolts or to cover the slope with wire mesh. It is generally preferable to remove loose rock and eliminate the hazard, but only if this will form a stable face and not undermine other potentially loose rock on the face. If the source of the rock falls is a zone of boulders in an erodible soil matrix that cannot be stabilised by bolting or effectively scaled, then a combination of ditch-containment structure may be more suitable. If the space at the toe of the slope for this work is limited, the only alternative may be to relocate or realign the facility.

When selecting and designing stabilisation measures that are appropriate for a site, geotechnical, construction and environmental issues must be considered. The geotechnical issues – geology, rock strength, ground water and stability analysis – are discussed in previous chapters. Construction and environmental issues, which can affect the costs and schedule of the work, must be addressed during the design phase of the project. Issues that are frequently important are equipment access, available work time during traffic closures and disposal of waste rock and soil.

Another factor to consider in the selection of stabilisation measures is the optimum level of work. For example, a minor scaling project will remove the loosest rock on the slope face, but, if the rock is susceptible to weathering, this work may have to be repeated every 3–5 years. Alternatively, a more comprehensive programme can be carried out using shotcrete and bolting, in addition to scaling. Although the initial cost of this second programme would be higher, it would be effective for a longer period, perhaps for 20–30 years. Alternative stabilisation programmes such as these, including the alternative of doing no work, can be compared using decision analysis. Decision analysis is a systematic procedure for evaluating alternative courses of action taking into account the likely range of construction costs and design life of the stabilisation work, as well as the probability and costs of rock falls occurring and causing accidents (Wyllie, McCammon and Brumund, 1979; Roberds, 1991; Roberds, Ho and Leroi, 2002; Wyllie, 2014a,b).

The following is a brief discussion of some construction issues that may have a significant influence on stabilisation work:

Blasting – damage to rock faces by excessively heavy blasting is a frequent cause of instability in the years following excavation of a slope. Methods of controlled blasting, such as pre-shearing and trim blasting as described in Chapter 13, can be used to excavate a slope to a specified line with minimal damage to the rock behind the face.

Topography – if the slope above the crest of a cut is steep, then stabilisation work that involves laying back the cut will have the effect of increasing the height of the cut. This increase in the cut height will require a larger catch ditch, and may result in additional stability problems, especially if the thickness of the layer of soil or weathered rock at the surface is substantial.

Construction access – determine the type of equipment that is likely to be required to carry out the work, and how this equipment will be used at the site. For example, if it is planned to excavate a substantial volume of rock in order to lay back a slope, then it is likely that airtrac drills and excavators will have to work on the slope. In steep terrain, it may be found that the construction of an access road for this equipment is costly and causes additional instability. Furthermore, a cut width of at least 5 m (16 ft) is required to provide sufficient working width for this equipment. Also, if stabilisation work is planned using large diameter rock bolts, then it is essential that suitable drilling equipment can access the site. For example, on steep faces, holes with a diameter larger than about 100 mm (4 in.) will have to be drilled with heavy equipment supported from a crane. Where it is not possible to use such heavy drilling equipment, it would be necessary to drill smaller diameter holes with handheld equipment, and install a greater number of smaller size bolts.

Construction costs – cost estimates for stabilisation work must take into account both the direct costs of the work on the slope, and indirect costs such as mobilisation, traffic control, waste disposal and environmental studies as discussed below. A significant cost issue on active transportation routes is the use of cranes to access the slope. If the work is done from a platform suspended from a crane located on the road, this may block two to three lanes of traffic. In contrast, traffic closures can be minimised by having the construction crews work off ropes secured above the crest of the slope.

Waste disposal – the least expensive method of disposing of waste rock produced by excavation and scaling operations in mountainous terrain is discarding rock down the slope below the site. However, disposing of waste rock in this manner has a number of drawbacks. First, a steeply sloping pile of loose rock may be a visual scar on the hillside, which can be difficult to vegetate. Second, the waste rock may become unstable if not adequately drained or keyed into the existing slope; if it fails, the material may move a considerable distance and endanger facilities located down slope. Third, where the site is located in a river valley, the discarded rock may fall in the river and have a deleterious effect on fish populations. In order to minimise these impacts, it is sometimes required that the excavated rock be hauled to designated, stable waste sites.

Another problem that may need to be addressed in the disposal of waste rock is acid-water drainage. In areas of North Carolina and Tennessee, for example, some argillite and schist formations contain iron-disulphides; percolation of water through fills constructed with this rock produces low pH, acidic runoff. One method that has been used to control this condition is to mix the rock with lime to neutralise the acid potential and then to place the blended material in the centre of the fill (Byerly and Middleton, 1981). Sometimes it is necessary to encase the rock–lime mixture in an impervious plastic membrane.

Aesthetics – a series of steep, high rock cuts above a highway may have a significant visual impact when viewed both by the road user and the local population. In scenic areas, it may be desirable to incorporate appropriate landscaping measures in the design of the rock cuts in order to minimise their visual impact (Norrish and Lowell, 1988). Examples of aesthetic treatments of rock faces include designing blasts to produce an irregular face with no traces of the blast holes, to recess the heads of rock bolts, and to colour and sculpt shotcrete so that it has the appearance of rock.

Dust, noise, ground vibration – many rock stabilisation operations can produce considerable noise and dust, and blasting has the additional risk of ground vibrations (see Section 13.5.1). Prior to letting the contract, consideration should be given to the acceptable ground vibration levels for the site conditions, so that necessary steps can be taken to limit their effects.

Biological and botanical effects – on some projects, steps may have to be taken to limit disturbance to wildlife and vegetation. Typical precautions that can be taken are to schedule work outside of specified 'windows' of animal activity, and to relocate protected plants outside the work area.

14.4 STABILISATION BY ROCK REINFORCEMENT

Figure 14.5 shows a number of reinforcement techniques that may be implemented to secure potentially loose rock on the face of a rock cut. The common feature of all these techniques is that they minimise relaxation and loosening of the rock mass that may take place as a

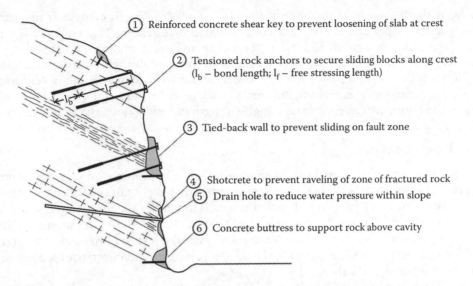

① Reinforced concrete shear key to prevent loosening of slab at crest

② Tensioned rock anchors to secure sliding blocks along crest
(l_b – bond length; l_f – free stressing length)

③ Tied-back wall to prevent sliding on fault zone

④ Shotcrete to prevent raveling of zone of fractured rock
⑤ Drain hole to reduce water pressure within slope

⑥ Concrete buttress to support rock above cavity

Figure 14.5 Rock slope reinforcement methods (Transportation Research Board, 1996).

result of excavation. Once relaxation has been allowed to take place, a loss of interlock occurs between the blocks of rock resulting in a significant decrease in the shear strength. Figure 5.13 illustrates the effect of installing rock bolts to maintain the interlock on high-roughness angle, second-order asperities. Once relaxation has taken place, it is not possible to reverse the process. For this reason, reinforcement of rock slopes is most effective if it is installed prior to excavation – a process known as pre-reinforcement.

14.4.1 Shear keys

Reinforced shear keys provide support for blocks of rock up to about 1 m thick, as well as zones of loose and weathered rock at the crest of the slope (Figure 14.5, Item 1). Shear keys are used where the support required is limited because of the small size of the blocks, and to prevent ravelling and loosening of closely fractured, weak rock. If bolts were to be installed in this rock, the ravelling would soon expose the head of the bolt resulting in loss of support. Shear keys comprise lengths of reinforcing steel about 25–32 mm (1–1.5 in.) diameter and about 1000 mm (3 ft) long fully grouted into holes about 500–750 mm (1.5–2 ft) deep drilled into stable rock. The holes are located close to the toe of the block to be supported, and are spaced at about 500–1000 mm (1.5–3 ft) depending on the support force required. Lengths of reinforcing steel about 6–10 mm (1/4 in.) diameter are then placed horizontally and secured to the exposed portion of the vertical bars. Finally, the reinforcing steel is fully encapsulated in shotcrete, or concrete poured in intimate contact with the rock.

Support provided by a shear key is equal to the shear strength of the vertical steel bars, and possibly the cohesion on the rock–concrete interface. Shear keys act as a resisting force in the limit equilibrium equation (7.31) that gives the factor of safety for a planar sliding block supported by a buttress with resistence, R_b (see Section 7.5.3). If the magnitude of the resisting force supplied by the buttress is R_b, then the factor of safety for a block with weight W support with a shear key is

$$FS = \frac{W \cdot \cos \psi_p \cdot \tan \phi}{W \sin \psi_p - R_k} \tag{14.3}$$

where ψ_p is the dip angle of the base of the block and ϕ the friction angle of this surface, assuming a dry slope. The factor of safety calculated by Equation 14.3 could be for a unit length of the slope, or a specified length, depending on how W and R_k are defined.

Shear keys on a much larger scale have been used for the stabilisation of dam foundations and abutments (Moore and Imrie, 1982). A tunnel was driven along a distinctly defined shear zone, with the excavation extending into sound rock on either side of the tunnel. The tunnel was then filled with concrete to create a high-strength inclusion along the sliding plane.

14.4.2 Rock anchors

Typical applications of rock anchors, as shown in Figure 14.5, Items 2 and 3, are to prevent sliding of blocks or wedges of rock on discontinuities dipping out of the face. It is important to note that the primary function of rock anchors is to modify the normal and shear forces acting on the sliding planes, and not to rely on the shear strength of steel where the anchor crosses this plane. In this chapter, the term 'rock anchor' refers to both rigid bars, and flexible cables that can be used in bundles; the design principles and construction methods are similar for both bars and cables.

Rock anchors may be fully grouted and untensioned, or anchored at the distal end and tensioned. The different applications of untensioned, pre-reinforcement bolts and tensioned anchors are shown in Figure 14.6 (see also Figure 4.14). Pre-reinforcement of an excavation may be achieved by installing fully grouted but untensioned bolts (dowels) at the crest of the cut prior to excavation. The fully bonded dowels prevent loss of interlock of the rock mass because the grouted bolts are sufficiently stiff to prevent movement on the discontinuities (Moore and Imrie, 1982; Spang and Egger, 1990). However, where blocks have moved and relaxed, it is generally necessary to install tensioned anchors to prevent further displacement and loss of interlock. The advantages of untensioned bolts are their lower-cost and quicker installation compared to tensioned anchors.

Tensioned rock anchors are installed across potential slide surfaces and bonded in sound rock beyond the surface. The application of a tensile force in the anchor, which is transmitted into the rock by a reaction plate at the rock surface, produces compression in the rock mass, and modifies the normal and shear stresses on the slide surface. Chapters 7, 8 and 10 each contain design information on the procedures for calculating both the anchor force and anchor orientation required to produce a specified factor of safety; as discussed in Section 7.4.1, the anchor orientation can be optimised to minimise the required force.

Once the anchor force and hole orientation requirements have been determined, the following nine steps are involved in an anchor installation (Littlejohn and Bruce, 1977; FHWA, 1982; British Standards Institute [BSI], 1989; Xanthakos, 1991; Wyllie, 1999; Post Tensioning Institute [PTI], 2006):

Step 1 – Drilling – determine the drill hole diameter and length that can be drilled at the site based on the available equipment, and the access for this equipment.

Step 2 – Bolt materials and dimensions – select the anchor materials and dimensions that are compatible with hole diameter, and the required anchoring force.

Step 3 – Corrosion – assess the corrosivity of the site, and apply an appropriate level of corrosion protection to the anchors.

Step 4 – Bond type – select either cement or resin grout or a mechanical anchor to secure the distal end of the anchor in the hole. Factors influencing this decision include the hole diameter, tensile load, anchor length, rock strength and speed of installation.

Step 5 – Bond length – based on the bonding type, hole diameter, anchor tension and rock strength, calculate the required bond length.

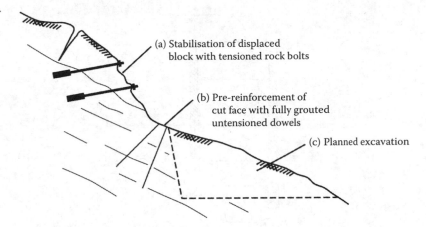

Figure 14.6 Reinforcement of a rock slope: (a) tensioned rock bolts in a displaced block; (b) fully grouted, untensioned dowels installed prior to excavation to pre-reinforce the rock (Transportation Research Board, 1996).

Step 6 – Total anchor length – calculate the total anchor length, which is the total of the bond length and free-stressing length. The free-stressing length should extend from the rock surface to the top of the bond zone, with the top of the bond zone being below the potential sliding plane.

Step 7 – Anchor pattern – layout the anchor pattern so that they are approximately evenly spaced on the face, and produce the required overall anchor force.

Step 8 – Waterproofing drill holes – check that no discontinuities exist in the bond zone into which grout could leak, and seal the open cracks if necessary.

Step 9 – Testing – set up a testing procedure that will verify whether the bonded length can sustain the design load, and that the full length of the free-stressing length is being tensioned.

Each of these nine steps is discussed in more detail as follows:

Step 1, Drilling – the diameter of the drill hole is partially determined by the available drilling equipment, but must also meet certain design requirements. The hole diameter should be large enough to allow the anchor to be inserted into the hole without driving or hammering, and allow full embedment in a continuous column of grout. A hole diameter significantly larger than the anchor will not materially improve the design and will result in unnecessary drilling costs and possibly excessive grout shrinkage. As a guideline, the diameter of the drill hole should be at least 2.5 times the diameter of the full anchor assembly with corrosion protection and grout tubes.

Percussion drilling – holes for rock anchors are usually drilled with percussion equipment that utilises a combination of impact and rotation of a tungsten carbide drill bit to crush the rock and advance the hole. The cuttings are removed by compressed air that is pumped down a hole in the centre of the rods and is exhausted up the annulus between the rods and the wall of the hole. Percussion drills are either pneumatic or hydraulic powered, and the hammer is either at the surface or down-the-hole (DTH). The advantages of percussion drills are their high-penetration rates, good availability and the slightly rough wall that is produced. Precautions that must be taken include minimising hole deviation by controlling the down pressure on the rods, and avoid losing the drill string in zones of weathered and broken rock.

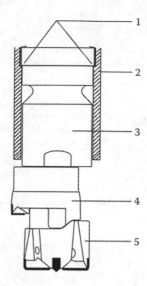

Figure 14.7 Tubex drill bit for advancing casing through soil and weathered rock. (1) Shoulder, (2) bit tube, (3) guide, (4) reamer and (5) pilot bit. (Courtesy: Sandvik Drilling.)

Where holes are to be drilled through an upper layer of soil, or intermediate zones of weathered rock in which collapse of the hole is possible, equipment is available that installs casing as the hole is advanced. Equipment manufactured by Tubex* uses a bit that expands, when torque is applied, to ream out the hole to a diameter slightly larger than the casing (Figure 14.7). At the completion of drilling, the drill rods and contracted bit can be withdrawn inside the casing. The maximum hole diameter for Tubex drills is 356 mm. Alternatively, drills manufactured by Klemm and Barber advance casing during drilling by applying thrust and torque to the casing, which is independent of the thrust and torque on the drill rods.

For holes drilled with handheld percussion drills, the limits for efficient operation are a maximum hole diameter of about 60 mm (2.5 in.) and a maximum length of about 6 m (20 ft). For track mounted percussion drills, the hole diameters range between approximately 75 and 150 mm (3–6 in.), and the maximum hole length for top hammer drills is approximately 60 m (200 ft), with the main limitation being excessive hole deviation. For DTH drills, the maximum hole length is several hundred metres. For hole diameters larger than about 150 mm (6 in.) in hard rock, a substantial increase in the size of the drilling equipment is required, and this equipment is usually used for vertical holes rather than near-horizontal holes.

Rotary drilling – for drilling in weak rock such as chalk and some shales, it is possible to use rotary drilling methods that include augers, drag bits and tri-cone bits; drill hole diameters range from 150 to 600 mm. These methods generally require that the hole be self-supporting, although when using hollow-stem augers, bentonite or cement grout can be circulated in the hole to stabilise the walls.

Diamond drilling is not usually used for anchor installation because of the cost, and the creation of polished walls of the hole that may have a low-bond strength with the bonding grout.

---

* Manufacturer's names are given as examples only, and are not intended as endorsements of their products.

Step 2, Anchor materials and dimensions – anchors are available as either deformed steel bars, or 7-wire strand cables. Figures 14.8 and 14.9 show typical features of these two types of anchor, both of which incorporate an optional corrosion protection system.

The strength of steel is defined by its ultimate (σ_{ult}) and yield strength (σ_y). In designing anchors, the usual limits on the loads applied to individual anchors are as follows:

- Design load $\leq [0.6\ \sigma_{ult}]$
- Lock-off load $\leq [0.7\ \sigma_{ult}]$
- Maximum test load $\leq [0.8\ \sigma_{ult}]$

Bar anchors – most bars are manufactured with a continuous, course thread that is resistant to damage, and allows the bars to be cut to any length to suit site conditions. Common steel grades used for thread bars are 517/690 MPa (75/88 ksi) and 835/1030 MPa (121/149 ksi): yield stress/ultimate stress, with a modulus of elasticity of 204.5 GPa. Bar diameters range from 19 to 57 mm (¾ to 2-¼ in.) and lengths of bar can be joined by threaded couplers. It is usual that only a single bar is installed in each hole, rather than groups of bars. If a tension force is applied to the bar, the head of the bar is secured with a reaction plate, washer and threaded nut.

Another type of bar anchor is a self-drilling product that comprises a hollow-core drill steel with a continuous coil thread, and a disposable bit. The anchor is used most commonly when drilling through broken rock and soil seams where the hole tends to collapse as soon as the drill steel is removed. When using this type of anchor, the drill steel is left in the hole when drilling is complete, following which cement grout is pumped down the centre hole to fill the annual space and encapsulate the anchor. It is also possible to use an adaptor with the drill that allows grout to be circulated down the hole while drilling; this approach may be used where compressed air circulation is not effective in removing the cuttings. Where the application of the anchor is reinforcement of fractured rock masses, rather than anchoring of defined blocks, then it is appropriate to install fully grouted, untensioned anchors. However, it is possible to install tensioned anchors by driving a smooth casing over the grouted bar to the depth of the top of the bond zone. The grout above the bond zone is washed out as the casing is driven to create a debonded, free-stressing length. Some trade names of self-drilling anchors are MAI and IBO, and they are available in diameters ranging from 25 to 51 mm, with ultimate strengths of 200–800 kN (45–180 kips), respectively.

Strand anchors – wire strand is manufactured by twisting together seven, 5 mm (0.2 in.) diameter steel wires to form a strand with a diameter of 12 mm (1/2 in.). Each strand has an ultimate tensile strength of 260 kN, and anchors with higher capacities can be produced by assembling individual strands into bundles; bundles as large as 94 strands have been used in improving the seismic stability of concrete dams. For slope stabilisation requiring shallow dip holes, the largest bundle may be about 12 strands. The strands are flexible, which facilitates handling in the field, but they cannot be coupled. When tension is applied to strand, the end exposed at the surface is secured with a pair of tapered wedges that grip the strand and fit tightly into a tapered hole in the reaction plate (see Figure 14.9, detail 1).

Step 3, Corrosion protection – corrosion protection for steel bar and strand anchors should be considered for all projects, including temporary installations if the site conditions are corrosive (King, 1977; Baxter, 1997). Even if anchors are not subject to corrosion at the time of installation, conditions may change in the future that must be accounted for in

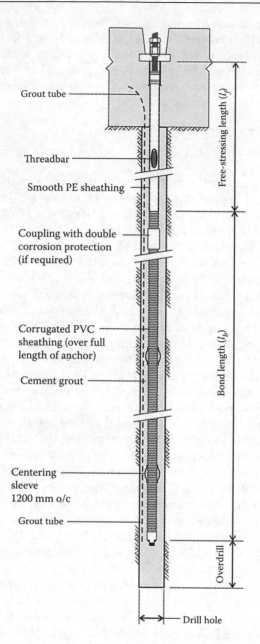

Figure 14.8 Typical threadbar rock anchor with double corrosion protection system comprising grouted corrugated plastic sleeve over full length of anchor, and smooth sheath on unbonded length (Class I corrosion protection). (Courtesy: DSI Anchor System.)

design. The following list describes conditions that will usually create a corrosive environment for steel anchors (Hanna, 1982; Post Tensioning Institute [PTI], 2006):

1. Soils and rocks that contain chlorides
2. Marine environments where they are exposed to sea water that contains chlorides and sulphates
3. Fully saturated clays with high-sulphate content

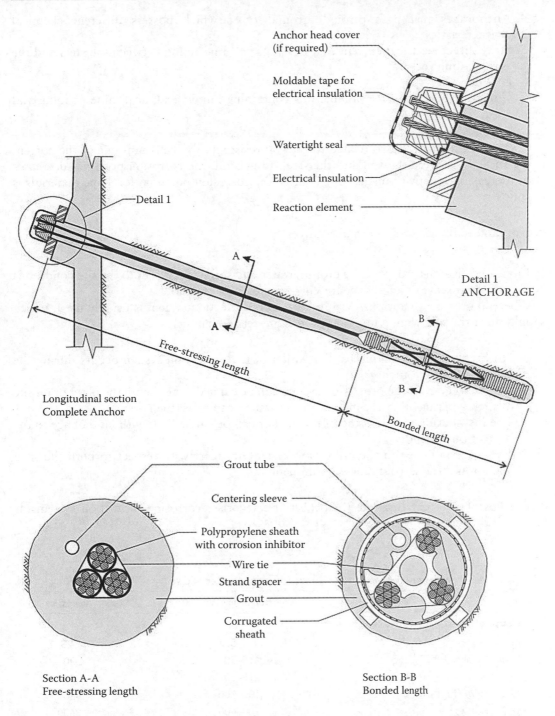

Anchor head cover
(if required)

Moldable tape for
electrical insulation

Watertight seal

Electrical insulation

Reaction element

Detail 1

A

A

Free-stressing length

Longitudinal section
Complete Anchor

B

B

Bonded length

Detail 1
ANCHORAGE

Grout tube

Centering sleeve

Polypropylene sheath
with corrosion inhibitor

Wire tie

Strand spacer

Grout

Corrugated
sheath

Section A-A
Free-stressing length

Section B-B
Bonded length

Figure 14.9 Typical multi-strand cable anchor with corrosion protection system comprising grouted corrugated plastic sleeve on bond length, and smooth greased sheath on unbounded length. (Courtesy: Lang Tendons Inc.)

4. Anchorages that pass through ground types which possess different chemical characteristics
5. Stray direct electrical current that develops galvanic action between the steel and the surrounding rock
6. Peat bogs
7. Cinder, ash or slag fills; organic fills containing humic acid; acid mine or industrial waste

Corrosion potential is also related to the soil resistivity by the magnitude of the current that can flow between the soil and the steel. In general, the corrosion potential decreases with increasing resistivity of the soil. That is, the corrosion potential for slope materials is as follows:

organic soil > clay > silt > sand > gravel

Table 14.5 lists the properties of ground water and soil with respect to the site conditions being non-aggressive or aggressive for corrosion of steel anchors.

Where aggressive conditions exist, a corrosion protection system is usually used, which should meet the following requirements for long-term reliability:

1. No break down, cracking or dissolution of the protection system occurs during the service life of the anchor.
2. The fabrication of the protection system can be carried out either in the plant or on site in such a manner that the quality of the system can be verified.
3. The installation and stressing of the anchor can be carried out without damage to the protection system.
4. The materials used in the protection system are inert with respect to both the steel anchor and the surrounding environment.

The Post Tensioning Institute (PTI) (2006) categorises corrosion protection systems as Class I and Class II as follows:

Table 14.5 Parameter limits for corrosiveness of ground water and soil

	Non-aggressive	Aggressive
Properties of ground water		
pH	6.5–5.5	<4.5
Lime-dissolving (CO_2), mg/L	15–30	>30
Ammonium (NH_4^+), mg/L	15–30	>30
Magnesium (Mg^{2+}), mg/L	100–300	>300
Sulphate (SO_4^{2-}), mg/L	200–600	>600
Properties of soil		
Resistivity (Ω), ohm/cm	2000–5000	<2000
pH	5–10	<5

Source: Transportation Research Board (TRB). 2002. Evaluation of metal tensioned systems in geotechnical applications. NCHRP Project No. 24-13, Washington, DC, 102 pages plus figures and appendices.

- *Class I* protection is used for permanent anchors in aggressive environments, or in non-aggressive environments where the consequences of failure are significant. Both the bond and free-stressing lengths of the tendon or bar are protected with either cement grout-filled encapsulation or an epoxy coating; the head of the anchor is also protected.
- *Class II* protection is used for temporary anchors in non-aggressive environments; protection is limited to grout on the bond length, a sheath on the free-stressing length, and protection of the head if exposed. Figures 14.8 and 14.9 show typical Class I corrosion protection systems for bar and strand anchors, respectively.

Based on these categories of corrosion, a decision tree has been developed to assess the vulnerability of rock anchors to corrosion and loss of anchorage capacity (Figure 14.10) (TRB, 2002). High-strength steel (ultimate tensile strength $\sigma_{ult} > 1000$ MPa [145 ksi]) is vulnerable to attack from hydrogen embrittlement and corrosion stress cracking. Generally, high-strength steel is used to manufacture wire strand elements ($\sigma_{ult} \approx 1700$–$1900$ MPa [250–276 ksi]), and strand anchors are more vulnerable to corrosion than bar anchors because of the larger surface area of steel.

It is also possible to estimate the service life of anchors based on the rate of corrosion by calculating the loss of element thickness over time. The service life t in years is given by

$$\ln(t) = \frac{\ln(X) - \ln(K)}{n} \tag{14.4}$$

where X is the loss of thickness or radius (μm), and K and n are constants (Table 14.6). Values for constant n may vary from 0.6 to 1.0 for a variety of site conditions, but it is usual to use a default value of $n = 1.0$.

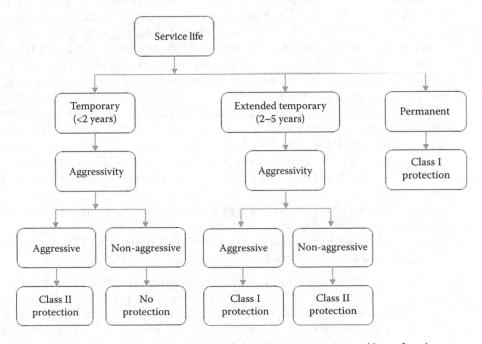

Figure 14.10 Decision tree assessing vulnerability of elements to corrosion and loss of anchorage capacity.

Table 14.6 Values of constants K and n for corrosion rate calculations

Parameter	Normal $\Omega = 2000-5000; pH = 5-10$	Aggressive $\Omega = 700-2000; pH = 5-10$	Very aggressive $\Omega = <700; pH < 5$
K (μm)	35	50	340

Ω is soil resistivity (ohm/cm).

The loss of thickness X is computed from the original radius r_o, and the critical radius r_{crit} that is the radius at which the yield stress is reached, at constant load, assuming that the working stress equals 0.6 times the yield stress. If the original cross-section area is A_o, then the value for r_{crit} due to the loss of cross-section due to corrosion is

$$r_{crit} = \sqrt{\frac{0.6 A_o}{\pi}} \qquad (14.5)$$

and

$$X = (r_o - r_{crit}) \qquad (14.6)$$

For example, for a 25 mm diameter bar installed in aggressive ground ($K = 50$ μm), the critical radius will be 10 mm (0.39 in.) and $X = 2500$ μm. For these conditions, the service life will be approximately 50 years.

The service life calculations discussed in this section refer, of course, to unprotected steel components. Methods of protecting the steel against corrosion include galvanising, applying an epoxy coating or encapsulating the steel in cement. Cement is commonly used for corrosion protection, primarily because it creates a high pH environment that protects the steel by forming a surface layer of hydrous ferrous oxide. In addition, cement grout is inexpensive, simple to install, has sufficient strength for most applications, and a long service life. Because of the brittle nature of grout and its tendency to crack, particularly when loaded in tension or bending, it is usual that the protection system comprises a combination of grout and a plastic (high-density polyethylene [HDPE]) sleeve. In this way, the grout produces the high pH environment around the steel, while the plastic sleeve provides protection against cracking. In order to minimise the formation of shrinkage cracks that reduce corrosion resistance of the grout, it is usual to use non-shrink grouts for all components of the installation. Figures 14.8 and 14.9 show examples of three-layer corrosion protection systems in which the steel is encapsulated in a grout-filled HDPE sheath, and the outer annulus, between the sheath and the rock, formed by the centring sleeves, is filled with a second grout layer.

For anchors with free-stressing lengths, it is particularly important that the head be protected from both corrosion and damage. This is because loss of the nut or wedges, or fracturing of the rock under the reaction plate, will result in loss of tension in the anchor even if the remainder of the anchor is entirely intact.

Step 4, Bond type – tensioned anchors comprise two portions – a bond length and a free-stressing length (Figures 14.8 and 14.9). In the bond length, the bar or strand is bonded by one of a variety of means to the surrounding rock. In the free-stressing length, the bar or strand is unbonded and is free to strain as tension is applied. Figure 14.5 shows that the bond zone is located in stable rock below the potential sliding plane so that, when the anchor is tensioned, stresses are applied to this plane to increase the factor of safety (see Section 7.5).

Methods of securing the distal end of an anchor in the drill hole include resin, mechanical and cement grout anchors. The selection of the appropriate anchor will depend on such factors as the required capacity of the anchor, speed of installation, strength of the rock in the anchor zone, access to the site for drilling and tensioning equipment, and the level of corrosion protection required. The following is a brief discussion of each of these anchorage methods:

Resin anchors comprise a plastic cartridge of about 25 mm (1 in.) in diameter and 200 mm (8 in.) long that contains a liquid resin and a hardener that set when mixed together (Figure 14.11). Setting times vary from about 1 min to as much as 90 min, depending on the reagents used. The setting time is also dependent on the temperature, with fast-setting resin hardening in about 4 min at a temperature of −5°C, and in about 25 s at 35°C.

The installation method consists of inserting a sufficient number of cartridges into the drill hole to fill the annular space around the bar. It is important that the hole diameter, in relation to the bar size, be within specified tolerances so that complete mixing of the resin is achieved when the bar is spun. This usually precludes the use of coupled anchors because the hole diameter to accommodate the coupling will be too large for complete resin mixing. The bar is spun as it is driven through the cartridges to mix the resin and form a rigid solid anchorage. The required speed of rotation is about 60 revolutions per minute, and spinning is continued for about 30 s after the bar has reached the end of the hole. It is preferable that threaded bar be rotated in the direction that augers the resin into the hole, particularly in up-holes.

The maximum bolt length is limited to about 12 m (40 ft) because most drills cannot rotate longer bars at sufficient speed to mix the resin. It is possible to install a tensioned, resin-grouted bolt by using a fast-setting (about 2 min) resin for the anchor, and a slower setting (30 min) resin for the remainder of the bar. The bolt is tensioned between the times that the fast and slow resins set.

Figure 14.11 Resin cartridges for anchoring rock bolts (Transportation Research Board, 1996).

The primary advantage of resin anchorage is the simplicity and speed of installation, with support of the slope being provided within minutes of spinning the bolt. The disadvantages are the limited length and tension capacity (~400 kN) of the bolt, and the fact that only rigid bars can be used. Furthermore, the resin is not as effective as cement grout for corrosion protection of the steel. Unlike cement grout, resin does not provide the high pH protective layer against corrosion, and it cannot be verified that the resin completely encapsulates the steel.

Mechanical anchors comprise a pair of steel platens that are pressed against the walls of the drill hole. The anchor is expanded by driving or torqueing a steel wedge between the platens.

The advantage of mechanical anchors is that installation is rapid, and tensioning can be carried out as soon as the anchor has been set. Grouting can then be carried out using a grout tube attached to the bar, or through the centre hole in the case of the bolt manufactured by the Williams Form Hardware and Rockbolt Co. The disadvantages of mechanical anchors are that they can be used only in medium-to-strong rock in which the anchor will grip, and the maximum working tensile load is about 200 kN. Mechanical anchors for permanent installations must always be fully grouted because the wedge will creep and corrode in time, resulting in loss of support.

Cement grout is the most common method of anchoring long-service-life rock anchors in a wide range of rock and soil conditions because cement is inexpensive, provides corrosion protection and installation is simple. Figures 14.8 and 14.9 show typical cement anchorage installations with centring sleeves on the anchor assembly to ensure complete encapsulation of the steel. The grout mix usually comprises non-shrink, unsanded cement with water, at a water:cement ratio in the range of 0.4–0.45. This ratio will produce a grout that can be pumped down a 12 mm (1/2 in.) diameter grout tube, yet produce a high-strength, continuous grout column with minimal bleed of water from the mix. Admixtures are sometimes added to the grout to reduce bleeding, and increase the viscosity of the grout. Grout is always placed with a grout tube extending to the distal end of the hole in order to displace air and water in the hole.

Step 5, Bond length – for cement and resin grout anchored bolts, the stress distribution along the bond length is highly non-uniform; the highest stress is concentrated in the proximal end of the bond zone, and ideally the distal end of the bond is unstressed (Farmer, 1975). However, it is found as a simplification that the required length of the bond zone can be calculated assuming that the shear stress at the rock–grout interface is uniformly distributed along the anchor. Based on this assumption, the average ultimate shear stress in the bond zone τ_{ult} is given by Equation 14.7

$$t_{ult} = \frac{T \cdot FS}{\pi \cdot d_h \cdot l_b}$$

(14.7)

or, the design bond length l_b is

$$l_b = \frac{T \cdot FS}{\pi \cdot d_h \cdot \tau_{ult}}$$

(14.8)

where T is the design tension force and d_h is the hole diameter. Values of τ_{ult} can be estimated from the uniaxial compressive strength (σ_i) of the rock in the anchor zone according to the following relationship (Littlejohn and Bruce, 1977):

$$\tau_{ult} = \frac{\sigma_i}{10} \tag{14.9}$$

Approximate ranges of average ultimate rock–grout bond stress (τ_{ult}) related to rock type are presented in Table 14.7. In the application of Equation 14.8, the minimum factor of safety (FS) is usually 2.0 for permanent anchors.

Step 6, Total anchor length – the total length of a tensioned anchor is the sum of the bond length as determined in steps 4 and 5, and the free-stressing length (see Figure 14.5). The free-stressing length extends from the proximal end of the bond zone to the head of the bolt, and the three components to this length are as follows. First, the distal end of the free-stressing length must be beyond the potential sliding surface so that the tensile force applied to the anchor is transferred to the rock at the sliding surface. If the location of this surface is precisely known, then the distal end of the free-stressing length could be 1–2 m below the sliding surface, while, if the sliding surface is a zone rather than a plane, this distance should be increased accordingly. The second component of the length extends from the sliding surface to the rock surface, and this length will depend on the slope geometry. The third length component is the distance from the rock surface to the head of the anchor where the bearing plate and nut are located. For strong rock, the bearing plate can bear directly on the rock (Figure 14.5, Item 2), whereas, in conditions where the stress under the bearing plate could crush the rock, a reinforced concrete or shotcrete reaction pad would be required (Figure 14.5, Item 3).

Step 7, Anchor pattern – the layout of the anchors on the face should be such that a reasonably uniform stress is applied to the sliding plane. This will require that the horizontal and vertical spacing be about equal. Also, the anchors should not be too close to the base of the slope where the thickness of rock above the slide plane is limited, or close to the crest where the anchor may pass through a tension crack. For a plane failure where the support force T is calculated per unit length of slope, then the required vertical spacing S_v for an installation comprising n horizontal rows is given by

$$S_v = \frac{B \cdot n}{T} \tag{14.10}$$

where B is the design tension force in each bolt.

Table 14.7 Typical average ultimate rock–grout bond stresses in cement grout anchorages

Rock type	Average ultimate bond stress, MPa (psi)
Granite, basalt	1.7–3.1 (250–450)
Dolomitic limestone	1.4–2.1 (200–300)
Soft limestone	1.0–1.4 (150–200)
Slates, hard shales	0.8–1.4 (120–200)
Soft shales	0.2–0.8 (30–120)
Sandstones	0.8–1.7 (120–250)
Weathered sandstone	0.7–0.8 (100–120)
Chalk	0.2–1.1 (30–155)
Weathered marl	0.15–0.25 (25–35)
Concrete	1.4–2.8 (200–400)

Source: Post Tensioning Institute (PTI). 2006. *Post-Tensioning Manual*, 6th edition. PTI, Phoenix, AZ, 70pp.

Step 8, Waterproofing drill hole – if the drill hole intersects open discontinuities in the bond length into which significant leakage of grout could occur, it will be necessary to seal these discontinuities before installing the anchor. The potential for grout leakage into the rock can be checked by filling the hole with water and applying an excess pressure of 35 kPa (5 psi). If, after allowing time for saturation of the rock mass around the hole, the water leakage over a 10-min period exceeds 9.5 L, then it is possible that grout in the bond zone will flow into the rock before it has time to set. Under these conditions, the hole is sealed with a low-fluid or sanded grout, and then is re-drilled after a setting time of less than about 15–24 h; if the grout is allowed to set fully, it is possible that the drill will wander from the original alignment and intersect the ungrouted discontinuity. A second water inflow test is then conducted. The procedure for testing, grouting and re-drilling is repeated until the hole is sealed.

Step 9, Testing – where tensioned rock anchors are installed, a procedure is required to check whether the anchor can sustain the full design load at the required depth, and that no loss of load occurs with time. It is found that an adequate setting time for cement grout before testing is a minimum of 3 days.

A testing procedure has been drawn up by the Post Tensioning Institute (PTI) (2006) and comprises the following four types of tests:

1. Performance test
2. Proof test
3. Creep test
4. Lift-off test

The performance and proof tests consist of a cyclic-testing sequence, in which the deflection of the head of the anchor is measured as the anchor is tensioned (Figure 14.12). The design load should not exceed 60% of the ultimate strength of the steel, and the maximum test load is usually 133% of the design load, which should not exceed 80% of the ultimate strength of the steel. As a guideline, performance tests are usually carried out on the first two to three anchors and on 2% of the remaining anchors, while proof tests are carried out on the remainder of the anchors. The testing sequences are as follows, where AL is an alignment load to take slack out of the anchor assembly and P is the design load (Figure 14.13a):

Figure 14.12 Test set-up for a tensioned multi-strand cable anchor comprising hydraulic jack load, and dial gauge on independent mount to measure anchor elongation. (Photograph by W. Capaul.)

Performance test:

AL, 0.25P
AL, 0.25P, 0.5P
AL, 0.25P, 0.5P, 0.75P
AL, 0.25P, 0.5P, 0.75P, 1.0P
AL, 0.25P, 0.5P, 0.75P, 1.0P, 1.2P
AL, 0.25P, 0.5P, 0.75P, 1.0P, 1.2P, 1.33P – hold for creep test*
AL, P – lock-off anchor, carry out lift-off test

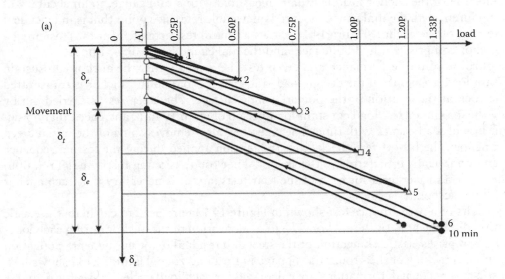

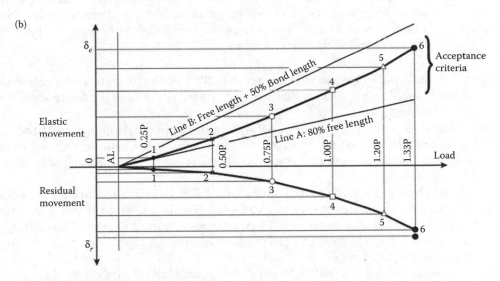

Figure 14.13 Results of performance test for tensioned anchor: (a) cyclic load/movement measurements; (b) load/elastic movement plot (Post-Tensioning Institute, 2006).

---

* *Creep test* – elongation measurements are made at 1–6 and 10 min. If the total creep exceeds 1 mm (0.04 in.) between 1 and 10 min, the load is maintained for an additional 50 min with elongation measurements made at 20–50 and 60 min.

Proof test:

AL, 0.25P, 0.5P, 0.75P, 1.0P, 1.2P, 1.33P – hold for creep test*
P – lock-off anchor, carry out lift-off test

The usual method of tensioning rock bolts is to use a hollow-core hydraulic jack that allows the load to be applied without bending the anchor and to be precisely measured, as well as cycling the load and holding it constant for the creep test. It is important that the hydraulic jack be calibrated before each project to ensure that the indicated load is accurate. The deflection of the anchor head is usually measured with a dial gauge, to an accuracy of about 0.05 mm, with the dial gauge mounted on a stable reference point that is independent of movement of the anchor. Figure 14.12 shows a typical test arrangement for tensioning a cable anchor comprising a hydraulic jack, and the dial gauge set up on a tripod.

The purpose of the performance and creep tests is to ensure that the anchor can sustain a constant load greater than the design load, and that the load in the anchor is transmitted into the rock at the location of the potential slide surface. The creep test is carried out by holding the maximum test load constant for a period of up to 10 min, and check that no significant loss of load occurs with time. The creep test also removes some of the initial creep in the anchor. The lift-off test checks that the tension applied during the testing sequence has been permanently transferred to the anchor. The Post Tensioning Institute (PTI) (2006) provides acceptance criteria for each of the four tests, and it is necessary that each anchor meets all the acceptance criteria.

The results of a performance test shown in Figure 14.13a are used to calculate the elastic elongation δ_e of the head of the anchor. The total elongation of the anchor during each loading cycle comprises elastic elongation of the steel and residual δ_r (or permanent) elongation due to minor cracking of the grout and slippage in the bond zone. Figure 14.13a shows how the elastic and residual deformations are calculated for each load cycle. Values for δ_e and δ_r at each test load, together with the PTI load-elongation acceptance criteria, are then plotted on a separate graph (Figure 14.13b). For both performance and proof tests, the four acceptance criteria for tensioned anchors are as follows:

First, the total elastic elongation is greater than 80% of the theoretical elongation of the stressing length – this ensures that the load applied at the head is being transmitted to the bond length.

Second, the total elastic elongation is less than the theoretical elongation of the stressing length plus 50% of the bond length – this ensures that load in the bond length is concentrated in the upper part of the bond and no significant shedding of load occurs to the distal end.

Third, for the creep test, the total elongation of the anchor head during the period of 1–10 min is not greater than 1 mm (0.04 in.) (Figure 14.14), or if this is not met, is less than 2 mm (0.08 in.) during the period of 6–60 min. If necessary, the duration of the creep test can be extended until the movement is less than 2 mm (0.08 in.) for one logarithmic cycle of time.

Fourth, the lift-off load is within 5% of the designed lock-off load – this checks that no loss of load has occurred during the operation of setting the nut or wedges, and releasing the pressure on the tensioning jack.

---

* *Creep test* – elongation measurements are made at 1–6 and 10 min. If the total creep exceeds 1 mm (0.04 in.) between 1 and 10 min, the load is maintained for an additional 50 min with elongation measurements made at 20–50 and 60 min.

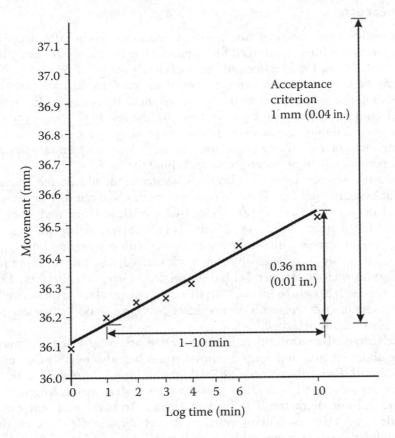

Figure 14.14 Results of creep test showing measured elongation over 10 min test period compared with acceptance criterion of a 1 mm (0.04 in.) elongation.

The working shear strength at the steel–grout interface of a grouted deformed bar is usually greater than the working strength at the rock–grout interface. For this reason, the required anchor length is typically determined from the stress level developed at the rock–grout interface.

14.4.3 Reaction wall

Figure 14.5, Item 3, shows an example where a sliding type failure is possible in closely fractured rock. If tensioned rock bolts are used to support this portion of the slope, the fractured rock may degrade and ravel from under the reaction plates of the anchors, and eventually the tension in the bolts will be lost. In these circumstances, a reinforced concrete wall can be constructed to cover the area of fractured rock, and then the holes for the rock anchors can be drilled through sleeves in the wall. Finally, the anchors are installed and tensioned against the face of the wall (see Figure 1.1). The wall acts as both a protection against ravelling of the rock, and a large reaction plate for the rock anchors. Where necessary, reinforced shotcrete can be substituted for concrete.

Since the purpose of the wall is to distribute the anchor loads into rock, the reinforcing for the wall should be designed such that no cracking of the concrete occurs under the concentrated loads of the anchor heads. It is also important that drain holes are drilled through the concrete to prevent buildup of water pressure behind the wall.

14.4.4 Shotcrete

Shotcrete is a pneumatically applied, fine-aggregate mortar that is usually placed in a 50–100 mm (4 in.) layer, and is often reinforced for improved tensile and shear strength (American Concrete Institute [ACI], 1995). Zones and beds of closely fractured or susceptible to weathering rock may be protected by applying a layer of shotcrete to the rock face (Figure 14.5, Item 4). The shotcrete will control both the fall of small blocks of rock, and progressive ravelling that could eventually produce unstable overhangs. However, shotcrete provides little support against sliding for the overall slope; its primary function is surface protection. Another component of a shotcrete installation is the provision of drain holes through the shotcrete to prevent buildup of water pressures behind the face.

Reinforcement – for permanent applications, shotcrete should be reinforced to reduce the risk of cracking and spalling. The two common methods of reinforcing are welded-wire mesh, or steel or polypropylene fibres. Welded-wire mesh is fabricated from light gauge (~3.5 mm [0.14 in.] diameter) wire on 100 mm (4 in.) centres, and is attached to the rock face on about 1–2 m centres with steel pins, complete with washers and nuts, grouted into the rock face. The mesh must be close to the rock surface, and fully encased in shotcrete, taking care that no voids are formed behind the mesh. On irregular surfaces, it can be difficult to attach the mesh closely to the rock. In these circumstances, the mesh can be installed between two layers of shotcrete, with the first layer creating a smoother surface to which the mesh can be more readily attached.

An alternative to mesh reinforcement is to use steel or polypropylene fibres that are a component of the shotcrete mix and form a reinforcement mat throughout the shotcrete layer (Morgan, McAskill, Richardson et al., 1989; Morgan, Heere, McAskill et al., 1999). The steel fibres are manufactured from high-strength carbon steel with a length of 30–38 mm (1-1/4 to 1-1/2 in.) and diameter of 0.5 mm (0.02 in.). To resist pull-out, the fibres have deformed ends or are crimped. The proportion of steel fibres in the shotcrete mix is about 60 kg/m^3 (100 lb/yd^3), while comparable strengths are obtained for mixes containing 6 kg of polypropylene fibres per cubic metres (10 lb/yd^3) of shotcrete. The principal function of fibres is to significantly increase the shear, tensile and post-crack strengths of the shotcrete compared to non-reinforced shotcrete (see Figure 14.16); shotcrete will tend to be loaded in shear and tension when blocks of fractured rock loosen behind the face.

The disadvantages of steel fibres are their tendency to rust at cracks in the shotcrete, and the hazard of the 'pin cushion' effect where persons pass close to the face; polypropylene fibres overcome both these disadvantages;

Mix design – shotcrete mixes comprise cement and aggregate (10–2.5 mm [1/2 to 1/8 in.] gravel and sand) together with admixtures (superplasticisers) to provide high early strengths. The properties of shotcrete are enhanced by the use of micro-silica ('silica fume') that is added to the mix as a partial replacement for cement. Silica fume is an ultrafine powder with a particle size approximately equal to that of smoke. When added to shotcrete, silica fume reduces rebound, allows thicknesses of up to 500 mm (18 in.) to be applied in a single pass, and covers surfaces with running water. The long-term strength is increased in most cases.

Shotcrete can be applied as either a wet mix or a dry mix. For wet-mix shotcrete, the components, including water, are mixed at a ready-mix concrete plant and the shotcrete is delivered to the site by ready-mix truck. This approach is suitable for sites with good road access and the need for large quantities. For dry-mix shotcrete, the dry components are mixed at the plant and then placed in 1 m^3 bags that have a valve in the bottom (Figure 14.15). At the site, the bags are discharged into the hopper on the pump and a pre-moisturiser adds 4% water to the mix. The mix is then pumped to the face where additional water is added

Figure 14.15 Dry-mix shotcrete process using bagged mix feeding a pump and pre-moisturiser.

through a ring valve at the nozzle. The advantages of the dry-mix process are its use in locations with difficult access, and where small quantities are being applied at a time. It is also useful to be able to adjust the quantity of water in the mix in areas where seepage on the face is variable.

Typical mixes for dry-mix and wet-mix silica fume, steel fibre-reinforced shotcrete are shown in Table 14.8 (Morgan, McAskill, Richardson et al., 1989).

Shotcrete strength – the strength of shotcrete is defined by three parameters that correspond to the types of loading conditions to which shotcrete may be subjected when applied to a slope. Typical values for these parameters are as follows:

1. Compressive strength of 20 MPa (2900 psi.) at 3 days and 30 MPa (4350 psi) at 7 days
2. First crack flexural strength of 4.5 MPa (650 psi) at 7 days
3. Toughness indices of $I_5 = 4$ and $I_{10} = 6$

The flexural strength and toughness indices are determined by cutting a beam with dimensions of 100 mm (4 in.) square in section and 350 mm (14 in.) long from a panel shot in the field, and testing the beam in bending. The test measures the deformation beyond the peak strength, and the method of calculating the I_5 and I_{10} toughness indices from these measurements is shown in Figure 14.16.

Surface preparation – the effectiveness of shotcrete is influenced by the condition of the rock surface to which it is applied – the surface should be free of loose and broken rock, soil, vegetation and ice. The surface should also be damp to improve the adhesion between

Table 14.8 Typical shotcrete mixes – wet mix and dry mix

Material	Dry-mix shotcrete		Wet-mix shotcrete	
	kg/m³	% dry materials	kg/m³	% dry materials
Cement, Type I	400	18.3	420	19.7
Silica fume	50	2.3	40	1.9
10 mm (1/2 in.) coarse aggregate	500	22.9	480	22.6
Sand	1170	53.7	1120	52.7
Steel fibres	60	2.8	60	2.7
Water reducer	–	–	2 L	0.1
Super plasticiser	–	–	6 L	0.3
Air entraining admixture	–	–	If required	If required
Water	170ᵃ	–	180	–
Total wet mass	2350	100	2300	100

Note: Conversion factor – 1 kg/m³ = 1.68 lb/yd³.

ᵃ Total water from pre-moisturiser and added at nozzle (based on saturated surface dry aggregate concept).

the rock and the shotcrete, and the air temperature should be above 5°C for the first 7 days when the shotcrete is setting, unless heating is applied during curing. Drain holes should be drilled through the shotcrete to prevent buildup of water pressure behind the face; the drain holes are usually about 0.5 m deep, and located on 1–2 m centres. In massive rock, the drain holes should be drilled before the shotcrete is applied, and located to intersect discontinuities that carry water. The holes are temporarily plugged with wooden pegs or rags while applying the shotcrete.

Aesthetics – a requirement on some civil projects is that shotcreted faces should have a natural appearance. That is, the shotcrete should be coloured to match the natural rock

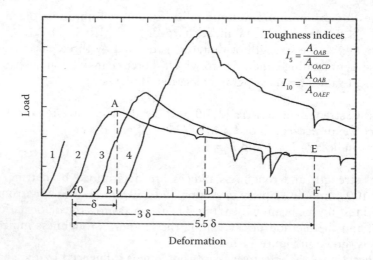

Figure 14.16 Load-deformation characteristics of steel fibre-reinforced shotcrete. Curve 1: without fibres; Curve 2: 1% vol. fibres; Curve 3: 2% vol. fibres; and Curve 4: 3% vol. fibres (American Concrete Institute [ACI], 1995).

colour, and the face sculpted to show a pattern of 'discontinuities'. This work is obviously costly, but the final appearance can be a very realistic replica of a rock face.

14.4.5 Buttresses

Where a rock fall or weathering has formed a cavity in the slope face, it may be necessary to construct a concrete buttress in the cavity to prevent further falls (Figure 14.5, Item 6). The buttress fulfils two functions: first, to retain and protect areas of weak rock, and second, to support the overhang. Buttresses should be designed so that the direction of thrust from the rock supports the buttress in compression. In this way, bending moments and overturning forces are minimised and heavy reinforcement of the concrete, or tiebacks anchored in the rock are not needed.

If the buttress is to prevent relaxation of the rock, it should be founded on a clean, sound rock surface. If this surface is not at right angles to the direction of thrust, then the buttress should be anchored to the base using steel pins to prevent sliding. Also, the top of the buttress should be poured so that it is in contact with the underside of the overhang. In order to meet this second requirement, it may be necessary to place the last pour through a hole drilled downwards into the cavity from the rock face, and to use a non-shrink agent in the mix.

14.4.6 Concrete grillages

A method of stabilisation for cuts slopes in weathered rock that is widely used in Asia is reinforced concrete grillages (Figure 14.17). Grillages are a network of concrete beams, oriented horizontally and vertically and spaced at about 3 m (10 ft), with a rock anchor at each

Figure 14.17 (**See colour insert.**) Construction of reinforced concrete grillage to support cut slope in weathered rock (near Niigata, Japan).

intersection point. The rock anchors, which serve two functions – to reinforce the rock slope and to secure the grillage to the rock slope, may be steel bars or strand cable. The length and load capacity of the rock anchors depends on the requirements for slope support. The open face between the beams is covered with wire mesh to hold loose rock fragments in place, and is kept free of shotcrete to facilitate drainage, supplemented with drain holes if necessary.

Grillages are usually constructed with shotcrete that is built up on the steel reinforcing cage, with workers using the trowels to shape the beams.

14.4.7 Drainage

As shown in Table 14.1, ground water in rock slopes is often a primary or contributory cause of instability, and a reduction in water pressure usually improves stability. This improvement can be quantified using the design procedures discussed in Chapters 7–10. Methods of controlling water pressure include limiting surface infiltration, and drilling horizontal drain holes or driving tunnels to create outlets for the water (Figure 14.18). Selection of the most appropriate method for the site will depend on such factors as the intensity of the rainfall or snow melt, the hydraulic conductivity of the rock and the dimensions of the slope.

Surface infiltration – in climates that experience intense rainfall that can rapidly saturate the slope and cause surface erosion, it is beneficial to stability to construct drains both behind the crest, and on benches on the face, to intercept the water (The Government of the Hong Kong – Civil Engineering Department, 2000). These drains are lined with masonry or concrete to prevent the collected water from infiltrating the slope, and are dimensioned to carry the expected peak design flows. The drains are also interconnected so that the water is discharged into the storm drain system or nearby water courses. Where the drains are on steep gradients, it is sometimes necessary to incorporate energy dissipation protrusions in the base of the drain to limit flow velocities. In climates with high rainfall and rapid vegetation growth, periodic maintenance will be required to keep the drains clear (see Figure 5.1).

Horizontal drain holes – an effective means of reducing the water pressure in many rock slopes is to drill a series of drain holes (inclined upwards at about 5°) into the face (see Figure 6.1). Since most of the ground water is contained in discontinuities, the holes should be aligned so that they intersect the discontinuities that are carrying the water. For the conditions shown in Figure 14.5, the drain holes are drilled at a shallow angle to intersect the more persistent discontinuities that dip out of the face. If the holes were drilled at a steeper angle, parallel to these discontinuities, drainage would be less effective.

Drain holes can be drilled to depths of several hundred metres, sometimes using drilling equipment that installs the perforated casing as the drill advances to prevent caving; perforations are sized to minimise infiltration of fines that are washed from fracture infillings. Also, it is common to drill a fan of holes from a single set up to minimise the drill moves (Cedergren, 1989).

No commonly used formulae is available to calculate the required spacing of drill holes, but as a guideline, holes are usually drilled on a spacing of about 3–10 m (10–30 ft), to a depth of about one-half to one-third of the slope height. Another aspect of the design of drain holes is disposal of the seepage water. If this water is allowed to infiltrate the toe of the slope, it may result in degradation of low-strength materials, or produce additional stability problems downstream of the drains. Depending on site conditions, it may be necessary to collect seepage water from the drains in a manifold and dispose of it at some distance from the slope.

Pumped wells – as shown in Figure 14.18, specific geologic features can be drained with vertical wells in which submersible electric pumps equipped with level switches are installed. The advantage of pumped wells is that they can be designed to provide a required amount of

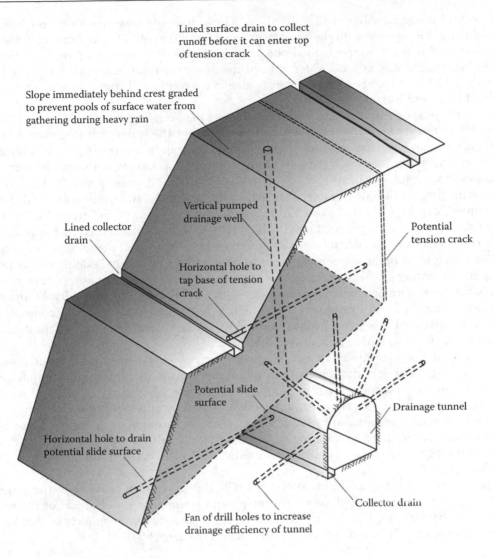

Figure 14.18 Slope drainage methods.

drawdown of the water table, and the performance can be monitored with nearby piezometers and measurements of discharge volume from the pump. The disadvantages of wells are that a drill hole with a diameter large enough for the pump is required, that usually has to be cased for stability. In addition, power has to be supplied to the site, and the entire installation may be lost if slope movement occurs.

Drainage tunnels – for large slides, it may not be possible to significantly reduce the water pressure in the slope with relatively short drain holes. In these circumstances, a drainage tunnel may be driven, preferably in stable rock below the slide plane, from which a series of drain holes are drilled up into the saturated rock. For example, the Downie Slide in British Columbia, Canada has an area of about 7 sq. km (2.7 sq. mile) and a thickness of about 250 m (820 ft). Stability of the slope was of concern when the toe was flooded by the construction of a dam downstream of the slide. A series of drainage tunnels with a total length of 2.5 km (9000 ft) were driven at an elevation just above the high-water level of the reservoir. From these tunnels, a total of 13,500 m (44,000 ft) of drain holes was drilled to reduce the ground water pressures within the slope. These drainage measures have been effective in reducing the water level in the slide by as much as 120 m (400 ft), and reducing the rate of movement from 10 mm/year to about 2 mm/year (0.4–0.08 in./year) (Forster, 1986). In a mining application, ground water control measures for the Chuquicamata pit in Chile include a 1200 m (3900 ft) long drainage tunnel in the south wall, and a number of pumped wells (Flores and Karzulovic, 2000).

Since the purpose of a drainage tunnel is to lower the overlying water table, and the water pressure in the slope, it is necessary that the volume of the water that is removed by the drainage system is greater than the surface infiltration rate plus the water that is removed from storage within the rock mass. That is, for rock formations with low porosity and low-specific storage, it is likely that transient conditions will develop where the head diminishes with time as the slope drains. The layout of the tunnels, and drain holes drilled from the tunnel, can be optimised using the 3D numerical modelling of the ground water flow. Because the flow within the rock mass will usually be strongly influenced by the geological structure, the tunnel and drain holes should be oriented to intersect the set(s) of discontinuities in which most of the water is expected to flow, such as persistent bedding planes rather than less persistent joints. In practice, it is often found that most of the flow occurs in a few discontinuities, with flow in some areas, while other areas may be entirely dry. In all cases, the calculated flow and drawdown values will be estimates because of the complex and uncertain relationship between ground water flow and structural geology, and the difficulty of obtaining representative conductivity values.

Because of the uncertainty in the influence of the geology on ground water flow, an important aspect of slope drainage is to install piezometers to monitor the effect of drainage measures on the water pressure in the slope. For example, one drain hole with a high flow may only be draining a small, permeable zone in the slope and monitoring may show that more holes would be required to lower the water table throughout the slope. Conversely, in low-conductivity rock, monitoring may show that a small seepage quantity that evaporates as it reaches the surface is sufficient to reduce the water pressure and improve stability conditions.

Theoretical modelling of drainage systems will also provide an estimate of the quantity of discharge water, which will be useful for planning pumping and disposal of the water. However, the planning should take into account the uncertainty of models in that actual flow rates are likely to differ from calculated values (Heuer, 1995).

14.4.8 'Shot-in-place' buttress

On landslides where the slide surface is a well-defined geological feature such as a continuous bedding surface, stabilisation may be achieved by blasting this surface to produce a 'shot-in-place' buttress (Aycock, 1981; Moore, 1986). The effect of the blasting is to disturb the rock surface and effectively increase its roughness, which increases the total friction angle. If the total friction angle is greater than the dip of the slide surface, then sliding may be halted. Fracturing and dilation of the rock may also help reduce water pressures on the slide surface.

The method of blasting involves drilling a pattern of holes to the slide surface and placing an explosive charge at this level that is just sufficient to break the rock, but not to cause more widespread damage. This technique requires that the drilling begin while it is still safe for the drills to access the slope, and before the rock becomes too broken for the drills to operate. Obviously, this stabilisation technique should be used with great caution because of the potential for exacerbating instability conditions, and probably should only be used in an emergency situation when no suitable alternatives are possible.

14.5 STABILISATION BY ROCK REMOVAL

Stabilisation of rock slopes can be accomplished by the removal of potentially unstable rock; Figure 14.19 illustrates typical removal methods including:

- Resloping zones of unstable rock
- Trim blasting of overhangs
- Scaling of individual blocks of rock

This section describes these methods, and the circumstances where removal should and should not be used. In general, rock removal is a preferred method of stabilisation because the work will eliminate the hazard, and no future maintenance will be required. However, removal should only be used where it is certain that the new face will be stable, and the risk of undermining the upper part of the slope is minimal. Area 2 in Figure 14.19 is an example of where rock removal should be carried out with care. It would be safe to remove the outermost loose rock, provided that the fracturing was caused by blasting and only extended to a shallow depth. However, if the rock mass is deeply fractured, continued scaling will soon develop a cavity that will undermine the upper part of the slope.

Removal of loose rock on the face of a slope is not effective where the rock is highly degradable, such as shale. In these circumstances, exposure of a new face will just start a new cycle of weathering and instability. For this condition, more appropriate stabilisation would be protection of the face with shotcrete.

14.5.1 Resloping and unloading

Where overburden or weathered rock occurs in the upper portion of a cut, it is often necessary to cut this material at an angle flatter than in the more competent underlying rock (Figure 14.19, Item 1). The design procedure for resloping and unloading starts with back

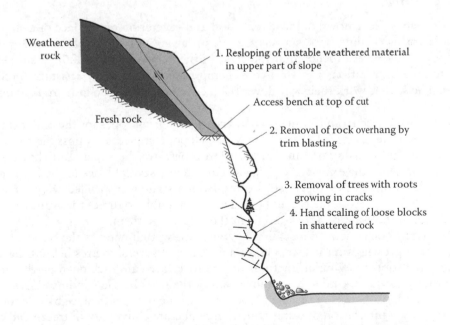

Weathered rock

1. Resloping of unstable weathered material in upper part of slope

Access bench at top of cut

Fresh rock

2. Removal of rock overhang by trim blasting

3. Removal of trees with roots growing in cracks

4. Hand scaling of loose blocks in shattered rock

Figure 14.19 Rock removal methods for slope stabilisation (Transportation Research Board, 1996).

analysis of the unstable slope. By setting the factor of safety of the unstable slope to 1.0, it is possible to calculate the rock mass strength parameters (see Section 5.4). This information can then be used to calculate the required reduced slope angle and/or height that will produce the necessary factor of safety.

Another condition that should be taken account of during design is weathering of the rock some years after construction, at which time resloping may be difficult to carry out. A bench can be left at the toe of the soil or weathered rock to provide a catchment area for minor slope failures and provide equipment access.

Resloping and unloading is usually carried out by excavating equipment such as excavators and bulldozers. Consequently, the cut width must be designed to accommodate suitable excavating equipment on the slope with no danger of collapse of the weak material in the slope while equipment is working; this width would usually be at least 5 m (16 ft). Safety for equipment access precludes the excavation of 'sliver' cuts in which the toe of the new cut coincides with that of the old cut.

14.5.2 Trimming

Failure or weathering of a rock slope may form an overhang on the face (Figure 14.19, Item 2), which could be a hazard if it were to fail. In these circumstances, removal of the overhang by trim blasting may be the most appropriate stabilisation measure. Section 13.4 discusses methods of controlled blasting that are applicable to situations where it is required to trim blast small volumes of rock with minimal damage to the rock behind the trim line.

Where the burden on a trim blast is limited, flyrock may be thrown a considerable distance because of the limited thickness of rock to contain the explosive energy. In these circumstances, appropriate precautions such as the use of blasting mats would be required to protect any nearby structures and power lines. Blasting mats are fabricated from rubber tires or conveyor belts chained or wired together.

14.5.3 Scaling

Scaling describes the removal of loose rock, soil and vegetation on the face of a slope using hand tools such as scaling bars, shovels and chain saws. On steep slopes, workers are usually supported by ropes, anchored at the crest of the slope (Figure 14.20). A suitable type of rope for these operations is a steel-core, hemp rope that is highly resistant to cuts and abrasion. The scalers work their way down the face to ensure that no loose rock above them is dislodged.

A staging suspended from a crane is an alternative to using ropes for the scalers to access the face. The crane is located at the base of the slope if no access is possible to the crest. The disadvantages of using a crane, rather than ropes, are the expense of the crane, and on highway projects, the extended outriggers can occupy several lanes of the highway with consequent disruption to traffic. Also, scaling from a staging suspended from a crane can be less safe than using ropes because the scalers are not able to direct the crane operator to move quickly in the event of a rock fall from the face above them.

An important component of a scaling operation in wet climates is the removal of trees and vegetation growing on the face, and to a distance of several metres behind the crest of the slope. Tree roots growing in fractures on the rock face can force opening the fractures and eventually cause rock falls. Also, movement of the trees by the wind produces leverage by the roots on loose blocks. The general loosening of the rock on the face by tree roots also permits increased infiltration of water which, in temperate climates, will freeze and expand

Figure 14.20 High-scaler suspended on rope and belt while removing loose rock on steep rock slope (Thompson River Canyon, British Columbia, Canada) (Transportation Research Board, 1996).

and cause further opening of the cracks. As shown in Table 14.1, approximately 0.6% of the rock falls on the California highway system can be attributed to root growth.

14.5.4 Rock removal operations

Where rock removal operations are carried out above active highways or railroads, or in urban areas, particular care must be taken to prevent injury or damage from falling rock. This will usually require that all traffic be stopped while rock removal is in progress, and until the slope has been made safe and the road has been cleared of debris. Where pipelines or cables are buried at the toe of the slope, it may be necessary to protect them, as well as pavement surfaces or rail track, from the impact of falling rock. Protection can usually be provided by depositing a cover of sand and gravel to a depth of about 1.5–2 m (4–6 ft), or placing rubber blast mats. If the rock removal operation is on an active transportation route where work is carried out in short 'work windows' when traffic is temporarily stopped, the time required to place and remove the protection material should be taken into account when planning work schedules.

14.6 PROTECTION MEASURES AGAINST ROCK FALLS

An effective method of minimising the hazard of rock falls is to let the falls occur and control the distance and direction in which they travel. Methods of rock fall control and protection of facilities at the toe of the slope include catchment ditches and barriers, wire mesh fences, mesh hung on the face of the slope and concrete rock sheds. A common feature of all these protection structures is their energy-absorbing characteristics in which the rock fall is either stopped over some distance, or is deflected away from the facility that is being protected. As described in this section, it is possible by the use of appropriate techniques to control rocks with diameters of as much as 2–3 m (6–10 ft), falling from heights of several hundred metres, and impacting with energies as high as 1 MJ. Rigid structures, such as reinforced concrete walls or fences with stiff attachments to fixed supports, are rarely appropriate for stopping rock falls.

14.6.1 Rock fall modelling

Selection and design of effective protection measures require the ability to predict rock fall behaviour. An early study of rock falls was made by Ritchie (1963), who drew up an empirical ditch design chart related to the slope dimensions (see Section 14.6.2). Since the 1980s, the prediction of rock fall behaviour was enhanced by the development of a number of computer programmes that simulate the behaviour of rock falls as they roll and bounce down slope faces (Wu, 1984; Descoeudres and Zimmerman, 1987; Spang, 1987; Hungr and Evans, 1988; Pfeiffer and Bowen, 1989; Pfeiffer, Higgins and Turner, 1990; Azzoni and de Freitas, 1995).

Figure 14.21 shows an example of the output from the rock fall simulation programme, *RocFall* (Rocscience). The cross-section shows the trajectories of 20 rock falls, one of which rolls out of the ditch. Figure 14.21b and c shows, respectively, the maximum bounce heights and total kinetic energy at intervals down the slope. The inputs for the programme comprise the slope and ditch geometry, the irregularity (roughness) of the face, the restitution coefficients of the slope materials, the mass and shape of the block, and the start location and velocity. The degree of variation in the shape of the ground surface is modelled by randomly varying the surface roughness for each of a large number of runs, which in turn produces a range of trajectories.

The results of analyses such as those shown in Figure 14.21, together with geological data on block sizes and shapes, can be used to estimate the dimensions of a ditch, or the optimum position, required height and energy capacity of a fence or barrier. In some cases, it may also be necessary to verify the design by constructing a test structure. Sections 14.6.2 to 14.6.5 describe types of ditches, fences and barriers, and the conditions in which they can be used.

It is considered that reliable rock fall simulation requires a 3D model of the slope because rock falls, similar to water flow, tend to follow gullies in the slope rather than fall down straight lines defined on a 2D section. Also, documentation of actual rock falls shows that the normal coefficient of restitution (e_N) is related to the angle of impact rather than the slope material. That is, $e_N \sim 0.1$ for steep (near vertical impacts) and $e_N > 1$ (i.e. up to ~3) for shallow impacts (impact angle <20°) (Wyllie, 2014a,b).

Benched slopes: the excavation of intermediate benches on rock cuts usually increases the rock fall hazard, and is therefore not recommended for most conditions. Benches can be a hazard where the crests of the benches fail due to blast damage, and the failed benches leave irregular protrusions on the face. Rock falls striking these protrusions tend to bounce away from the face and land a considerable distance from the base. Where the narrow benches fill with debris, they will not be effective in catching rock falls. It is rarely possible to remove this debris because of the hazard to equipment working on narrow, discontinuous benches.

There are, however, two situations where benched slopes are a benefit to stability. First, in horizontally bedded sandstone/shale/coal sequences, the locations and vertical spacing of the benches are often determined by the lithology. Benches are placed at the top of the least resistant beds, such as coal or clay shale, which weather quicker (Wright, 1997). With this configuration, the more resistant lithology is not undermined as the shale weathers (Figure 14.22). The width of intermediate benches may vary from 6 to 8 m (20–26 ft), and the face angle depends on the durability of the rock. For example, shales with a slake durability index of 50–79 are cut at angles of 43° (1.33H:1V) and heights up to 9 m (30 ft), while massive sandstone and limestone may be excavated at face angles as steep as 87°(1/20H:1V) and

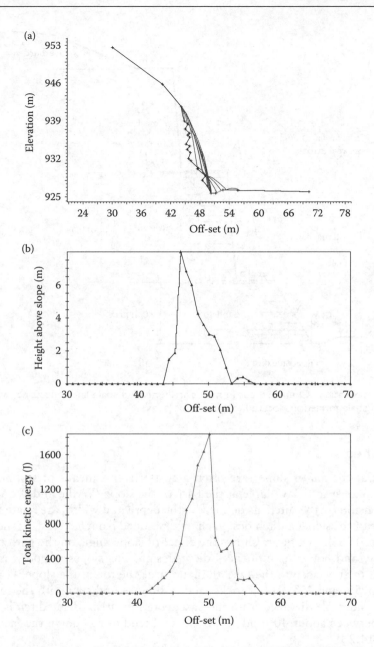

Figure 14.21 Example of analysis of rock fall behaviour: (a) trajectories of 20 rock falls; (b) variation in vertical bounce heights along the slopes; (c) variation in total kinetic energy along the slope.

heights up to 15 m (50 ft). Figure 14.22 also shows a 4.5 m (15 ft) wide bench at the base of
the overburden slope to contain minor sloughing and provide access for cleaning.

A second application for benched slopes is in tropical areas with deeply weathered
rock and periods of intense rain. In these conditions, concrete-lined drainage ditches running along each bench and down the slope face are essential to collect runoff and prevent
scour and erosion of the soil and weak rock (The Government of the Hong Kong – Civil
Engineering Department, 2000). Examples of benched slopes in weathered rock are shown
in Figures 9.22 and 9.23.

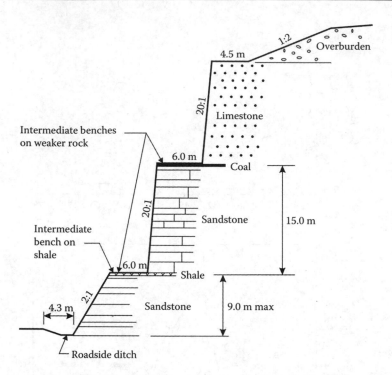

Figure 14.22 Configuration of benched cut in horizontally bedded shale and sandstone, with weaker coal and shale formations located at base of cut faces.

14.6.2 Ditches

Catch ditches at the toe of slopes are often a cost-effective means of stopping rock falls, provided adequate space is available at the base of the slope (Wyllie and Wood, 1981). The required dimensions of the ditch, as defined by the depth and width, are related to the height and face angle of the slope; a ditch design chart developed from field tests is shown in Figure 14.23 (Ritchie, 1963). The figure shows the effect of slope angle on the path that rock falls tend to follow, and how this influences ditch design. For slopes steeper than about 75°, the rocks tend to stay close to the face and land near the toe of the slope. For slope angles between about 55° and 75°, falling rocks tend to bounce and spin with the result that they can land a considerable distance from the base; consequently, a wide ditch is required. For slope angles between about 40° and 55°, rocks will tend to roll down the face and into the ditch (Figure 14.23).

To update the work carried out by Ritchie, a comprehensive study of rock fall behaviour and the capacity of catchment areas has been carried out by the Oregon Department of Transportation (2001). This study examined rock fall from heights of 12, 18 and 24 m (40, 60 and 80 ft) on slopes with five increments of face angles ranging from vertical to 45° (1V:1H). The catchment areas at the toe of the slope were planar surfaces with inclinations of 76° (1/4H:1V), 80° (1/6H:1V) and horizontal to simulate unobstructed highway shoulders. The tests observed both the distance of the first impact point from the toe, and the roll-out distance; these distances provide ratios on: (first impact distance/slope height) and (roll-out distance/slope height) that are useful for ditch design. The report includes design charts that show, for all combinations of slope geometry, the relationship between the percent rock retained and the width of the catchment area.

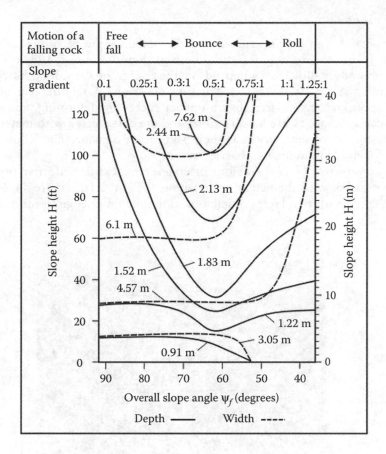

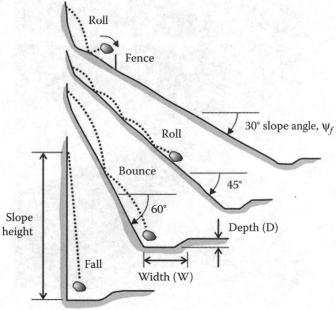

Figure 14.23 Ditch design chart for rock fall catchment (Ritchie, 1963).

14.6.3 Barriers

A variety of barriers can be constructed to either enhance the performance of excavated ditches, or to form catchment zones at the base of slopes (Andrew, 1992a). The required type of barrier and its dimensions depend on the energy of the falling blocks, the slope dimensions and the availability of construction materials. A requirement of all barriers is flexibility upon impact. Barriers absorb impact energy by deforming, and systems with high-impact energy capacity are both flexible, and are constructed with materials that can withstand the impact of sharp rocks without significant damage. The following is a brief description of some commonly used barriers.

Gabions and concrete blocks: Gabions or concrete blocks are effective protection barriers for falling rock with diameters up to about 0.75 m (2 ft). Figure 14.24a shows an example of a ditch with two layers of gabions along the outer edge forming a 1.5 m (5 ft) high barrier.

(a)

(b)

Figure 14.24 Rock fall containment structures: (a) rock catch ditch with 1.5 m high gabion along outer edge (Kicking Horse River Canyon, British Columbia); (b) barrier constructed with MSE wall and wire-rope fence on top of wall (Interstate 40 near Asheville, North Carolina). (Courtesy: North Carolina Department of Transportation.)

The function of a barrier is to form a 'ditch' with a vertical outer face that traps rolling rock. Barriers are particularly useful at the toe of slopes flatter than about 55° where falling rock rolls and spins down the face with shallow trajectories. Such rocks may land in a ditch at the base of the slope but can roll up the sloping outer side; a vertical barrier will help to trap such falls.

Gabions are rock-filled, wire mesh baskets, typically measuring 0.91 m by 0.91 m (3 ft) in cross-section, that are often constructed on-site with local waste rock. Rock fill properties are usually of dimensions ranging from 75 to −100 mm (3–4 in.), are not susceptible to weathering and contain no fines to enhance drainage; the unit mass of rock fill is about 1700 kg·m^{-3} (2870 lb/yd^3).

Advantages of gabions are the ease of construction on steep hillsides and where the foundation is irregular, and their capacity to sustain considerable impact from falling rock. However, gabions are susceptible to damage by impacts of rock and maintenance equipment, and repair costs can become significant. Barriers constructed with precast concrete blocks with similar dimensions to gabions are also used on transportation systems for rock fall containment. Although concrete blocks are somewhat less resilient than gabions, they have the advantages of wide availability and rapid installation. In order for concrete blocks to be effective, flexibility must be provided by allowing movement at the joints between the blocks. In contrast, mass concrete walls are rigid and tend to shatter on impact (Wyllie, 2014a,b).

Geofabric–soil barriers (mechanically stabilised earth [MSE]): barriers have been constructed using geofabric and soil layers, each about 0.6 m (2 ft) thick, built up to form a barrier which may be as high as 4 m (13 ft) (Threadgold and McNichol, 1984). By wrapping the fabric around each layer, it is possible to construct a barrier with steep front and back faces; the face subject to impact can be protected from damage with such materials as railway ties, gabions, rubber tires and replaceable fabric bags (Figure 14.25). The capacity of a barrier of this type to stop rock falls depends on its overall mass in relation to the impact energy such that the entire barrier is not displaced or overturned. Other capacity issues are plastic deformation of the fill in the area of the impact, and the shear resistance of the compacted fill (Grimod and Giacchetti, 2013; Wyllie, 2014a,b).

Full-scale tests and computer models of MSE barriers show that about 80%–85% of the impact energy is absorbed by plastic deformation of the fill around the impact point, and a further 15%–20% is absorbed by shear displacement within the body of the barrier (Figure 14.25b). The dimensions in Figure 14.25b need to be defined when designing an MSE barrier. That is, the barrier needs to have enough mass to resist impact due to sliding or overturning, and develop shear resistance through the body of the barrier to prevent punching failure.

Testing of prototype barriers by the Colorado Department of Transportation has shown that limited shear displacement occurs on the fabric layers, and that they can withstand high-energy impacts without significant damage (Barrett and White, 1991). Also, a 4 m (13 ft) high geofabric and soil barrier successfully withstood impact from boulders with volumes of up to 13 m^3 (17 yd^3) and impact energies of 5000 kJ on Niijima Island in Japan, and a similar 1.8 m wide geofabric barrier stopped rock impacts delivering 950 kJ of energy (Protec Engineering, 2002).

A disadvantage of MSE barriers such as those shown in Figure 14.25 is that a considerable space is required for both the barrier and the catchment area behind it.

14.6.4 Rock catch fences and attenuators

Extensive testing of fences and nets suitable for installation on steep rock faces, in ditches and on talus run-out zones has been carried out by the manufacturers, research organisations

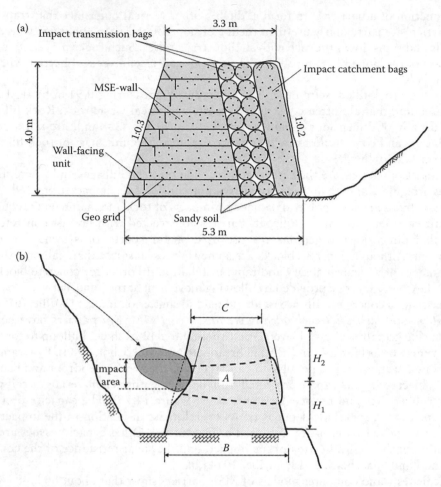

Figure 14.25 Rock fall barriers constructed with soil and geofabric, and a variety of facings: (a) A 4 m (13 ft) high wall with impact energy capacity of 5000 kJ (Protec Engineering, 2002). (b) Mechanism of impact energy dissipation in MSE barrier due to plastic deformation of fill around the impact point and horizontal shear displacement (Grimod and Giacchetti, 2013).

and purchasers such as departments of transportation (Smith and Duffy, 1990; Barrett and White, 1991; Duffy and Haller, 1993). Nets are also being used in open-pit mines for rock fall control (Brawner and Kalejta, 2002; Wyllie, 2014a,b). A design suitable for a particular site depends on the topography, anticipated impact loads, trajectory height and foundation conditions. A common feature of all these designs is their ability to withstand impact energy from rock falls due to their construction without any rigid components. When a rock impacts a net, the mesh is deformed which then engages in energy-absorbing components over an extended time of collision. This deformation significantly increases the capacity of these components to stop rolling rock and allows the use of light, low-cost elements in construction.

Wire-rope nets: nets with energy absorption capacity ranging from 40 to 8000 kJ have been developed as proprietary systems by a number of manufacturers, principally in Europe and Japan. The components of these nets are a series of steel I-beam posts on about 6 m (20 ft) centres, anchored to the foundation with grouted bolts, and guy cables anchored on the slope (Figure 14.26). The bases of the posts may be either fixed as shown in Figure 14.26,

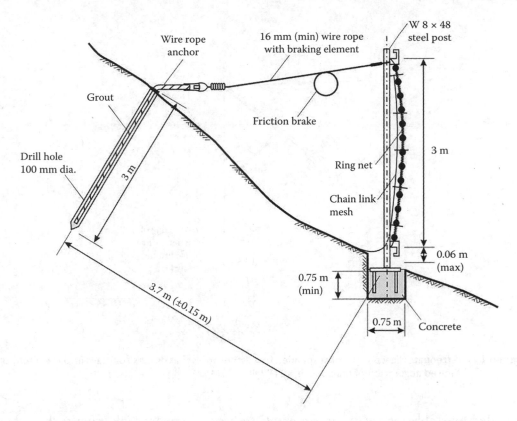

Figure 14.26 Rock fall fence with steel posts supporting flexible wire mesh net (Transportation Research Board, 1996).

or hinged. Posts with fixed bases, which are required where it is not possible to support the posts with guy wires, are fabricated with heavier components than hinged posts. Additional flexibility is provided by incorporating the braking elements on the cables supporting the nets and on the guy cables. It has been found that nets are also effective in containing debris flows because the water rapidly drains from the debris material and its mobility is diminished.

Typically, the mesh is a two-layer system comprising a 50 mm (2 in.) chain link mesh, and either woven steel wire-rope mesh or interlocking 0.31 m (1 ft) diameter steel rings, each of which is interlocked with four adjacent rings. The rings are fabricated from 3 mm (0.11 in.) diameter high tensile steel wire, and the number of wires in each ring varies between 5 and 19 depending on the design energy capacity of the net. The wire-rope and net dimensions will vary with expected impact energies and block sizes. The mesh is attached with steel shackles to a continuous perimeter wire rope that runs through brackets at the top and bottom of each post.

An important design consideration for nets is the deformation of the net that occurs during impact, and the need to provide adequate clearance between the deformed net and the protected structure.

Rock fall attenuators: attenuators comprise steel nets freely suspended from a series of posts and an upper support cable such that rock falls are redirected, rather than stopped, by the net (Figure 14.27). If the net is open at the base, then the rocks fall into the ditch where they can be readily removed. Where rocks fall down a narrow gully or chute

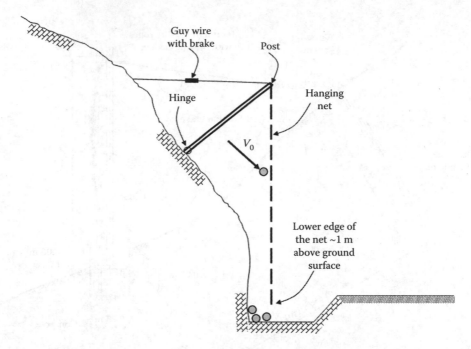

Figure 14.27 Attenuator fence with net suspended from steel posts that directs rock falls into the ditch with limited absorption of impact energy (Wyllie, 2014a,b).

bounded by stable rock walls, it is possible to hang a support cable from rock anchors from which the net, spanning the gully, is suspended (Andrew, 1992b; Wyllie, 2014a,b). The primary advantage of attenuators, compared to fences such as those shown in Figure 14.26, is that only a portion of the rock fall impact energy is absorbed by the net and the remainder of the energy is retained in the moving block of rock that is stopped when it impacts the ground. Because attenuators only absorb a portion of the impact energy, they can be constructed with lighter materials than required for the type of fence shown in Figure 14.26, where the entire impact energy is absorbed by the structure. Another feature of attenuators is that they are self-cleaning. That is, the lower edge of the net is about 1 m (3 ft.) above the ground level so that rocks fall into the ditch and do not accumulate behind the net.

Maintenance requirements and worker safety of fence systems should be considered in design. A properly designed system should not need frequent repairs if the impacts are within the tolerances of design energies. However, cleaning of accumulated rock is necessary for any system. Typically, fixed barriers such as geofabric walls require room behind them for cleaning operations. In contrast, woven wire rope and ring nets do not have this requirement because of their modular design, allowing the nets to be cleaned from the front by removing or lifting each panel in turn.

14.6.5 Draped mesh

Wire mesh hung on the face of a rock slope can be an effective method of containing the rock falls close to the face and preventing them from bouncing on to the road (Ciarla, 1986). Because the mesh absorbs some of the energy of the falling rock, the required dimensions of the ditch at the base of this slope are considerably reduced from those shown in Figure 14.23. Chain link mesh is a suitable method for controlling rock falls with dimensions less than about 0.6–1 m (2–3 ft), and woven wire rope or ring nets are suitable for rock with dimensions up to about 1.3 m (4.3 ft). For installations covering a high slope where the weight of the lightweight mesh may exceed its strength, the mesh can be reinforced with interwoven wire rope. In all cases, the upper edge of the mesh or net should be placed close to the source of the rock fall so that the blocks have little momentum when they impact the mesh.

Installation options for draped mesh are to either let it hang freely from a support cable along the crest as described in the previous paragraph, or to pin it to the face with a pattern of bolts. It is usually preferable to NOT use intermediate bolts for two conditions. First, if the rock degrades rapidly, the fallen rock will accumulate behind the bolted mesh forming 'baskets' of rock that can break suddenly and be a hazard to vehicles, for example. Under these circumstances, it is preferable to let the rock fall down the slope behind the mesh. Second, if a catchment area is available at the base of the slope, then rock falls can accumulate in the ditch where it can be removed by maintenance crews.

Conditions for which the installation of pinned mesh is appropriate are where the rock is reasonably strong, and no catchment area exists at the base of the slope. For these circumstances, pinning the mesh tightly to the slope provides a resisting force that helps to hold potentially loose blocks of rock on the face.

14.6.6 Warning fences

Fences and warning signals that are triggered by falling rock are often used to protect railroads, and occasionally highways. The warning fence consists of a series of posts and cantilever cross-arms, which support rows of wires spaced about 0.5 m (1.5 ft) apart. The wires are connected to a signal system that shows a red light if a wire is broken. The signal light is located far enough from the rock slope that the traffic has time to stop, and then to proceed with caution before it reaches the rock fall location. Warning systems can also be incorporated into rock fall fences (Section 14.6.4), so that a second level of protection is provided in the event of a large fall that exceeds the energy capacity of the fence.

Warning fences are most applicable on transportation systems where traffic is light enough to accommodate occasional closures of the line. However, the use of warning fences as a protection measure has a number of disadvantages. The signal lights must be located a considerable distance from the slope compatible with stopping distance, and falls may occur after the traffic has passed the light. Also, false alarms can be caused by minor falls of rock or ice, and maintenance costs can be significant.

14.6.7 Rock sheds

In areas of extreme rock fall hazard where stabilisation of the slope would be very costly, construction of a rock shed may be justified. Figure 14.28 shows two alternate configurations for sheds depending on the path of the falling rock. Where the rock falls have a steep trajectory, the shed has a horizontal roof and is usually constructed with precast-reinforced concrete beams and columns, with a layer of energy-absorbing material such as gravel on the roof (Figure 14.28a).

Where the slope angle is shallow enough that rock falls tend to roll down the face, it is possible to construct sheds with sloping roofs that are designed to deflect rolling rock. Figure 14.28b shows three sheds that deflect rock falls over a single track railway. Because

(a) (b)

Figure 14.28 Typical rock shed construction: (a) Reinforced concrete structures with horizontal roof covered with layer of gravel. (Photograph courtesy of Dr. H. Yoshida, Kanazawa, Japan.) (b) Sheds constructed with timber and reinforced concrete with sloping roofs that deflect rock falls over the railway (White Canyon, Thompson River, British Columbia, Canada). (Courtesy of Canadian National Railway.)

these sheds do not sustain direct impact, they are of much lighter construction than that in Figure 14.28a, and have no protective layer on the roof. However, these deflection structures extend a considerable distance up the slope, compared with the direct impact shed that is just wide enough for vehicle clearance.

Extensive research on the design of rock sheds has been carried out in Japan, where tens of kilometres of sheds have been constructed to protect both railways and highways (Yoshida, Ushiro, Masuya et al., 1991; Ishikawa, 1999), and also in Switzerland (Vogel, Labiouse and Masuya, 2009). Much of the research in Japan has involved full-scale testing in which concrete blocks incorporating 3D accelerometers and angular rate sensors have been dropped on prototype sheds that are fully instrumented to measure deceleration in the boulder, and the induced stresses and strains in the major structural components. Objectives of the tests are to determine the effectiveness of various cushioning materials in absorbing and dispersing the impact energy, and to assess the influence of the flexibility of the structure on the maximum impact that can be sustained without damage.

A critical feature in the design of concrete sheds is the weight and energy absorption characteristics of the cushioning material. Ideally, the cushion should both absorb energy by compression, and disperse the point impact energy so that the energy transmitted into the structure occurs over a larger area. Furthermore, the cushion should remain intact after impact so that it does not need to be replaced. The effectiveness of the cushion material can be expressed as the difference between the kinetic energy of impact by boulder impact, and the force that is absorbed by the structure. As the result of the testing, an empirical equation has been developed that relates the kinetic energy of the moving boulder at the moment of impact to the equivalent static force that is induced, through the cushioning layer, into the shed. The parameters that make up the equation to calculate the force are

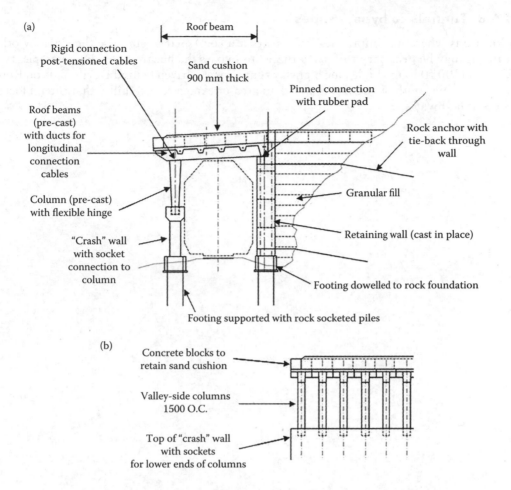

(a)

Roof beam

Rigid connection
post-tensioned cables

Sand cushion
900 mm thick

Pinned connection
with rubber pad

Roof beams
(pre-cast)
with ducts for
longitudinal
connection
cables

Rock anchor with
tie-back through
wall

Column (pre-cast)
with flexible hinge

Granular fill

"Crash" wall
with socket
connection to
column

Retaining wall (cast in place)

Footing dowelled to rock foundation

Footing supported with rock socketed piles

(b)

Concrete blocks to
retain sand cushion

Valley-side columns
1500 O.C.

Top of "crash" wall
with sockets
for lower ends of columns

Figure 14.29 Rock fall shed to protect single track railway constructed with precast concrete roof beams and columns. (a) Cross-section and (b) longitudinal section (Yoshida, Ushiro, Masuya et al., 1991).

the weight of the rock fall and its diameter, the fall height, the properties of the cushioning material expressed as the Lame parameter, and the thickness of the cushioning material (Wyllie, 2014a,b).

Gravel is the most commonly used cushioning material because it is inexpensive and widely available. However, the disadvantage of gravel is its weight, and a point will occur where the dead load of the gravel weight exceeds the rock fall live impact loading. Rubber tires have also been used for cushioning, but it is found that they are highly compressible with little energy absorption. A viable alternative to gravel is reinforced Styrofoam that is an effective energy-absorbing material with low unit weight, which allows for some saving in the dimensions of the structure (Mamaghani, Yoshida and Obata, 1999). The disadvantage of Styrofoam is its cost compared to gravel.

Concrete sheds incorporate flexible, energy-absorbing features, in addition to the cushion on the roof, that include coupling the roof beams with post-tensioned cables that allow each beam to deflect independently and dissipate the energy over the width of several beams. Also, the connection between the roof beams and the outside columns supporting the roof, and the columns themselves, incorporates flexible hinges (Figure 14.29).

14.6.8 Tunnels to bypass slides

In locations where it is impractical to construct a rock shed or stabilise the slope by other means, it may be necessary to drive a tunnel to bypass the hazard zone. For example, twin 1280 m (4200 ft) long, single lane highway tunnels were driven behind Devils Slide on Route 1 in California south of Pacifica to avoid an area of extensive instability that often blocked the coast highway.

Chapter 15

Movement monitoring

15.1 MOVEMENT OF SLOPES AND OPERATIONAL SAFETY

Many rock slopes move to varying degrees during the course of their operational lives. Such movement indicates that the slope is in a quasi-stable state, but this condition may continue for many years, or even centuries, without failure occurring. However, in other cases, initial minor slope movement may be a precursor for accelerating movement followed by collapse of the slope. Because of the unpredictability of slope behaviour, movement-monitoring programmes can be of value in managing slope hazards, and provide information that is useful for the design of remedial work. Examples of civil slopes where long-term movement monitoring has been carried out, with telemetry to remotely access the readings, are Turtle Mountain in Alberta, Canada, the site of the Frank Slide (Read, Langenberg, Cruden et al., 2005; Dehls, Farina, Martin et al., 2010), and the Debeque Slide in Colorado (Gaffney, White and Ellis, 2002).

Slope movement is the most common in open-pit mines, and many mines continue to operate safely with moving slopes that are carefully monitored to warn of deteriorating stability conditions. Other slopes that undergo long-term movement are landslides that may creep for hundreds of years, resulting in accumulative movement of tens of metres. Such movement may comprise an approximately uniform creep rate, on which may be superimposed short periods of more rapid movement resulting from such events as earthquakes, unusually high-precipitation periods and human activities. Human activities that can be detrimental to slope stability include excavation of the base, and changing the ground water conditions by dam filling or irrigation.

This chapter describes common methods of monitoring movement of rock slopes, and interpretation of the results. It is considered that monitoring programmes are most appropriate for actively excavated slopes such as open-pit mines and quarries which have a limited operational life and where a carefully managed, ongoing survey operation can be set up. The survey can identify accelerating movement of the slope and allow measures to be taken to minimise the risk by moving operations away from the active slide. Figure 15.1 shows an example of an open-pit slope, where careful monitoring identified the increasing rate of movement indicative of deteriorating stability, which allowed the actual collapse to be photographed. Several cases of slope monitoring at open pits where mining continued for several months below the moving slope have been well documented. Eventually, the rate of movement increased rapidly indicating that stability conditions were deteriorating and operations were halted shortly before the slope failed (Kennedy and Neimeyer, 1970; Brawner, Stacey and Stark, 1975; Wyllie and Munn, 1979; Broadbent and Zavodni, 1982). Movement of natural slopes has also been monitored to predict failure time as described in Section 15.7.1 (Fukuzono, 1985; Mufundirwa et al., 2010; Guthrie and Nicksiar, 2016).

Figure 15.1 Photographs of slope failure in an open-pit mine; failure was predicted as the result of slope monitoring: (a) horizontal movement of 22.9 m (75 ft) prior to collapse; (b) and (c) toppling block a few seconds after photograph (a). (Photograph by P. F. Stacey; Brawner, Stacey and Stark, 1975.)

Monitoring may also be suitable for large landslides that threaten facilities such as reservoirs, transportation systems and residential areas. The weaknesses of such programmes are that they may require sophisticated monitoring and telemetry equipment that will be costly to operate and maintain for long periods, that is, years or decades. Also, it may be difficult to identify deteriorating stability conditions that will clearly show a need to evacuate the site. It is considered that where the risk to lives and property is significant, remediation is preferred to long-term monitoring.

This chapter includes a discussion of monitoring of open-pit mines because most measurement of actively moving slopes is carried out at mines and it is from these results that experience has been developed on monitoring methods and interpretation of the results.

15.2 SELECTION OF MONITORING SYSTEMS

The following is a discussion on issues related to the reliability and cost of movement measurements that should be considered when planning slope-monitoring programmes. Abellán, Oppikofer, Jaboyedoff et al. (2014) provide a useful discussion on the accuracy and reproducibility of slope-monitoring methods.

15.2.1 Visual inspections

Despite the value, and increasing adaption of sophisticated instrumentation discussed in this chapter, it is the author's opinion that visual inspections of slopes are an essential part of slope management programmes. That is, careful observations of slopes will often identify possible movement conditions such as slight cracking, fresh faces and minor rock falls before movement can be detected by the instrumentation. Furthermore, subtle changes with time can be observed by trained and experienced personnel. This approach is consistent with the general rule that 'simple is better'.

15.2.2 Movement magnitude and extent

Slopes in strong rock where stability is governed by geological structure may only move a few centimetres before failure, and if large, intact blocks of rock are moving, their movement is likely to be representative of the overall slope movement. For these conditions, placement of a few reflector prisms on the slope and monitoring their positions with precise geodetic survey equipment should provide reliable information on slope movement.

In contrast, slopes in weathered or closely jointed rock may move tens of metres prior to failure. Furthermore, if the surficial rock comprises loose blocks that are toppling or rotating independently of the main slope movement, it may be necessary to use a remote system to scan the entire moving slope, and nearby stable areas, to identify overall movement rate.

15.2.3 Measurement accuracy and reproducibility

The slope movement-monitoring method that is selected must be able to measure displacement to the required level of accuracy and reproducibility consistent with the expected type of movement described in Section 15.2.2. That is, accuracy is the comparison between the actual position of a reference point on the slide, and its measured distance. For example, geodetic surveying of prism positions from a stable base station under ideal conditions will detect 5–10 mm (0.2–0.4 in.) of slope movement.

A separate issue is the actual performance of the instrumentation and the reproducibility of the results. For example, with light detection and ranging (LiDAR) systems, errors will develop where the sight distance is greater than the design range of the equipment, in rain, fog or dusty conditions and where the reflectivity of the slope surface is poor. Changes in these conditions with time will produce differences in the measured position of the slope. An overall comment on LiDAR scans is that a dense point cloud that produces a detailed image of the slope may not be an accurate representation of the actual rock surface (Abellán, Oppikofer, Jaboyedoff et al., 2014). One method of checking reproducibility is to make scans, a few days apart, of a known stable area and check the results to see the magnitude of any apparent movement.

Further comments on reflectorless survey methods are that reproducibility is improved if permanent reference points are fixed to stable locations on the slope, and if these reference points are then located relative to a local or Universal Transverse Mercator (UTM) coordinates. These points will be of great value in accurately overlaying point clouds from successive surveys; detection of movement areas of the slope requires that three-dimensional vectors be calculated between comparable points within each cloud.

15.2.4 Site environment

Site conditions such as snow cover, rain, fog, dust, varying humidity and temperature along the sight line and vegetation that will change with time can interfere with measurements to varying degrees. For example, LiDAR scans are scattered by rain and fog, and cannot penetrate snow, although it is possible to filter out vegetation so that the ground surface is measured. For reflectorless survey methods such as LiDAR, the reflectivity of the rock surface may also be an issue that influences the accuracy of the measurements, and it will not be possible to measure 'occlusion' areas such as sharp changes in slope attitude that cannot be observed by the line of sight.

Another factor in designing monitoring systems may be a reliable power supply – although monitoring systems can often be powered with solar cells, these may not be suitable for locations that are often in shadow, during short days in winter, and where the panels are covered in snow. For systems where electrical cables are used, an occasional issue is animals chewing through the insulation and damaging exposed electrical cables.

15.2.5 Safety, site access

If the slope is very steep or unstable, it may not be safe for personnel to access monitoring stations on the slope. In these circumstances, remote monitoring systems such as radar or LiDAR scanning would be preferable to crack width measurements, for example.

15.2.6 Measurement frequency

In situations where the slope is moving rapidly, it may be necessary to make measurements daily, or even hourly, and immediately process the results to produce update movement–time plots. In these circumstances, measurements made by infrequent satellite orbits (i.e. interferometric synthetic aperture radar [InSAR]) or aerial photogrammetry would not be suitable compared to land-based scanning.

15.2.7 Automation, telemetry

Many monitoring methods can incorporate systems to automatically read the instruments at programmed intervals, analyse the data and transmit the results to remote locations for interpretation. These systems can also incorporate alarms that are triggered if pre-set movement thresholds are exceeded (Baker, 1991). It is considered that it is preferable, initially at least, to set up a simple system and ensure that it provides reliable and accurate results, and not put emphasis on expensive technology that has a large manpower commitment to install and maintain.

15.2.8 Monitoring costs

Costs for purchase, installation and maintenance of monitoring equipment, and for personnel time and data interpretation can rapidly escalate, particularly if telemetry is used to receive the results at a location remote from the slope. This may be acceptable for a short-term, high-risk situation, such as a temporary highway closure while stabilisation work is carried out, but if the monitoring duration is open ended, the funds may be better spent on remediation work.

On civil projects, the cost of monitoring should take into account that the work may be carried out by outside personnel with rented equipment and be required for an indefinite period. It is the author's experience that sophisticated, long-term monitoring of civil slopes is too costly for most owners of these facilities. The exception is the operators of large dams where the consequence of failure is extreme, and for these conditions, long-term monitoring is usually carried out.

In contrast, slope monitoring for open-pit mines is often the responsibility of in-house personnel with equipment that can be used for other mine operations, and may only be carried out for a short time that active slope movement is occurring. Under these circumstances, the cost of monitoring may only be a small portion of the overall operating costs.

In the case of sub-surface monitoring of a slide compared to surface measurements, the costs will be influenced by the cost of drilling holes in which to install the instruments. If holes can be drilled with available blast hole drills, then the drilling cost will be negligible in comparison to drilling diamond core holes; diamond coring may only be necessary if information on site geology is required as part of the slope investigation programme.

15.2.9 Technical developments

Many monitoring systems are undergoing rapid advancements with respect to both hardware and software to interpret the results. Therefore, this chapter describes a range of systems that are commonly used to monitor slope movement, but does not provide details such as costs and level of accuracy because these numbers are expected to be soon outdated.

15.3 TYPES OF SLOPE MOVEMENT

In setting up a movement-monitoring programme, it is useful to have an understanding of the type of movement that is occurring. This information can be used to select appropriate instrumentation for the site, and assist in interpretation of the results. For example, if the slope were undergoing a toppling failure, then crack width monitors at the crest would provide direct measurement of horizontal movement. In comparison, if an inclinometer were to be installed in a toppling failure, it may not be certain that it extended to a depth below the zone of movement, which would result in erroneous readings (see Section 15.5.3). Furthermore, the type of movement is related to the failure mechanism and this information can be used to ensure that an appropriate type of stability analysis is used. That is, outward and downward movement at the crest and bulging at the toe would indicate a plane or circular failure, whereas horizontal movement at the crest would only be more indicative of a toppling failure.

The following is a discussion on types of slope movement, and their implications for slope stability.

15.3.1 Initial response

When a slope is first excavated or exposed, a period of initial response occurs as a result of elastic rebound, relaxation and/or dilation of the rock mass due to changes in stress induced by the excavation (Zavodni, 2000). This initial response will occur most commonly in open-pit mines, where the excavation rate is relatively rapid. In comparison, the exposure of slopes by the retreat of glaciation or gradual steepening of slopes due to river erosion at the toe will occur over time periods that may be orders of magnitude longer than that of mines. However, the cumulative strain of such slopes can be considerable. Elastic rebound strain takes place without the development of a definite sliding surface, and is likely the result of dilation and shear of existing discontinuities within the rock mass.

Martin (1993) reports on the initial response measurements of three open-pit mines, which showed that total displacement varied from 150 mm (6 in.) in a strong massive rock mass at Palabora in South Africa, to more than 500 mm (20 in.) in highly fractured and altered rock at the Goldstrike Mine in Nevada. The rates of movement during initial response periods decreased with time and eventually showed no movement. Based on the monitoring carried out at Palabora, the following relationship has been established between the rate of movement V (mm/day) and the time t (days):

$$V = A \cdot e^{-b \cdot t} \tag{15.1}$$

where A and b are parameters that are a function of the rock mass properties, the slope height and angle, the mining rate, external influences and the ultimate failure mechanism. The reported values of A range from 0.113 to 2.449, while values for b range from 0.0004 to 0.00294.

The critical property of the relationship in Equation 15.1 is that the rate of movement decreases with time, indicating that the slope is not at risk from failure.

Another characteristic of initial response type of movement is that can occur within a large volume of rock. For example, during the steepening of the 150 m (500 ft) deep Berkeley Pit from a slope angle of 45° to an angle of 60°, movement measurements in two adits showed that rebound occurred at a distance of up to 120 m (400 ft) behind the face at the toe of the slope (Zavodni, 2000). This rebound and relaxation mechanism has been modelled using the FLAC and UDEC codes (Itasca Consulting Group Inc., 2012) with the objective of predicting such behaviour in similar pits (see Chapter 12 for descriptions of FLAC and UDEC).

15.3.2 Regressive and progressive movement

Following a period of initial response and then possible stability, slope 'failure' would be indicated by the presence of tension cracks at or near the crest of the slope. The development of such cracks is evidence that the movement of the slope has exceeded the elastic limit of the rock mass. However, it is possible that operations can safely continue under these conditions with the implementation of a monitoring system. Eventually, an 'operational slope failure' may develop, which can be described as a condition where the rate of displacement exceeds the rate at which the slide material can be safety excavated (Call, 1982).

A means of identifying either plastic strain of the rock mass or operational failure is to distinguish between regressive and progressive time–displacement curves (Figure 15.2a). A regressive failure (curve A) is one that shows short-term decelerating displacement cycles if disturbing events external to the slope, such as blasting or water pressure, are removed. Conversely, a progressive failure (curve B) is one that displaces at an increasing rate, with the increase in rate often being algebraic to the point of collapse, unless stabilisation measures are implemented. Correct interpretation of the curves is valuable in understanding the slope failure mechanism, and in predicting the future performance of the slope.

Figure 15.2 also shows geological conditions that are commonly associated with these types of time–displacement curves. Where the slope contains discontinuities that dip out of the face, but at a shallow angle that is flatter than the friction angle of these surfaces

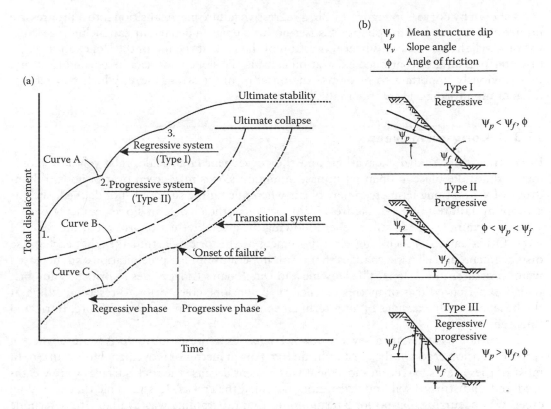

Figure 15.2 Types of slope movement: (a) typical regressive and progressive displacement curves; (b) structural geological conditions corresponding to types of slope movement (Broadbent and Zavodni, 1982).

(Type I), then it is usual that some external stimuli such as blasting or water pressures will be required to initiate movement. The onset of movement indicates that the factor of safety of the slope has dropped just below 1.0, but with a reduction of the external stimuli, the factor of safety will increase and the rate of movement will begin to reduce. In the case of water pressures causing movement, the opening of tension cracks and dilation of the rock mass may temporarily result in the water pressures diminishing, but as pressures gradually build up, another cycle of movement may start. Another condition associated with regressive movement is stick-slip behaviour, which is related to the difference between the static and dynamic coefficients of friction on rock surfaces (Jaeger and Cook, 1976).

Operations can be continued below slopes experiencing regressive movement, but it is necessary that excavation be conducted for short periods with frequent pullbacks, with care being taken to identify the transition to a progressive failure (Zavodni, 2000). As shown in Figure 15.2, geological conditions that may be associated with progressive failure are discontinuities that dip out of the face at a steeper angle than the friction angle (Type II). Also, a slide surface on which the shear strength gradually diminishes with displacement may experience progressive failure. The duration of the progressive stage of a failure has varied from 4 days to 45 days, with no obvious correlation between the time and the site conditions (Zavodni and Broadbent, 1980). However, more rapid failure would be expected where the slide surface is well defined, such as a persistent bedding plane.

As shown by curve C in Figure 15.2b, a regressive failure may transition into a progressive failure and rapidly lead to collapse. Causes of this change in behaviour can include the excavation daylighting a sliding surface, breakup of the rock at the toe of the slope, an increase in water pressure, or continued excavation causing the slope to accelerate beyond recovery. It is obviously important to recognise the onset of progressive failure, which will require diligent monitoring and careful analysis of the results.

15.3.3 Long-term creep

In contrast to the rapid excavation, and the consequent large-scale, relatively fast movements that take place in open-pit mines, mountain slopes may creep over periods of hundreds of years. Long-term creep may occur where the failure surface is not defined, such as a toppling failure (Type III, Figure 15.2b), or where the change in slope geometry is very slow, for example, due to stress relief following the glacial retreat or erosion at the toe by a river. Other causes of such long-term movement are historical earthquakes that each cause displacement, and climatic changes that result in periods of high precipitation and increased water pressures in the slope. The Downie and Dutchman's Ridge Slides in British Columbia, which experienced tens of metres of ancient, downslope creep prior to reservoir filling at the base, are both examples of long-term creep (Moore and Imrie, 1982; Moore, Imrie and Enegren, 1997).

The author has examined several dozen landslides in western North America, where a series of tension cracks at the crest indicate that tens of metres of movement has occurred. In most of these cases, no recent movement was evident because the rock surfaces were weathered and undisturbed soil and vegetation had filled the cracks. It is possible that very slow creep was occurring, but no long-term monitoring programme was available to determine if this was the case. In one case in Alaska, comparison of historic photographs in the local museum showed no substantive change in the appearance of the slope over a period of 120 years. From these observations, it has been concluded that the presence of tension cracks does not necessarily indicate that the risk of collapse is imminent. However, the hazard may be significant if recent movement is observed such as disturbance to the soil and movement of blocks of rock, or a change to the forces acting on the slope is proposed, such as excavation at the toe for highway construction, for example.

15.4 SURFACE-MONITORING METHODS

This section describes common procedures for making surface measurements of slope movement. With respect to using surface rather than sub-surface monitoring, it is noted that surface measurements can only be used, where the surface movement accurately represents the overall movement of the slope. Surface monitoring may be ineffective where the surficial rock is fragmented and loose, and/or the slope is heavily vegetated.

15.4.1 Crack width monitors

Tension cracks are an almost universal feature of slope movement, and crack width measurements are often a reliable and inexpensive means of monitoring movement. Figure 15.3 shows two methods of measuring crack widths. The simplest procedure is to install pins on either side of the crack and measure the distance between them with a steel tape (Figure 15.3a). If pairs of pins are installed on either side of the crack, then the diagonal distance

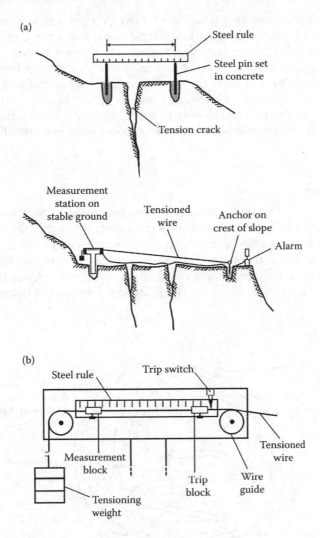

Figure 15.3 Measurement of tension crack width: (a) measurement of distance between steel pins; (b) wire extensometer with trip switch to warn of excessive movement (Wyllie and Munn, 1979).

can also be measured to check the transverse displacement. The maximum practical distance between the pins is probably 2 m.

Figure 15.3b shows a wire extensometer that can be used to measure the total movement across a series of cracks over a distance of as much as 20 m (60 ft). The measurement station is located on stable ground beyond the cracks, and the cable extends to a pin located on the crest of the slope. The cable is tensioned by the weight, and movement is measured by the position of the steel block threaded on the cable. If the movement exceeds the length of the steel rule, the cable can be extended by moving the counterweight and resetting the steel block to the left end of the rule. The wire extensometer can also incorporate a warning system comprising a second steel block threaded on the cable that is set at a selected distance from a trip switch. If the movement exceeds this pre-set limit, the trip switch is triggered and an alarm is activated. As movement occurs, it is necessary to reset the position of the front block, with the distance from the trip switch being determined by the rate of movement. The selection of an appropriate distance from the trip switch is important in order that the alarm provides a warning of deteriorating stability conditions, while not triggering false alarms that result in operators losing confidence in the value of the monitoring.

The main limitations of crack width monitoring are that the upslope pin or reference point must be on stable ground, and that it is necessary that people access the crest of the slide to make the measurements. This work could be hazardous where the slope is moving rapidly. These limitations can be overcome to some degree by automating the system using vibrating wire strain gauges and data loggers to automatically read the distance and record the measurements.

15.4.2 Surveying

On large slides where access to the slope is hazardous and/or frequent and precise measurements and rapid analysis of the results are needed, surveying using EDM (electronic distance measurement) equipment may be a suitable monitoring method (Vamosi and Berube, 1987; ACG, 1998). The three usual components of a survey system are as follows (Figure 15.4). First, one or several reference points are required on stable ground, but that can be viewed from the instrument stations closer to the slide. Second, a number of instrument stations are

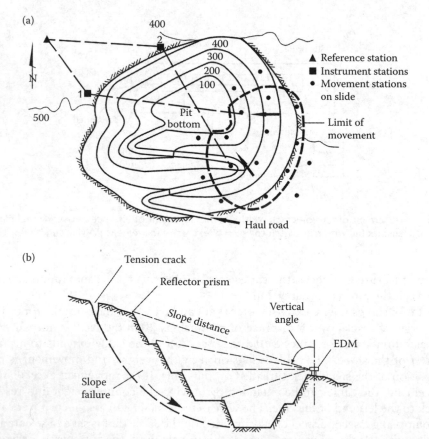

Figure 15.4 Survey system to remotely measure slope movement: (a) typical arrangement of reference, instrument and monitoring stations; (b) measurement of vertical angle and distance to determine vertical displacement (Wyllie and Munn, 1979).

set up on reasonably stable ground at locations from which the slide is visible. If the coordinate positions of the movement stations are to be measured, then the instrument stations should be arranged such that they form an approximately equilateral triangle. Third, stations (reflectors) are set up on, and possibly just outside, the slide area that are then located relative to the instrument stations. The stations on the slide can be reflectors used on heavy equipment, or survey prisms, depending on the sight distance and the accuracy required.

It is preferable that the measurement direction be in the likely direction of movement so that the distance readings approximate the actual slide movement. For example, in Figure 15.4a, it is preferable to measure stations on the north portion of the slide from instrument station 1, and those on the south side from instrument station 2.

The survey arrangement shown in Figure 15.4 can be used to measure at the desired frequency or level of accuracy. For example, for slow-moving slides, the readings may be made every few weeks or months, while for a rapidly moving slide above an active mining operation, an automated system can be set up that takes a series of readings at pre-set intervals, and records and plots the results. Also, quick checks of stability can be made by making distance measurements only, and these can be followed up with triangulation measurements to determine the coordinates of each station at less frequent intervals. Figure 15.4b shows that measurement of the vertical angle and the slope distance allows the vertical displacement to be measured, which is of value in determining the mechanism of failure (see Section 15.5).

The use of individual targets located on the slope is only effective if the movement of the targets is representative of the movement of the overall slope. If it is only possible to place the targets on loose blocks of rock that are sliding or toppling independently of the main slope displacement, then the monitoring may be misleading, and it will be necessary to adopt a monitoring method that measures the overall slope movement as discussed below.

15.4.3 Tiltmeters

It is possible to measure the tilt of a feature to a resolution of about 10 arc seconds using a tiltmeter. These measurements involve bolting or gluing a base plate to the rock face on which the tiltmeter is precisely mounted. The instrument can either be permanently mounted on the face so that readings can be made at any time, or can be located on the mounting plate just when readings are being made.

The advantage of tiltmeters is rapid and precise measurements of the tilt, from which an assumed movement can be calculated. The disadvantages are that the instrument is costly, and it may be difficult to find a small portion of the rock face, the movement of which is representative of the slope movement. It is considered that the primary application of tiltmeters is on structures such as dams and retaining walls rather than rock slopes.

15.4.4 Photogrammetry and digital photography

Aerial photography has long been used in photogrammetry to prepare topographic maps, and comparison of sets of photographs over periods of several years or decades can be useful to compare large-scale changes in slope movement with time. The development of digital imaging has allowed photogrammetric mapping to be carried out both more rapidly and accurately so that it may be used for measuring slope movement.

Another use of digital imaging is to take successive images, or sets of images, to prepare mosaics of slope faces, and then to compare the images to detect areas of movement; the images can be taken by consumer-grade digital cameras. Another approach is to mount the camera on a drone (unmanned aerial vehicle, UAV) to obtain views of the slope that are not possible from the ground, where, for example, trees obscure the slope or a body of water is located at the base of the slope.

Movement-monitoring involves the use of algorithms to analyse three-dimensional point clouds to differentiate between outward movement (accumulation) and detachment (rock falls). Information of the analysis of point clouds, which is undergoing rapid development, is reported in sites such as www.rockbench.org and www.3D-landslide.com.

15.4.5 Light detection and ranging (LiDAR)

Laser imaging using LiDAR scanners involves making a precise three-dimensional map of the entire face. The system involves directing the laser at the slope, selecting the area to be scanned and the density of the scan, and then the laser rapidly and automatically makes a large number of closely spaced scans to cover the area. The result is a dense, accurate three-dimensional point cloud that can be processed to produce a contour map. By making a series of such maps over time, the position of the face can be compared between each scan, and the location and magnitude of the movement measured. Monitoring of movement over time requires that the point clouds from each series of scans be precisely overlaid so that three-dimensional vectors of movement of the same points over time can be calculated. At present (2016), it is possible to overlay the point clouds, without reference stations on stable ground, by working with points on the slope that are outside the slide area such that a comparable points on the slide area can be identified. Another feature of this monitoring is that after the identification of an area of movement, its coordinates can be entered into the scanner and the laser beam will then point to this location on the slope (or tunnel) so that a ground inspection can be carried out.

LiDAR scans in reflectorless mode may not be as accurate as radar scans (see Section 15.4.6) unless they are calibrated against targets on the slope that have been located (registered) by first-order surveys.

LiDAR scans can be made from the ground (terrestrial, TLS) or from the air (aerial, ALS) depending on requirements for sight lines and access to the slope. With the use of appropriate software, it is possible to integrate TLS and ALS scans, or other topographic mapping, to prepare complete maps of faces that may not be visible from a single location.

15.4.6 Radar scanning

Radar scanning is now routinely used for monitoring open-pit slopes because rapid scans of large slope areas can be made in near real time without the use of reflectors, and the results are not affected by atmospheric conditions; movement can be detected to an accuracy of about 1 mm. The scans produce a two-dimensional model of the slope so radar monitoring is sometimes combined with topographic surveys to produce more detailed three-dimensional models. The results of the scans are transmitted to a central office, where they are processed to produce plots showing areas of outward (accumulation) and detachment (rock falls), from which plots of slope movement against time can be prepared.

Available radar scanners include systems that can be carried in backpacks and rapidly set up at remote locations with the optimum line-of-site view of the slide, and powered with solar panels (Lowry, Gomez, Zhou et al., 2013). Other systems need to be towed by a small truck, so road access is required to the monitoring location.

15.4.7 Global positioning system

The global positioning system (GPS) is based on satellites orbiting the earth, with positioning being achieved by timing signals transmitted by the satellites to ground receivers.

A minimum of four satellites with unobstructed lines of sight to the monitoring station are required to obtain three-dimensional positioning. GPS may be a suitable method of monitoring slope movement where the slide covers a large area, and where the measurement of movement at discrete locations is acceptable. Stations can be set up on the slide and their coordinates measured with the GPS unit in real time at any desired frequency, 24 hours a day, regardless of weather or lighting conditions.

The level of accuracy of a standard GPS system may be about 1–2 m (3–6 ft) that is usually adequate for locating mine equipment, but additional measures are required to improve the accuracy for slope monitoring. For example, greater accuracy can be achieved by setting up a base station on stable ground outside the slide area, and accurately determining its position. The GPS readings on the slide are then referenced to the coordinates of the base station (differential GPS). Accuracy of 10–20 mm (0.4–0.8 in.) will require a system known as real-time kinetics (RTK), where corrections to the readings are made every few seconds. Accuracy may be affected where measurements are made in steep canyons such that satellite geometry is not ideal.

The advantages of GPS monitoring are the low cost and ease of set up, but the disadvantages are that movement is only measured at the location of the GPS units, and safe access to the reference points will be required when taking measurements. Permanent GPS points can be set up on the face to eliminate the need to access the slope, but this is more costly.

15.4.8 Synthetic aperture radar (InSAR)

A technique for precise monitoring of movement over large areas is to use radar satellite remote-sensing techniques. This technique is known as InSAR and involves capturing a radar image of the ground surface, which is then compared with images taken at a different time to obtain relative ground movement. Significant features of this technique are that the image can cover an area as large as 2500 km^2, relative movements can be measured in the range of 5–25 mm, and the measurements are independent of the weather, cloud cover and daylight.

These attributes mean that InSAR is ideally suited to precise movement monitoring of large areas over long time periods, with no need to set up reference points on the ground. However, some limitations of the technique are that the frequency of the measurements are governed by the interval between satellite orbits over the site, which presently (2015) is about once every 10–45 days, and the processing of the data can take another few days and be costly. Also, movement is most accurately monitored in the vertical direction, so less reliable information may be provided of sub-horizontal movement of steep rock faces.

Ground-based GB-InSAR measurement equipment is also available where a fixed base station is set up that incorporates a horizontal rail about 1.5 m (5 ft) along on which the unit moves as it scans the face. This equipment has been used to monitor movement of the rock face on Turtle Mountain in southwest Alberta that is the location of the 1903 Frank Slide that had a volume of about 30 million cubic metres (40 million cubic yards) (Dehls, Farina, Martin et al., 2010).

15.5 SUB-SURFACE-MONITORING METHODS

Sub-surface measurement of slope movement is often a useful component of a monitoring programme in order to provide a more complete picture of the slope behaviour. In cases where surface monitoring is not feasible, sub-surface measurements will be the only measurements available. The main purpose of these measurements is to locate the slide surface or surfaces, determine the slide volume and monitor the rate of movement. In some cases, the holes used to install movement-monitoring equipment can also be used to measure water pressure.

15.5.1 Borehole probes

One of the simplest sub-surface-monitoring methods is the borehole probe comprising a length of reinforcing steel about 2 m (6 ft) long that is lowered down the drill hole on a length of rope. If the hole intersects a moving slide plane, the hole will be displaced at this level and it will no longer be possible to pull the bar past this point. Similarly, a probe can be lowered down the hole, and in this way, both the top and bottom of the slide plane can be located.

The advantages of the probe are the low cost and simplicity, but it will provide little information on the rate of movement.

15.5.2 Time-domain reflectometry

Time-domain reflectometry (TDR) is another means of locating a sliding surface, which can also monitor the rate of movement (Kane and Beck, 1996). This method involves grouting into a borehole a coaxial cable comprising inner and outer metallic conductors separated by an insulating material. When a voltage pulse waveform is sent down the cable, it will be reflected at any point where a change in the distance between the conductors has occurred. The reflection occurs because the change in distance alters the characteristic impedance of the cable. Movement of a sliding plane that causes a crimp or kink in the cable will be sufficient to change the impedance, and the instrumentation can detect the location of the movement.

The primary advantages of TDR are that the cable is inexpensive so that it can be sacrificed in a rapidly moving slide. Also, the readings can be made in a few minutes from a remote location either by extending the cable to a safe location off the slide, or by telemetry.

15.5.3 Inclinometers

Inclinometers are instruments ideally suited to long-term, precise monitoring of the position of a borehole over its entire length. By making a series of readings over time, it is also possible to monitor the rate of movement. The components of the inclinometer are a plastic casing with four longitudinal grooves cut in the inside wall, and a probe that is lowered down the casing on an electrical cable with graduated depth markings (Figure 15.5a). The probe contains two accelerometers, aligned so that they measure the tilt of the probe in two mutually perpendicular directions, usually one parallel to the slope face, and the other at right angles to the face. The probe is also equipped with a pair of wheels that run in the grooves in the casing and maintain the rotational stability of the probe.

It is also possible to install a string of in-place inclinometer probes over the entire length of the drill hole to measure the hole deviation in three dimensions, and to determine changes in deviation with time. The readings can be made at the surface, or remotely via telemetry.

The first requirement of accurate monitoring is to extend the borehole below the depth of movement so that readings made from the lower end of the hole are referenced to a stable base. Precautions are also needed during installation of the casing to maintain the vertical alignment of the grooves and prevent spiralling. Readings are made by lowering a probe to the end of the hole and then raising it in increments equal to the length of the wheelbase L of the probe. At each depth increment, the tilt angle ψ is measured. Figure

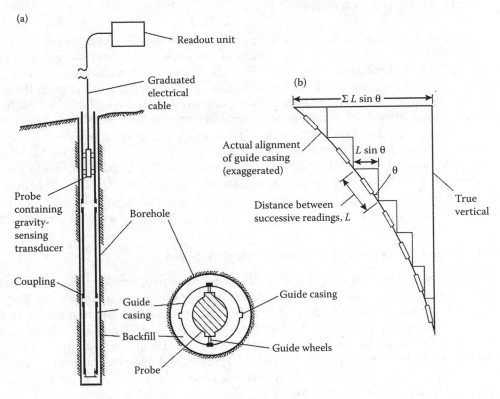

Figure 15.5 Inclinometer for measuring borehole deflection: (a) arrangement of grooved casing and inclinometer probe; (b) principle of calculating deflection from tilt measurements (Dunnicliff, 1993).

15.5b shows the procedure for calculating the displacement ($L \sin \psi$) for each increment, and the total displacement at the top of the hole $\Sigma(L \sin \psi)$. A check of the results is usually made by rotating the probe by 180° and taking a second set of readings. Another precaution is to allow time during the readings for the probe to reach temperature equilibrium in the hole.

15.6 MICROSEISMIC MONITORING

Slope movement is often associated with fracturing of rock on the slide surface, and within the body of the slide, that produces microseismic noise. These events can be detected by geophones located in boreholes drilled close to or into the slide. Deteriorating stability conditions may be indicated by increasing frequency of the events. By installing a number of geophones in widely spaced holes, it is possible to calculate the location of the event using triangulation.

Microseismic monitoring can be useful for conditions where access to the slide surface is dangerous, and can complement surface-monitoring methods such as radar and LiDAR scanning. Geophones can be installed in holes outside the slide area so that they are not damaged by slide movement, and can incorporate data loggers, a power supply and telemetry to remotely record the events (Read, Langenberg, Cruden et al., 2005).

In general, the most common application of microseismic monitoring is in deep underground mines to detect and locate rock bursts.

15.7 DATA INTERPRETATION

Interpretation of the movement data is an essential part of monitoring operations in order to quickly identify acceleration or deceleration of the slope that indicates deteriorating or improving stability conditions, respectively. This allows appropriate action to be taken with respect to the safety and economics of the operation. The importance of updating the plots as the readings are made is that failures have occurred within days of cracks first being observed (Stacey, 1996). Unfortunately, instances have occurred of careful monitoring measurements being recorded, but because they were not plotted, acceleration of the slope that was a clear precursor to failure was not recognised.

This section describes a number of procedures for interpreting monitoring results to provide information on both stability conditions, and the mechanism of failure. With respect to the mechanism of failure, this information is useful in designing stabilisation measures, which require that the appropriate method of analysis be used, such as planar or toppling.

15.7.1 Time–movement and time–velocity plots

The monitoring programme, whether carried out by surveying, crack width gauges, GPS, satellite scans or inclinometers, will provide readings of movement against time. Plots of this data are fundamental to understanding the mechanism of the slope movement, and possibly predicting the time of failure. Empirical methods of predicting the time of failure are as follows. First, as discussed below, semi-log plots of the displacement rate against time can be used to find the velocity at the time of collapse. Second, plotting the inverse displacement rate against time shows that the rate converges to zero as the slope approaches failure (Federico et al., 2015; Guthrie and Nicksiar, 2016).

The following is a discussion of a monitoring programme in an open-pit coal mine, where a movement-monitoring programme was used to safely mine below a moving slope for most of the life of the pit (Wyllie and Munn, 1979).

Figure 15.6a shows a cross-section of the final pit and the slope above the pit, as well as the lithology and geological structure of this slope. Soon after mining commenced at the 1870 m elevation in March 1974, a toppling failure was initiated in the overturned siltstone beds at the crest of the pit. When mining on the 1840 m bench, a series of cracks formed on the 1860 m bench and a movement-monitoring programme was set up in February 1975 using a combination of wire extensometers (Figure 15.3b), and surveying of prisms. This monitoring system was used to control the mining operation with the objective of mining back to the final wall so that the coal could eventually be mined to the bottom of the pit. Figure 15.6b shows the sensitivity of the slope movement to mining at the toe of the toppling slope, and typical regressive behaviour as soon as the shovel was pulled back. This experience was used to establish the criterion, based on hourly movement readings, that mining would be halted as soon as the rate of movement reached 25 mm (1 in.) per hour. When this rate reduced to 15 mm (0.6 in.) per day over a period of about 10 days, mining recommenced. Using this control procedure, mining continued towards the final depth of about 1700 m with the slope moving at an average rate of 6 mm (0.25 in.) per day.

In April 1976, the slope started to accelerate, and over the next two months, a total movement of about 30 m (100 ft) occurred on the mountain above the pit and the maximum velocity reached almost 1 m/day (3 ft/day) (Figure 15.6c and d). The acceleration on the slope movement plots gave an adequate warning of deterioration stability conditions and mining was abandoned. The area with the greatest movement was on the toppling beds along the crest of the pit, and in early June 1976, two separate slope failures occurred with a total volume of 570,000 m³ (750,000 yd³). After the failures, the monitoring system was

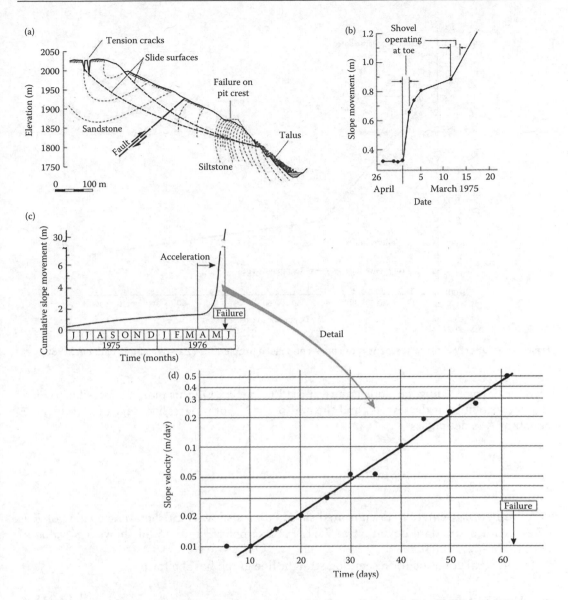

Figure 15.6 Movement monitoring at open-pit coal mine: (a) cross-section of pit and hillside showing geology and extent of slope failure; (b) regressive slope movement when mining at 1840 m level; (c) slope movement over 13 months leading to slope failure; (d) slope velocity over two months prior to failure (Wyllie and Munn, 1979).

re-established, which showed that the rate of movement was gradually decreasing, and after a month it was decided to restart mining at the bottom of the pit. This decision was also based on borehole probe measurements, which showed that the circular slide surfaces, associated with the toppling at the pit crest, daylighted in the upper part of the pit slope (Figure 15.6a). Therefore, mining at the base of the pit would have little effect on stability.

The type of movement-monitoring data shown in Figures 15.6c and d has been analysed to help predict the time of failure once the progressive stage of slope movement has developed (Zavodni and Broadbent, 1980). Figure 15.7 shows the semi-log time–velocity plot in

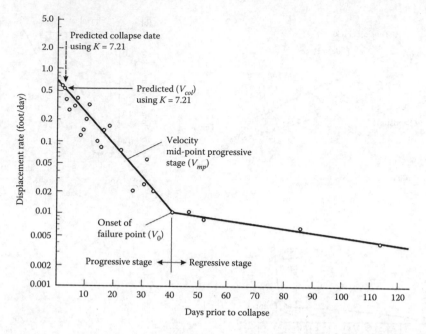

Figure 15.7 Liberty Pit displacement rate curve and failure predictions (Zavodni and Broadbent, 1980).

feet per day preceding the slope failure at the Liberty Pit. On this plot, it is possible to identify the velocities at the start V_0 and the mid-point V_{mp} of progressive stage of movement. A
constant K is defined as

$$K = \frac{V_{mp}}{V_0} \tag{15.2}$$

A study of six carefully documented slope failures shows that the average value of K is
−7.21, with a standard deviation of 2.11. For example, Figure 15.6d shows a K value of
about −7 (−0.07/0.01).

The general equation for a semi-log straight line graph has the form

$$V = C \cdot e^{S \cdot t} \tag{15.3}$$

where V is the velocity, C is the intercept of the line on the time axis, e is the base of the
natural logarithm, S is the slope of the line and t is the time. Therefore, the velocity at any
time is given by

$$V = V_0 \cdot e^{S \cdot t} \tag{15.4}$$

Combining Equations 15.2 and 15.4 gives the following relationship for the velocity at
collapse V_{col}:

$$V_{col} = K^2 \cdot V_0 \tag{15.5}$$

The use of Equation 15.5 in conjunction with a time–velocity plot allows an estimation to be made of the time of collapse. For example, from Figure 15.6d where $K = -7$ and $V_0 = 0.01$ m per day, the value of V_{col} is 0.49 m per day. Extrapolation of the velocity–time line shows that this rate will occur at about 61 days, which is very close to the actual day of collapse.

It is likely that the rate at which collapse occurs depends to some degree on the geological conditions on the sliding plane. For example, failure is likely to occur faster if sliding is occurring on a well-defined structural feature such as a fault that daylights on the face, compared to that shown in Figure 15.6a where the slide plane developed by fracture through intact rock. The reason for this is that in the case of a fault, the shear strength will diminish from peak to residual with relatively little movement compared to that of failure through the rock mass.

15.8 SLOPE FAILURE MECHANISMS

Figure 15.8 shows a number of methods of analysing movement-monitoring results that may help to identify the mechanism of slope failure. This information can be useful in applying the appropriate method of stability analysis, and in the design of stabilisation measures.

Figure 15.8a shows a combined displacement and velocity plot, which shows that the acceleration of the slope stopped after day 5. This change in behaviour is clearly evident on the velocity plot where the velocity is constant after day 5. In comparison, on the movement plot, the change in gradient is not so obvious. This slope movement would be typical of regressive-type instability.

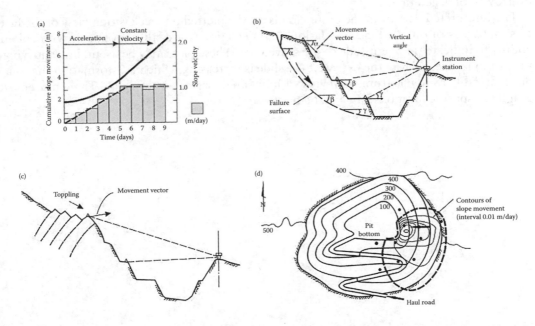

Figure 15.8 Interpretation of movement-monitoring data: (a) displacement and velocity plots show the onset of regressive movement; (b) movement vectors showing circular failure mechanism; (c) movement vectors showing toppling failure mechanism; (d) slope velocity contours show extent of slope movement (Wyllie and Munn, 1979).

Figure 15.8b shows the magnitude and dips of movement vectors for survey stations on the crest, mid-height and toe of the slide. The dip angles approximately equal the dip of the underlying sliding surface, and in this instance indicate that a circular failure is taking place in which the sliding surface is steep near the crest and near horizontal near the base. This information would also show the location of the base of the slide, which may not be base of the slope, as was the case for the slide shown in Figure 15.6a. Figure 15.8c shows a movement vector for a typical toppling failure in which the stations located on the overturning beds at the crest may move outwards, as well as upwards by a small amount, while little movement occurs below the crest.

Figure 15.8d shows contours of slope velocity plotted on a plan of the pit. These plots show both the extent of the slide and the area(s) of most rapid movement. Such plots, when regularly updated, can be useful in identifying an increase in the dimensions of the slide, and/or a change in the most rapidly moving area. This information will assist in the planning of mining operations, for example, to complete mining in the south east corner of the pit before failure occurs. Also, if the haul road is on a relatively slowly moving part of the slide, this may allow time to develop new access to the pit.

15.9 INTERPRETATION OF POINT CLOUDS

Slope movement monitoring using scanning methods, such as LiDAR, radar and InSAR, and digital images, requires that point clouds generated by each successive scan be overlayed so that equivalent points within the scans can be compared and the distance between them measured to produce three-dimensional vectors of movement. Matching of point cloud data is enhanced if reference stations are located on the slope. Factors that may influence the position of the point clouds are differing site conditions over time such as rain, snow cover or growth of vegetation.

The interpreted data from the point clouds can be plotted on an elevation view of the face to show areas of outward (accumulation) and depletion (rock fall) movement, with colour coding to indicate the magnitude of the movement. These plots may be useful in identifying, for example, areas where the extent of the slide is increasing, which is information that may not be readily available from surveying of individual prisms on the slope. Time–movement graphs of specific locations can also be prepared.

Chapter 16

Civil engineering applications

16.1 INTRODUCTION

When a slope above an important civil engineering structure is found to be unstable, an urgent decision is commonly required on effective and economical remedial measures. Evidence of potential instability includes open tension cracks behind the crest, movement at the toe, failures of limited extent in part of the slope, or failure of an adjacent slope in similar geology. Whatever the cause, once doubt has been cast upon the stability of an important slope, it is essential that its overall stability should be investigated and, if necessary, appropriate remedial measures implemented. This chapter describes six rock slopes on civil engineering projects, in a variety of geological and climatic conditions, and the slope stabilisation measures that were implemented. For each example, information is provided on the geology, rock strength and ground water conditions, as well as the stability analysis, design of remedial work and construction issues.

Case studies 1 to 4 describe rock slopes containing well-defined geological structure that controlled stability, forming planar (studies 1 and 2), wedge (study 3) and toppling (study 4) instability. In contrast, case studies 4 and 5 describe slopes where the structure orientation is not well defined and the sliding planes form circular surfaces through the rock mass.

The purpose of these case studies is to describe the application of the investigation and design techniques described in the previous chapters of the book.

16.2 CASE STUDY 1 – HONG KONG: CHOICE OF REMEDIAL MEASURES FOR PLANE FAILURE (BY DR. EVERT HOEK, VANCOUVER, CANADA)

16.2.1 Site description

A 60 m (200 ft) high cut had an overall face angle of 50°, made up of three, 20 m (65 ft) high benches with face angles of 70° (Figure 16.1). A small slide in a nearby slope had caused attention to be focussed on this particular cut, and concern had been expressed that a major slide could occur resulting in serious damage to an important civil engineering structure at the foot of the cut. An assessment was required of the short- and the long-term stability of the cut, and recommendations for appropriate remedial measures, should these prove necessary. No previous geological or engineering studies had been carried out on this cut, and no boreholes were known to exist in the area. The site was in an area of high-rainfall intensity and low seismicity. A horizontal seismic coefficient k_H of $0.08 \cdot g$ had been suggested as the maximum acceleration to which this cut was likely to be subjected.

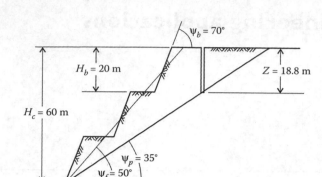

Figure 16.1 Case study 1 – geometry assumed for two-dimensional plane failure analysis of the slope.

16.2.2 Geology

The cut was in slightly weathered granite containing several sets of steeply dipping joints, as well as sheet jointing that dipped at 35° and formed the natural slopes in the area. Faced with this problem and having no geological or engineering data from which to work, the first task was to obtain a representative sample of structural geology to establish the most likely failure mode. Time would not allow for a drilling programme, so the collection of structural data had to be based upon surface mapping, which was reasonable because of the extensive rock exposure in the cut face and natural slopes.

The structural mapping identified the geometrical and structural geology features listed in Table 16.1.

16.2.3 Rock shear strength

Because no information was available on the shear strength of the sheet joints forming the potential sliding surface, the strength values used in design were estimated from previous experience of the stability of slopes in granite. Figure 5.21 is a summary of shear-strength values developed primarily from back analysis of slope failures; point '11' most closely represented the strength of the rock at the site. Based on this experience, it was considered that even heavily kaolinised granites exhibit friction values in the range of 35–45° because of the angular nature of the mineral grains. The cohesion of these surfaces was likely to be variable depending on the degree of weathering of the surface, and the persistence of the joints; a cohesion range of 50–200 kPa (7.2–29 psi) was selected.

Table 16.1 Orientation of slope and joint sets shown in Figures 16.1 and 16.2

Feature	Dip (°)	Dip direction (°)
Overall slope face	50	200
Individual benches	70	200
Sheet joint	35	190
Joint set J1	80	233
Joint set J2	80	040
Joint set J3	70	325

16.2.4 Ground water

No information on ground water conditions was available. However, since the site was in an area that experienced periods of intense rainfall, it was expected that significant transient ground water pressures would develop in the slope following these events.

16.2.5 Stability analysis

The stereoplot of the data in Table 16.1 is shown in Figure 16.2, including a friction circle of 35°. Note that, although the three joint sets provided a number of steep release surfaces which would allow blocks to separate from the rock mass, none of their lines of intersection, which are circled in Figure 16.2, fall within the zone designated as potentially unstable (refer to Figure 8.3b). On the other hand, the great circle representing the sheet joints passes through the zone of potential instability. Furthermore, the dip direction of the sheet joints is close to that of the cut face, so the most likely failure mode was a plane slide on the sheet joints in the direction indicated in Figure 16.2.

The stability check carried out in Figure 16.2 suggested that both the overall cut and the individual benches were potentially unstable, and it was therefore clearly necessary to carry out further analysis of both.

The two steeply dipping joint sets J1 and J2 were oriented approximately parallel to the slope face, and it was likely that tension cracks would form behind the crest of the cut on these discontinuities. One possible failure mode was that illustrated as Model I in Figure 16.3; this theoretical model assumed that a tension crack occurred in the dry state in the most critical position (refer to Figure 7.6), and that this crack filled to depth z_w with water during a period of exceptionally heavy rain. A simultaneous earthquake subjected the slope to ground motion that was simulated with a horizontal seismic coefficient k_H of $0.08 \cdot g$, generating a force of $(k_H \cdot W)$, where W was the weight of the sliding block. The factor of

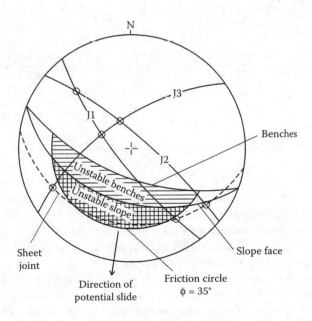

Figure 16.2 Case study 1 – stereoplot of geometric and geological data for slope shown in Figure 16.1.

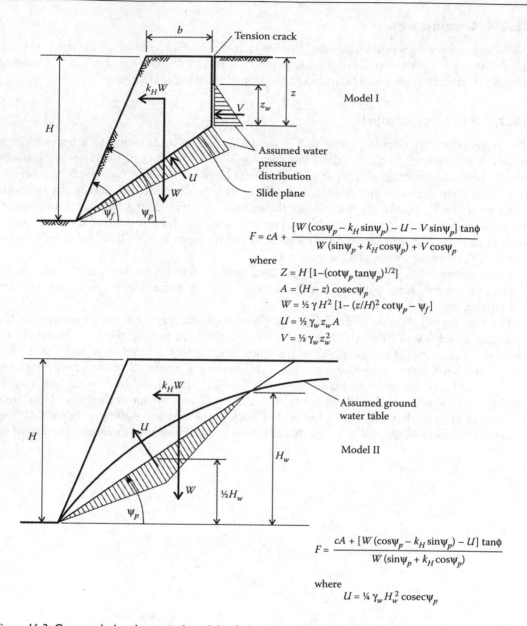

$$F = cA + \frac{[W(\cos\psi_p - k_H \sin\psi_p) - U - V\sin\psi_p]\tan\phi}{W(\sin\psi_p + k_H\cos\psi_p) + V\cos\psi_p}$$

where
$$Z = H\,[1-(\cot\psi_p\,\tan\psi_p)^{1/2}]$$
$$A = (H - z)\,\mathrm{cosec}\,\psi_p$$
$$W = \tfrac{1}{2}\,\gamma H^2\,[1-(z/H)^2\cot\psi_p - \psi_f]$$
$$U = \tfrac{1}{2}\,\gamma_w z_w A$$
$$V = \tfrac{1}{2}\,\gamma_w z_w^2$$

$$F = \frac{cA + [W(\cos\psi_p - k_H\sin\psi_p) - U]\tan\phi}{W(\sin\psi_p + k_H\cos\psi_p)}$$

where
$$U = \tfrac{1}{4}\,\gamma_w H_w^2\,\mathrm{cosec}\,\psi_p$$

Figure 16.3 Case study 1 – theoretical models of plane slope failures.

safety of this slope with the inclusion of a pseudo-static horizontal earthquake loading is given by the equations in Figure 16.3 (refer to Section 11.6).

To allow for the possible presence of substantial sub-surface water, an alternative theoretical model was proposed. This is illustrated as Model II in Figure 16.3 and, again this model includes the pseudo-static earthquake loading.

Having decided upon the most likely failure mode and having proposed one or more theoretical models to represent this failure mode, a range of possible slope parameter values was substituted into the factor of safety equations to determine the sensitivity of the slope to the different conditions to which it was likely to be subjected. Table 16.2 summarises the input

Table 16.2 Input data for case study 1 plane stability analysis

Parameter	Parameter value
Cut height	$H_c = 60$ m
Overall slope angle	$\psi_f = 50°$
Bench face angle	$\psi_b = 70°$
Bench height	$H_b = 20$ m
Failure plane angle	$\psi_p = 35°$
Distance to tension crack (slope)	$b_s = 15.4$ m
Distance to tension crack (bench)	$b_b = 2.8$ m
Rock density	$\gamma_r = 25.5$ kN/m³
Water density	$\gamma_w = 9.81$ kN/m³
Seismic coefficient	$k_H = 0.08$ g

data. The factors of safety of the slopes were calculated by substituting these values into the equations in Figure 16.3, as follows:

$$\text{Overall cut Model I} \quad FS = \frac{80.2 \cdot c + \left(18{,}143 - 39.3 \cdot z_w - 2.81 \cdot z_w^2\right) \cdot \tan\phi}{14{,}995 + 4.02 \cdot z_w^2}$$

$$\text{Overall cut Model II} \quad FS = \frac{104.6 \cdot c + \left(20{,}907 - 4.28 \cdot H_w^2\right) \cdot \tan\phi}{17{,}279}$$

$$\text{Individual benches Model I} \quad FS = \frac{17.6 \cdot c + \left(2815 - 86.3 \cdot z_w - 2.81 \cdot z_w^2\right) \cdot \tan\phi}{2327 + 4.02 \cdot z_w^2}$$

$$\text{Individual benches Model II} \quad FS = \frac{34.9 \cdot c + \left(4197 - 4.28 \cdot H_w^2\right) \cdot \tan\phi}{3469}$$

One of the most useful studies of the factor of safety equations was to find the shear strength which would have to be mobilised for failure (i.e. $FS = 1.0$). These analyses examined the overall cut and the individual benches, for a range of water pressures. Figure 16.4 gives the results of the study and the numbered curves on this plot represent the following conditions:

Curve 1	Overall cut, Model I: dry	$z_w = 0$
Curve 2	Overall cut, Model I: saturated	$z_w = z = 14$ m (46 ft)
Curve 3	Overall cut, Model II: dry	$H_w = 0$
Curve 4	Overall cut, Model II: saturated	$H_w = 60$ m (200 ft)
Curve 5	Individual bench, Model I: dry	$z_w = 0$
Curve 6	Individual bench, Model II: saturated	$z_w = z = 9.9$ m (32.5 ft)
Curve 7	Individual bench, Model II: dry	$H_w = 0$
Curve 8	Individual bench, Model II: saturated	$H_w = H = 20$ m (65.6 ft)

The reader may feel that a consideration of all these possibilities is unnecessary, but it is only coincidental that, because of the geometry of this particular cut, the shear

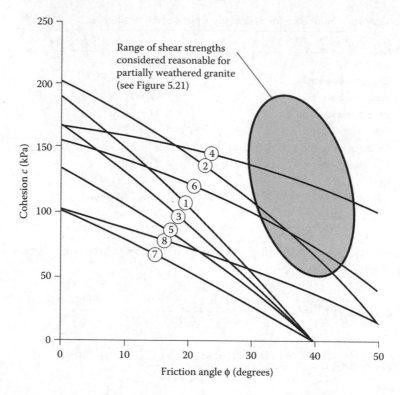

Figure 16.4 Case study 1 – shear strength mobilised for failure of slope considered.

strength values found happen to fall reasonably close together. In other cases, one of the conditions may be very much more critical than the others, and it would take consider-able experience to detect this condition without going through the calculations required to produce Figure 16.4.

The elliptical area in Figure 16.4 surrounds the range of shear strengths considered rea-sonable for partially weathered granite. As discussed in Section 16.2.3, these values are based on the plot given in Figure 5.21. Figure 16.4 shows that when the cut is fully saturated and subject to earthquake loading (curves 2, 4 and 6), the likely available shear strength along the sliding surfaces would be exceeded by the driving forces acting on the sliding sur-face, and failure would be possible. In contrast, the dry slopes are stable (curves 1, 3, 5, 7). Considering the rate of weathering of granite in tropical environments over the operational life of the slope, with a consequent reduction in available cohesive strength, these results indicated that the cut was unsafe and that the steps should be taken to increase its stability.

Four basic methods for improving the stability of the cut were considered:

1. Reduction of cut height
2. Reduction of cut face angle
3. Drainage
4. Reinforcement with tensioned anchors

In order to compare the effectiveness of these different methods, it was assumed that the sheet joint surface had a cohesion of 100 kPa (14.5 psi) and a friction angle of 35°. The increase in factor of safety for a reduction in slope height, slope angle and water level was found by altering one of the variables at a time in the equations in Figure 16.3. The influence

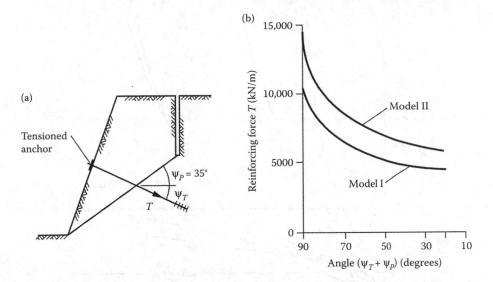

Figure 16.5 Slope stabilisation with tensioned rock anchors: (a) orientation of tensioned anchors; (b) total reinforcing force required for a factor of safety of 1.5.

of reinforcing the cut was obtained by modifying these equations to include a bolting force as shown in Equation 7.25.

$$\text{Model I:}\quad FS = \frac{c \cdot A + (W \cdot (\cos\psi_p - k_H \cdot \sin\psi_p) - U - V \cdot \sin\psi_p + T \cdot \sin(\psi_T + \psi_p)) \cdot \tan\phi}{W \cdot (\sin\psi_p + k_H \cdot \cos\psi_p) + V \cdot \cos\psi_p - T \cdot \cos(\psi_T + \psi_p)}$$

(16.1)

$$\text{Model II:}\quad FS = \frac{c \cdot A + (W \cdot (\cos\psi_p - k_H \cdot \sin\psi_p) - U + T \cdot \sin(\psi_T + \psi_p)) \cdot \tan\phi}{W \cdot (\sin\psi_p + k_H \cdot \cos\psi_p) - T \cdot \cos(\psi_T + \psi_p)}$$

(16.2)

where T is the total reinforcing force (kN/m) applied by the anchors, and ψ_T is the plunge or inclination of this force below the horizontal (Figure 16.5a).

Figure 16.6 compares the different methods that were considered for increasing the stability of the overall cut. In each case, the change is expressed as a percentage of the total range of each variable: $H = 60$ m (200 ft), $\psi_f = 50°$, $z_w/z = 1$, $H_W = 60$ m (200 ft). However, the variation of the reinforcing force is expressed as a percentage of the weight of the wedge of rock being supported. In calculating the effect of the reinforcement, it was assumed that the anchors are installed horizontally, that is, $\psi_T = 0°$. The influence of the anchor inclination ψ_T on the reinforcing load required to produce a factor of safety of 1.5 is shown in Figure 16.5b. This shows that the required bolting force can be approximately halved by installing the bolts horizontally ($\psi_T = 0°$, or $(\psi_T + \psi_p) = 35°$), rather than normal to the plane ($\psi_T = 55°$, or $(\psi_T + \psi_p) = 90°$). As discussed in Section 7.5.1, the generally optimum angle for tensioned rock anchors is given by Equation 7.26. In practice, cement grouted anchors are installed at about 10–15° below the horizontal to facilitate grouting.

16.2.6 Stabilisation options

The following is a discussion on the stabilisation options analysed in Figure 16.6.

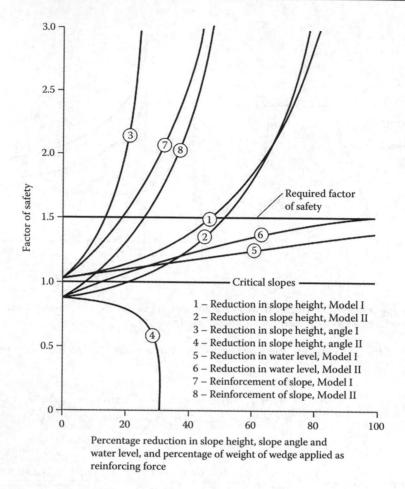

Figure 16.6 Case study 1 – comparison between alternative methods of increasing stability of overall slope.

Reduce height – curves 1 and 2 show that reduction in cut height is not an effective solution to the problem. In order to achieve the required factor of safety of 1.5, the slope height would have to be reduced by 50%. If this solution were to be adopted, it would be more practical to excavate the entire slope since most of the volume of rock to be excavated is contained in the upper half of the slope.

Reduce face angle – reducing the angle of the cut face would be an effective stabilisation measure as shown by line 3. The required factor of safety of 1.5 is achieved for a reduction of less than 25% of the slope angle. That is, the slope angle should be reduced from 50° to 37.5°. This finding is generally true and a reduction in the face angle is often an effective remedial measure. In the case of slopes under construction, using a flatter slope is always a prime choice for improving stability. However, a practical consideration is the difficulty of excavating a 'sliver' cut because of the limited access for equipment on the narrow, lower part of the cut.

Curve 4 (reduction of face angle for slope without tension crack) is an anomaly and demonstrates that calculations can sometimes produce unrealistic results. The reduction in factor of safety shown by this curve is a result of the reduction in the weight of the sliding block as the face angle is reduced. Since the water pressure on the sliding surface remains constant, the effective stress acting on the sliding surface decreases and hence the frictional component of the resisting force decreases. When a very thin sliver of rock remains, the water pressure will 'float' it off the slope. The problem with this analysis lies in the assumption that the block is completely impermeable and that the water remains trapped beneath the sliding surface. In fact, the block would break up long before it floated and hence the water pressure acting on the sliding surface would be dissipated.

Drainage – curves 5 and 6 show that drainage would not be an effective stabilisation option for both slope models. In neither case is a factor of safety of 1.5 achieved. This is something of a surprise since drainage is usually one of the most cost-effective drainage measures. The reasons for the poor performance of drainage in this case are due to the combination of the slope geometry and the shear strength of the failure surface.

Anchoring – curves 7 and 8 show that reinforcing the cut by means of tensioned anchors with a force equal to 5000 kN/m (39 kips/ft) of slope length would achieve a factor of safety of 1.5, assuming the anchors are installed just below the horizontal. In other words, reinforcement of an 100 m (330 ft) length of slope would require the installation of 500 anchors, each with a capacity of 1 MN (225 kips).

The two most attractive options for long-term remediation were reinforcement using tensioned cables or bar anchors, or reduction of the slope face angle. Reinforcement was rejected because of the high cost, and the uncertainty of long-term corrosion resistance of the steel anchors. The option finally selected was to reduce the face angle to 35° by excavating the entire block down to the sheet joints forming the sliding surface. This effectively removed the problem. Since good-quality aggregate is always needed in Hong Kong, it was decided to work the slope as a quarry. It took several years to organise this activity and during this time, the water levels in the slope were monitored with piezometers. Although the road was closed twice during this period, no major problems occurred and the slope was finally excavated back to the sliding plane.

16.3 CASE STUDY 2 – CABLE ANCHORING PLANE FAILURE

16.3.1 Site description

A potential rock fall hazard had developed on a steep, 38 m (125 ft) high-rock face located above a highway (Figure 16.7), and a stabilisation programme was required that could be implemented with minimal interruption to traffic. The rock falls could range from substantial failures with volumes of several hundred cubic metres formed by widely spaced, persistent joints, to falls of crushed and fractured rock with dimensions of tens of centimetres. The rock falls were a hazard to traffic because of the limited sight distance, and the 2 m (6 ft) wide ditch between the toe of the rock cut and the highway shoulder had limited capacity to contain rock falls (see Figure 14.22).

The evidence of instability was a series of tension cracks that had opened as the result of down-slope movement on the sheet joints dipping out of the face; this movement had created a layer of crushed rock on each of the sheet joints.

A survey of the slope, together with structural geological mapping of the face, was used to determine the shape and dimensions of the major blocks, and the orientations of major joint sets. It was decided that diamond drilling would not be required to obtain geological data because of the excellent exposure of the rock in the face.

16.3.2 Geology

The cut had been excavated in a very strong, course-grained, fresh, blocky granite containing four sets of joints that were generally orthogonally oriented. The predominant structural geology features were two sets of joints dipping out of the face – sheet joints dipping at an

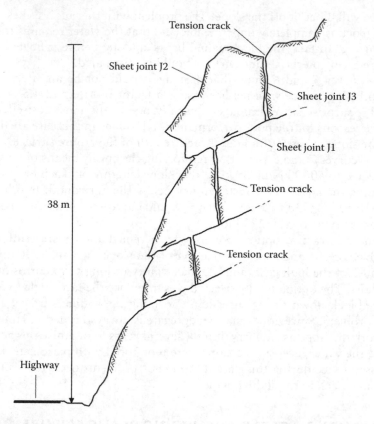

Figure 16.7 Case study 2 – cross-section of slope showing movement along sheet joints and location of tension cracks.

angle of about 20–30° (J1), and a second set (J2) dipping at 45–65°. A third set (J3) dipped steeply into the face, and a fourth set (J4) also steeply dipping, was oriented at right angles to the face. Figures 16.8a and b show, respectively, the contoured and great circle stereoplots of the surface-mapping data. The sheet joints had continuous lengths of several tens of metres, and the spacings of all the joints were in the range of 5–8 m (16–26 ft).

The blocks were sliding on the sheet joints (J1) with joint set J3 forming a series of tension cracks behind the face, while set J4 formed release surfaces at the sides of the blocks (Figure 16.7). The mapping located tension cracks that were up to 150 mm (6 in.) wide, as well as sheared rock on the sheet joints at the base of the each major block. The substantial volumes of the blocks were the result of the persistence and wide spacing of the discontinuities.

Since the dip directions of the slope face and the sheet joints were within 20° of each other, this geometry formed a plane failure as illustrated in Figures 2.16 and 2.17.

16.3.3 Rock shear strength

The blocks shown in Figure 16.7 were sliding on the continuous sheet joints because the rock was very strong, and fracturing through intact rock was not possible. Therefore, the stability of the blocks was partially dependent on the shear strength of the sheet joints, which were smooth and slightly undulating, and the only infilling was slight weathering of the surfaces. For these conditions, the cohesion was zero and the shear strength comprised only friction.

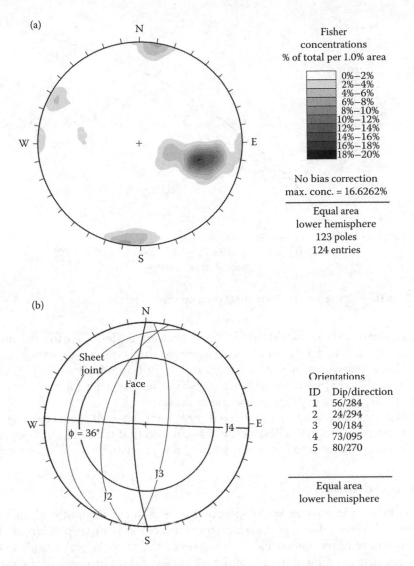

Figure 16.8 Case study 2 – stereonet of structural geology of slope: (a) contoured plot showing pole concentrations; (b) great circles of joint sets, slope face and friction circle.

The total friction angle of the sheet joints, comprising the friction (ϕ_r) of the granite and surface roughness (*i*), was determined as follows. Block samples of rock containing open, but undisturbed, sheet joints were collected at the site and 100 mm (4 in.) square test pieces were cut from the block. Direct shear tests were then conducted to determine the friction angle of the granite; each sample was tested four times using normal stresses σ of 150, 300, 400 and 600 kPa (21.8, 43.6, 58, 87.2 psi) (refer Figure 5.16). These normal stress test values were based on the likely stress acting on the lower sheet joint, calculated from

$$\sigma = \gamma_r \cdot H \cdot \cos \psi_p \qquad (16.3)$$

For a rock unit weight (γ_r) of 26 kN/m³ (165 lb/ft³), a rock thickness (*H*) of 25 m (82 ft) on a plane dipping (ψ_p) at 25°, the normal stress is 600 kPa (87.2 psi).

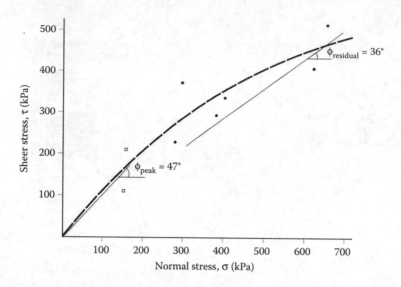

Figure 16.9 Case study 2 – results of direct shear tests on sheet joints in granite.

The direct shear tests showed that the peak friction angle (ϕ_{peak}) for the initial test at $\sigma = 150$ kPa (21.8 psi) was 47°, but at higher stresses, this diminished as the result of shearing on the sliding surface to a residual angle ($\phi_{residual}$) of 36° (Figure 16.9). Because of the displacement that had already taken place along the *in situ* sheet joints, it was considered appropriate to use the residual friction angle in design.

Examination of the rock along the sheared sheet joints showed that the rock was crushed with little intact rock contact. Therefore, it was decided that it would not be appropriate to add a roughness component to friction angle, and a friction angle of 36° was used in design.

16.3.4 Ground water

The site was located in an area with high precipitation in the form of both rain and snow. The level of the water table was assessed from observations of seepage from sheet joints exposed at the base of the lowest block. It was expected that the water table would generally be low because the dilated nature of the rock mass would promote drainage. However, during heavy precipitation events, it was likely that high, transient water pressures would develop and this was accounted for in the design.

It was assumed for design that water would accumulate in the tension crack to depth z_w, and that water forces would be generated both in the tension crack (V) and along the sliding plane (U) (Figure 16.10).

16.3.5 Earthquakes

The site was located in a seismically active area where the design horizontal seismic coefficient was $k_H = 0.15 \cdot g$. This coefficient was used in the pseudo-static stability analysis of the slope.

16.3.6 Stability analysis

The nominal static factor of safety of individual blocks sliding on the sheet joints ($\phi = 36°$) dipping at 25° was about ($FS = \tan \Phi / \tan \psi_p = \tan 36 / \tan 25 = 1.56$). However, the shear

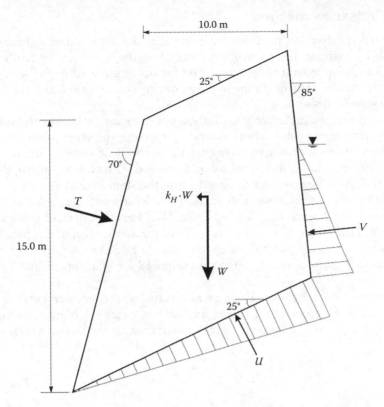

Figure 16.10 Case study 2 – cross-section of block used in design to model the assemblage of rock blocks.

movement along the sheet joints and the corresponding pattern of tension cracks behind the face shown in Figure 16.7 indicated that, under certain conditions, the factor of safety had diminished to approximately 1.0. It was considered that the cause of the movement was a combination of water pressures and ice jacking on the joints, seismic ground motions over geologic time and blast damage during construction. Also, movement could have been progressive over time, and as shearing occurred, crushing of rock asperities along the sliding surfaces reduced the friction angle.

The stability of the sliding blocks was studied using a plane stability model in which it was assumed that the cross-section was uniform at right angles to the slope face, and that sliding took place on a single plane dipping out of the face. In order to apply this model to the actual slope, a simplifying assumption was made in which the three blocks were replaced by a single equivalent block that had the same weight as the total of the three blocks and the same stability characteristics.

The shape and dimensions of the equivalent single block were defined by the following parameters (Figure 16.10):

Sliding plane, dip $\psi_p = 25°$; tension crack, dip $\psi_t = 85°$; slope face, dip $\psi_f = 70°$; upper slope, dip $\psi_s = 25°$; height of face, $H = 18$ m (60 ft); distance of tension crack behind crest, $b = 10$ m (33 ft). For this slope geometry, the depth of the tension crack is 13 m (43 ft).

Stability analysis of this block showed that the factor of safety was approximately 1.0 when the water in the tension crack was about 2.5 m (8.2 ft) deep, and a pseudo-static seismic coefficient of $0.15 \cdot g$ was applied. The static factor of safety for these conditions was 1.44, and reduced to 1.17 when the water level in the tension crack was 50% of the crack depth ($z_w = 6.5$ m (21.3 ft)).

16.3.7 Stabilisation method

Two alternative stabilisation methods were considered for the slope: either to remove the unstable rock by blasting and then, if necessary bolt the new face, or reinforce the existing slope by installing tensioned anchors. The factors considered in the selection were the requirement to maintain traffic on the highway during construction, and the long-term reliability of the stabilised slope.

It was decided that the preferred stabilisation option was to reinforce the slope by installing a series of tensioned rock anchors extending through the sheet joints into sound rock. The advantages of this alternative were that the method of stability analysis was clearly defined, and that the anchor installation could proceed with minimal disruption to traffic.

The rock anchoring system was designed using the slope model shown in Figure 16.10. For static conditions and the tension crack half filled with water ($z_w = 6.5$ m (21 ft)), it was calculated that an anchoring force of 550 kN/m (37.7 kips/ft) length of slope was necessary to increase the static factor of safety to 1.69. With the application of the pseudo-static seismic coefficient, the factor of safety was approximately 1.1. The anchors were installed at an angle of 15° below the horizontal, which was required for efficient drilling and grouting of the anchors.

The arrangement of anchors on the face was dictated by the requirements to reinforce each of the three blocks, to intersect the sheet joints and to locate the bond zone for the anchors in sound rock (Figure 16.11). Because of the limited area on the face in which anchors could

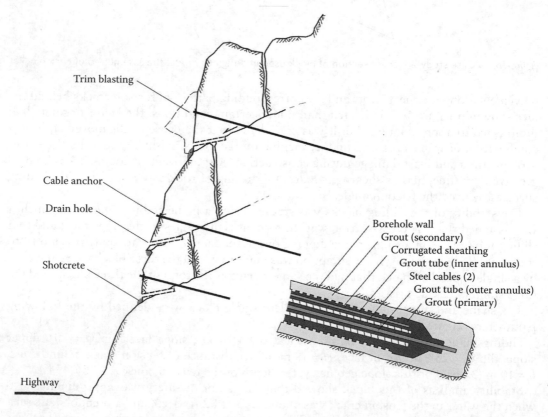

Figure 16.11 Case study 2 – cross-section of stabilised slope showing layout of cable anchors, and the trim blast, shotcrete and drain holes; detail shows lower end of cable anchors with arrangement of grout tubes.

be installed, it was necessary to minimise the number of anchors. This was achieved using steel strand cables, because of their higher tensile strength compared to rigid bars. A further advantage of the cables was that they could be installed in a hole drilled with a light drill that would be set up on the slope without the support of a heavy crane that would block traffic. Also, the installation would be facilitated because cable bundles were lighter than bars, and could be installed as a single length without the use of couplings.

The details of the anchor design that met these design and construction requirements were as follows: Working tensile load of two-strand, 15 mm (0.6 in.) diameter, seven-wire strand anchor at 50% of ultimate tensile strength = 248 kN.

For three rows of anchors arranged as shown in Figure 16.11, the total support force, ($T = 3 \cdot 248 = 744$ kN) (167 kips). Therefore, the required horizontal spacing between the vertical rows (Equation 7.27):

$$\text{Spacing} = \frac{\text{Supplied anchor force by three rows of anchors}}{\text{Required anchor force for factor of safety of 1.5}}$$
$$= \frac{744 \, \text{kN}}{550 \, \text{kN/m}}$$
$$\approx 1.5 \, \text{m}$$

The required bond length for the anchors was calculated using Equation 14.8 for a drill hole diameter of 80 mm, an allowable shear strength of the rock–grout bond of 1000 kPa (145 psi) (see Table 14.8) and an anchor tension of 248 kN. The calculated bond length was 1 m (3 ft), but the actual bond length used was 2 m (6 ft) to allow for loss of grout in fracture rock in the bond zones, and to ensure that the steel–grout bond strength was not exceeded.

In addition to the cable anchors, which were required to prevent large-scale instability, stabilisation measures to minimise the risk of surficial rock falls that could be a hazard to traffic included hand scaling of loose rock, shotcreting of fractured zones, trim blasting of an overhang and drilling drain holes (Figure 16.11).

16.3.8 Construction issues

The following is a brief description of a number of issues that were addressed during construction to accommodate the site conditions actually encountered:

- Drilling was carried out with a down-the-hole hammer drill, without the use of casing.
- The thrust and rotation components for the drill were mounted on a frame that was bolted to the rock face, with a crane only being used to move the equipment between holes. This arrangement allowed drilling to proceed with minimal disruption to highway traffic.
- Grouting of the anchor holes to the surface was not possible because the grout often flowed into open fractures behind the face. In order to ensure that the 2 m (6 ft) long bond zones were fully grouted, the lower portion of each hole was filled with water and a well sounder was used to monitor the water level. Where seepage into fractures occurred, the holes were sealed with cement grout and then redrilled, following which a further water test was carried out.
- Corrosion protection of the anchors was provided with a corrugated plastic sheath that encased the steel cables, with cement grout filling the annular spaces inside and outside the sheath. In order to facilitate handling of the cable assemblies on the steep

rock face, the grouting was only carried out once the anchors had been installed in the hole. This involved two grout tubes and a two-stage grouting process as follows. First, grout (primary) was pumped down the tube contained within the plastic sheath to fill the sheath and encapsulate the cables. Second, grout (secondary) was pumped down the tube sealed into the end cap of the sheath to fill the annular space between the sheath and the borehole wall.

- Testing of the anchors to check the load capacity of the bond zone was carried out using the procedures discussed in Section 14.4.2, *Step 9 – Testing* (Post Tensioning Institute, 2006).

16.4 CASE STUDY 3 – STABILITY OF WEDGE IN BRIDGE ABUTMENT

16.4.1 Site description

This case study describes the stability analysis of a bridge abutment in which the geological structure formed a wedge in the steep rock face on which the abutment was founded (Figure 16.12). The analysis involved defining the shape and dimensions of the wedge, the shear strength of the two sliding planes, and the magnitude and orientation of a number of external forces. The stability of the wedge was examined under a combination of load conditions, and the anchoring force was calculated to produce a factor of safety against sliding of at least 1.5.

The site was located in an area subject to both high precipitation and seismic ground motions. The bridge was a tensioned cable structure with the cables attached to a concrete

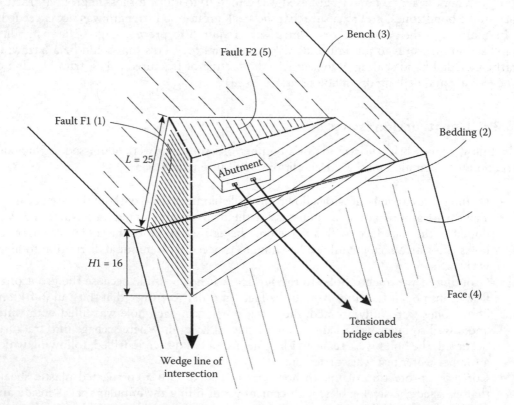

Figure 16.12 Case study 3 – view of wedge in bridge abutment showing five planes forming the wedge.

reaction block located on a bench cut into the rock face. The cables exerted an outward force on the abutment (15° below the horizontal) along the axis of the bridge. The structural geology of the site comprised bedding and two sets of faults that together formed wedge-shaped blocks in the slope below the abutment. The stability of the slope was examined using the wedge stability analysis method to determine the static and seismic factors of safety, with and without rock anchors. Figure 16.12 is a sketch of the abutment showing the shape of the wedge and the orientation of the bridge (Q) force. The anchors were installed in the upper surface of the abutment, inclined at an angle of 45° below the horizontal, and oriented 180° from the direction of the line of intersection. In Figure 16.12, the five planes forming the wedge are numbered according to the system shown in Figure 8.18a.

16.4.2 Geology

The rock was slightly weathered, strong, massive sandstone with the bedding dipping at an angle of 22° to the west (orientation 22/270). The site investigation identified a persistent bedding plane at a depth of 16 m (52 ft) below the bench level that contained a weak shale interbed (plane 1). This plane (plane 1) formed the flatter of the two sliding planes forming the wedge block. The slope also contained two sets of faults with orientations 80/150 (F1) and 85/055 (F2). The faults were planar and contained crushed rock and fault gouge, and were likely to have continuous lengths of tens of metres. Fault F1 formed the second sliding plane (plane 2), on the left side of the wedge (Figure 16.12). Fault F2 formed the tension crack at the back of the wedge (plane 5), and was located at a distance of 25 m (82 ft) behind the slope crest, measured along the outcrop of fault F1.

Figure 16.13 is a stereonet showing the orientations of the great circles of the three discontinuity sets, and the slope face (orientation 78/220), and upper bench (orientation 02/230).

16.4.3 Rock strength

The stability analysis required shear-strength values for both the F1 fault and the bedding. The fault was likely to be a continuous plane over the length of the wedge, for which the

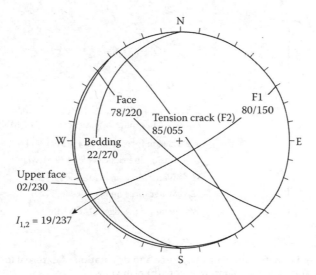

Figure 16.13 Case study 3 – stereonet of five planes forming wedge in bridge abutment shown in Figure 16.12.

shear strength of the crushed rock and gouge would comprise friction with no significant cohesion. The shear strength of the bedding plane was that of the shale interbed. The shear strength of both materials was determined by laboratory testing using a direct shear test machine (see Figure 5.16).

The direct shear tests carried out on fault infilling showed friction angles averaging 25° with zero cohesion, and for the shale the friction angle was 20° and the cohesion was 50 kPa (7.3 psi). Although both the fault and the bedding were undulating, it was considered that the effective roughness of these surfaces would not be incorporated in the friction angle because shearing was likely to take place entirely within the weaker infilling, and not on the rock surfaces.

16.4.4 Ground water

This area was subject to periods of intense rain that was likely to flood the bench at the crest of the slope. Based on these conditions, it was assumed for the analysis that maximum water pressures would be developed on the planes forming the wedge.

16.4.5 Seismicity

The seismic coefficient for the site was $k_H = 0.1 \cdot g$. The stability analysis used the pseudo-static method in which the product of the seismic coefficient, the gravity acceleration and the weight of the wedge was assumed to produce a horizontal force acting out of the slope along the line of intersection of the wedge.

16.4.6 External forces

The external forces acting on the wedge comprised water forces on planes 1, 2 and 5, the seismic force, the bridge load and the rock anchors. Figure 16.14 shows the external forces in plan and section views.

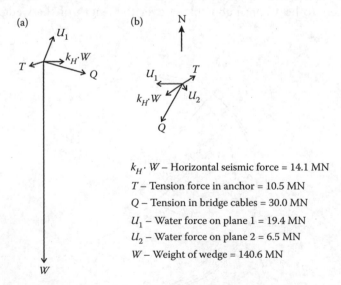

$k_H \cdot W$ – Horizontal seismic force = 14.1 MN

T – Tension force in anchor = 10.5 MN

Q – Tension in bridge cables = 30.0 MN

U_1 – Water force on plane 1 = 19.4 MN

U_2 – Water force on plane 2 = 6.5 MN

W – Weight of wedge = 140.6 MN

Figure 16.14 Case study 3 – sketch showing magnitude and orientation of external forces on wedge: (a) section view along the line of intersection; (b) plan view.

The water forces were the product of the areas of planes 1 and 2 and the water pressure distribution. The seismic force was the product of the horizontal seismic coefficient and the weight of the wedge. The analysis procedure was to run the stability analysis to determine the weight of the wedge (volume multiplied by rock unit weight), from which the seismic force was calculated.

For the bridge, the structural load on the abutment due to the tensioned cables had a magnitude of 30 MN, and trend and plunge values of 210° and 15°, respectively. The trend coincided with the bridge axis that was not at right angles to the rock face, and the plunge coincided with the sag angle of the catenary created by the sag in the cables.

The rock anchors were installed in the upper surface of the bench and extended through the bedding plane into stable rock to apply normal and shear (up-dip) forces to the bedding plane.

16.4.7 Stability analysis

The stability of the abutment was analysed using the comprehensive wedge analysis procedure described in Appendix III, and the computer program SWEDGE version 4.01 by Rocscience Inc. The input data required for this analysis comprised the shape and dimensions of the wedge, the rock properties and the external forces acting on the wedge. Values of these input parameters, and the calculated results, were as follows:

1. *Wedge shape and dimensions:* The shape of the wedge was defined by five surfaces with orientations as shown in Figure 16.13:

Plane 1 (bedding):	22°/270°
Plane 2 (fault F1):	80°/150°
Plane 3 (upper slope):	02°/230°
Plane 4 (face):	78°/220°
Plane 5 (tension crack, fault F2):	85°/055°

    The orientation of the line of intersection between planes 1 and 2 was calculated to be

    Line of intersection: 18.6°/237°

    The dimensions of the wedge were defined by two length parameters:

    Height, $H1$ (vertical height from line of intersection to crest):   16 m
    Length, L (length along plane 1 from crest to tension crack):   25 m

2. *Rock properties:* The rock properties comprised the shear strengths of planes 1 and 2, and the rock unit weight:

Bedding with shale interbed:	$c_1 = 50$ kPa (7.3 psi), $\phi_1 = 20°$
Fault F1:	$c_2 = 0$, $\phi_2 = 35°$
Unit weight of rock:	$\gamma_r = 0.026$ MN/m³ (165 lb/ft³)
Unit weight of water:	$\gamma_w = 0.01$ MN/m³ (62.4 lb/ft³)

3. *External forces:* The magnitude and orientation of the external forces were as follows. Water forces acted normal to each plane and were calculated to have the following values, for fully saturated conditions:

$U_1 = 18.8$ MN

$U_2 = 6.13$ MN

$U_5 = 1.49$ MN

The wedge weight acted vertically and was calculated (from the wedge volume and the rock unit weight) to have magnitude:

$W = 141$ MN

The horizontal component of the seismic force acted in the direction along the line of intersection and had magnitude:

$k_H \cdot W = 0.1 \cdot W$
$\qquad = 14.1$ MN oriented at $0°/237°$

The bridge force Q acted along the centre line of the bridge at an angle of $15°$ below the horizontal:

$Q = 30$ MN oriented at $15°/210°$

The factor of safety of the abutment with no reinforcement provided by tensioned anchors was as follows:

1. $FS = 2.58$ – dry, static, $Q = 0$
2. $FS = 2.25$ – saturated, static, $Q = 0$
3. $FS = 1.73$ – saturated, $k_H = (0.1 \cdot g)$, $Q = 0$
4. $FS = 1.32$ – saturated, static, $Q = 30$ MN
5. $FS = 1.10$ – saturated, $k_H = (0.1 \cdot g)$, $Q = 30$ MN

It was considered that the factors of safety for load conditions (4) and (5) were inadequate for a structure critical to the operation of the facility, and that the minimum required static and seismic factors of safety should be 1.5 and 1.25, respectively. These factors of safety were achieved, with the bridge load applied, by the installation of tensioned anchors (tension load T), which gave the following results:

1. $FS = 1.54$ – saturated, static, $T = 10.5$ MN, $\psi_T = 15°$, $\alpha_T = 056°$ (parallel to the line of intersection)
2. $FS = 1.26$ – saturated, $k_H = (0.1 \cdot g)$, $T = 10.5$ MN, $\psi_T = 15°$, $\alpha_T = 056°$

It was found that the factor of safety for the reinforced wedge could be optimised by varying the orientation of the anchors. If the trend of the anchors was between the trends of the line of intersection and the bridge load (i.e. $\alpha_T = 035°$), it was possible to reduce the anchor force required to achieve the required factor of safety to 8.75 MN.

It is noted that the discussion in this case study only addressed the stability of the wedge, and did not discuss the method of attaching the tensioned bridge cables to the rock wedge. Also, it was assumed that all the external forces acted through the centre of gravity of the wedge so that no moments were generated.

16.5 CASE STUDY 4 – STABILISATION OF TOPPLING FAILURE

16.5.1 Site description

A rock slope above a railway was about 25 m (82 ft) high, and the rock forming the slope was a blocky granite in which a toppling failure was occurring (Wyllie, 1980). The movement of the upper toppling block was crushing the rock at the base and causing rock falls that were a hazard to railway operations (Figures 16.15 and 16.16). The site was in a high-precipitation climate, with a moderate risk of seismic ground motions. Stabilisation measures were undertaken to limit the rock fall hazards, and to prevent additional toppling motion.

16.5.2 Geology

The granite at the site was fresh and very strong, and contained three well-defined sets of joints with orthogonal orientations. The most prominent set (J1) dipped at about 70°, with the strike at right angles to the railway alignment. The second set (J2) had the same strike

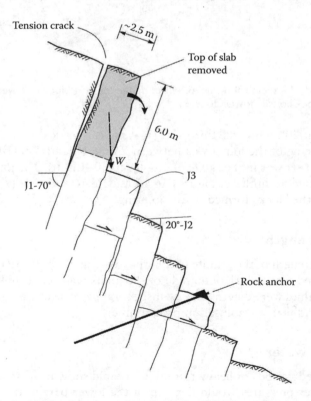

Figure 16.15 Case study 4 – idealised configuration of toppling slabs showing excavation and bolting.

Figure 16.16 Case study 4 – sketch showing extent of upper toppling block removed by blasting, and location of rock bolts in lower slope.

but dipped at about 20°, while the third set (J3) was near vertical with the strike parallel to the track. The spacing of the joints was between 2 m (6 ft) and 3 m (10 ft), and the persistence of the J1 joint set was in the range of 10–40 m (30–130 ft). The joints were planar but rough, and contained no infilling. Figures 16.15 and 16.16 show sketches of the slope and the dimensions of the blocks formed by the jointing.

16.5.3 Rock strength

The compressive strength of the granite was in the range of 50–100 MPa (7250–14,500 psi), and it was estimated that the friction angle of the joints was between 40° and 45° with no cohesion. These values were determined by inspection because of the limited time available to assess the site and plan a stabilisation programme.

16.5.4 Ground water

The site experienced periods of heavy rainfall and rapid snow melt, so it was expected that transient high-water pressures would develop in the lower part of the slope. In the upper part of the slope, it was unlikely that water pressures would occur because water would not collect in the open tension cracks exposed in the face.

16.5.5 Stability conditions

The uniform spacing and orientation of the J1 joints formed a series of slabs in the slope that were approximately 2.5 m (8 ft) wide and had vertical heights of as much as 20 m (66 ft). The slabs dipped at about 70° so the centre of gravity of the slabs lay outside the base when the height exceeded about 6 m; this was a necessary condition for toppling (see Figure 1.11). As shown in Figure 16.15, the upper slab had an exposed face about 7 m (23 ft) high and toppling of this slab had opened a tension crack about 200 mm (8 in.) wide along the J1 joint set. As the upper block toppled, it generated thrust forces on the lower slabs. The short length of these lower slabs meant that their centres of gravity were well inside their bases so toppling did not occur. However, the thrust was great enough to cause the lower blocks to slide on the J2 joint set. This set dipped at 20° and had a friction angle of about 40°; limit equilibrium analysis of the sliding blocks showed that the thrust force required to cause sliding was equal to about 50% of the weight of the block. This shear displacement caused some fracturing and crushing of the rock that was the source of the rock falls.

The mechanism of instability at the site was essentially identical to the theoretical toppling mechanism discussed in Chapter 10 and shown in Figure 10.8. That is, the tall, upper slabs toppled and caused the lower, shorter slabs to slide. Possible stabilisation options for these conditions included reducing the height of the toppling slabs so that the centre of gravity lay inside the base, or installing a support force in the sliding slabs at the base. These two measures were adopted, with the combined effect of reducing the tendency for the upper slabs to topple, and preventing movement of the lower slabs.

16.5.6 Stabilisation method

The following three stabilisation measures were undertaken to reduce the rock fall hazard and improve the long-term stability of the slope:

- Hand scaling was carried out on the face above the railway to remove loose rock. This work included the removal of all trees growing in open cracks in the rock because these had contributed to the loosening of the blocks of rock on the face.
- A row of bolts was installed through one of the lower slabs. This work was done prior to excavation at the crest in order to prevent any further movement due to blasting vibrations.
- Blasting was used to remove the upper 6 m (20 ft) of the top slab. The blasting was carried out in stages in order to limit blast vibrations in the lower slope and allow additional bolts to be installed if further movement occurred. The blasting pattern comprised 6 m (20 ft) long holes on about 0.6 m (2 ft) centres, with three rows being detonated on each blast. A light explosive charge of 0.4 kg/m^3 (0.67 lb/yd^3) was used by placing spacers between the sticks of explosive in the blast holes.

16.6 CASE STUDY 5 – CIRCULAR FAILURE ANALYSIS OF EXCAVATION FOR ROCK FALL DITCH

16.6.1 Site description

As the result of a series of rock falls from a rock face above a railway, a programme was undertaken to improve stability conditions (Figure 16.17). The initial stabilisation work involved selective hand scaling of loose rock and bolt installation, but it was found that this only provided an improvement for one or two years before new rock falls occurred as

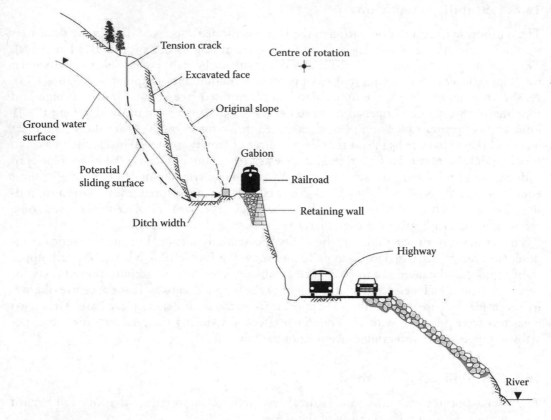

Figure 16.17 Case study 5 – geometry of slope above railway. Sketch shows dimensions of ditch after excavation of slope, and shape of potential circular sliding surface through the rock mass.

the rock weathered and blocks loosened on joint surfaces. Rock falls were a potential hazard because the curved alignment and 2 km stopping distance meant that trains could not be brought to a halt if a rock fall was observed. In order to provide long-term protection against rock falls, it was decided to excavate the face to create a ditch that was wide enough to contain substantial falls from the new face. This work involved a drilling and blasting operation to cut back the slope to a face angle of 75°, and constructing a gabion wall along the outer edge of the ditch that acted as an energy-absorbing barrier to contain rock falls (Wyllie and Wood, 1981).

The railway and highway were located on benches cut into a rock slope above a river, with steep rock faces above and below the upper bench on which the railway was located; a 30 m (100 ft) length of the track was supported by a masonry retaining wall (Figure 16.17). The original cut above the railway was about 30 m (100 ft) high at a face angle of 60°, and the 2 m (6.6 ft) wide ditch at the toe of the slope was not adequate to contain rock falls (see Figure 14.22). Blasting used to excavate the original slope had caused moderate damage to the rock behind the face.

16.6.2 Geology

The cut was in medium strong, slightly to moderately weathered volcanic tuff containing joints spaced at about 0.5–2 m (1.5–6 ft), and lengths up to 3 m (10 ft). A consistent set of joints had a near vertical dip and a strike at about 45° to the strike of the cut face. However, the orientations of the other joints were variable over short distances. Many of the joints had calcite infillings that had a low-cohesive strength.

Because of the variable orientations and limited persistence of the joints throughout the length of the cut, structurally controlled instability of the overall rock slope was limited.

16.6.3 Ground water

Because of the low precipitation in the area, it was assumed that the ground water level would have little influence on overall slope stability. The assumed ground water table is shown in Figure 16.17.

16.6.4 Rock shear strength

An important design issue for the project was the stability of the overall cut face above the railway, and whether it could be cut back safely to create a rock fall catch ditch. The rock strength relevant to this design was that of the rock mass because potential failure surfaces would pass partially through intact rock, and partially along any low-persistence joints oriented approximately parallel to this surface. It was not possible to either test samples with diameters of several metres that would be representative of the rock mass, or to determine the proportions of intact rock and joint plane that would form the sliding surface in the slope. Therefore, two empirical methods as described below were used to estimate the cohesion and friction angle of the rock mass.

16.6.5 Circular stability analysis

Because no geological structure existed that would form a sliding surface, it was likely that instability would take the form of a shallow circular failure, as described in Chapter 9 and shown in Figure 16.17.

The first method of estimating the rock mass strength was to carry out a back analysis of the existing 30 m (100 ft) high cut above the railway, which involved the following steps. First, no evidence of instability was observed of the overall slope, which had been standing for over 100 years, or nearby natural slopes in the same rock type. Second, these slopes had probably been subject in the past to earthquakes and occasional periods of high-water pressure. Therefore, a factor of safety in the range of 1.5–2.0 was assumed for the existing slope. Third, as discussed in Section 16.4.3, the water table was in the lower part of the slope (see Figure 9.4), and it was appropriate to use chart number 2 to perform stability analyses (Figure 9.7). Fourth, for blocky rock with no significant clay on the joint surfaces, a friction angle of 35° was estimated; the rock unit weight was 26 kN/m^3. Using these data, for the 30 m (100 ft) high slope at a face angle of 60°, it was possible to use the circular failure design chart to calculate the rock mass cohesion as approximately 150 kPa (21.8 psi) (for $FS = 1.75$; $\tan \phi/FS = 0.40$; $c/(\gamma \cdot H \cdot FS) = 0.11$). Figure 5.21 was used as an additional guideline in selecting shear-strength values.

As a comparison with the back analysis method of determining rock mass strength, the Hoek–Brown strength criterion (see Section 5.5) was used to calculate a friction angle of 38° and a cohesion of about 180 kPa (26.1 psi) (input parameters: $\sigma_i = 40$ MPa (5800 psi)); $GSI = 45$; $m_i = 10$; $D = 0.9$ (poor blasting in original construction; slope height – 30 m) based on the program RocData version 4.014 (Rocscience Inc.).

The two sets of strength values are reasonably close, but the difference illustrates the uncertainty in determining rock mass strengths, and the need to carry out sensitivity analyses to evaluate the possible influence on this range in strengths on stability.

16.6.6 Ditch and slope design

The two principle design issues for the project were the dimensions of the ditch to contain rock falls, and the stability of the slope excavated to create the ditch.

Ditch – the required depth and width of the ditch to contain rock falls is related to both the height and slope angle of the cut face as illustrated in Figure 14.22 (Ritchie, 1963). These design recommendations show that the required ditch dimensions are reduced for a proposed face angle of 75°, compared to the existing 60° face. Another factor in the ditch design was the face angle of the outside face of the ditch. If this face is steep and constructed with energy-absorbing material, then rocks that land in the base of the ditch are likely to be contained. However, if the outer face has a gentle slope, they may roll out of the ditch.

For a 30 m (100 ft) high-rock face at an angle of 75°, the required ditch dimensions were a depth of 2 m (6.6 ft) and a base width of 7 m (23 ft). In order to reduce the excavation volume, the ditch was excavated to a depth of 1 m (3 ft), and a 2 m (6.6 ft) high gabion wall was placed along the outer side of the excavation to create a vertical, energy-absorbing barrier.

Slope stability – the stability of the excavated slope was examined using circular chart number 2 (see Figure 9.7). The proposed excavation would increase the face angle from 60° to 75° without increasing the height of 30 m (100 ft) significantly, and the rock mass strength and the ground water conditions in the new slope would be identical to those in the existing slope. Chart number 2 showed that the factor of safety of the new slope was about 1.3 ($c/(\gamma \cdot H \tan \phi) = 0.275$; $\tan \phi/FS \approx 0.2$). Figure 16.15 shows the approximate location of the potential tension crack, and sliding surface with the minimum factor of safety, determined using Figure 9.8 ($X = -0.9 \cdot H$; $Y = H$; $b/H = 0.15$).

16.6.7 Construction issues

The excavation was by drill and blast methods because the rock was too strong to be broken by rippers. The following are some of the issues that were addressed during construction:

The blasting was carried out in 4.6 m (15 ft) high lifts using vertical holes. The 'step-out' required at the start of each bench to allow clearance for the head of the drill was 1.2 m (4 ft), so the overall slope angle was 75°. The production holes were 63 mm (2-1/2 in) diameter on a 1.5 m (5 ft) square pattern and the powder factor was 0.3 kg/m³ (0.5 lb/yd³).

Controlled blasting was used on the final face to minimise the blast damage to the rock behind the face. The final line holes were spaced at 0.6 m (2 ft) and charged with decoupled, low-velocity explosive at a load factor of 0.3 kg/m (0.2 lb/ft) of hole length. The final line holes were detonated last in the sequence (cushion blasting) because the limited burden precluded pre-shear blasting (see Section 13.4).

The detonation sequence of the rows in the blast was at right angles to the face in order to limit the throw of blasted rock on to the railway and highway and minimise closure times.

The track was protected from the impact of falling rock by placing a 1 m (3 ft) thick layer of gravel on the track before each blast. This could be quickly removed to allow operations of the train.

Near the bottom of the cut, it was necessary to protect from blast damage the masonry-retaining wall supporting the track. This was achieved by controlling the explosive weight per delay so that the peak particle velocity of the vibrations in the wall did not exceed 100 mm/s (4 in/s) (see Section 13.5).

16.7 CASE STUDY 6 – PASSIVE REINFORCEMENT OF SLOPE IN CLOSELY FRACTURED ROCK MASS (BY NORMAN I. NORRISH, SEATTLE, WASHINGTON)

16.7.1 Site description

An existing slope, being cut back to widen a transportation route, required excavation of new cuts up to 28 m (90 ft) high at a face angle of 76°. The natural slope above the crest of the cut was at an angle up to 40°. A 76° cut face angle was required to limit the cut height in the steep terrain, and to minimise the width of the right of way required for construction (Figure 16.18).

For safe excavation of cut, and long-term stability, it was necessary to install support from the top down as the cut was excavated. This case study describes the design and installation of the fully grouted, untensioned steel bars (passive dowels) that were used to support the cut.

16.7.2 Geology

The project geology comprised lava and pyroclastic flows, forming basalt and tuff formations with significant variability of the rock composition over short distances. In addition, low-grade metamorphism and surficial weathering contributed to the variability of the rock

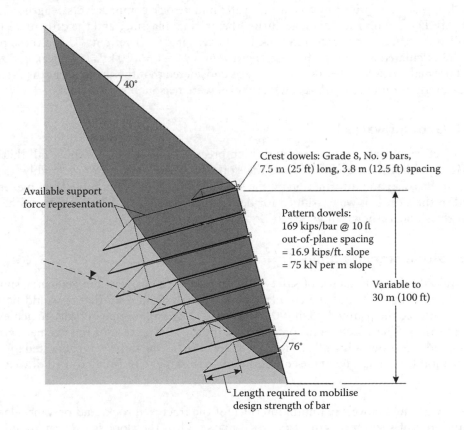

Figure 16.18 Case study 6 – slope section showing potential circular slip surface, and locations and depths of fully grouted, steel dowels.

characteristics. As is common in these volcanic formations, the discontinuities had low persistence and were generally randomly oriented. The rock was moderately to highly weathered (Grade I to II), and the intact rock was moderately to extremely weak. In drill core, the RQD was less than 50%.

16.7.3 Rock strength

The shear strength used in design was that of the rock mass comprising both fragments of intact rock and the randomly oriented, low-persistence joints. Because it was not practical to measure the strength of the rock mass in the laboratory, the shear strength was estimated from back analysis of the existing slopes that were assumed to be marginally stable, and by the Hoek–Brown strength criterion as described in Section 5.5.

The back analysis was carried out on a 15 m (50 ft) high slope at a face angle of 74° with a history of localised slope failures, where the factor of safety was estimated to be 1.06. Assuming a circular-type sliding surface, with the water table in the lower part of the slope, circular failure chart 2 (refer to Figure 9.7) was used for the back analysis. For a friction angle of this rock mass of 35°, the calculated cohesion was 45 kPa (6.5 psi); this strength is in reasonable agreement with the (c–ϕ) values related to the rock characteristics shown in Figure 5.21.

For the Hoek–Brown rock mass strength, the parameters were $GSI = 35$ for 'blocky/disturbed/seamy' rock and 'fair/poor' discontinuity surface condition; $m_i = 18$ for volcanic rock masses comprised of basalt and tuff; intact rock compressive strength = 21 MPa (3000 psi); $D = 0.9$ for rock damage during the original blasting, and the difficulty of using controlled blasting in the fractured rock to be excavated in sliver cuts. Using these parameters, the calculated rock mass shear-strength values were $c = 120$ kPa (17.4 psi), $\varphi = 32°$ at a typical normal stress level for the slope. It was considered that the shear strengths calculated by back analysis and the Hoek–Brown criterion were reasonably consistent.

16.7.4 Ground water

The project was located in an area with high precipitation and heavy snow fall that could result in rapid infiltration of water into the slope. However, because of the highly fractured nature of the rock, and information on the level of the ground water table from piezometers located in the slopes, it was assumed for design that the water table would be in the lower part of the slope as shown in Figure 16.18.

16.7.5 Slope stability analysis

The required minimum factor of safety for the excavated cut with the geometry shown in Figure 16.18 was 1.30, and the circular stability analysis showed that it would be necessary to reinforce the planned 28 m (90 ft) high, 76° face angle slope in order to achieve this factor of safety. The selected reinforcement method was a pattern of fully cement grouted, untensioned steel dowels installed, starting at the crest, as the slope was excavated in a series of horizontal lifts. The advantages of reinforcing the slope with fully grouted dowels are as follows:

- Fully grouted dowels maintain interlock of the fractured rock, and prevent relaxation and loss of shear strength of the rock mass. When the slope is excavated and strain occurs in the rock, the dowel capacity is mobilised only to the extent that the rock mass deforms. Such rock mass deformation tensions the dowels and mobilises their strength at the location required to resist the movement, provided that the dowels are installed as the excavation proceeds.

- The full support function of the dowels is maintained even if ravelling of rock occurs at the face under the plates. This is in contrast to end-anchored, tensioned bolts where loss of rock under the face plate results in the bolts being detensioned such that they provide no slope support.
- The cost of installation of fully grouted, untensioned dowels is less than that of tensioned rock bolts because of the simpler and more rapid installation procedure.

Fully grouted dowels are discussed in Section 7.4.2 (reinforcement with fully grouted untensioned dowels) and Section 14.4.2 (rock anchors).

The stability analysis was carried out using the program Slide (Rocscience Inc.), with the rock shear strength and ground water parameters discussed in the previous paragraphs. The fully grouted dowels were modelled as soil nails with a working bond strength of 310 kPa (45 psi). It was found that the required factor of safety was achieved with 45 mm (1.75 in.) diameter, solid, steel dowels installed in a pattern of 3.8 m (12.5 ft) vertical spacing, and 3 m (10 ft) horizontal spacing. These bolts had a yield load of 750 kN (169 kips), and a bond length of 6 m (20 ft) was required to develop the full yield load in the bar. The total dowel lengths ensured that the bond zone was beyond the potential critical sliding surface.

16.7.6 Construction method

The slope was excavated by drill and blast methods in 5 m (16 ft) lifts using trim blasting to minimise damage to the rock behind the face. Two rows of fully grouted dowels, with epoxy coating for corrosion protection, were installed in each lift. The drill holes were inclined at 15° below the horizontal to facilitate cement grouting, and the hole diameter was 125 mm (5 in.) to accommodate the bar together with the centering sleeves and grout tube.

In addition to the main pattern of dowels, 7.5 m (25 ft) long, 28 mm (1-1/4 in.) diameter dowels spaced at 4 m (13 ft) were installed along the crest before starting excavation to support the loose and weathered rock along the crest. This installation was an essential safety measure.

It was found that the drill holes for the dowels would stay open without casing so solid bars were used, with couplings as necessary to join bar segments. The lengths were suited to the standard lengths supplied by the mill to minimise wastage. The bars were centered in the drill holes with centering sleeves, and grout was placed upwards from the distal end of the bar through a grout tube attached to the bar.

After installing the dowels, the face was covered with a 100 mm (4 in.) thick layer of steel fibre-reinforced shotcrete to control surficial ravelling of the fractured rock. The plates and nuts on the exposed ends of the bolts were embedded in shotcrete to help hold the shotcrete on the steep face.

At the completion of the project, 25 m (75 ft) deep drain holes, spaced at 7.5 m (25 ft), were drilled in two rows at the base of the slope.

Appendix I: Stereonets for hand plotting of structural geology data

I.1 INTRODUCTION

Analysis of the orientation of structural geology data involves first, plotting poles representing the dip and dip direction of each discontinuity. This plot will help to identify discontinuity sets, for which both the average orientation and the scatter (dispersion) can be calculated. The second step in the analysis is to plot great circles representing the average orientation of each set, major discontinuities such as faults, and the dip and dip direction of the cut face. Hand plotting of structural data can be carried out on the stereonets provided in this appendix. Further details of the plotting procedures are provided in Chapter 2, Section 2.5.

I.2 PLOTTING POLES

Poles can be plotted on the polar stereonet (Figure I.1) on which the dip direction is indicated on the periphery of the circle, and the dip is measured along radial lines with zero degrees at the centre. It should be noted that the stereonet shown in Figure I.1 is the lower hemisphere plot in which the dip direction scale starts at the bottom of the circle and increases in a clockwise direction, with the north arrow corresponding to the dip direction of 180°. The reason for setting up the scale in this manner is that if the field readings, as measured with a structural compass, are plotted directly on the stereonet, the poles are correctly plotted on the lower hemisphere plot (see Section 2.5.2).

The procedure for plotting poles is to lay a sheet of tracing paper on the printed polar net and mark the north direction and each quadrant position around the edge of the outer circle. A mark is then made to show the pole that represents the orientation of each discontinuity as defined by its dip (radial distance from the centre) and dip direction (circumferential position). Poles for shallow dipping discontinuities lie close to the centre of the circle, and poles of steeply dipping discontinuities lie close to the periphery of the circle, while poles for planes dipping to the north lie on the lower half of the circle and poles for planes dipping to the south lie on the upper half of the circle.

I.3 CONTOURING POLE CONCENTRATIONS

Concentrations of pole orientations can be identified using a counting net such as that shown in Figure I.2. The Kalsbeek net is made up of mutually overlapping hexagons, each with an area of 1/100 of the full area of the stereonet. Contouring is performed by overlaying the counting net on the pole plot and counting the number of poles in each hexagon; this number is marked on the net. These numbers of poles are converted into percentages by dividing the number of poles in each hexagon by the total number of poles and multiplying by 100.

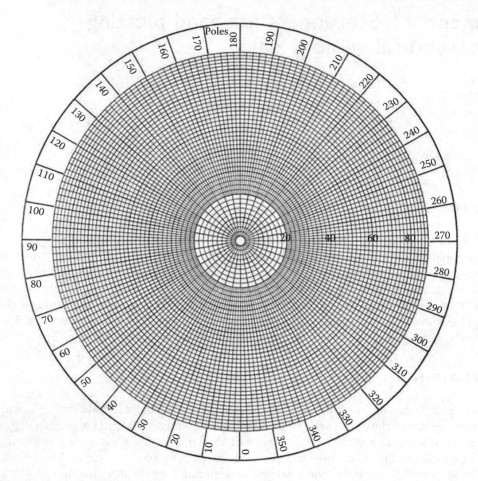

Figure I.1 Polar net for plotting poles to planes.

Contours of pole concentrations can be developed by interpolating between the percentage values in each hexagon (see Section 2.5.3).

I.4 PLOTTING GREAT CIRCLES

Great circles are plotted on the equatorial net (Figure I.3), but they cannot be plotted directly on this net because the true dip can only be scaled off the horizontal axis. The plotting procedure for great circles consists of the following steps (see Section 2.5.4):

1. Lay a piece of tracing paper on the net with a thumb tack through the centre point so that the tracing paper can be rotated on the net.
2. Mark the north direction of the net on the tracing paper.
3. Locate the dip direction of the plane on the scale around the circumference of the net and mark this point on the tracing paper. Note that the dip direction scale on the equatorial net for plotting great circles starts at the north point at the top of the circle and increases in a clockwise direction.
4. Rotate the tracing paper until the dip direction mark coincides with one of the horizontal axes of the net, that is, the 90° or 180° points of the dip direction scale.

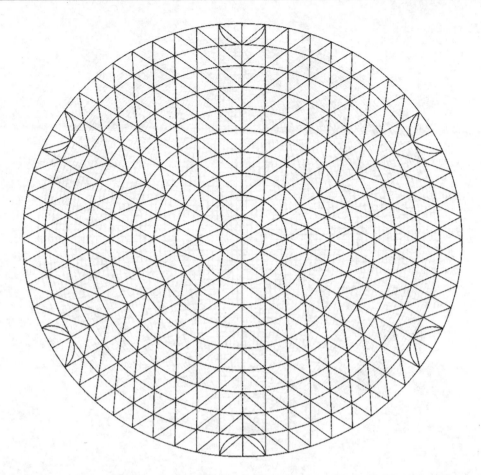

Figure I.2 Kalsbeek counting net for contouring pole concentrations.

5. Locate the arc on the net corresponding to the dip of the plane and trace this arc on to the paper. Note that a horizontal plane has a great circle at the circumference of the net, and a vertical plane is represented by a straight line passing through the centre of the net.
6. Rotate the tracing paper so that the two north points coincide and the great circle is oriented correctly.

I.5 LINES OF INTERSECTION

The intersection of two planes is a straight line that defines the direction of sliding of a wedge formed by these two planes. The procedure for determining the orientation of the line of intersection between two planes is as follows (see Section 2.5.5):

1. Locate the line of intersection between the two planes which is represented by the point at which the two great circles intersect.
2. Draw a line from the centre of the net through the point of intersection and extend it to the circumference of the net.

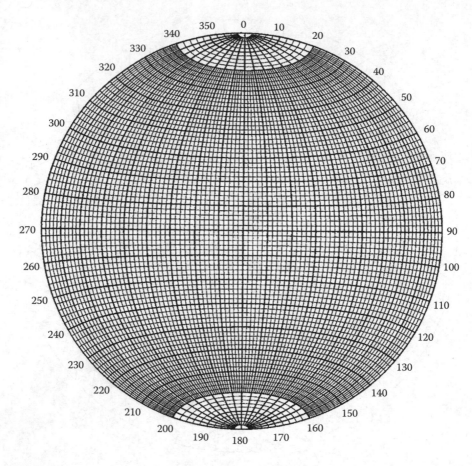

Figure I.3 Equal-area equatorial net for plotting poles and great circles.

3. The trend of the line of intersection is given by the position where the line drawn in step 2 intersects the scale on the circumference of the net.

4. Rotate the tracing paper until the line drawn in step 2 lies over one of the horizontal axes of the net (dip directions 90° or 180°). The plunge of the line of intersection is read off the scale on the horizontal axis, with a horizontal plunge having a point of intersection at the circumference and a vertical plunge at the centre of the net.

Appendix II: Quantitative description of discontinuities in rock masses

II.I INTRODUCTION

This appendix provides details of the parameters used in geological mapping and diamond drilling for the quantitative description of rock masses. The information provided is based entirely on the procedures drawn up by the International Society of Rock Mechanics (ISRM, 1981a,b, 2007), and which are discussed in more detail in Chapter 4 of this book. The objectives of using the ISRM procedures for geological mapping and drill core logging are as follows. First, these procedures are quantitative so that each parameter is measured and the results can be used either directly, or interpreted, in design. Second, the use of standardised procedures allows different personnel to work to the same standards, and to produce comparable information.

The following is a description of the parameters that describe the rock mass, together with tables listing values used to quantify these parameters. Also provided are mapping forms that can be used to record both geological mapping data and oriented core logging. Further information on geological characterisation and methods of data collection are discussed in Chapter 4.

II.2 ROCK MASS CHARACTERISATION PARAMETER

Figure II.1 illustrates the parameters that characterise the rock mass, and Figure II.2 shows how they can be divided into six classes related to the rock material and its strength, the discontinuity characteristics, infilling properties, the dimensions and shape of the blocks of rock and ground water conditions. Each of the parameters is discussed below.

II.2.1 Rock material description

II.2.1.1 A: Rock type

The value of including the rock type in describing a rock mass is that this defines the process by which the rock was formed. For example, sedimentary rocks such as sandstone usually contain well-ordered sets of discontinuities because they are laid down in layers, and are medium-to-low strength because they have usually only been subject to moderate heating and compression. Also, the rock type gives an indication of the properties of the rock mass from general experience of their engineering performance. For example, granite tends to be strong, massive and resistant to weathering, in comparison to shale which is often weak and fissile, and can weather rapidly when exposed to wetting and drying cycles.

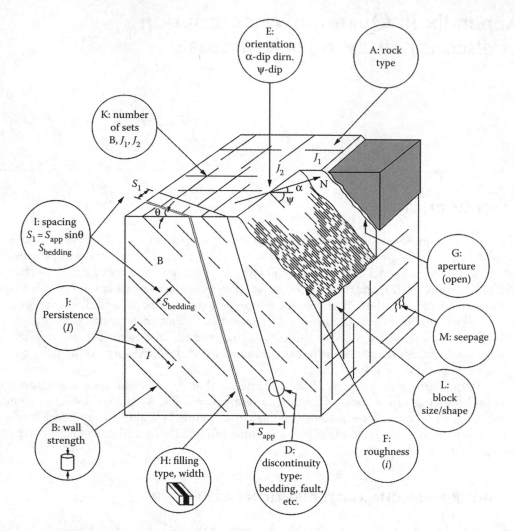

Figure II.1 Diagram illustrating rock mass properties (Wyllie, 1999).

Table II.1 shows a procedure for defining the rock type. This procedure involves identifying three primary characteristics of rock as follows:

1. Colour, as well as whether light or dark minerals predominate
2. Texture or fabric ranging from crystalline, granular or glassy
3. Grain size that can range from clay particles to gravel (Table II.2)

II.2.1.2 B: Rock strength

The compressive strength of the rock, comprising both the intact blocks of rock and the discontinuity surfaces, is an important component of shear strength and deformability, especially if the surfaces are in direct rock-to-rock contact as in the case of unfilled joints. Slight shear displacement of individual joints caused by shear stresses within the rock mass often

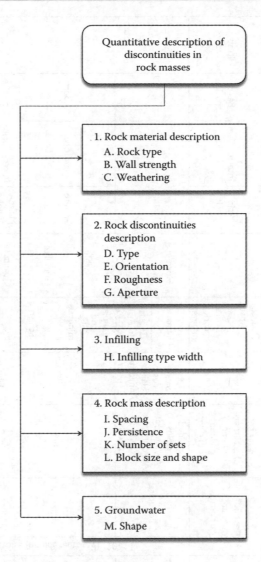

Figure II.2 Parameters describing rock mass characteristics.

results in small asperity contact areas. Where the concentrated stresses locally approach or exceed the compression strength of the rock wall materials, asperity damage occurs. The surface strength is quantified in the determination of shear strength as the joint compressive strength (*JCS*) as discussed in Section 4.4.3.

Table II.3 defines ranges of rock and soil strength with a corresponding grade (R6 to S1) related to simple field identification procedures.

II.2.1.3 C: Weathering

- Rock masses are frequently weathered near the ground surface, and are sometimes altered by hydrothermal processes. Weathering (and alteration) is generally more pronounced on rock exposed on the discontinuity surfaces than in the interior of rock

Table II.I Rock type

Genetic group		Detrital sedimentary		Chemical organic	Metamorphic		Igneous			
Usual structure		Bedded	Bedded (Pyroclastic)		Foliated	Massive	Massive			
Composition		Grains of rock, quartz, feldspar and minerals / At least 50% of grains are of carbonate	At least 50% of grains are of fine-grained volcanic rock		Quartz, feldspars, micas, acicular dark minerals		Light coloured minerals are quartz, feldspar, mica and feldspar-like minerals			Dark minerals
Grain size class	Grain size (mm)						Acid rocks	Intermediate rocks	Basic rocks	Ultrabasic rocks
Very coarse grained	60	Grains are of rock fragments Rounded grains: Conglomerate								
Coarse grained	2	Angular grains: Breccia — Calcirudite	Rounded grains Agglomerate; Angular gains Volcanic breccia		Migmatite	Hornfels	Granite	Diorite (Pegmatite)	Gabbro	Pyroxenite and peridotite
Medium grained	0.06	Sandstone: Grains are mainly mineral fragments Quartz Sandstone: 95% quarts, voids empty or cemented Arkose: 75% quarts, up to 25% feldspar: voids empty of cemented Argillarceous sandstone: 75% quartz, 15% + fine detrital material — Calcarenite	Tuff	Saline rocks Halite Anhydrite Gypsum	Gneiss alternate layers of granular and flakey minerals	Marble	Micro-granite	Micro-diorite	Dolerite	Serpentine
	0.002			Chert	Schist	Granulite				
Fine grained	0.06	Mudstone shale: fissile mudstone Siltstone 50% fine-grained particles — Calcisiltite	Fine-grained Tuff	Flint	Phyllite	Quartzite	Rhyolite	Andesite	Basalt	
Very fine grained	0.002	Claystone 50% very fine-grained particles Calcareous mudstone — Calcilutite	Very fine-grained Tuff	Coal Others	Slate	Amphibolite				
					Mylonite					
Glassy							Obsideian and pitchstone			Tachylyte

Limestone (undifferentiated) (20)

Source: Adapted from Geological Society Engineering Group Working Party. 1977. Knill, J. L. et al., eds., Quart. J. Eng. Geol. Hydrogeol., 3(1):1–24.

Table II.2 Grain size scale

Grain size description	Grain size dimension
Boulders	200–600 mm (7.9–23.6 in.)
Cobbles	60–200 mm (2.4–7.9 in.)
Coarse gravel	20–60 mm (0.8–0.24 in.)
Medium gravel	6–20 mm (0.2–0.8 in.)
Fine gravel	2–6 mm (0.1–0.2 in.)
Coarse sand	0.6–2 mm (0.02–0.1 in.)
Medium sand	0.2–0.6 mm (0.008–0.02 in.)
Fine sand	0.06–0.2 mm (0.002–0.008 in.)
Silt, clay	<0.06 mm (<0.002 in.)

blocks because water flow that initiates weathering occurs in the discontinuities. This water flow results in the rock strength on the discontinuity surfaces being less than that of the fresher rock found in the interior of the rock blocks. A description of the state of weathering or alteration, both for the rock material and the rock mass, is therefore an essential part of the description of the rock mass.

• The two primary weathering processes are mechanical disintegration, and chemical decomposition including solution. Generally, both mechanical and chemical effects

Table II.3 Classification of rock material strengths

Grade	Description	Field identification	Approximate range of compressive strength	
			MPa	psi
R6	Extremely strong rock	Specimen can only be chipped with geological hammer	>250	>36,000
R5	Very strong rock	Specimen requires many blows of geological hammer to fracture it	100–250	15,000–36,000
R4	Strong rock	Specimen requires more than one blow with a geological hammer to fracture it	50–100	7000–15,000
R3	Medium strong rock	Cannot be scraped or peeled with a pocket knife; specimen can be fractured with single firm blow of geological hammer	25–50	3500–7000
R2	Weak rock	Can be peeled with a pocket knife; shallow indentations made by firm blow with point of geological hammer	5–25	725–3500
R1	Very weak rock	Crumbles under firm blows with point of geological hammer; can be peeled by a pocket knife	1–5	150–725
R0	Extremely weak rock	Indented by thumbnail	0.25–1	35–150
S6	Hard clay	Indented with difficulty by thumbnail	>0.5	>70
S5	Very stiff clay	Readily indented by thumbnail	0.25–0.5	35–70
S4	Stiff clay	Readily indented by thumb but penetrated only with great difficulty	0.1–0.25	15–35
S3	Firm clay	Can be penetrated several inches by thumb with moderate effort	0.05–0.1	7–15
S2	Soft clay	Easily penetrated several inches by thumb	0.025–0.05	4–7
S1	Very soft clay	Easily penetrated several inches by fist	<0.025	<4

Table II.4 Weathering and alteration grades

Grade	Term	Weathering description
I	Fresh	No visible sign of rock material weathering; perhaps slight discolouration on major discontinuity surfaces
II	Slightly weathered	Discolouration indicates weathering of rock material and discontinuity surfaces. All the rock material may be discoloured by weathering and may be somewhat weaker externally than in its fresh condition
III	Moderately weathered	Less than half of the rock material is decomposed and/or disintegrated to a soil. Fresh or discoloured rock is present either as a continuous framework or as corestones
IV	Highly weathered	More than half of the rock material is decomposed and/or disintegrated to a soil. Fresh or discoloured rock is present either as a discontinuous framework or as corestones
V	Completely weathered	All rock material is decomposed and/or disintegrated to soil. The original mass structure is still largely intact
VI	Residual soil	All rock material is converted to soil. The mass structure and material fabric are destroyed. A large decrease in volume occurs, but the soil has not been significantly transported

act together, depending on climatic regime and lithology, but chemical weathering is usually dominant. Mechanical weathering results in opening of discontinuities, the formation of new discontinuities by rock fracture, the opening of grain boundaries and the fracture or cleavage of individual mineral grains. Chemical weathering results in discolouration of the rock and leads to the eventual decomposition of silicate minerals to clay minerals; some minerals, notably quartz, resist this action and may survive unchanged. Solution is an aspect of chemical weathering which is particularly important in the case of carbonate and saline minerals (Chapter 3).

- The relatively thin 'skin' of surface rock that affects shear strength and deformability can be tested by means of simple index tests. The apparent uniaxial compression strength can be estimated both from the Schmidt hammer tests and from scratch/geological hammer tests, since the later have been approximately calibrated against a large body of test data as shown in Table II.3.
- Mineral coatings will affect the shear strength of discontinuities, to a marked degree if the surfaces are planar and smooth. The type of mineral coatings should be described where possible. Samples should be taken when in doubt.

Table II.4 defines grades of rock weathering.

II.2.2 Discontinuity description

II.2.2.1 D: Discontinuity type

- The discontinuity type is useful in the description of the rock mass because each type has properties that influence the behaviour of the rock mass. For example, faults can have lengths of several kilometres and contain low-strength gouge, while joints lengths usually do not exceed a few metres and they often contain no infilling. Section 4.3.3 describes the characteristics of the most common types of discontinuities which include faults, bedding, foliation, joints, cleavage and schistosity.

II.2.2.2 E: Discontinuity orientation

- The orientation of a discontinuity in space is described by the dip (ψ) of the line of the steepest declination measured from horizontal, and by the dip direction (α) measured clockwise from true north. Example: (45°/025°).
- The orientation of discontinuities relative to an engineering structure largely controls the possibility of unstable conditions or excessive deformations developing. The importance of orientation increases when other conditions for deformation are present, such as low-shear strength and a sufficient number of discontinuities or joint sets for sliding to occur.
- The mutual orientation of discontinuities will determine the shape of the individual blocks comprising the rock mass.

II.2.2.3 F: Roughness

- The roughness of a discontinuity surface is a potentially important component of its shear strength, especially in the case of undisplaced and interlocked features (e.g. unfilled joints). The importance of surface roughness declines as aperture, or infilling thickness, or the degree of previous displacement increases.
- The roughness can be characterised by waviness, and the unevenness or asperities. Waviness describes large-scale undulations which, if interlocked and in contact, cause dilation during shear displacement since they are too large to be sheared off. Unevenness or asperities describe small-scale roughness. Asperities will tend to be damaged during shear displacement if the ratio of the rock strength on the discontinuity surface to the normal stress level is low, in which case little dilation will occur on these small-scale features.
- Both waviness and asperities can be measured in the field, and asperities can be measured as a component of a direct shear test.
- Roughness can be sampled by linear profiles taken parallel to the direction of sliding, that is, parallel to the dip (dip vector). In cases where sliding is controlled by two intersecting discontinuity planes, the direction of potential sliding is parallel to the line of intersection of the planes.
- The purpose of all roughness sampling is to estimate or calculate shear strength and dilation. Presently available methods of interpreting roughness profiles and estimating shear strength include measuring the i value (or inclination) of the irregularities, or the joint roughness coefficient (JRC) of the surface (Figure II.3). The contribution of the surface roughness to the total friction angle of a surface is discussed in more detail in Section 4.4.3.

Descriptive terms that can be used to define roughness are a combination of small-scale features (several centimetres dimensions): *rough, smooth, slickensided* and larger-scale features (several metres dimensions): *stepped, undulating, planar*. These terms can be combined to describe decreasing levels of roughness shown in Table II.5.

II.2.2.4 G: Aperture

- Aperture is the perpendicular distance separating the adjacent rock surfaces of an open discontinuity, in which the intervening space is air or water filled. Aperture is thereby distinguished from the width of a filled discontinuity. Discontinuities that have been filled (e.g. with clay) also come under this category if the filling material has been washed out locally.

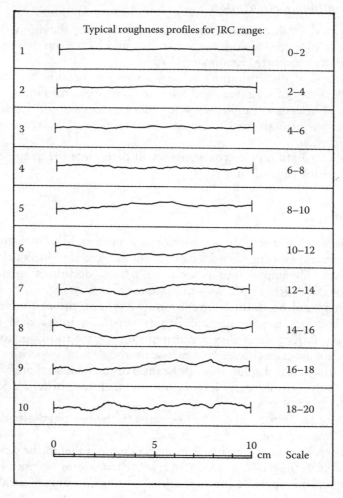

Figure II.3 Roughness profiles and corresponding range of JRC values (International Society for Rock Mechanics, 1981a).

Table II.5 Descriptive terms for roughness

Roughness grade	Roughness description
I	Rough, stepped
II	Smooth, stepped
III	Slickensided, stepped
IV	Rough, undulating
V	Smooth, undulating
VI	Slickensided, undulating
VII	Rough, planar
VIII	Smooth, planar
IX	Slickensided, planar

Table II.6 Aperture dimensions

Dimensions (mm)	Aperture description	
<0.1	Very tight	'Closed' features
0.1–0.25	Tight	
0.25–0.5	Partly open	
0.5–2.5	Open	'Gapped' features
2.5–10	Moderately wide	
>10	Wide	
10–100	Very wide	'Open' features
100–1000	Extremely wide	
>1000	Cavernous	

- Large aperture can result from shear displacement of discontinuities having appreciable roughness and waviness, from tensile opening, outwash of infillings and solution. Steeply dipping discontinuities that have opened in tension as a result of valley erosion or glacial retreat may have 'cavernous' apertures.
- In most sub-surface rock masses, apertures will probably be less than half a millimetre, compared to the 'wide' width of some outwash or extension varieties. Unless discontinuities are exceptionally smooth and planar, it will not be of great significance to the shear strength that a 'closed' feature is 0.1 mm or 1.0 mm wide. However, hydraulic conductivity is influenced by minor changes in aperture.
- Unfortunately, visual observation of small apertures is inherently unreliable since, with the possible exceptions of drilled holes and bored tunnels, visible apertures are likely to be disturbed by blasting or surface weathering effects. Aperture can be measured indirectly by hydraulic conductivity testing.
- Apertures are recorded from the point of view of both their loosening and conducting capacity. Joint water pressure, inflow of water and outflow of storage products (both liquid and gas) will all be affected by aperture.

Aperture can be described by the terms listed in Table II.6.

II.2.3 Infilling description

II.2.3.1 H: Infilling type and width

- Infilling is the term for material separating the adjacent rock surfaces of discontinuities, for example, calcite, chlorite, clay, silt, fault gouge, breccia, etc. The perpendicular distance between the adjacent rock surfaces is termed the *width* of the filled discontinuity, as opposed to the *aperture* of a gapped or open feature.
- Due to the enormous variety of occurrences, filled discontinuities display a wide range of physical behaviour, in particular as regards their shear strength, deformability and hydraulic conductivity. Short- and long-term behaviour may be quite different such that it is easy to be misled by favourable short-term conditions.
- The wide range of physical behaviour depends on many factors of which the following are probably the most important:
  - Mineralogy of filling material (Table II.1)
  - Grading or particle size (Table II.2)
  - Over-consolidation ratio
  - Water content and conductivity (Table II.12)

- Previous shear displacement
- Surface roughness (Figure II.3 and Table II.5)
- Width (Table II.6)
- Fracturing or crushing of surface rock
- Every attempt should be made to record the above factors, using the quantitative descriptions where possible, together with sketches and/or colour photographs of the most important occurrences. Certain index tests are suggested for a closer investigation of major discontinuities considered to be a threat to stability. In special cases, the results of these field descriptions may warrant recommendation for large-scale *in situ* testing, possibly in the case of slopes containing a critical structural feature above an important facility.

II.2.4 Rock mass description

II.2.4.1 I: Spacing

- The spacing of adjacent discontinuities largely controls the size of individual blocks of intact rock. Several closely spaced sets tend to give conditions of low-rock mass cohesion, whereas those that are widely spaced are much more likely to yield interlocking conditions. These effects depend upon the persistence of the individual discontinuities.
- In exceptional cases, a close spacing may change the mode of failure of a rock mass from plane or wedge, to circular or even to flow (e.g. a 'sugar cube' shear zone in quartzite). With exceptionally close spacing, the orientation is of little consequence as failure may occur through rotation or rolling of the small rock pieces.
- As in the case of orientation, the importance of spacing increases when other conditions for deformation are present, that is, low-shear strength and a sufficient number of discontinuities or joint sets for sliding to occur.
- The spacing of individual discontinuities and associated sets has a strong influence on the mass hydraulic conductivity and seepage characteristics. In general, the hydraulic conductivity of any given set will be inversely proportional to the spacing, if individual joint apertures are comparable.

Spacing can be described by the terms listed in Table II.7.

II.2.4.2 J: Persistence

- Persistence implies the areal extent or size of a discontinuity within a plane. It can be crudely quantified by observing the discontinuity trace lengths on the surface of exposures. It is one of the most important rock mass parameters, but one of the most difficult to quantify.

Table II.7 Joint spacing dimensions

Joint spacing description	Spacing (mm)
Extremely close spacing	<20
Very close spacing	20–60
Close spacing	60–200
Moderate spacing	200–600
Wide spacing	600–6000
Very wide spacing	2000–6000
Extremely wide spacing	>6000

Table II.8 Joint persistence dimensions

Joint persistence description	Dimensions (m)
Very low persistence	<1
Low persistence	1–3
Medium persistence	3–10
High persistence	10–20
Very high persistence	>20

- The discontinuities of one particular set will often be more continuous than those of the other sets. The minor sets will therefore tend to terminate against the primary features, or they may terminate in solid rock.
- In the case of rock slopes, it is of the greatest importance to attempt to assess the degree of persistence of those discontinuities that are unfavourably orientated for stability. The degree to which discontinuities persist beneath the adjacent rock blocks without terminating in solid rock or terminating against other discontinuities determines the degree to which failure of intact rock would be involved in eventual failure. Perhaps more likely, it determines the degree to which 'down-stepping' would have to occur between adjacent discontinuities for a slip surface to develop. Persistence is also of the greatest importance to tension crack development behind the crest of a slope.
- Frequently, rock exposures are small compared to the area or length of persistent discontinuities, and the real persistence can only be guessed. Less frequently, it may be possible to record the dip length and the strike length of exposed discontinuities and thereby estimate their persistence along a given plane through the rock mass using the probability theory (see Section 4.5). However, the difficulties and uncertainties involved in the field measurements will be considerable for most rock exposures.

Persistence can be described by the terms listed in Table II.8.

II.2.4.3 K: Number of sets

- The mechanical behaviour of a rock mass and its appearance will be influenced by the number of sets of discontinuities that intersect one another. The mechanical behaviour is especially affected since the number of sets determines the extent to which the rock mass can deform without involving failure of the intact rock. The number of sets also affects the appearance of the rock mass due to the loosening and displacement of blocks in both natural and excavated faces (Figure II.4).
- The number of sets of discontinuities may be an important feature of rock slope stability, in addition to the orientation of discontinuities relative to the face. A rock mass containing a number of closely spaced joint sets may change the potential mode of slope failure from translational or toppling to rotational/circular.
- In the case of tunnel stability, three or more sets will generally constitute a 3D block structure having a considerable more 'degrees of freedom' for deformation than a rock mass with less than three sets. For example, a strongly foliated phyllite with just one closely spaced joint set may give equally good tunnelling conditions as a massive granite with three widely spaced joint sets. The amount of overbreak in a tunnel will usually be strongly dependent on the number of sets.

The number of joint sets occurring locally (e.g. along the length of a tunnel) can be described according to the scheme shown in Table II.9.

Major singular discontinuities that are not part of a joint set should be recorded individually.

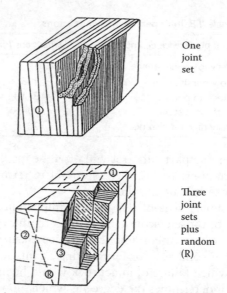

One
joint
set

Three
joint
sets
plus
random
(R)

Figure II.4 Examples that demonstrate the effect of the number of joint sets on the mechanical behaviour and appearance of a rock mass (International Society for Rock Mechanics, 1981a).

Table II.9 Number of joint sets

Joint set grade	Description of joint sets
I	Massive, occasional random joints
II	One joint set
III	One joint set plus random
IV	Two joint sets
V	Two joint sets plus random
VI	Three joint sets
VII	Three joint sets plus random
VIII	Four or more joint sets
IX	Crushed rock, earth like

II.2.4.4 L: Block size and shape

- Block size is an important indicator of rock mass behaviour. Block dimensions are determined by discontinuity spacing, number of sets and persistence of the discontinuities delineating potential blocks.
- The number of sets and the orientation determine the shape of the resulting blocks, which can take the approximate form of cubes, rhombohedra, tetrahedrons, sheets, etc. However, regular geometric shapes are the exception rather than the rule since the joints in any one set are seldom consistently parallel. Jointing in sedimentary rocks usually produces the most regular block shapes.
- The combined properties of block size and interblock shear strength determine the mechanical behaviour of the rock mass under given stress conditions. Rock masses composed of large blocks tend to be less deformable, and in the case of underground construction, develop favourable arching and interlocking. In the case of slopes, a small block size may cause the potential mode of failure to resemble that of soil (i.e.

Table II.10 Block dimensions

Block dimensions	J_v (joints/m³)
Very large blocks	<1.0
Large blocks	1–3
Medium-sized blocks	3–10
Small blocks	10–30
Very small blocks	>30

Table II.11 Block size and shape for rock masses

Rock mass grade	Description of rock masses
I	Massive – few joints or very wide spacing
II	Blocky – approximately equidimensional
III	Tabular – one dimension considerably smaller than the other two
IV	Columnar – one dimension considerably larger than the other two
V	Irregular – wide variations of block size and shape
VI	Crushed – heavily jointed to 'sugar cube'

circular/rotational), instead of the translational or toppling modes of failure usually associated with discontinuous rock masses. In exceptional cases, 'block' size may be so small that flow occurs, as with a 'sugar cube' shear zones in quartzite.

- Rock quarrying and blasting efficiency are related to the *in situ* block size. It may be helpful to think in terms of a block size distribution for the rock mass, in much the same way that soils are categorised by a distribution of particle sizes.
- Block size can be described either by means of the average dimension of typical blocks (block size index I_b), or by the total number of joints intersecting a unit volume of the rock mass (volumetric joint count J_v).

Table II.10 lists descriptive terms that give an impression of the corresponding block size. Values of $J_v > 60$ would represent crushed rock, typical of a clay-free crushed zone.

Rock masses. Rock masses can be described by the adjectives given in Table II.11, to give an impression of block size and shape (Figure II.5).

II.2.5 Ground water

II.2.5.1 M: Seepage

- Water seepage through rock masses results mainly from flow through water conducting discontinuities ('secondary' hydraulic conductivity). However, in the case of certain sedimentary rocks, such as poorly indurated sandstone, the 'primary' hydraulic conductivity of the rock material may be significant such that a proportion of the total seepage occurs through the pores. The rate of seepage is proportional to the local hydraulic gradient and to the relevant directional conductivity, proportionality being dependent on laminar flow. High-velocity flow through open discontinuities may result in head losses due to turbulence.
- The prediction of ground water levels, likely seepage paths and approximate water pressures, may often give advance warning of stability or construction difficulties. The field

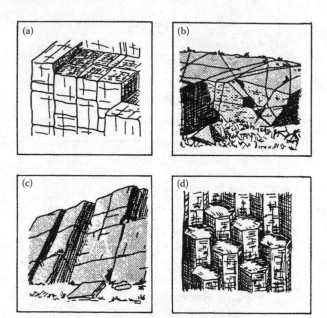

Figure II.5 Sketches of rock masses illustrating (a) blocky; (b) irregular; (c) tabular and (d) columnar block shapes (International Society for Rock Mechanics, 1981a).

description of rock masses must inevitably precede any recommendation for field conductivity tests so these factors should be carefully assessed at the early stages of the investigation.

- Irregular ground water levels and perched water tables may be encountered in rock masses that are partitioned by persistent impermeable features such as dykes, clay-filled discontinuities or low-conductivity beds. The prediction of these potential flow barriers and associated irregular water tables is of considerable importance, especially for projects where such barriers might be penetrated at depth by tunnelling, resulting in high-pressure inflows.

- Water seepage caused by drainage into an excavation may have far-reaching consequences in cases where a sinking ground water level would cause settlement of nearby structures founded on overlying clay deposits.

- The approximate description of the local hydrogeology should be supplemented with detailed observations of seepage from individual discontinuities or particular sets, according to their relative importance to stability. A short comment concerning recent precipitation in the area, if known, will be helpful in the interpretation of these observations. Additional data concerning ground water trends, and rainfall and temperature records will be useful supplementary information.

- In the case of rock slopes, the preliminary design estimates will be based on assumed values of effective normal stress. If, as a result of field observations, one has to conclude that pessimistic assumptions of water pressure are justified, such as a tension crack full of water and a rock mass that does not drain readily, then this will clearly influence the slope design. So also will the field observation of rock slopes where high-water pressures can develop due to seasonal freezing of the face that blocks drainage paths.

Seepage from individual unfilled and filled discontinuities or from specific sets exposed in a tunnel or in a surface exposure can be assessed according to the descriptive terms in Tables II.12 and II.13.

In the case of an excavation such as a tunnel that acts as a drain for the rock mass, it is helpful if the flow into individual sections of the structure is described. This should ideally be performed immediately after excavation since groundwater levels, or the rock mass storage, may be depleted rapidly. Descriptions of seepage quantities are given in Table II.14.

- A field assessment of the likely effectiveness of surface drains, inclined drill holes, or drainage tunnels should be made in the case of major rock slopes. This assessment will depend on the orientation, spacing and apertures of the relevant discontinuities.
- The potential influence of frost and ice on the seepage paths through the rock mass should be assessed. Observations of seepage from the surface trace of discontinuities may be misleading in freezing temperatures. The possibility of ice-blocked drainage paths should be assessed from the points of view of surface deterioration of a rock excavation, and of overall stability.

Table II.12 Seepage quantities in unfilled discontinuities

Seepage grade	Seepage in unfilled discontinuities
I	Discontinuity is very tight and dry, water flow along it does not appear possible
II	Discontinuity is dry with no evidence of water flow
III	Discontinuity is dry but shows evidence of water flow, that is, rust staining
IV	Discontinuity is damp but no free water is present
V	Discontinuity shows seepage, occasional drops of water, but no continuous flow
VI	Discontinuity shows a continuous flow of water – estimate L/min and describe pressure, that is, low, medium, high

Table II.13 Seepage quantities in filled discontinuities

Seepage grade	Seepage in filled discontinuities
I	Filling materials are heavily consolidated and dry, significant flow appears unlikely due to very low conductivity
II	Filling materials are damp, but no free water is present
III	Filling materials are wet, occasional drops of water
IV	Filling materials show signs of outwash, continuous flow of water – estimate L/min
V	Filling materials are washed out locally, considerable water flow along out wash channels – estimate L/min and describe pressure, that is, low, medium, high
VI	Filling materials are washed out completely, very high-water pressures experienced, especially on first exposure – estimate L/min and describe pressure

Table II.14 Seepage quantities in tunnels

Seepage grade	Tunnel seepage quantity
I	Dry walls and roof, no detectable seepage
II	Minor seepage, specify dripping discontinuities
III	Medium inflow, specify discontinuities with continuous flow (estimate L/min/10 m length of excavation)
IV	Major inflow, specify discontinuities with strong flows (estimate L/min/10 m length of excavation)
V	Exceptionally high inflow, specify source of exceptional flows (estimate L/min/10 m length of excavation)

II.3 FIELD MAPPING SHEETS

The two mapping sheets included with this appendix provide a means of recording the qualitative geological data described in this appendix.

Sheet 1 – Rock mass description sheet describes the rock material in terms of its colour, grain size and strength, the rock mass in terms of the block shape, size, weathering and the number of discontinuity sets and their spacing.

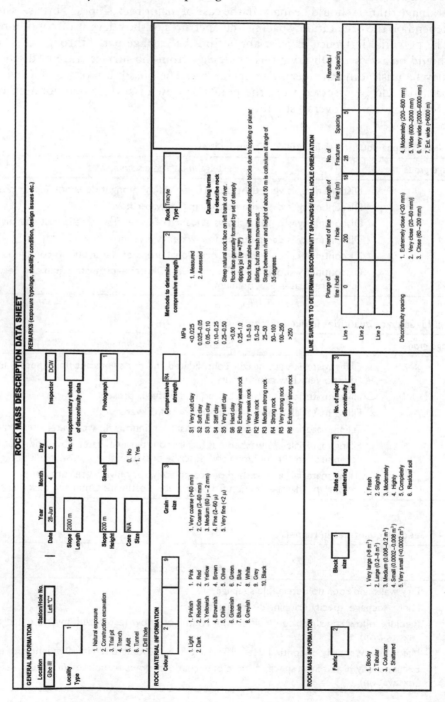

Sheet 2 – Discontinuity survey data sheet describes the characteristics of each discontinuity in terms of its type, orientation, persistence, aperture/width, filling, surface roughness and water flow. This sheet can be used for recording both outcrop (or tunnel) mapping data and oriented core data (excluding persistence and surface shape).

The sheets also include important information on the location, date and mapping conditions.

DISCONTINUITY SURVEY DATA SHEET

GENERAL INFORMATION

Location | Station / Hole No. | Date: Day / Month / Year | Inspector | Discontinuity data sheet no. | of

NATURE AND ORIENTATION OF DISCONTINUITY

Columns: Chainage or Depth | Type | Dip | Dip Direction | Persistence | Termination | Aperture/Width | Nature of Filling | Strength of Filling | Surface Roughness | Surface Shape | Waviness Wavelength | Waviness Amplitude | JRC | Water Flow | Spacing | Remarks

Type
- 0. Fault zone
- 1. Fault
- 2. Joint
- 3. Cleavage
- 4. Schistosity
- 5. Shear
- 6. Fissure
- 7. Tension Crack
- 8. Foliation
- 9. Bedding

Persistence
- 1. Very low persistence — <1 m
- 2. Low persistence — 1–3 m
- 3. Medium persistence — 3–10 m
- 4. High persistence — 10–20 m
- 5. Very high persistence — >20 m

Termination
- 0. Neither end visible
- 1. One end visible
- 2. Both ends visible

Aperture/width
- 1. Very tight (<0.1 mm)
- 2. Tight (0.1–0.25 mm)
- 3. Partly open (0.25–0.5 mm)
- 4. Open (0.5–2.5 mm)
- 5. Moderately wide (2.5–10 mm)
- 6. Wide (>10 mm)
- 7. Very wide (1–10 cm)
- 8. Extremely wide (10–100 cm)
- 9. Cavernous (>1 m)

Surface shape
- 1. Stepped
- 2. Undulating
- 3. Planar

Nature of filling
- 1. Clean
- 2. Surface staining
- 3. Non-cohesive
- 4. Inactive clay or clay matrix
- 5. Swelling clay or clay matrix
- 6. Cemented
- 7. Chlorite, talc or gypsum
- 8. Other - specify

Surface roughness
- 1. Rough
- 2. Smooth
- 3. Polished
- 4. Slickensided

Compressive strength of infilling

MPa
- S1 Very soft clay — <0.025
- S2 Soft clay — 0.025–0.05
- S3 Firm clay — 0.05–0.10
- S4 Stiff clay — 0.10–0.25
- S5 Very stiff clay — 0.25–0.50
- S6 Hard clay — >0.50
- R0 Extremely weak rock — 0.25–1.0
- R1 Very weak rock — 1.0–5.0
- R2 Weak rock — 5.0–25
- R3 Medium strong rock — 25–50
- R4 Strong rock — 50–100
- R5 Very strong rock — 100–250
- R6 Extremely strong rock — >250

Spacing
- 1. Extremely close spacing — <20 mm
- 2. Very close spacing — 20–60 mm
- 3. Close spacing — 60–200 mm
- 4. Moderate spacing — 200–600 mm
- 5. Wide spacing — 600–2000 mm
- 6. Very wide spacing — 2000–6000 mm
- 7. Extremely wide spacing — >6000 mm

Water flow (open)
- 0. The discontinuity is very tight and dry; water flow along it does not appear possible.
- 1. The discontinuity is dry with no evidence of water flow.
- 2. The discontinuity is dry but shows evidence of water flow, i.e. rust staining, etc.
- 3. The discontinuity is damp but no free water is present.
- 4. The discontinuity shows seepage, occasional drops of water, but no continuous flow.
- 5. The discontinuity shows a continuous flow of water (Estimate l/min and describe pressure, i.e. low, medium, high).

Water flow (filled)
- 6. The filling materials are heavily consolidated and dry; significant flow appears unlikely due to very low permeability.
- 7. The filling materials are damp, but no free water is present.
- 8. The filling materials are wet; occasional drops of water.
- 9. The filling materials show signs of outwash, continuous flow of water (estimate litres/minute).
- 10. The filling materials are washed out locally, considerable water flow along out-wash channels (estimate litres/minute and describe pressure, i.e. low, medium, high).

Appendix III: Comprehensive solution wedge stability

III.I INTRODUCTION

This appendix presents the equations and procedure to calculate the factor of safety for a wedge failure as discussed in Chapter 8. This comprehensive solution includes the wedge geometry defined by five surfaces, including a sloped upper surface and a tension crack, water pressures, different shear strengths on each slide plane and up to two external forces (Figure III.1). External forces that may act on a wedge include tensioned anchor support, foundation loads and earthquake ground motions. The forces are vectors defined by their magnitude, and their plunge and trend. If necessary, several force vectors can be combined to meet the limit of two forces. It is assumed that all forces act through the centre of gravity of the wedge so no moments are generated, and no rotational slip or toppling occurs.

III.2 ANALYSIS METHODS

The equations presented in this appendix are identical to those in Appendix 2 of *Rock Slope Engineering Edition 3* (Hoek and Bray, 1981). These equations have been found to be versatile and capable of calculating the stability of a wide range of geometric and geotechnical conditions. However, the analysis has two limitations that are discussed in Section III.3.

As an alternative to the comprehensive analysis presented in this appendix, a shorter analysis is available that can be used for a more limited set of input parameters. Sections 8.3 and 8.4 in Chapter 8 show calculation procedures for a wedge formed by planes 1, 2, 3 and 4 shown in Figure III.1, but with no tension crack. The shear strength is defined by different cohesions and friction angles on planes 1 and 2, and the water pressure condition assumed is that the slope is saturated. However, no external forces can be incorporated in the analysis.

Also included in Chapter 8 is a set of charts for rapid calculation of the factor of safety for wedges for dry, friction only conditions with no tension crack.

III.3 ANALYSIS LIMITATIONS

For the comprehensive stability analysis presented in this appendix, the following two limitations are noted. First, a geometric limitation is related to the relative inclinations of upper slope (plane 3) and the line of intersection, and second a simplified procedure is used for modifying water pressures. The following is a discussion of these two limitations.

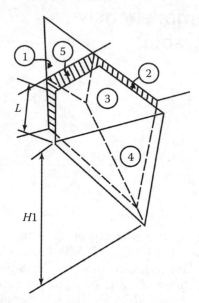

Figure III.I Wedge geometry used for comprehensive stability analysis.

Wedge geometry – for a wedge with steep upper slope (plane 3), and a line of intersection that has a shallower dip than the upper slope (i.e. $\psi_3 > \psi_i$), no intersection occurs between the plane and the line and the programme will terminate with the error message 'Tension crack invalid' (see Equations III.50 to III.53). The reason for this error message is that the calculation procedure is to first calculate the dimensions of the overall wedge from the slope face to the apex (intersection of the line of intersection with plane 3). Then, the dimensions of a wedge between the tension crack and the apex are calculated. Finally, the dimensions of the wedge between the face and the tension crack are found by subtracting the overall wedge from the upper wedge (see Equations III.54 to III.57).

However, for the wedge geometry where ($\psi_3 > \psi_i$), a wedge can still be formed if a tension crack (plane 5) is present; it is possible to calculate the factor of safety for these conditions using a different set of equations (e.g. Swedge, Rocscience Inc.).

Water pressure – the analysis incorporates the average values of the water pressure on the sliding planes (u_1 and u_2), and on the tension crack (u_5). These values are calculated assuming that the wedge is fully saturated. That is, the water table is coincident with the upper surface of the slope (plane 3), and that the pressure drops to zero where planes 1 and 2 intersect the slope face (plane 4). These pressure distributions are simulated as follows. Where no tension crack exists, the water pressures on planes 1 and 2 are given by ($u_1 = u_2 = \gamma_w \cdot H_w/6$), where H_w is the vertical height of the wedge defined by the two ends of the line of intersection. The second method allows for the presence of a tension crack and gives ($u_1 = u_2 = u_5 = \gamma_w \cdot H_{5w}/3$), where H_{5w} is the depth of the bottom vertex of the tension crack below the upper ground surface. The water forces are then calculated as the product of these pressures and the areas of the respective planes.

To calculate stability of a partially saturated wedge, the reduced pressures are simulated by reducing the unit weight of the water, γ_w. That is, if it is estimated that the tension crack is one-third filled with water, then a unit weight of ($\gamma_w/3$) is used as the input parameter. It is considered that this approach is adequate for most purposes because water levels in slopes are variable and difficult to determine precisely.

III.4 SCOPE OF SOLUTIONS

The solution described below is for computation of the factor of safety for translational slip of a tetrahedral wedge formed in a rock slope by two intersecting discontinuities (planes 1 and 2), the upper ground surface (plane 3), the slope face (plane 4) and a tension crack (plane 5). The solution allows for water pressures on the two slide planes and in the tension crack, and for different strength parameters on the two slide planes. Plane 3 may have a different dip direction to that of plane 4. The influence of an external load E and a cable tension T are included in the analysis. Supplementary sections are provided for the examination of the minimum factor of safety for a given external load, and for minimising the anchoring force required for a given factor of safety.

The solution allows for the following conditions:

- Interchange of planes 1 and 2
- The possibility of one of the planes overlying the other
- The situation where the crest overhangs the toe of the slope (in which case $\eta = -1$)
- The possibility of contact being lost on either plane ($N_1, N_2 < 0$)

III.5 NOTATIONS

The wedge geometry is illustrated in Figure III.1; the following input data are required:

ψ, α = dip, dip direction of plane, or plunge, trend of force
$H1$ = slope height referred to plane 1
L = distance of tension crack from crest, measured along the trace of plane 1
c = cohesion of each slide plane
ϕ = angle of friction of each slide plane
γ = unit weight of rock
γ_w = unit weight of water
T = anchor tension
E = external load
$\eta = -1$ if face is overhanging and
$\quad$ +1 if face does not overhang

Other terms used in the solution are:

FS = factor of safety against sliding along line of intersection, or on plane 1 or plane 2
A = area of sliding plane or tension crack
W = weight of wedge
V = water thrust on tension crack (plane 5)
N_a = total normal force of plane 1 $\left.\rule{0pt}{2.5em}\right\}$ when contact is maintained on plane 1 only
S_a = shear force on plane 1
Q_a = shear resistance on plane 1

FS_1 = factor of safety
N_b = total normal force on plane 2 $\left.\rule{0pt}{2.5em}\right\}$ when contact is maintained on plane 2 only
S_b = shear force on plane 2
Q_b = shear resistance on plane 2

FS_2 = factor of safety

N_1, N_2 = effective normal reactions

S = total shear force on planes 1 and 2

Q = total shear resistance on planes 1 and 2

} when contact is maintained on both planes 1 and 2

FS_3 = factor of safety

N_1', N_2', S', etc. = values of N_1, N_2, S, etc. when $T = 0$

N_1'', N_2'', S'', etc. = values of N_1, N_2, S, etc. when $E = 0$

$\vec{a}$ = unit normal vector for plane 1

$\vec{b}$ = unit normal vector for plane 2

$\vec{d}$ = unit normal vector for plane 3

$\vec{f}$ = unit normal vector for plane 4

$\vec{f_5}$ = unit normal vector for plane 5

$\vec{g}$ = vector in direction of intersection line of 1, 4

$\vec{g_5}$ = vector in direction of intersection line of 1, 5

$\vec{i}$ = vector in direction of intersection line of 1, 2

$\vec{j}$ = vector in direction of intersection line of 3, 4

$\vec{j_5}$ = vector in direction of intersection line of 3, 5

$\vec{k}$ = vector in plane 2 normal to $\vec{i}$

$\vec{l}$ = vector in plane 1 normal to $\vec{i}$

R = magnitude of vector $\vec{i}$

G = square of magnitude of vector $\vec{g}$

G_5 = square of magnitude of vector $\vec{g_5}$

Note: The computed value of V is negative when the tension crack dips away from the toe of the slope, but this does not indicate a tensile force.

III.6 SEQUENCE OF CALCULATIONS

The following sections list the sequence of calculations required to determine the factor of safety for three slope conditions related to the external loads T and E. In addition, a worked example is provided showing how to find the optimum orientation of a tensioned rock anchor to minimise the required support force to achieve a specified factor of safety.

III.6.1 Calculation of factor of safety when forces T and E are either zero, or completely specified in magnitude and direction

1. Components of unit vectors in directions of normals to planes 1–5, and of forces T and E.

$$(a_x, a_y, a_z) = (\sin\psi_1 \cdot \sin\alpha_1, \sin\psi_1 \cdot \cos\alpha_1, \cos\psi_1) \tag{III.1}$$

$$(b_x, b_y, b_z) = (\sin\psi_2 \cdot \sin\alpha_2, \sin\psi_2 \cdot \cos\alpha_2, \cos\psi_2) \tag{III.2}$$

$$(d_x, d_y, d_z) = (\sin\psi_3 \cdot \sin\alpha_3, \sin\psi_3 \cdot \cos\alpha_3, \cos\psi_3) \tag{III.3}$$

$$(f_x, f_y, f_z) = (\sin\psi_4 \cdot \sin\alpha_4, \ \sin\psi_4 \cdot \cos\alpha_4, \ \cos\psi_4) \tag{III.4}$$

$$(f_{5x}, f_{5y}, f_{5z}) = (\sin\psi_5 \cdot \sin\alpha_5, \ \sin\psi_5 \cdot \cos\alpha_5, \ \cos\psi_5) \tag{III.5}$$

$$(t_x, t_y, t_z) = (\cos\psi_t \cdot \sin\alpha_t, \ \cos\psi_t \cdot \cos\alpha_t, \ -\sin\psi_t) \tag{III.6}$$

$$(e_x, e_y, e_z) = (\cos\psi_e \cdot \sin\alpha_e, \ \cos\psi_e \cdot \cos\alpha_e, \ -\sin\psi_e) \tag{III.7}$$

2. Components of vectors in the direction of the lines of intersection of various planes.

$$(g_x, g_y, g_z) = (f_y \cdot a_z - f_z \cdot a_y), \quad (f_z \cdot a_x - f_x \cdot a_z), \quad (f_x \cdot a_y - f_y \cdot a_x) \tag{III.8}$$

$$(g_{5x}, g_{5y}, g_{5z}) = (f_{5y} \cdot a_z - f_{5z} \cdot a_y), \quad (f_{5z} \cdot a_x - f_{5x} \cdot a_z), \quad (f_{5x} \cdot a_y - f_{5y} \cdot a_x) \tag{III.9}$$

$$(i_x, i_y, i_z) = (b_y \cdot a_z - b_z \cdot a_y), \quad (b_z \cdot a_x - b_x \cdot a_z), \quad (b_x \cdot a_y - b_y \cdot a_x) \tag{III.10}$$

$$(j_x, j_y, j_z) = (f_y \cdot d_z - f_z \cdot d_y), \quad (f_z \cdot d_x - f_x \cdot d_z), \quad (f_x \cdot d_y - f_y \cdot d_x) \tag{III.11}$$

$$(j_{5x}, j_{5y}, j_{5z}) = (f_{5y} \cdot d_z - f_{5z} \cdot d_y), \quad (f_{5z} \cdot d_x - f_{5x} \cdot d_z), \quad (f_{5x} \cdot d_y - f_{5y} \cdot d_x) \tag{III.12}$$

$$(k_x, k_y, k_z) = (i_y \cdot b_z - i_z \cdot b_y), \quad (i_z \cdot b_x - i_x \cdot b_z), \quad (i_x \cdot b_y - i_y \cdot b_x) \tag{III.13}$$

$$(l_x, l_y, l_z) = (a_y \cdot i_z - a_z \cdot i_y), \quad (a_z \cdot i_x - a_x \cdot i_z), \quad (a_x \cdot i_y - a_y \cdot i_x) \tag{III.14}$$

3. Numbers proportional to cosines of various angles.

$$m = g_x \cdot d_x + g_y \cdot d_y + g_z \cdot d_z \tag{III.15}$$

$$m_5 = g_{5x} \cdot d_x + g_{5y} \cdot d_y + g_{5z} \cdot d_z \tag{III.16}$$

$$n = b_x \cdot j_x + b_y \cdot j_y + b_z \cdot j_z \tag{III.17}$$

$$n_5 = b_x \cdot j_{5x} + b_y \cdot j_{5y} + b_z \cdot j_{5z} \tag{III.18}$$

$$p = i_x \cdot d_x + i_y \cdot d_y + i_z \cdot d_z \tag{III.19}$$

$$q = b_x \cdot g_x + b_y \cdot g_y + b_z \cdot g_z \tag{III.20}$$

$$q_s = b_x \cdot g_{5x} + b_y \cdot g_{5y} + b_z \cdot g_{5z} \tag{III.21}$$

$$r = a_x \cdot b_x + a_y \cdot b_y + a_z \cdot b_z \tag{III.22}$$

$$s = a_x \cdot t_x + a_y \cdot t_y + a_z \cdot t_z \tag{III.23}$$

$$v = b_x \cdot t_x + b_y \cdot t_y + b_z \cdot t_z \tag{III.24}$$

$$w = i_x \cdot t_x + i_y \cdot t_y + i_z \cdot t_z \tag{III.25}$$

$$s_e = a_x \cdot e_x + a_y \cdot e_y + a_z \cdot e_z \tag{III.26}$$

$$v_e = b_x \cdot e_x + b_y \cdot e_y + b_z \cdot e_z \tag{III.27}$$

$$w_e = i_x \cdot e_x + i_y \cdot e_y + i_z \cdot e_z \tag{III.28}$$

$$s_5 = a_x \cdot f_{5x} + a_y \cdot f_{5y} + a_z \cdot f_{5z} \tag{III.29}$$

$$v_5 = b_x \cdot f_{5x} + b_y \cdot f_{5y} + b_z \cdot f_{5z} \tag{III.30}$$

$$w_5 = i_x \cdot f_{5x} + i_y \cdot f_{5y} + i_z \cdot f_{5z} \tag{III.31}$$

$$\lambda = i_x \cdot g_x + i_y \cdot g_y + i_z \cdot g_z \tag{III.32}$$

$$\lambda_5 = i_x \cdot g_{5x} + i_y \cdot g_{5y} + i_z \cdot g_{5z} \tag{III.33}$$

$$\varepsilon = f_x \cdot f_{5x} + f_y \cdot f_{5y} + f_z \cdot f_{5z} \tag{III.34}$$

4. Miscellaneous factors.

$$R = \sqrt{(1 - r^2)} \tag{III.35}$$

$$\rho = \frac{1}{R^2} \cdot \frac{n \cdot q}{|n \cdot q|} \tag{III.36}$$

$$\mu = \frac{1}{R^2} \cdot \frac{m \cdot q}{|m \cdot q|} \tag{III.37}$$

$$\nu = \frac{1}{R} \cdot \frac{p}{|p|} \tag{III.38}$$

$$G = g_x^2 + g_y^2 + g_z^2 \tag{III.39}$$

$$G_5 = g_{5x}^2 + g_{5y}^2 + g_{5z}^2 \tag{III.40}$$

$$M = (G \cdot p^2 - 2 \cdot |m \cdot p| \cdot \lambda + m^2 \cdot R^2)^{1/2} \tag{III.41}$$

$$M_5 = (G_5 \cdot p^2 - 2 \cdot |m_5 \cdot p| \cdot \lambda_5 + m_5^2 \cdot R^2)^{1/2} \tag{III.42}$$

$$h = \frac{H1}{|g_z|} \tag{III.43}$$

$$h_5 = \frac{M \cdot h - |p| \cdot L}{M_5} \tag{III.44}$$

$$B = \frac{\left(\tan^2 \phi_1 + \tan^2 \phi_2 - 2 \cdot \left(\dfrac{\mu \cdot r}{\rho} \right) \cdot \tan \phi_1 \cdot \tan \phi_2 \right)}{R^2} \tag{III.45}$$

5. Plunge and trend, respectively, of line of intersection of planes 1 and 2.

$$\psi_i = \arcsin(v \cdot i_z) \tag{III.46}$$

$$\alpha_i = \arctan\left(\frac{-v \cdot i_x}{-v \cdot i_y} \right) \tag{III.47}$$

The term $-v$ should not be cancelled out in Equation III.47, since this is required to determine the correct quadrant when calculating values for dip direction, α_i.

6. Check on wedge geometry.

No wedge is formed, terminate computation

$$\begin{cases} \text{if } (p \cdot i_z) < 0, \text{ or} & \text{(III.48)} \\[2ex] \text{if } (\eta \cdot q \cdot i_z) < 0 & \text{(III.49)} \end{cases}$$

Tension crack invalid, terminate computation

$$\begin{cases} \text{if } (\varepsilon \cdot \eta \cdot q_5 \cdot i_z) < 0, \text{ or} & \text{(III.50)} \\[2ex] \text{if } h_5 < 0, \text{ or} & \text{(III.51)} \\[2ex] \text{if } \left[\left| \dfrac{m_5 \cdot h_5}{m \cdot h} \right| \right] > 1, \text{ or} & \text{(III.52)} \end{cases}$$

$$\text{if } \left[\left| \frac{n \cdot q_5 \cdot m_5 \cdot h_5}{n_5 \cdot q \cdot m \cdot h} \right| \right] > 1 \tag{III.53}$$

7. Areas of faces and weight of wedge.

$$A_1 = \frac{|m \cdot q| \cdot h^2 - |m_5 \cdot q_5| \cdot h_5^2}{2 \cdot |p|} \tag{III.54}$$

$$A_2 = \frac{\left((|q| \cdot m^2 \cdot h^2)/(|n|) - (|q_5| \cdot m_5^2 \cdot h_5^2)/(|n_5|)\right)}{2 \cdot |p|} \qquad \text{(III.55)}$$

$$A_5 = \frac{|m_5 \cdot q_5| \cdot h_5^2}{2 \cdot |n_5|} \qquad \text{(III.56)}$$

$$W = \frac{\gamma\left((q^2 \cdot m^2 \cdot h^3)/(|n|) - (q_5^2 \cdot m_5^2 \cdot h_5^3)/(|n_5|)\right)}{6 \cdot |p|} \qquad \text{(III.57)}$$

8. Water pressures.
   a. With no tension crack

$$u_1 = u_2 = \frac{\gamma_w \cdot h \cdot |m_5 \cdot i_z|}{6 \cdot |p|} \qquad \text{(III.58)}$$

   b. With tension crack

$$u_1 = u_2 = u_5 = \frac{\gamma_{w \cdot h_5 \cdot |h_5|}}{3 \cdot d_z} \qquad \text{(III.59)}$$

$$V = u_5 \cdot A_5 \cdot \eta\left(\frac{\varepsilon}{|\varepsilon|}\right) \qquad \text{(III.60)}$$

9. Effective normal reactions on planes 1 and 2 assuming contact on both planes.

$$N_1 = \rho\{W \cdot k_z + T(r \cdot v - s) + E(r \cdot v_e - s_e) + V(r \cdot v_5 - s_5)\} - u_1 \cdot A_1 \qquad \text{(III.61)}$$

$$N_2 = \mu\{W \cdot l_z + T(r \cdot s - v) + E(r \cdot s_e - v_e) + V(r \cdot s_5 - v_5)\} - u_2 \cdot A_2 \qquad \text{(III.62)}$$

10. Factor of safety when $N_1 < 0$ and $N_2 < 0$ (contact is lost on both planes).

$$FS = 0 \qquad \text{(III.63)}$$

11. If $N_1 > 0$ and $N_2 < 0$, contact is maintained on plane 1 only and the factor of safety is calculated as follows:

$$N_a = W \cdot a_z - T \cdot s - E \cdot s_e - V \cdot s_5 - u_1 \cdot A_1 \cdot r \qquad \text{(III.64)}$$

$$S_x = (T \cdot t_x + E \cdot e_x + N_a \cdot a_x + V \cdot f_{5x} + u_1 \cdot A_1 \cdot b_x) \qquad \text{(III.65)}$$

$$S_y = (T \cdot t_y + E \cdot e_y + N_a \cdot a_y + V \cdot f_{5y} + u_1 \cdot A_1 \cdot b_y) \qquad \text{(III.66)}$$

$$S_z = (T \cdot t_z + E \cdot e_z + N_a \cdot a_z + V \cdot f_{5z} + u_1 \cdot A_1 \cdot b_z) + W \qquad \text{(III.67)}$$

$$S_a = \left(S_x^2 + S_y^2 + S_z^2\right)^{1/2} \tag{III.68}$$

$$Q_a = (N_a - u_1 \cdot A_1) \cdot \tan\phi_1 + c_1 \cdot A_1 \tag{III.69}$$

$$FS_1 = \left(\frac{Q_a}{S_a}\right) \tag{III.70}$$

12. If $N_1 < 0$ and $N_2 > 0$, contact is maintained on plane 2 only and the factor of safety is calculated as follows:

$$N_b = (W \cdot b_z - T \cdot v - E \cdot v_e - V \cdot v_5 - u_2 \cdot A_2 \cdot r) \tag{III.71}$$

$$S_x = -(T \cdot t_x + E \cdot e_x + N_b \cdot b_x + V \cdot f_{5x} + u_2 \cdot A_2 \cdot a_x) \tag{III.72}$$

$$S_y = -(T \cdot t_y + E \cdot e_y + N_b \cdot b_y + V \cdot f_{5y} + u_2 \cdot A_2 \cdot a_y) \tag{III.73}$$

$$S_z = -(T \cdot t_z + E \cdot e_z + N_b \cdot b_z + V \cdot f_{5z} + u_2 \cdot A_2 \cdot a_z) + W \tag{III.74}$$

$$S_b = \left(S_x^2 + S_y^2 + S_z^2\right)^{1/2} \tag{III.75}$$

$$Q_b = (N_b - u_2 \cdot A_2) \cdot \tan\phi_2 + c_2 \cdot A_2 \tag{III.76}$$

$$FS_2 = \left(\frac{Q_b}{S_b}\right) \tag{III.77}$$

13. If $N_1 > 0$ and $N_2 > 0$, contact is maintained on both planes and the factor of safety is calculated as follows:

$$S = v(W \cdot i_z - T \cdot w - E \cdot w_e - V \cdot w_5) \tag{III.78}$$

$$Q = N_1 \cdot \tan\phi_1 + N_2 \cdot \tan\phi_2 + c_1 \cdot A_1 + c_2 \cdot A_2 \tag{III.79}$$

$$FS_3 = \left(\frac{Q}{S}\right) \tag{III.80}$$

III.6.2 Calculation of orientation of external load E of given magnitude that produces the minimum factor of safety

1. Evaluate N_1'', N_2'', S'', Q'', FS_3'' by use of Equations III.61, III.62, III.78 to III.80 with $E = 0$.
2. If $N_1'' < 0$ and $N_2'' < 0$, even before E is applied. Then, $FS = 0$, terminate computation.

3. $D = \left\{ (N_1'')^2 + (N_2'')^2 + 2\left(\dfrac{mn}{|mn|} \right) \cdot N_1'' \cdot N_2'' \cdot r \right\}^{1/2}$ (III.81)

$\psi_e = \arcsin\left\{ \left(-\dfrac{1}{G}\left(\dfrac{m}{|m|} \right) \cdot N_1'' \cdot a_z + \dfrac{n}{|n|} \cdot N_2'' \cdot b_z \right) \right\}$ (III.82)

$a_e = \arctan\left\{ \dfrac{\dfrac{m}{|m|} \cdot N_1'' \cdot a_x + \dfrac{n}{|n|} \cdot N_2'' \cdot b_x}{\dfrac{m}{|m|} \cdot N_1'' \cdot a_y + \dfrac{n}{|n|} \cdot N_2'' \cdot b_y} \right\}$ (III.83)

If $E > D$, and E is applied in the direction ψ_e, α_e, or within a certain range encompassing this direction, then contact is lost on both planes and $FS = 0$. Terminate calculation.

4. If $N_1'' > 0$ and $N_2'' < 0$, assume contact on plane 1 only after application of E. Determine S_x'', S_y'', S_z'', S_a'', Q_a'', FS_1'' from Equations III.65 to III.70 with $E = 0$. If $FS_1'' < 1$, terminate computation. If $FS_1'' > 1$:

$FS_1 = \dfrac{S_a'' \cdot Q_a'' - E\{(Q_a'')^2 + ((S_a'')^2 - E^2)\tan^2 \phi_1\}^{1/2}}{(S_a'')^2 - E^2}$ (III.84)

Orientation of force E:

$\psi_{e1} = \arcsin\left(\dfrac{S_z''}{S_a''} \right) - \arctan\left(\dfrac{\tan\phi_1}{(FS_1)} \right)$ (III.85)

$\alpha_{e1} = \arctan\left(\dfrac{S_x''}{S_y''} \right) + 180$ (III.86)

5. If $N_1'' < 0$ and $N_2'' > 0$, assume contact on plane 2 only after application of E. Determine S_x'', S_y'', S_z'', S_b'', Q_b'', FS_2'' from Equations III.72 to III.77 with $E = 0$. If $FS_2'' < 1$, terminate computation. If $FS_2'' > 1$:

$FS_2 = \dfrac{S_b'' \cdot Q_b'' - E\{(Q_b'')^2 + ((S_b'')^2 - E^2)\tan^2 \phi_2\}^{1/2}}{(S_b'')^2 - E^2}$ (III.87)

Orientation of force E:

$\psi_{e2} = \arcsin\left(\dfrac{S_z''}{S_b''} \right) - \arctan\left(\dfrac{\tan\phi_2}{(FS_2)} \right)$ (III.88)

$$\alpha_{e2} = \arctan\left(\frac{S_x''}{S_y''}\right) + 180° \tag{III.89}$$

6. If $N_1'' > 0$ and $N_2'' > 0$, assume contact on both planes after application of E.
   If $FS_3'' < 1$, terminate computation.
   If $FS_3'' > 1$:
   Note: parameter v is defined by Equation III.38.

$$FS_3 = \frac{S'' \cdot Q'' - E\{(Q'')^2 + B((S'')^2 - E^2)\}^{1/2}}{(S'')^2 - E^2} \tag{III.90}$$

$$\chi = \sqrt{B + (FS_3)^2} \tag{III.91}$$

$$e_x = -\frac{((FS_3) \cdot v \cdot i_x - \rho \cdot k_x \cdot \tan\phi_1 - \mu \cdot l_x \cdot \tan\phi_2)}{\chi} \tag{III.92}$$

$$e_y = -\frac{((FS_3)vi_y - \rho \cdot k_y \cdot \tan\phi_1 - \mu \cdot l_y \cdot \tan\phi_2)}{\chi} \tag{III.93}$$

$$e_z = -\frac{((FS_3) \cdot v \cdot i_z - \rho \cdot k_z \cdot \tan\phi_1 - \mu \cdot l_z \cdot \tan\phi_2)}{\chi} \tag{III.94}$$

Orientation of force E:

$$\psi_{e3} = \arcsin(e_z) \tag{III.95}$$

$$\alpha_{e3} = \arctan\left(\frac{e_x}{e_y}\right) \tag{III.96}$$

Compute s_e and v_e using Equations III.26 and III.27

$$N_1 = N_1'' + E \cdot \rho(r \cdot v_e - s_e) \tag{III.97}$$

$$N_2 = N_2'' + E \cdot \mu(r \cdot s_e - v_e) \tag{III.98}$$

Check that $N_1 \geq 0$ and $N_2 \geq 0$

III.6.3 Minimum cable or bolt tension T_{min} required to raise the factor of safety to a specified value of factor of safety

1. Evaluate N_1', N_2', S', Q' by means of Equations III.61, III.62, III.78 and III.79 with $T = 0$.
2. If $N_2' < 0$, contact is lost on plane 2 when $T = 0$. Assume contact on plane 1 only, after application on T. Evaluate S_x', S_x', S_z', S_a' and Q_a' using Equations III.65 to III.69 with $T = 0$.

$$T_1 = \frac{((FS)S_a' - Q_a')}{\sqrt{(FS)^2 + \tan^2 \phi_1}}$$ (III.99)

Orientation of force T:

$$\psi_{t1} = \arctan\left(\frac{\tan\phi_1}{(FS)}\right) - \arcsin\left(\frac{S_z'}{S_a'}\right)$$ (III.100)

$$\alpha_{t1} = \arctan\left(\frac{S_x'}{S_y'}\right)$$ (III.101)

3. If $N_1' < 0$, contact is lost on plane 1 when $T = 0$. Assume contact on plane 2 only, after application of T. Evaluate S_x', S_y', S_z', S_b' and Q_b' using Equations III.72 to III.76 with $T = 0$.

$$T_2 = \frac{((FS)S_b' - Q_b')}{\sqrt{(FS)^2 + \tan^2 \phi_2}}$$ (III.102)

Orientation of force T:

$$\psi_{t2} = \arctan\left(\frac{\tan\phi_2}{(FS)}\right) - \arcsin\left(\frac{S_z'}{S_b'}\right)$$ (III.103)

$$\alpha_{t2} = \arctan\left(\frac{S_x'}{S_y'}\right)$$ (III.104)

4. All cases. No restrictions on values of N_1' and N_2'. Assume contact on both planes after application of T.

$$\chi = \sqrt{((FS)^2 + B)}$$ (III.105)

$$T_3 = \frac{((FS)\,S' - Q')}{\chi}$$ (III.106)

$$t_x = \frac{((FS)\cdot v\cdot i_x - \rho\cdot k_x\cdot \tan\phi_1 - \mu\cdot l_x\cdot \tan\phi_2)}{\chi}$$ (III.107)

$$t_y = \frac{((FS)\cdot v\cdot i_y - \rho\cdot k_y\cdot \tan\phi_1 - \mu\cdot l_y\cdot \tan\phi_2)}{\chi}$$ (III.108)

$$t_z = \frac{((FS)\cdot v\cdot i_z - \rho\cdot k_z\cdot \tan\phi_1 - \mu\cdot l_z\cdot \tan\phi_2)}{\chi}$$ (III.109)

Orientation of force T:

$$\psi_{t3} = \arcsin(-t_z) \tag{III.110}$$

$$\alpha_{t3} = \arctan\left(\frac{t_x}{t_y}\right) \tag{III.111}$$

Compute s and v using Equations III.23 and III.24.

$$N_1 = N_1' + T_3 \cdot \rho(r \cdot v - s) \tag{III.112}$$

$$N_2 = N_2' + T_3 \cdot \mu(r \cdot s - v) \tag{III.113}$$

If $N_1 < 0$ or $N_2 < 0$, ignore the results of this section.
If $N_1' > 0$ and $N_2' > 0$, $T_{min} = T_3$
If $N_1' > 0$ and $N_2' < 0$, T_{min} = smallest of T_1, T_3
If $N_1' < 0$ and $N_2' > 0$, T_{min} = smallest of T_2, T_3
If $N_1' < 0$ and $N_2' < 0$, T_{min} = smallest of T_1, T_2, T_3

III.6.4 Worked examples

This section provides four worked examples showing calculations for the factor of safety of a wedge for saturated and drained slopes, and the method of calculating the optimum orientation of tensioned anchors T for a given value of factor of safety.

The parameters defining the wedge are as follows (see Figure III.1):

Plane	1	2	3	4	5
ψ	45	70	12	65	70
α	105	235	195	185	165

$H1 = 100$ ft, $L = 40$ ft, $c_1 = 500$ lb/ft^2, $c_2 = 1000$ lb/ft^2
$\eta = +1$ – face not overhanging
$\phi_1 = 20°$, $\phi_2 = 30°$, $\gamma = 160$ lb/ft^3

1. Saturated slope with no external loads.
   $u_1 = u_2 = u_5$ (saturated) from Equation III.59; $T = E = 0$.
   $(a_x, a_y, a_z) = (0.683, -0.183, 0.707)$
   $(b_x, b_y, b_z) = (-0.769, -0.539, 0.342)$
   $(d_x, d_y, d_z) = (-0.054, -0.201, 0.978)$
   $(f_x, f_y, f_z) = (-0.079, -0.903, 0.423)$
   $(f_{5x}, f_{5y}, f_{5z}) = (0.243, -0.908, 0.342)$
   $(g_x, g_y, g_z) = (-0.561, 0.345, 0.631)$
   $(g_{5x}, g_{5y}, g_{5z}) = (-0.579, 0.0616, 0.575)$
   $(i_x, i_y, i_z) = (-0.319, 0.778, 0.509)$
   $(j_x, j_y, j_z) = (-0.798, 0.055, -0.033)$

$(j_{5x}, j_{5y}, j_{5z}) = (-0.819, -0.256, -0.098)$
$(k_x, k_y, k_z) = (0.540, -0.283, 0.770)$
$(l_x, l_y, l_z) = (-0.643, -0.573, 0.473)$
$m = 0.578$
$m_5 = 0.582$
$n = 0.574$
$n_5 = 0.735$
$p = 0.359$
$q = 0.462$
$q_5 = 0.609$
$r = -0.185$
$s_5 = 0.574_3$
$v_5 = 0.419$
$w_5 = -0.609$
$\lambda = 0.768$
$\lambda_5 = 0.525$
$\varepsilon = 0.945$
$R = 0.983$
$\rho = 1.036$
$\mu = 1.036$
$\nu = 1.018$
$G = 0.832$
$G_5 = 0.670$
$M = 0.334$
$M_5 = 0.440$
$h = 158.45$
$h_5 = 87.52$
$B = 0.563$

Line of intersection orientation:

$\psi_i = 31.20°$
$\alpha_i = 157.73°$

$\left.\begin{array}{l} p \cdot i_z > 0 \\ n \cdot q \cdot i_z > 0 \end{array}\right\}$ Wedge is formed

$\left.\begin{array}{l} \varepsilon \cdot \eta \cdot q_5 \cdot i_z > 0 \\ h_5 > 0 \\ \dfrac{|m_5 \cdot h_5|}{|m \cdot h|} = 0.55 < 1 \end{array}\right\}$ Tension crack valid

$\dfrac{|n \cdot q_5 \cdot m_5 \cdot h_5|}{|n_5 \cdot q \cdot m \cdot h|} = 0.57 < 1$

Areas of planes, weight of wedge and water forces:

$A_1 = 5565.01 \text{ ft}^2$
$A_2 = 6428.1 \text{ ft}^2$
$A_5 = 1846.6 \text{ ft}^2$
$W = 2.8272 \times 10^7 \text{ lb}$
$u_1 = u_2 = u_5 = 1084.3 \text{ lb/ft}^2; \ V = 2.0023 \times 10^6 \text{ lb}$

$N_1 = 1.5171 \times 10^7 \, \text{lb}$
$N_2 = 5.7892 \times 10^6 \, \text{lb}$ }Both positive, therefore contact on planes 1 and 2

$S = 1.5886 \times 10^7 \, \text{lb}$
$Q = 1.8075 \times 10^7 \, \text{lb}$

Factor of safety:
   $FS = 1.14$

2. Dry slope with no external loads.
   $u_1 = u_2 = u_5 = 0$ (dry); $T = E = 0$.
   As in (1) above except as follows:
   $V = 0$
   $N_1 = 2.2565 \times 10^7 \, \text{lb}$
   $N_2 = 1.3853 \times 10^7 \, \text{lb}$ }Both positive, therefore contact on both planes 1 and 2

   $S = 1.4644 \times 10^7 \, \text{lb}$
   $Q = 2.5422 \times 10^7 \, \text{lb}$

Factor of Safety, $FS_3 = 1.74$
   This shows that the effect of draining the slope is to increase the factor of safety from 1.14 to 1.74.

3. Dry slope with external load E

   $u_1 = u_2 = u_5 = 0$ (dry); $T = 0$; $E = 8 \times 10^6 \, \text{lb}$

   Find the value of the minimum factor of safety, and the orientation of this force.

   Values of N_1'', N_2'', S'', Q'', FS_3'' as given in (2).
   $N_1'' > 0$, $N_2'' > 0$, $FS_3'' > 1$, continue calculation.
   $B = 0.563$

Minimum factor of safety, $FS_3 = 1.04$
   $\chi = 1.280$
   $e_x = 0.121$
   $e_y = -0.992$
   $e_z = 0.028$

Orientation of force E:
   $\psi_{e3} = -1.62°$ – *plunge of force E (upwards)*
   $\alpha_{e3} = 173.03°$ – *trend of force E (out of face)*
   $N_1 = 1.9517 \times 10^7 \, \text{lb}$
   $N_2 = 9.6793 \times 10^6 \, \text{lb}$ }Both positive, therefore contact maintained on both planes

   This shows that the factor of safety is a minimum of 1.04 when the external force E, acts upwards and out of the face at the calculated orientation.

4. Saturated slope with support force T at minimum value for factor of safety of 1.5.
   As in (1), except find optimum orientation of minimum support force T required to increase the factor of safety to 1.5.
   N_1', N_2', S' and Q' – as given in (1).

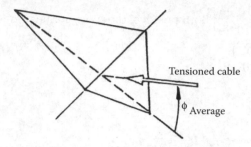

Figure III.2 Optimum orientation of support force (tensioned cable or bolt) to minimise force to achieve specified factor of safety.

$\chi = 1.6772$
$T_3 = 3.4307 \times 10^6 \, \text{lb} \quad - T_{\min}$ *(minimum cable tension)*
$t_x = -0.182$
$t_y = 0.975$
$t_z = 0.121$
$\psi_{t3} = -6.98°$ *– plunge of cable (upwards)*
$\alpha_{t3} = 349.43°$ *– trend of cable (into face)*

Note that the optimum plunge and trend of the anchor are approximately:
$\psi_{t3} \approx \psi_i + 180° - 1/2 \, (\phi_1 + \phi_2) = 31.2 + 180 - 25 = -6.2°$ (upwards)
and $\alpha_{t3} \approx \alpha_i \pm 180° = 157.73 + 180 = 337.73°$ (into face)

That is, the best direction in which to install an anchor to reinforce a wedge is:
   Anchor aligned with the line of intersection of the two planes, viewed from the bottom of the slope, and inclined at the average friction angle to the line of intersection (Figure III.2).

Note that for cement anchored bolts or cables, grouting is facilitated by installations inclined about 15° below the horizontal.

Appendix IV: Conversion factors

Imperial unit	SI unit	SI unit symbol	Conversion factor (Imperial to SI)	Conversion factor (SI to imperial)
Length				
Mile	kilometre	km	1 mile = 1.609 km	1 km = 0.6214 mile
Foot	metre	m	1 ft = 0.3048 m	1 m = 3.2808 ft
	millimetre	mm	1 ft = 304.80 mm	1 mm = 0.003 281 ft
Inch	millimetre	mm	1 in = 25.40 mm	1 mm = 0.039 37 in
Area				
Square mile	square kilometre	km^2	$1\ mile^2 = 2.590\ km^2$	$1\ km^2 = 0.3861\ mile^2$
	hectare	ha	$1\ mile^2 = 259.0\ ha$	$1\ ha = 0.003\ 861\ mile^2$
Acre	hectare	ha	1 acre = 0.4047 ha	1 ha = 2.4710 acre
	square metre	m^2	$1\ acre = 4047\ m^2$	$1\ m^2 = 0.000\ 247\ 1\ acre$
Square foot	square metre	m^2	$1\ ft^2 = 0.092\ 90\ m^2$	$1\ m^2 = 10.7639\ ft^2$
Square inch	square millimetre	mm^2	$1\ in^2 = 645.2\ mm^2$	$1\ mm^2 = 0.001\ 550\ in^2$
Volume				
Cubic yard	cubic metre	m^3	$1\ yd^3 = 0.7646\ m^3$	$1\ m^3 = 1.3080\ yd^3$
Cubic foot	cubic metre	m^3	$1\ ft^3 = 0.028\ 32\ m^3$	$1\ m^3 = 35.3147\ ft^3$
	litre	L	$1\ ft^3 = 28.32\ L$	$1\ L = 0.035\ 31\ ft^3$
Cubic inch	cubic millimetre	mm^3	$1\ in^3 = 16\ 387\ mm^3$	$1\ mm^3 = 61.024 \times 10^{-6}\ in^3$
	cubic centimetre	cm^3	$1\ in^3 = 16.387\ cm^3$	$1\ cm^3 = 0.061\ 02\ in^3$
	litre	L	$1\ in^3 = 0.016\ 39\ L$	$1\ L = 61.02\ in^3$
Imperial gallon	cubic metre	m^3	$1\ gal = 0.004\ 55\ m^3$	$1\ m^3 = 220.0\ gal$
	litre	L	1 gal = 4.546 L	1 L = 0.220 gal
Pint	litre	L	1 pt = 0.568 L	1 L = 1.7598 pt
US gallon	cubic metre	m^3	$1\ US\ gal = 0.0038\ m^3$	$1\ m^3 = 264.2\ US\ gal$
	litre	L	1 US gal = 3.8 L	1 L = 0.264 US gal
Mass				
Ton	tonne	t	1 ton = 0.9072 tonne	1 tonne = 1.1023 ton
Ton (2000 lb) (US)	kilogram	kg	1 ton = 907.19 kg	1 kg = 0.001 102 ton
Ton (2240 lb) (UK)			1 ton = 1016.0 kg	1 kg = 0.000 984 ton
Kip	kilogram	kg	1 kip = 453.59 kg	1 kg = 0.002 204 6 kip
Pound	kilogram	kg	1 lb = 0.4536 kg	1 kg = 2.204 6 lb

(Continued)

Imperial unit	SI unit	SI unit symbol	Conversion factor (Imperial to SI)	Conversion factor (SI to imperial)
Mass density				
Ton per cubic yard (2000 lb) (US)	kilogram per cubic metre	kg/m^3	1 ton/yd^3 = 1186.55 kg/m^3	1 kg/m^3 = 0.000 842 8 ton/yd^3
	tonne per cubic metre	t/m^3	1 ton/yd^3 = 1.1866 t/m^3	1 t/m^3 = 0.8428 ton/yd^3
Ton per cubic yard (2240 lb) (UK) pound per cubic foot	kilogram per cubic metre	kg/cm^3	1 ton/yd^3 = 1328.9 kg/m^3	1 kg/m^3 = 0.000 75 ton/yd^3
			1 lb/ft^3 = 16.02 kg/m^3	1 kg/cm^3 = 0.062 42 lb/ft^3
	tonne per cubic metre	t/m^3	1 lb/ft^3 = 0.01602 t/m^3	1 t/m^3 = 62.42 lb/ft^3
Pound per cubic inch	gram per cubic centimetre	g/cm^3	1 lb/in^3 = 27.68 g/cm^3	1 g/cm^3 = 0.036 13 lb/in^3
	tonne per cubic metre	t/m^3	1 lb/in^3 = 27.68 t/m^3	1 t/m^3 = 0.036 13 lb/in^3
Force				
Ton force (2000 lb) (US)	kilonewton	kN	1 tonf = 8.896 kN	1 kN = 0.1124 tonf (US)
Ton force (2240 lb) (UK)			1 tonf = 9.964 KN	1 kN = 0.1004 tonf (UK)
Kip force	kilonewton	kN	1 kipf = 4.448 kN	1 kN = 0.2248 kipf
Pound force	newton	N	1 lbf = 4.448 N	1 N = 0.2248 lbf
Tonf/ft (2000 lb) (US)	kilonewton per metre	kN/m	1 tonf/ft = 29.189 kN/m	1 kN/m = 0.034 26 tonf/ft (US)
Tonf/ft (2240 lb) (UK)	kilonewton per metre		1 tonf/ft = 32.68 kN/m	1 kN/m = 0.0306 tonf/ft (UK)
Pound force per foot	newton per metre	N/m	1 lbf/ft = 14.59 N/m	1 N/m = 0.068 52 lbf/ft
Flow rate				
Cubic foot per minute	cubic metre per second	m^3/s	1 ft^3/min = 0.000 471 9 m^3/s	1 m^3/s = 2118.880 ft^3/min
	litre per second	L/s	1 ft^3/min = 0.4719 L/s	1 L/s = 2.1189 ft^3/min
Cubic foot per second	cubic metre per second	m^3/s	1 ft^3/s = 0.028 32 m^3/s	1 m^3/s = 35.315 ft^3/s
	litre per second	L/s	1 ft^3/s = 28.32 L/s	1 L/s = 0.035 31 ft^3/s
Gallon per minute	litre per second	L/s	1 gal/min = 0.075 77 L/s	1 L/s = 13.2 gal/min
Pressure, stress				
Ton force per square foot (2000 lb) (US)	kilopascal	kPa	1 tonf/ft^2 = 95.76 kPa	1 kPa = 0.01044 ton f/ft^2
Ton force per square foot (2240 lb) (UK)			1 tonf/ft^2 = 107.3 kPa	1 kPa = 0.00932 ton/ft^2
Pound force per square foot	pascal	Pa	1 lbf/ft^2 = 47.88 Pa	1 Pa = 0.020 89 lbf/ft^2
	kilopascal	kPa	1 lbf/ft^2 = 0.047 88 kPa	1 kPa = 20.89 lbf/ft^2
Pound force per square inch	pascal		1 lbf/in^2 = 6895 Pa	1 Pa = 0.000 1450 lbf/in^2
	kilopascal	kPa	1 lbf/in^2 = 6.895 kPa	1 kPa = 0.1450 lbf/in^2

(Continued)

Imperial unit	SI unit	SI unit symbol	Conversion factor (Imperial to SI)	Conversion factor (SI to imperial)
Weight density[a]				
Pound force per cubic foot	kilonewton per cubic metre	kN m^{-3}	1 lbf/ft^3 = 0.157 kN m^{-3}	1 kN m^{-3} = 6.37 lbf/ft^3
Energy				
Foot lbf	joule	J	1 ft lbf = 1.356 J	1 J = 0.7376 ft lbf

[a] Assuming a gravitational acceleration of 9.807 m/s^2.

References

Abellán, A., Oppikofer, T., Jaboyedoff, M., Rosser, N. J., Lim, M. and Lato, M. J. 2014. Terrestrial laser scanning of rock slope instabilities. *Earth Surface Processes Landforms*, 39(1):80–97.

Abrahamson, N. A. 2000. State of the practice of seismic hazard evaluation. *Proc. Geoeng2000*, Melbourne, Australia, pp. 659–85.

Adhikary, D. P., Dyskin, A. V., Jewell, R. J. and Stewart, D. P. 1997. A study of the mechanism of flexural toppling failures of rock slopes. *Rock Mech. Rock Eng.*, 30(2):75–93.

Adhikary, D. P. and Guo, H. 2002. An orthotropic cosserat elasto-plastic model for layered rocks. *Rock Mech. Rock Eng.*, 35(3):161–70.

Alzo'ubi, A. M. 2009. *The effect of tensile strength on the instability of rock slopes*, PhD thesis, University of Alberta, Canada.

American Association of State Highway and Transportation Officials (AASHTO) 1984. *A Policy on Geometric Design of Highways and Streets*. AASHTO, Washington, DC.

American Association of State Highway and Transportation Officials (AASHTO) 2012. *LRFD Bridge Design Specifications, Customary US Units*. AASHTO, Washington, DC, Contract No.: LRFDUS-6.

American Concrete Institute (ACI) (Revised 1995). Specifications for materials, proportioning and application of shotcrete. ACI Report 506.2-95.

Anderson, R. A. and Schuster, R. L. 1970. Stability of clay shales interbedded with Columbia River basalt. *Proceedings of the 8th Annual Engineering Geology and Soils Engineering Symposium*, Pocatello, Idaho, pp. 273–84.

Andrew, R. D. 1992a. Restricting rock falls. *Civil Eng. ASCE*, Washington, DC, October, 62(10):66–7.

Andrew, R. D. 1992b. *Selection of Rock Fall Mitigation Techniques Based on Colorado Rock Fall Simulation Program*. Transportation Research Record 1343, Transportation Research Board, National Research Council, Washington, DC, pp. 20–2.

Ashby, J. 1971. *Sliding and toppling modes of failure in models and jointed rock slopes*. MSc thesis, London University, Imperial College, UK.

Athanasiou-Grivas, D. 1979. Probabilistic evaluation of safety of soil structures. *J. Geotech. Eng. ASCE*, 105(GT9):109–15.

Athanasiou-Grivas, D. 1980. A reliability approach to the design of geotechnical systems. Rensselaer Polytechnic Institute Research Paper, *Transportation Research Board Conference*, Washington, DC.

Atkinson, L. C. 2000. The role and mitigation of groundwater in slope stability. In: *Slope Stability in Surface Mines*, Chapter 9, Society of Mining Engineers of AIME, Englewood, CO, pp. 89–96.

Atlas Powder Company. 1987. *Explosives and Rock Blasting*. Atlas Powder Company, Dallas, TX.

Australian Center for Geomechanics (ACG). 1998. Integrated monitoring systems for open pit wall deformation. X. Ding, S. B. Montgomery, M. Tsakiri, C. F. Swindells and R. J. Jewell, eds., MERIWA Project M236, Report ACG:1005-98.

Australian Drilling Industry (ADI). 1996. *Drilling: Manual of Methods, Applications and Management*, 4th edition. Australian Drilling Industry Training Committee, CRC Press LLC, Boca Raton, FL, 615pp.

Aycock, J. H. 1981. Construction problems involving shale in a geologically complex environment – State Route 32, Grainger County, Tennessee. *Proceedings of the 32nd Highway Geology Symposium*, Gatlinburg, TN.

Azzoni, A. and de Freitas, M. H. 1995. Experimentally gained parameters decisive for rock fall analysis. *Rock Mech. Rock Eng.*, 28(2):111–24.

Baecher, G. B., Lanney, N. A. and Einstein, H. 1977. Statistical description of rock properties and sampling. *Proceedings of the 18th U.S. Symposium on Rock Mechanics*. Johnson Publishing Co., Keystone, CO.

Baker, D. G. 1991. Wahleach power tunnel monitoring. In: *Field Measurements in Geotechnics*. CRC/Balkema, Leiden, Netherlands, Rotterdam, pp. 467–79.

Baker, W. E. 1973. *Explosives in Air.* University of Texas Press, Austin, TX.

Balmer, G. 1952. A general analytical solution for Mohr's envelope. *Am. Soc. Test. Mat.*, 52:1260–71.

Bandis, S. C. 1993. Engineering properties and characterization of rock discontinuities. In: *Comprehensive Rock Engineering: Principles, Practice and Projects*, J. A. Hudson, ed., Pergamon Press, Oxford, Vol. 1, pp. 155–83.

Barrett, R. K. and White, J. L. 1991. Rock fall prediction and control. *Proceedings of the National Symposium on Highway and Railway Slope Maintenance*, Association of Engineering Geology, Chicago, pp. 23–40.

Barton, N. and Bandis, S. 1983. Effects of block size on the shear behaviour of jointed rock. *Issues in Rock Mechanics – Proceedings of the 23rd U.S. Symposium on Rock Mechanics*, Berkeley, CA, Society of Mining Engineers of AIME, Englewood, CO, pp. 739–60.

Barton, N. R. 1971. *A model study of the behavior of excavated slopes.* PhD thesis, University of London, Imperial College, 520pp.

Barton, N. R. 1973. Review of a new shear strength criterion for rock joints. *Eng. Geol.*, 7:287–322.

Barton, N. R. 1974. A review of the shear strength of filled discontinuities in rock. Norwegian Geotechnical Institute, Pub. No. 105.

Baxter, D. A. 1997. Rockbolt corrosion under scrutiny. *Tunnels Tunneling Int.*, July, 35–8.

Beale, G. and Read, J. 2013. *Guidelines for Evaluating Water in Pit Slope Stability.* CSIRO Publishing, Melbourne, Australia.

Bieniawski, Z. T. 1976. Rock mass classification in rock engineering. In: *Exploration for Rock Engineering, Proceedings of Symposium*, Z. T. Bieniawski, ed., CRC/Balkema, Cape Town, Leiden, Netherlands, Vol. 1, pp. 97–106.

Birch, J. S. 2008. Using 3DM analyst mine mapping suite for underground mapping. In: *Laser and Photogrammetric Methods for Rock Tunnel Characterization*, F. Tonon, ed. 42nd U.S. Rock Mechanics Symposium, San Francisco, CA.

Bishop, A. W. 1955. The use of the slip circle in the stability analysis of earth slopes. *Geotechnique*, 5:7–17.

Bishop, A. W. and Bjerrum, L. 1960. The relevance the triaxial test to the solution of stability problems. *Proceedings of the ASCE Conference Shear Strength of Cohesive Soils*, Boulder, CO, pp. 437–501.

Black, W. H., Smith, H. R. and Patton, F. D. 1986. Multi-level ground water monitoring with the MP system. *Proceedings of the NWWA-AGU Conference on Surface and Borehole Geophysical Methods and Groundwater Instrumentation*, Denver, CO, pp. 41–61.

Bobet, A. 1999. Analytical solutions for toppling failure (Technical Note). *Int. J. Rock Mech. Min. Sci.*, 36:971–80.

Branner, J. C. 1895. Decomposition of rocks in Brazil. *Geol. Soc. Am. Bull.*, 7(1):255–314.

Brawner, C. O. and Kalejta, J. 2002. Rock fall control in surface mining with cable ringnet fences. *Proceedings of the Annual Meeting*, Society of Mining Engineers of AIME, Phoenix, AZ, 10pp.

Brawner, C. O., Pentz, D. L. and Sharp, J. C. 1971. Stability studies of a footwall slope in layered coal deposit. *13th Symposium on Rock Mechanics – Stability of Rock Slopes*, University of Illinois, Urbana, ASCE, pp. 329–65.

Brawner, C. O., Stacey, P. F. and Stark, R. 1975. A successful application of mining with pitwall movement. *Proceedings of the Canadian Institute of Mining, Annual Western Meeting*, Edmonton, October, 20pp.

Brawner, C. O. and Wyllie, D. C. 1975. Rock slope stability on railway projects. *Proceedings of the American Railway Engineering Association, Regional Meeting*, Vancouver, BC.

Bray, J. D. and Travasarou, T. 2009. Pseudostatic coefficient for use in simplified seismic slope stability evaluation. *J. Geotech. Geoenviron. Eng.*, 135(9):1336–40.

Brideau, M.-A. 2010. *Three-dimensional kinematic controls on rock slope stability conditions.* PhD thesis, Simon Fraser University, Burnaby, Canada.

British Standards Institute (BSI). 1989. *British Standard Code of Practice for Ground Anchorages, BS 8081:1989.* BSI, 2 Park Street, London, W1A 2BS, 176pp.

Broadbent, C. D. and Zavodni, Z. M. 1982. Influence of rock structure on stability. In: *Stability in Surface Mining*, Chapter 2, C. O. Brawner, ed., Society of Mining Engineers of AIME, Englewood, CO, Vol. 3.

Brock, R. W., Grount, F. F. and Swanson, C. O. 1943. Weathering of igneous rocks near Hong Kong. *Geol. Soc. Am. Bull.*, 54(6):717–38.

Brooker, E. W. and Anderson, I. H. 1979. Rock mechanics in damsite location. *Proceedings of the 5th Canadian Rock Mechanics Symposium*, Toronto, Canada, 75–90.

Brown, E. T. 1970. Strength of models of rock with intermittent joints. *J. Soil Mech. Found. Eng. Div. ASCE*, 96(SM6):1935–49.

Byerly, D. W. and Middleton, L. M. 1981. Evaluation of the acid drainage potential of certain Precambrian rocks in the Blue Ridge Province. *32nd Annual Highway Geology Symposium*, Gatlinburg, TN, pp. 174–85.

Byrne, R. J. 1974. *Physical and numerical models in rock and soil slope stability.* PhD thesis, James Cook University of North Queensland, Australia.

Cala, M., Kowalski, M. and Stopkowicz, A. 2014. The three-dimensional (3D) numerical stability analysis of Hyttemalmen open-pit. *Arch. Min. Sci.*, 59(3):609–20.

California Department of Conservation, C. G. S. 2008. *Guidelines for Evaluating and Mitigating Seismic Hazards in California.* California Geological Survey, Sacramento. Special Publication 117A.

California Division of Mines and Geology (CDMG). 1997. *Guildelines for Evaluating and Mitigating Seismic Hazards in California.* Califiornia Div. of Mines and Geology. Special Report 117. www/consrv.ca.gov/dmg/pubs/sp/117/

Call, R. D. 1982. Monitoring pit slope behaviour. In: *Stability in Surface Mining*, Chapter 9, C. O. Brawner, ed., Society of Mining Engineers of AIME, Englewood, CO, Vol. 3, pp. 42–62.

Call, R. D. 1992. Slope stability. In: *SME Mining Engineering Handbook*. Society of Mining Engineers of AIME, Englewood, CO, Vol. 1, pp. 881–96.

Call, R. D., Saverly, J. P. and Pakalnis, R. 1982. A simple core orientation device. In: *Stability in Surface Mining*, C. O. Brawner, ed., Society of Mining Engineers of AIME, Englewood, CO, pp. 465–81.

Canada Department of Energy, Mines and Resources. 1978. *Pit Slope Manual.* DEMR, Ottawa, Canada.

Canadian Geotechnical Society. 1992. *Canadian Foundation Engineering Manual.* BiTech Publishers Ltd., Vancouver, Canada.

Canadian Standards Association. 1988. *Design of Highway Bridges.* CAN/CSA-56-88, Rexdale, Ontario.

Carvajal, H. E. M., Restrepo, P. A. I. and Azevedo, G. F. 2012. Landslide risk management in Medellin, Colombia. In: *Extreme Rainfall Induced Landslides*, W. A. Lacerda, E. M. Palmeira, A. L. Netto and M. Ehrlich, eds., Oficina de Textos, Brazil, pp. 299–323.

Carvalho, J. L., Kennard, D. T. and Lorig, L. 2002. Numerical analysis of the east wall of Toquepala Mine, Southern Andes of Peru. In: *EUROCK 2002*, C. Dinas da Gama and L. Ribeiro e Sousa, eds., Sociedade Portuguesa de Geotecnica, Maderia, Portugal, pp. 615–25.

Casagrande, L. 1934. Näherungsverfahren zur Ermittlung der Sickerung in geschütteten Dämmen auf underchläassiger Sohle. *Die Bautechnik*, Heft 15.

Caterpillar Inc. 2015. *Caterpillar Performance Handbook*, 29th Edition. Peoria, IL, 1014 pages.

Caterpillar Tractor Co. 2015. *Caterpillar Performance Handbook*, 32nd edition. Caterpillar Tractor Co, Peoria, IL.

Cavers, D. S. 1981. Simple methods to analyze buckling of rock slopes. *Rock Mech.*, 14(2):87–104.

Cedergren, H. R. 1989. *Seepage, Drainage and Flow Nets*, 3rd edition. John Wiley & Sons, New York, 465pp.

Chen, Z. 1995a. Keynote Lecture: Recent developments in slope stability analysis. *Proceedings of the 8th International Congress on Rock Mechanics*, ISRM, Tokyo, Japan, Vol. 3, pp. 1041–8.

Chen, Z. editor in chief. 1995b. Transactions of the stability analysis and software for steep slopes in China. Vol. 3.1: Rock classification, statistics of database of failed and natural slopes. China Institute of Water Resources and Hydroelectric Power Research (in Chinese).

Cheng, Y. and Liu, S. 1990. Power caverns of the Mingtan Pumped Storage Project, Taiwan. In: *Comprehensive Rock Engineering*, J. A. Hudson, ed., Pergamon Press, Oxford, Vol. 5, pp. 111–32.

Christensen Boyles Corp. 2000. *Diamond Drill Products – Field Specifications*. CBC, Salt Lake City, UT.

Ciarla, M. 1986. Wire netting for rock fall protection. *Proceedings of the 37th Annual Highway Geology Symposium*, Helena, Montana, Montana Department of Highways, pp. 100–18.

C.I.L. 1983. *Blaster's Handbook*. Canadian Industries Limited, Montréal, Québec.

Clark, I. 1979. *Practical Geostatistics*, 1st edition. Elsevier Applied Science, Amsterdam.

Clayton, M. A. 2014. *Characterization and analysis of the Mitchell Creek Landslide: A large-scale rock slope instability in north-western British Columbia*. MSc thesis, Simon Fraser University, Burnaby, Canada.

Clayton, M. A. and Stead, D. 2015. A discontinuum numerical modelling investigation of failure mechanisms at the Mitchell Creek landslide, B.C., Canada. *13th International Congress of Rock Mechanics*, Montreal, Canada, Canadian Institute of Mining, Metallurgy and Petroleum.

Cluff, L. S., Hansen, W. R., Taylor, C. L., Weaver, K. D., Brogan, G. E., Idress, I, M., McClure, F. E. and Bayley, J. A. 1972. Site evaluation in seismically active regions – An interdisciplinary team approach. *Proceedings of the International Conference on Microzonation for Safety, Construction, Research and Application*, Seattle, WA, Vol. 2, pp. 9-57–9.87.

Coates, D. F., Gyenge, M. and Stubbins, J. B. 1965. Slope stability studies at Knob Lake. *Proceedings of the Rock Mechanics Symposium*, Toronto, pp. 35–46.

Colog Inc. 1995. *Borehole Image Processing System* (BIP). Golden, Colorado, and Raax Co. Ltd., Australia.

Commonwealth Scientific and Industrial Research Organization CSIRO. 2001. *SIROJOINT Geological Mapping Software*. CSIRO Mining and Exploration, Pullenvale, Queensland, Australia.

Commonwealth Scientific and Industrial Research Organization (CSIRO). 2009. *Guidelines for Open Pit Slope Design*, J. Read and P. F. Stacey, eds., CRC/Balkema, Leiden, Netherlands, 496pp.

Corominas, J. 1996. The angle of reach as a mobility index for small and large landslides. *Can. Geotech. J.*, 33:260–71.

Costa, J. E. and Schuster, R. L. 1988. The formation and failure of natural dams. *Geological Society of America Bulletin*, 100(7):1054–68.

Coulomb, C. A. 1773. Sur une application des règles de Maximis et Minimis a quelques problèmes de statique relatifs à l'Architechture. *Acad. Roy. des Sciences Memoires de math. et de physique par divers savans*, 7:343–82.

Cruden, D. M. 1977. Describing the size of discontinuities. *Int. J. Rock Mech. Min. Sci. Geomech. Abstr.*, 14:133–7.

Cruden, D. M. 1997. Estimating the risks from landslides using historical data. In: *Proceedings of the International Workshop on Landslide Risk Assessment*, D. M. Cruden and R. Fell, eds., Honolulu, HI, CRC/Balkema, Leiden, Netherlands, pp. 177–84.

Cundall, P. A. 1971. A computer model for simulating progressive, large scale movements in blocky rock systems. *Proceedings of the International Symposium on Rock Fracture*, Nancy, France, paper 11–8.

Cundall, P. A. and Damjanac, B. 2009. A comprehensive 3D model for rock slopes based on micromechanics. *Proceedings of the Slope Stability 2009 Conference*, Santiago, Chile, 10pp.

Davies, J. N. and Smith, P. L. P. 1993. Flexural toppling of siltstones during a temporary excavation for a bridge foundation in North Devon. In: *Comprehensive Rock Engineering*, J. Hudson, ed., Chapter 31, Pergamon Press, Oxford, UK, Vol. 5, pp. 759–75.

Davis, G. H. and Reynolds, S. J. 1996. *Structural Geology of Rocks and Regions*, 2nd edition. John Wiley & Sons, New York, 776pp.

Davis, S. N. and DeWiest, R. J. M. 1966. *Hydrogeology.* John Wiley & Sons, New York.

Dawson, E. M., Roth, W. H. and Drescher, A. 1999. Slope stability analysis by strength reduction. *Géotechnique*, 49(6):835–40.

de Freitas, M. H. and Watters, R. J. 1973. Some field examples of toppling failure. *Géotechnique*, 23(4):495–513.

Deere, D. U. and Miller, R. P. 1966. *Engineering classification and index properties of intact rock.* Technical Report No. AFWL-TR-65-116, Air Force Weapons Laboratory, Kirkland Air Force Base, New Mexico.

Deere, D. U. and Patton, F. D. 1971. Slope stability in residual soils. *4th PanAmerican Conference on Soil Mechanics and Foundation Engineering*, San Juan, Puerto Rico.

Dehls, J. F., Farina, P., Martin, D. and Froese, C. 2010. Monitoring Turtle Mountain using ground-based synthetic aperture radar (GB-InSAR). *GEO2010 – 63rd Canadian Geotechnical Conference*, Calgary, Alberta, Canadian Geotechnical Society (CGS).

Dershowitz, W. S. and Einstein, H. H. 1988. Characterizing rock joint geometry with joint system models. *Rock Mech. Rock Eng.*, 20(1):21–51.

Dershowitz, W. S., Lee, G., Geier, J., Foxford, T., LaPointe, P. and Thomas, A. 2004. *FracMan – Interactive Discrete Feature Data Analysis, Geometric Modeling, and Exploration Simulation*, User documentation. Golder Associates Inc., Seattle, WA.

Dershowitz, W. S., Lee, G., Geier, J., Hitchcock, S. and LaPointe, P. 1994. *FRACMAN Interactive Discrete Feature Data Analysis, Geometric Modeling, and Exploration Simulation*. User documentation V. 2.4, Golder Associates Inc., Redmond, WA.

Descoeudres, F. and Zimmerman, T. 1987. Three-dimensional calculation of rock falls. *Proceedings of the International Conference Rock Mechanics*, Montreal, Canada.

Deutsch, C. V. 2002. *Geostatistical Reservoir Modeling.* Oxford University Press, Oxford, UK, 400pp.

Dowding, C. H. 1985. *Blast Vibration Monitoring and Control.* Prentice-Hall, Upper Saddle River, NJ.

Duffy, J. D. and Haller, B. 1993. Field tests of flexible rock fall barriers. *Proceedings of the International Conference on Transportation Facilities through Difficult Terrain*, Aspen, CO, CRC/Balkema, Leiden, Netherlands, pp. 465–73.

Duncan, J. M. 1996. *Landslides: Investigation and Mitigation.* Transportation Research Board, Special Report 247, Washington, DC, Chapter 13, Soil Slope Stability Analysis, pp. 337–71.

Dunnicliff, J. 1993. *Geotechnical Instrumentation for Monitoring Field Performance*, 2nd edition. John Wiley & Sons, New York, 577pp.

Duran, J. and Douglas, L. H. 1999. Do rock slopes designed with empirical rock mass strength stand up? *9th Congress of the International Society of Rock Mechanics*, Paris, pp. 87–90.

Dyno Nobel. 2013. Product technical information. Dyno Nobel, http://www.dynonobel.com.

Eberhardt, E., Stead D. and Coggan, J. S. 2004. Numerical analysis of initiation and progressive failure in natural rock slopes – The 1991 Randa rockslide. *Int. J. Rock Mech.*, 41:69–87.

Einstein, H. H. 1993. Modern developments in discontinuity analysis. In: *Comprehensive Rock Engineering*, J. Hudson, ed., Chapter 9, Pergamon Press, Oxford, UK, Vol. 3, pp. 193–213.

Einstein, H. H., Veneziano, D., Baecher, G. B. and O'Reilly, K. J. 1983. The effect of discontinuity persistence on slope stability. *Int. J. Rock Mech. Min. Sci. Geomech. Abstr.*, 20:227–36.

Elmo, D. 2006. *Evaluation of a hybrid FEM/DEM approach for determination of rock mass strength using a combination of discontinuity mapping and fracture mechanics modelling, with particular emphasis on modelling of jointed pillars.* PhD thesis, University of Exeter, Exeter, UK.

Elmo, D., Clayton, C., Rogers, S., Beddoes, R. and Greer, S. 2011. Numerical simulations of potential rock bridge failure within a naturally fractured rock mass. *Proceedings of the 2011 International Symposium on Slope Stability in Mining and Civil Engineering*, Vancouver, pp. 18–21.

Erban, P. J. and Gill, K. 1988. Consideration of the interaction between dam and bedrock in a coupled mechanic-hydraulic FE program. *Rock Mech. Rock Engineering*, 21(2):99–118.

European Committee for Standardization. 1995. *Eurocode 1: Basis of Design and Actions on Structures – Part 1. Basis of Design*. Central Secretariat, Brussels.

Farmer, I. W. 1975. Stress distribution along a resin grouted anchor. *Int. J. Rock Mech. Geomech. Abstr.*, 12(11):347–51.

Fecker, E. and Rengers, N. 1971. Measurement of large scale roughness of rock planes by means of profilometer and geological compass. *Proceedings of the Symposium on Rock Fracture*, Nancy, France, Paper 1–18.

Federal Highway Administration (FHWA). 1982. *Tiebacks*. U.S. Department of Transportation, Washington, DC, Contract No. FHWA/RD-82/047.

Federal Highway Administration (FHWA). 1991. *Rock Blasting and Overbreak Control*. U.S. Department of Transportation, Washington, DC, Contract No. DTFW 61-90-R-00058.

Federal Highway Administration (FHWA). 1993. *Rockfall Hazard Mitigation Methods*. U.S. Department of Transportation, Washington, DC, Publication No. FHWA SA-95-085, Washington, DC.

Federal Highway Administration (FHWA). 1998a. *Geotechnical Earthquake Engineering*. U.S. Department of Transportation, Washington, DC, Contract No. FHWA-HI-99-012, 265pp.

Federal Highway Administration (FHWA). 1998b. *Rock Slopes*. U.S. Department of Transportation, Washington, DC, Publication No. FHWA-HI-99-007, FHWA, Washington, DC.

Federal Highway Administration (FHWA). 2011. *LRFD seismic analysis and design of transportation geotechnical features and structural foundations*. NHI Course No. 130094 Reference Manual. J. E. Kavazanjian, J.-N. J. Wang, G. R. Martin, A. Shamsabadi, I. P. Lam and S. E. Dickenson, eds., U.S. Department of Transportation, Washington, DC, Report No.: FHWA-NHI-11-032.

Federico, A., Popescu, M. and Murianni, A. 2015. Temporal prediction of landslide occurence: A possibility or a challenge? *Italian J. of Engineering Geology and Environment*, 1:41–60.

Fekete, S., Diederichs, M., Lato, M. and Grimstad, E. 2008. HD laser scanning in active tunnels: Challenges, solutions, applications. *Tunnelling Association of Canada, Annual Conference*, Toronto.

Fell, R. 1994. Landslide risk assessment and acceptable risk. *Can. Geotech. J.*, 31(2):261–72.

Fenton, G. A., Naghibi, F., Dundas, D., Bathurst, R. J. and Griffiths, D. V. 2015. Reliability-based geotechnical design in 2014 Canadian Highway Bridge Design Code. *Can. Geotech. J.*, 53(2):236–51.

Fisher, B. R. 2009. *Improved characterization and analysis of biplanar dip slope failures to limit model and parameter uncertainty in the determination of setback distances*. PhD thesis, University of British Columbia, Vancouver, BC.

Fisher, B. R. and Eberhardt, E. 2012. Assessment of parameter uncertainty associated with dip slope stability analyses as a means to improve site investigations. *J. Geotech. Geoenviron. Eng.*, 138(2):166–73.

Fleming, R. W., Spencer, G. S. and Banks, D. C. 1970. *Empirical study of the behaviour of clay shale slopes*. U.S. Army Nuclear Cratering Group Technical Report, No. 15.

Flores, G. and Karzulovic, A. 2000. The role of the geotechnical group in an open pit: Chuquicamata Mine, Chile. *Proceedings of the Slope Stability in Surface Mining*, Society of Mining Engineers of AIME, Englewood, CO, pp. 141–52.

Fookes, P. G. and Sweeney, M. 1976. Stabilization and control of local rock falls and degrading rock slopes. *Quart. J. Eng. Geol.*, 9:37–55.

Forster, J. W. 1986. *Geological problems overcome at Revelstoke, Part 2*. Water Power and Dam Construction, August.

Fredlund, M. 2014. Is there a movement toward three-dimensional slope stability analyses? *Geo-Strata*, 18(4):22–5.

Freeze, R. A. and Cherry, J. A. 1979. *Groundwater*. Prentice-Hall, Upper Saddle River, NJ, 604pp.

Frohlich, O. K. 1955. General theory of the stability of slopes. *Geotechnique*, 5:37–47.

Fukuzono, T. 1985. Method to predict failure time of slope collapse using inverse of surface moving velocity by precipitation, *Landslide*, 22:8–13 (in Japanese).

Gaffney, S. P., White, J. and Ellis, W. L. 2002. Instrumentation of the Debeque Canyon Landslide at Interstate 70 in West Central Colorado. *2002 Denver Annual Meeting*, Sunday, October 27, 2002, Colorado Convention Center, The Geological Society of America (GSA).

Gao, F. Q. and Stead, D. 2014. The application of a modified Voronoi logic to brittle fracture modelling at the laboratory and field scale. *Int. J. Rock Mech. Min. Sci.*, 68:1–14.

Geological Society Engineering Group Working Party. 1977. Report on the logging of rock cores for engineering purposes. J. L. Knill, C. R. Cratchley, K. R. Early, R. W. Gallois, J. D. Humphreys and J. Newbery, eds., *Quart. J. Eng. Geol. Hydrogeol.*, 3(1):1–24.

Geomechanica Inc. 2015. Irazu, http://www.geomechanica.com/services-slope-stab.html.

Gischig, V. S. 2011. *Kinematics and failure mechanisms of the Randa rock slope instability (Switzerland).* PhD thesis, ETH, Zurich, Switzerland.

Glass, C. E. 2000. The influence of seismic events on slope stability. In: *Proceedings of the Slope Stability in Surface Mining*, W. A. Hustrulid, M. K. McCarter and D. J. A. Van Zyl, eds., Society of Mining Engineers of AIME, Englewood, CO, pp. 97–105.

Golder, H. Q. 1972. The stability of natural and man-made slopes in soil and rock. In: *Geotechnical Practice for Stability in Open Pit Mining*, C. O. Brawner and V. Milligan, eds., Society of Mining Engineers of AIME, Englewood, CO, pp. 79–85.

Goldich, S. S. 1938. A study in rock weathering. *J. Geol.*, 46(1):17–58.

Goodman, R. E. 1964. The resolution of stresses in rock using stereographic projection. *Int. J. Rock Mech. Min. Sci.*, 1:93–103.

Goodman, R. E. 1970. The deformability of joints. In: *Determination of the In Situ Modulus of Deformation of Rock*. American Society for Testing and Materials Publications, Washington, DC, No. 477, pp. 174–96.

Goodman, R. E. 1976. *Methods of Geological Engineering in Discontinuous Rocks*. West Publishing Co., St. Paul, MN, 472pp.

Goodman, R. E. 1980. *Introduction to Rock Mechanics*. John Wiley & Sons, New York, 478pp.

Goodman, R. E. and Bray, J. 1976. Toppling of rock slopes. *ASCE, Proceedings of the Specialty Conference on Rock Engineering for Foundations and Slopes*, Boulder, CO, Vol. 2, pp. 201–34.

Goodman, R. E. and Shi, G. 1985. *Block Theory and Its Application to Rock Engineering*. Prentice-Hall, Upper Saddle River, NJ.

Griffiths, J. S., Stokes, M., Stead, D. and Giles, D. 2012. Landscape evolution and engineering geology: Results from IAEG Commission 22. *Bull. Eng. Geol. Environ.*, 71:605–36.

Grimod, A. and Giacchetti, G. 2013. Protection from high energy impacts using reinforced soil embankments: Design and experiences. In: *Landslide Science and Practice*, C. Margottini, P. Canuti and K. Sassa, eds., Springer-Verlag, Rome, pp. 189–96.

Grøneng, G., Lu, M., Nilsen, B. and Jenssen, A. K. 2010. Modelling of time-dependent behaviour of the basal sliding surface of the Åknes rockslide area in western Norway. *Eng. Geol.*, 114:412–22.

Guthrie, R. H. and Nicksiar, M. 2016. Time of failure – practical improvements of an analytical tool. *Proc. GeoVancouver 2016 Conf.*, Canadian Geotechnical Society, Vancouver, September, paper 3742, pp. 7.

Haar, M. E. 1962. *Ground Water and Seepage*. McGraw-Hill Co., New York.

Hagan, T. N. 1975. Blasting physics – What the operator can use in 1975. *Proceedings of the Australian Institute of Mining and Metallurgy, Annual Conference*, Adelaide, Part B, pp. 369–86.

Hagan, T. N. and Bulow, B. 2000. Blast designs to protect pit walls. In: *Slope Stability in Surface Mining*. Society Mining Engineers of AIME, Englewood, CO, pp. 125–30.

Haines, A. and Terbrugge, P. J. 1991. Preliminary estimate of rock slope stability using rock mass classification. *7th Congress of the International Society of Rock Mechanics*, Aachen, Germany, pp. 887–92.

Hamdi, P., Stead, D., Elmo, D., Töyrä, J. and Stöckel, B-M. 2014. Use of an integrated finite/discrete element method-discrete fracture network approach (FDEM-DFN) in characterizing surface subsidence associated with sub-level caving. *Proceedings of the DFNE 2014*, Vancouver, Canada, Paper DFN 2014-268, 8pp.

Hamel, J. V. 1970. The Pima Mine slide, Pima County, Arizona. *Geological Society of America, Abstracts with Programs*, 2(5):335.

Hamel, J. V. 1971a. The slide at Brilliant cut. *Proceedings of the 13th Symposium on Rock Mechanics*, Urbana, IL, pp. 487–510.

Hamel, J. V. 1971b. Kimberley Pit slope failure. *Proceedings of the 4th Pan-American Conference on Soil Mechanics and Foundation Engineering*, Puerto Rico, Vol. 2, pp. 117–27.

Hammett, R. D. 1974. *A study of the behaviour of discontinuous rock masses*. PhD thesis, James Cook University of North Queensland, Australia.

Han, C. 1972. *Technique for obtaining equipotential lines of ground water flow in slopes using electrically conducting paper*. MSc thesis, University of London (Imperial College), UK.

Hanna, T. H. 1982. *Foundations in Tension – Ground Anchors*. Trans Tech Publications/McGraw-Hill Book Co., Clausthal-Zellerfeld, West Germany.

Harp, E. L. and Jibson, R. W. 2002. Anomalous concentrations of seismically triggered rock falls in Pacoima Canyon: Are they caused by highly susceptible slopes or local amplification of seismic shaking? *Bull. Seismol. Soc. Am.*, 92(8):3180–9.

Harp, E. L., Jibson, R. W., Kayen, R. E., Keefer, D. S., Sherrod, B. L., Collings, B. D., Moss, R. E. S. and Sitar, N. 2003. Landslides and liquefaction triggered by the M7.9 Denali Fault earthquake of 3 November 2002. *GSA Today (Geological Society of America)*, August, 4–10.

Harp, E. L. and Noble, M. A. 1993. An engineering rock classification to evaluate seismic rock fall susceptibility and its application to the Wasatch Front. *Bull. Assoc. Eng. Geol.*, XXX(3):293–319.

Harp, E. L. and Wilson, R. C. 1995. Shaking intensity thresholds for rock falls and slides: Evidence from the 1987 Whittier Narrows and Superstition Hills earthquake strong motion records. *Bull. Seismol. Soc. Am.*, 85(6):1739–57.

Harr, M. E. 1977. *Mechanics of Particulate Matter – A Probabilistic Approach*. McGraw-Hill, New York, 543pp.

Harries, G. and Merce, J. K. 1975. The science of blasting and its use to minimize costs. *The Australasian Institute of Mining and Metallurgy Conference (AusIMM)*, June 1975, Adelaide.

Havaej, M. 2015. *Characterisation of high rock slopes using an integrated numerical modelling – Remote sensing approach*. PhD thesis, Simon Fraser University, Canada.

Havaej, M., Coggan, J., Stead, D. and Elmo, D. 2016a. A combined remote sensing–numerical modelling approach to the stability analysis of Delabole Slate Quarry, *Rock Mech. Rock Eng.*, 49(4):1227–45.

Havaej, M., Stead, D., Coggan, J. and Elmo, D. 2016b. Application of discrete fracture networks in the stability analysis of Delabole Slate Quarry, Cornwall, UK, *Proc. ARMA 2016*, Houston, TX, Paper#40, 10pp.

Havaej, M., Stead, D., Eberhardt, E. and Fisher, B. 2014a. Characterization of bi-planar and ploughing failure mechanisms in footwall slopes using numerical modelling. *Eng. Geol.*, 178:109–20.

Havaej, M., Stead, D., Mayer, J. and Wolter, A. 2014b. Modelling the relation between failure kinematics and slope damage in high rock slopes using a lattice scheme approach, *Proc. ARMA 2014*, Minneapolis, Paper #14-7374, 8pp.

Hemphill, G. B. 1981. *Blasting Operations*. McGraw-Hill Inc., New York, 258pp.

Hencher, S. R. and Richards, L. R. 1989. Laboratory direct shear testing of rock discontinuities. *Ground Eng.*, March, 24–31.

Hendron, A. J. and Patton, F. D. 1985. *The Vajont slide: A geotechnical analysis based on new geological observations of the failure surface*. Waterways Experiment Station Technical Report, U.S. Army Corps of Engineers, Vicksburg, MS, Vol. 1, p. 104.

Heuer, R. E. 1995. Estimating ground water flow in tunnels. *Proceedings of the Rapid Excavation and Tunneling Conference*, San Francisco, pp. 41–80.

Hiltunen, D. R., Hudyma, N. and Quigley, T. P. 2007. Ground proving three seismic refraction tomography programs. Transportation Research Board. *J. Transport. Res. Board*, 2016, 110–20.

Hocking, G. 1976. A method for distinguishing between single and double plane sliding of tetrahedral wedges. *Int. J. Rock Mech. Min. Sci.*, 13(7):225–6.

Hoek, E. 1968. Brittle failure of rock. In: *Rock Mechanics in Engineering Practice*, K. G. Stagg and O. C. Zienkiewicz, eds., John Wiley & Sons, London, pp. 99–124.

Hoek, E. 1970. Estimating the stability of excavated slopes in opencast mines. *Trans. Inst. Min. Metall.*, London, 79:A109–32.

Hoek, E. 1974. Progressive caving induced by mining an inclined ore body. *Trans. Int. Min. Metall.*, London, 83:A133–39.

Hoek, E. 1983. Strength of jointed rock masses, 23rd Rankine Lecture. *Géotechnique*, 33(3):187–223.

Hoek, E. 1990. Estimating Mohr–Coulomb friction and cohesion values from the Hoek–Brown failure criterion. *Int. J. Rock Mech. Min. Sci. Geomech. Abstr.*, 12(3):227–29.

Hoek, E. 1994. Strength of rock and rock masses. *ISRM News J.*, 2(2):4–16.

Hoek, E. and Bray, J. 1981. *Rock Slope Engineering*, 3rd edition. Institute of Mining and Metallurgy, London, UK.

Hoek, E., Bray, J. and Boyd, J. 1973. The stability of a rock slope containing a wedge resting on two intersecting discontinuities. *Quart. J. Eng. Geol.*, 6(1):22–35.

Hoek, E. and Brown, E. T. 1980a. Empirical strength criterion for rock masses. *J. Geotech. Eng Div. ASCE*, 106(GT9):1013–35.

Hoek, E. and Brown, E. T. 1980b. *Underground Excavations in Rock*, Institute of Mining and Metallurgy, London, UK, 527pp.

Hoek, E. and Brown, E. T. 1988. The Hoek–Brown failure criterion – A 1988 update. In: *Proceedings of the 15th Canadian Rock Mechanics Symposium*, J. C. Curran, ed., Department of Civil Engineering, University of Toronto, Toronto, pp. 31–8.

Hoek, E. and Brown, E. T. 1997. Practical estimates of rock mass strength. *Int. J. Rock Mech. Min. Sci. Geomech. Abstr.*, 34(8):1165–86.

Hoek, E., Carranza-Torres, C. and Corkum, B. 2002. Hoek–Brown Failure Criterion – 2002 edition. *Proceedings of the North American Rock Mechanics Society Meeting*, Toronto, Canada, July, pp. 267–73.

Hoek, E., Hutchinson, J., Kalenchuk, K. and Diederichs, M. 2009. Influence of *in situ* stresses on open pit design. In: *Guidelines for Open Pit Slope Design, App. 3*, J. Read and P. Stacey, eds., CSIRO, Brisbane, Australia.

Hoek, E., Kaiser, P. K. and Bawden, W. F. 1995. *Support of Underground Excavations in Hard Rock*. CRC/Balkema, Leiden, Netherlands, 215pp.

Hoek, E. and Marinos, P. 2000. Predicting Tunnel Squeezing. Tunnels and Tunnelling International. Part 1 – November 2000, Part 2 – December, 2000.

Hoek, E., Marinos, P. and Benissi, M. 1998. Applicability of the Geological Strength Index (GSI) classification for very weak and sheared rock masses. The case of the Athens Schist Formation. *Bull. Eng. Geol. Environ.*, 57(2):151–60.

Hoek, E. and Richards, L. R. 1974. *Rock Slope Design Review*. Golder Associates Report to the Principal Government Highway Engineer, Hong Kong, 150pp.

Hoek, E., Wood, D. F. and Shah, S. 1992. A modified Hoek–Brown criterion for jointed rock masses. In: *Proceedings of the Rock Characterization, Symposium of International Society of Rock Mechanics: Eurock'92*, J. A. Hudson, ed., British Geotechnical Society, London, pp. 209–14.

Hofmann, H. 1972. Kinematische Modellstudien zum Boschungs-problem in regalmassig geklüfteten Medien. Veröffentlichungen des Institutes für Bodenmechanik und Felsmechanik. Kalsruhe, Hetf 54.

Hong Kong Geotechnical Engineering Office. 2000. *Highway Slope Manual*. Civil Engineering Department, Govt., Hong Kong, Special Administrative Region, 114pp.

Hosseinian, A., Rasouli, V., Utikar, R. 2010. Fluid flow response of JRC exemplar profiles. In: *EUROCK 2010*, T. F. G. London, ed., *ISRM International Symposium*, Lausanne, Switzerland.

Huat, B. B. K., Toll, D. G. and Prasad, A. 2012. *Handbook of Tropical Residual Soils Engineering*. CRC Press, Boca Raton, FL, 536pp.

Hudson, J. A. and Priest, S. D. 1979. Discontinuities and rock mass geometry. *Int. J. Rock Mech. Min. Sci. Geomech. Abstr.*, 16:336–62.

Hudson, J. A. and Priest, S. D. 1983. Discontinuity frequency in rock masses. *Int. J. Rock Mech. Min. Sci. Geomech. Abstr.*, 20:73–89.

Huitt, J. L. 1956. Fluid flow in a simulated fracture. *J. Am. Inst. Chem. Eng.*, 2:259–64.

Hungr, O. 1987. An extension of Bishop's simplified method of slope stability to three dimensions. *Geotechnique*, 37:113–7.

Hungr, O. 1995. A model for the runout analysis of rapid flow slides, debris flows, and avalanches. *Can. Geotech. J.*, 32:610–23.

Hungr, O. 2005. Entrainment of material by debris flows. In: *Debris-Flow Hazards and Related Phenomena*, Chapter 7, M. Jakob and O. Hungr, eds., Praxis Springer, Berlin, pp. 305–24.

Hungr, O. 2014. *North Peak of Turtle Mountain, Frank, Alberta – Runout analyses of two potential landslides*, Report submitted to AER-Alberta Geological Survey, 19pp.

Hungr, O. and Amann, F. 2011. Limit equilibrium of asymmetric laterally constrained rockslides. *Int. J. Rock Mech. Min. Sci.*, 48(5):748–58.

Hungr, O. and Evans, S. G. 1988. Engineering evaluation of fragmental rock fall hazards. *Proceedings of the 5th International Symposium on Landslides*, Lausanne, Switzerland, July, pp. 685–90.

Hungr, O. and Evans, S. G. 1996. Rock avalanche runout prediction using a dynamic model. In: *Landslides*, Senneset, ed., Balkema, Rotterdam, pp. 233–38, ISBN 9054108185.

Hungr, O., Evans, S. G. and Hazzard, J. 1999. Magnitude and frequency of rock falls and rock slides along the main transportation corridors of south-western British Columbia. *Can. Geotech. J.*, 36:224–38.

Hungr, O., Morgan, G. C. and Kellerhals, R. 1984. Quantitative analysis of debris torrent hazards for design or remedial measures. *Can. Geotech. J.*, 21:663–77.

Husid, R. L. 1969. Analisis de terremotos: Analysis general. Revista del IDEM. No. 8, Santiago, Chile, 21–42.

Hutchinson, J. N. 1970. Field and laboratory studies of a fall in upper chalk cliffs at Joss Bay, Isle of Thanet. In: *Proceedings of the Roscoe Memorial Symposium*, R. H. G. Parry, ed., Cambridge University, March 29–31.

IAEG Commission on Landslides. 1990. Suggested nomenclature for landslides. *Bull. Int. Assoc. Eng. Geol.*, 41:13–6.

International Society of Explosives Engineers (ISEE). 2015. *Blaster's Handbook*, 18th edition. ISEE, Cleveland, OH, 742pp.

International Society for Rock Mechanics (ISRM). 1981a. *Suggested Methods for the Quantitative Description of Discontinuities in Rock Masses*. E. T. Brown, ed., Pergamon Press, Oxford, UK, 211pp.

International Society of Rock Mechanics (ISRM). 1981b. Basic geological description of rock masses. *Int. J. Int. J. Rock Mech. Min. Sci. Geomech. Abstr.*, 18:85–110.

International Society of Rock Mechanics (ISRM). 1985. Suggested methods for determining point load strength. *Int. J. Rock Mech. Min. Sci. Geomech. Abstr.*, 22(2):53–60.

International Society of Rock Mechanics (ISRM). 1991. Suggested methods of blast vibration monitoring. (co-ordinator: C. H. Dowding). *Int. J. Rock Mech. Min. Sci. Geomech. Abstr.*, 129(2):143–216.

International Society for Rock Mechanics (ISRM). 2007–2014. *Suggested Methods for Rock Characterization, Testing and Monitoring*, R. Ulusay, ed., Springer-Verlag, Switzerland, 293pp.

Isaaks, E. H. and Srivastava, R. M. 1989. *An Introduction to Applied Geostatistics*, 1st edition. Oxford University Press, Oxford, UK, 592pp.

Ishikawa, N. 1999. Recent progress on rock shed studies in Japan. *Proceedings of the Joint Japan-Swiss Scientific Seminar on Impact Loading by Rock Falls and Design of Protection Measures, Japan Society of Civil Engineers*, Kanazawa, Japan, pp. 1–6.

Itasca Consulting Group, Inc. 2012. *FLAC3D — Fast Lagrangian Analysis of Continua in Three-Dimensions, Ver. 5.0*. Minneapolis, MN.

Itasca Consulting Group, Inc. 2013. *3DEC – Three-Dimensional Distinct Element Code, Ver. 5.0*. Minneapolis, MN.

Itasca Consulting Group, Inc. 2014. *PFC2D – Particle Flow Code in Two Dimensions, Ver. 5.0*. Minneapolis, MN.

Itasca Consulting Group, Inc. 2014a. *PFC Suite – Particle Flow Code in Two and Three Dimensions, Ver. 5.0*. Minneapolis, MN.

Itasca Consulting Group, Inc. 2014b. *UDEC – Universal Distinct Element Code, Ver. 6.0 User's Manual*. Minneapolis, MN.

Itasca Consulting Group, Inc. 2015. *KUBRIX®, Ver. 15.0.* Minneapolis, MN.

Itasca Consulting Group, Inc. 2016. *FLAC – Fast Lagrangian Analysis of Continua, Ver. 8.0.* Minneapolis, MN.

Jacob, C. E. 1950. Flow of ground water. In: *Engineering Hydraulics*, H. Rouse, ed., John Wiley & Sons, New York, pp. 321–86.

Jaegar, J. C. 1970. The behaviour of closely jointed rock. *Proceedings of the 11th Symposium on Rock Mechanics*, Berkeley, CA, 57–68.

Jaeger, J. C. and Cook, N. G. W. 1976. *Fundamentals of Rock Mechanics*, 2nd edition. Chapman & Hall, London, UK, 585pp.

Janbu, N. 1954. Application of composite slide circles for stability analysis. *Proceedings of the European Conference on Stability of Earth Slopes*, Stockholm, Vol. 3, pp. 43–9.

Janbu, N., Bjerrum, L. and Kjaernsli, B. 1956. Soil mechanics applied to some engineering problems (in Norwegian with English summary). Norwegian Geotech. Inst., Publication 16.

Japan Ministry of Construction. 1983. *Reference Manual on Erosion Control Works (in Japanese).* Japan Erosion Control Department, Tokyo, Japan.

Jefferies, M., Lorig, L. and Alvarez, C. 2008. Influence of rock-strength spatial variability on slope stability. In: *Continuum and Distinct Element Numerical Modeling in Geo-Engineering – 2008*, R. Hart, C. Detournay and P. Cundall, eds., CRC Press, Boca Raton, FL, Paper 01–05.

Jennings, J. E. 1970. A mathematical theory for the calculation of the stability of open cast mines. In: *Proceedings of the Symposium on Planning of Open Pit Mines*, P. W. J. van Rensberg, ed., Johannesburg, South Africa, CRC/Balkema, Leiden, Netherlands, pp. 87–102.

Jibson, R. W. 1993. *Predicting Earthquake-Induced Landslide Displacements Using Newark's Sliding Block Analysis.* Transportation Research Record 1411, Transportation Research Board, Washington, DC, pp. 9–17.

Jibson, R. W. 2007. Regression models for estimating coseismic landslide displacement. *Eng. Geol.*, 91(2–4):209–18.

Jibson, R. W. 2011. Methods for assessing the stability of slopes during earthquakes – A retrospective. *Eng. Geol.*, 122(1–2):43–50.

Jibson, R. W. 2013. Mass-movement causes: Earthquakes. In: *Treatise on Geomorphology*, R. Marston and M. Stoffel, eds., Elsevier Inc., San Diego, 223–9.

Jibson, R. W. and Harp, E. L. 1995. *Inventory of landslides triggered by the 1994 Northridge, California Earthquake.* Department of the Interior, U.S.G.S., Open–File 95-213, 17pp.

Jibson, R. W., Harp, E. L. and Michael, J. A. 1998. *A method for producing digital probabilistic seismic landslide hazard map: An example from the Los Angeles, California Area.* Department of the Interior, U.S.G.S., Open–File Report 98-113, 17pp.

Jibson, R. W., Harp, E. L., Schulz, W. and Keefer, D. K. 2004. Landslides triggered by the 2002 Denali fault, Alaska, earthquake and the inferred nature of the strong shaking. *Earthquake Spectra*, 20(3):669–91.

Jibson, R. W., Rathje, E. M., Jibson, M. W. and Lee, Y. W. 2013. SLAMMER: Seismic LAndslide Movement Modeled using Earthquake Records. *U.S. Geological Survey Techniques and Methods 12-B1*, 1.1 edition, United States Department of the Interior, USGS.

John, K. W. 1970. Engineering analysis of three-dimensional stability problems utilizing the reference hemisphere. *Proceedings of the 2nd Congress – International Society of Rock Mechanics*, September 21–26, International Society for Rock Mechanics, Belgrade, Yugoslavia, Vol. 2, pp. 314–21.

Kalenchuk, K. S. 2010. *Multi-dimensional analysis of large, complex slope instability.* PhD thesis, Queen's University, Canada.

Kane, W. F. and Beck, T. J. 1996. Rapid slope monitoring. *Civil Eng. ASCE*, 66(6):56–8.

Kazerani, R. and Zhao J. 2010. Micromechanical parameters in bonded particle method for modelling of brittle material failure. *Int. J. Numer. Anal. Methods Geomech.*, 34:1877–95.

Keefer, D. L. 1984. Landslides caused by earthquakes. *Geol. Soc. Am. Bull.*, 95(4):406–21.

Keefer, D. L. 1992. The susceptibility of rock slopes to earthquake-induced failure. In: *Proceedings of the 35th Annual Meeting of the Association of Engineering Geologists*, M. L. Stout, ed., Ass. of Engineering Geologists, Long Beach, CA, pp. 529–38.

Kemeny, J. 2003. The time-dependent reduction of sliding cohesion due to rock bridges along discontinuities: A fracture mechanics approach. *Rock Mech. Rock Eng.*, 36(1):27–38.

Kennedy, B. A. and Neimeyer, K. E. 1970. Slope monitoring systems used in the prediction of a major slope failure at the Chuquicamata Mine, Chile. In: *Proceedings of the Symposium on Planning Open Pit Mines*, P. W. J. van Rensberg, ed., Johannesburg, South Africa, CRC/Balkema, Leiden, Netherlands, pp. 215–25.

Kiersch, G. A. 1963. Vajont reservoir disaster. *Civil Eng.*, 34(3):32–9.

Kikuchi, K., Kuroda, H. and Mito, Y. 1987. Stochastic estimation and modeling of rock joint distribution based on statistical sampling. *6th International Congress on Rock Mechanics*, Montreal, CRC/Balkema, Leiden, Netherlands, pp. 425–8.

King, R. A. 1977. *A review of soil corrosiveness with particular reference to reinforced earth.* Supplementary Report No. 316, Transport and Road Research Laboratory, Crowthorne, UK.

Kobayashi, Y. Harp, E. L. and Hagawa, T. 1990. Simulation of rock falls triggered by earthquakes. *Rock Mech. Rock Eng.*, 23(1):1–20.

Konya, C. J. and Walter, E. J. 1991. *Rock Blasting and Overbreak Control.* U.S. Department of Transportation, Federal Highway Administration, Contract No.: FHWA-HI-92-001.

Kreyszig, E. 1976. *Advanced Engineering Mathematics.* John Wiley & Sons, New York, 898pp.

Kulatilake, P. H. S. 1988. State of the art in stochastic joint geometry modeling. In: *Proceedings of the 29th U.S. Symposium on Rock Mechanics*, P. A. Cundall, J. Sterling and A. M. Starfield, eds., CRC/Balkema, Leiden, Netherlands, pp. 215–29.

Kulatilake, P. H. S. and Wu, T. H. 1984. Estimation of the mean length of discontinuities. *Rock Mech. Rock Eng.*, 17(4):215–32.

Ladegaard-Pedersen, A. and Dally, J. W. 1975. *A review of factors affecting damage in blasting.* Report to the National Science Foundation. Mechanical Engineering Department, University of Maryland, 170pp.

Lambe, W. T. and Whitman, R. V. 1969. *Soil Mechanics.* John Wiley & Sons, New York.

Langefors, U. and Kihlstrom, B. 1973. *The Modern Technique of Rock Blasting*, 2nd edition. John Wiley & Sons, New York, 405pp.

Lau, J. S. O. 1983. The determination of true orientations of fractures in rock cores. *Can. Geotech. J.*, 20:221–7.

Lee, J. and Green, R. A. 2008. Predictive relations for significant durations in stable continental regions. *14th World Conference on Earthquake Engineering*, October, Beijing, China.

Lee, S.-G. 2012. Lessons learned from extreme rainfall-induced landslides in South Korea. In: *Extreme Rainfall Induced Landslides – An International Perspective*, W. A. Lacerda, E. M. Palmeira, A. L. C. Netto and M. Ehrlich, eds., Brazil: Oficina de Textos, Brazil, pp. 141–59.

Ley, G. M. M. 1972. *The properties of hydrothermally altered granite and their application to slope stability in open cast mining.* MSc thesis, University of London, Imperial College, UK.

Leyshon, P. R. and Lisle, R. J. 1996. *Stereographic Projection Techniques in Structural Geology.* Butterworth and Heinemann, Oxford, UK, 104pp.

Lin, J-S. and Whitman, R. V. 1986. Earthquake induced displacement of sliding blocks. *J. Geotech. Eng. Div. ASCE*, 112(1):44–59.

Ling, H. I. and Cheng, A. H-D. 1997. Rock sliding induced by seismic forces. *Int. J. Rock Mech. Min. Sci.*, 34(6):1021–9.

Lisjak, A. and Grasselli, G. 2011. Rock slope stability under dynamic loading using a combined finite-discrete element approach. *Proceedings of the Pan-Am CGS Geotechnical Conference*, Toronto.

Littlejohn, G. S. and Bruce, D. A. 1977. *Rock Anchors – State of the Art.* Foundation Publications Ltd., Brentwod, Essex, UK.

Liu, K.-S. and Tsai, Y.-B. 2005. Attenuation relationships of peak ground acceleration and velocity for crustal earthquakes in Taiwan. *Bull. Seismol. Soc. Am.*, 95(3):1045–58.

Londe, P. 1965. Une method d'analyse a trois dimensions de la stabilite d'une rive rocheuse. *Annales des Ponts et Chaussees, Paris*, 1(1):37–60.

Londe, P., Vigier, G. and Vormeringer, R. 1969. Stability of rock slopes – Graphical methods. *J. Soil Mech. Found. Eng. Div. ASCE*, 96(SM4):1411–34.

Londe, P., Vigier, G. and Vormeringer, R. 1970. Stability of rock slopes – A three-dimensional study. *J. Soil Mech. Found. Eng. Div. ASCE*, 95(SM1):235–62.

Lorig, L. and Varona, P. 2001. Practical slope-stability analysis using finite-difference codes. In: *Slope Stability in Surface Mining*, W. A. Hustrulid, M. J. McCarter and D. J. A. Van Zyl, eds., Society of Mining Engineers of AIME, Englewood, CO, pp. 115–24.

Lorig, L. and Varona, P. 2004. Numerical Analysis. In *Rock Slope Engineering*, 4th edition, Chapter 10, D. C. Wyllie and C. W. Mah, eds., Taylor & Francis, London, UK, 431pp.

Lorig, L., Stacey, P. and Read, J. 2009. Slope design methods. In: *Guidelines for Open Pit Slope Design*, Chapter 10, J. Read and P. F. Stacey, eds., CSIRO Publishing, Brisbane, Australia, pp. 237–64.

Louis, C. 1969. *A study of ground water flow in jointed rock and its influence on the stability of rock masses*. Doctoral thesis, University of Karlsruhe (in German). English translation: Imperial College Rock Mechanics Research Report No. 10, 50pp.

Lowry, B., Gomez, F., Zhou, W., Mooney, M. A., Held, B. and Grasmick, J. 2013. High resolution displacement monitoring of a slow velocity slide using ground based radar interferometry. *Eng. Geol.*, 166:160–9.

Lutton, R. J. and Banks, D. C. 1970. Study of clay shales along the Panama Canal, Report 1, East Culebra and West Culebra Cuts and the model slope. The U.S. Army Corps of Engineers, Vicksburg, Mississippi. Report S-70-0, 285 pages.

Mahtab, M. A. and Yegulalp, T. M. 1982. A rejection criterion for definition of clusters in orientation data. *Proceedings of the 22nd Symposium on Rock Mechanics*, Berkeley, CA, Society of Mining Engineers of AIME, Englewood, CO, pp. 116–24.

Maini, Y. N. 1971. *In situ parameters in jointed rock – Their measurement and interpretation*, PhD thesis, University of London, UK.

Mamaghani, I. H. P., Yoshida, H. and Obata, Y. 1999. Reinforced expanded polystyrene stryrofoam covering rock sheds under impact of falling rock. *Proceedings of the Joint Japan-Swiss Scientific Seminar on Impact Loading by Rock Falls and Design of Protection Measures*, Japan Society of Civil Engineers, Kanazawa, Japan, pp. 79–89.

Marinos, P. and Hoek, E. 2000. GSI – A geologically friendly tool for rock mass strength estimation. *Proceedings of the GeoEng2000 Conference*, Melbourne, Australia.

Marinos. P. and Hoek, E. 2001. Estimating the geotechnical properties of heterogeneous rock masses such as flysch. Accepted for publication in *Bull. Int. Assoc. Eng. Geol.*, 60:85–91.

Markland, J. T. 1972. *A useful technique for estimating the stability of rock slopes when the rigid wedge sliding type of failure is expected*. Imperial College Rock Mechanics Research Report No. 19, 10pp.

Marsal, R. J. 1967. Large scale testing of rockfill materials. *J. Soil Mech. Found. Div. ASCE*, 93(SM2):27–44.

Marsal, R. J. 1973. Mechanical properties of rock fill. In: *Embankment Dam Engineering, Casagrande Volume*, John Wiley & Sons, New York, 109–200.

Martin, D. C. 1993. *Time-dependent deformation of rock slopes*, PhD thesis, University of London, UK.

Mas Ivars, D., Pierce, M. E., Darcel, C., Reyes-Montes, J., Potyondy, D. O., Young, P. and Cundall, P. A. 2011. The synthetic rock mass approach for jointed rock mass modelling. *Int. J. Rock Mech. Min. Sci.*, 48:219–44. doi: 10.1016/j.ijrmms.2010.11.014.

Massey, C. I., McSaveney, M. J., Taig, T. et al. 2014. Determining rock fall risk in Christchurch using rockfalls triggered by the 2010–2011 Canterbury earthquake sequence. *Earthquake Spectra*, 30(1):155–81.

Mayer, J. M. 2015. *Applications of uncertainty theory to rock mechanics and geotechnical mine design*, MSc thesis, Simon Fraser University, Canada, 237pp.

Mayer, J. M., Stead, D., de Bruyn, I. and Nowak, M. 2014. A sequential Gaussian simulation approach to modelling rock mass heterogeneity. In: *Proceedings of the ARMA Geomechanics Symposium*, Minneapolis, MN.

McCauley, M. L., Works, B. W. and Naramore, S. A. 1985. *Rockfall mitigation*. Report FHWA/CA/TL-85/12. FHWA, U.S. Department of Transportation, Washington, DC.

McDougall, S., Boultbee, N., Hungr, O., Stead, D. and Schwab, J. W. 2006. The Zymoetz River landslide, British Columbia, Canada: Description and dynamic analysis of a rock slide–debris flow. *Landslides*, 3(3):195–204.

McDougall, S. and Hungr, O. 2004. A model for the analysis of rapid landslide runout motion across three-dimensional terrain. *Can. Geotech. J.*, 41:1084–97.

McDougall, S., McKinnon, M. and Hungr, O. 2012. Developments in landslide runout prediction. In *Landslides: Types, Mechanisms and Modelling*, Chapter 16, J. Clague and D. Stead, eds., Cambridge University Press, UK, pp. 187–95.

McGuffey, V., Athanasiou-Grivas, D., Iori, J. and Kyfor, Z. 1980. *Probabilistic Embankment Design – A Case Study*. Transportation Research Board, Washington, DC.

McKinnon, M. 2010. *Landslide runout: Statistical analysis of physical characteristics and model parameters*, MSc thesis, University of British Columbia, Canada.

McKinstry, R., Floyd, J. and Bartley, D. 2002. Electronic detonator performance evaluation at Barrick Goldstrike Mines Inc. *J. Explosives Eng.*, International Society of Explosives Engineers, Cincinnati, OH, May/June, 12–21.

McMahon, B. K. 1982. *Probabilistic Design in Geotechnical Engineering*. Australian Mineral Foundation, AMF Course 187/82, Sydney.

McNeel, R. and Associates. 2016. *Rhino 5 for Windows*. Robert McNeel & Associates, Seattle.

Mearz, N. H., Franklin, J. A. and Bennett, C. P. 1990. Joint roughness measurements using shadow profilometry. *Int. J. Rock Mech. Min. Sci. Geomech. Abstr.*, 27(5):329–43.

Merrien-Soukatchoff, V., Korini, T. and Thoraval, A. 2011. Use of an integrated discrete fracture network code for stochastic stability analyses of fractured rock masses. *Rock Mech. Rock Eng.*, 45(2):159–81.

Meyerhof, G. G. 1984. Safety factors and limit states analysis in geotechnical engineering. *Can. Geotech. J.*, 21:1–7.

Middlebrook, T. A. 1942. Fort Peck slide. *Proceedings of the ASCE*, Vol. 107, Paper 2144, pp. 723–64.

Ministry of Construction, Japan. 1983. *Reference Manual on Erosion Control Works (in Japanese)*, Erosion Control Department, Tokyo, Japan, 386pp.

Mitchell, J. K. 1976. *Fundamentals of Soil Behavior*. John Wiley & Sons, New York. 422pp.

Mohr, O. 1900. Welche Umstände bedingen die Elastizitätsgrenze und den Bruch eines Materials. *Z. Ver. dt. Ing.*, 44:1524–30; 1572–7.

Moore, D. P. and Imrie, A. S. 1982. Rock slope stabilization at Revelstoke damsite. *Transactions of the 14th International Congress on Large Dams*, ICOLD, Paris, Vol. 2, pp. 365–85.

Moore, D. P., Imrie, A. S. and Enegren, E. G. 1997. Evaluation and management of Revelstoke reservoir slopes. *International Commission on Large Dams, Proceedings of the 19th Congress*, Florence, Italy, Q74, R1, pp. 1–23.

Moore, H. 1986. Construction of a shot-in-place rock buttress for landslide stabilization. *Proceedings of the 37th Highway Geology Symposium*, Helena, MT, 21pp.

Morgan, D. R., Heere, R., McAskill, N. and Chan, C. 1999. Comparative evaluation of system ductility of mesh and fibre reinforced shotcretes. *Proceedings of the Conference on Shotcrete for Underground Support VIII*, Engineering Foundation (New York) Campos do Jordao, Brazil, 23pp.

Morgan, D. R., McAskill, N., Richardson, B. W. and Zellers, R. C. 1989. A comparative study of plain, polypropylene fiber, steel fiber, and wire mesh reinforced shotcretes. *Transportation Research Board, Annual Meeting*, Washington, DC, 32pp (plus appendices).

Morgenstern, N. R. 1971. The influence of ground water on stability. *Proceedings of the 1st Symposium on Stability in Open Pit Mining*, Vancouver, Canada, Society of Mining Engineers of AIME, Englewood, CO, pp. 65–82.

Morgenstern, N. R. 1992. The role of analysis in the evaluation of slope stability. In: *Proceedings of the 6th International Symposium on Landslides*, D. H. Bell, ed., Christchurch, NZ, A. A. Balkema, Rotterdam, Vol. 3, pp. 1615–29.

Morgenstern, N. R. and Price, V. E. 1965. The analysis of the stability of general slide surfaces. *Geotechnique*, 15:79–93.

Morris, A. J. and Wood, D. F. 1999. Rock slope engineering and management process on the Canadian Pacific Railway. *50th Highway Geology Symposium*, Roanoke, Virginia.

Morriss, P. 1984. Notes on the probabilistic design of rock slopes. *Australian Mineral Foundation*, Notes for course on rock slope engineering, Adelaide, April.

Mufundirwa, A., Fujii, Y. and Kodama, J. 2010. A new practical method for prediction of geomechanical failure-time, *Int. J. Rock Mech. Mining Sci.*, 47(7):1079–1090.

Muller, L. 1968. New considerations of the Vajont slide. *Felsmechanik und engenieurgeologie*, 6(1): 1–91.

Munjiza, A., Owen, D. R. J. and Bicanic, N. 1995. A combined finite-discrete element method in transient dynamics of fracturing solids. *Eng. Comput.*, 12:145–74.

Nahon, D. B. 1991. *Introduction to the Petrology of Soils and Chemical Weathering*. John Wiley & Sons, New York.

Narendranathan, S., Thomas, R. D. H. and Neilsen, J. M. 2013. The effect of slope curvature in rock mass shear strength derivations for stability modelling of foliated rock masses. In: *Proceedings of the International Symposium on Slope Stability in Open Pit Mining and Civil Engineering*, P. Dight, ed., Australian Centre for Geomechanics, Brisbane, pp. 719–32.

Newmark, N. M. 1965. Effects of earthquakes on dams and embankments. *Geotechnique*, 15(2):139–60.

Nonveiller, E. 1965. The stability analysis of slopes with a slide surface of general shape. *Proceedings of the 6th International Conference Soil Mechanics and Foundation Engineering*, Montreal, Vol. 2, p. 522.

Norrish, N. I. and Lowell, S. M. 1988. Aesthetic and safety issues for highway rock slope design. *Proceedings of the 39th Annual Highway Geology Symposium*, Park City, Utah.

Office of Surface Mining (OSM). 2001. Use of explosives: Preblasting surveys. U.S. Department of the Interior, Surface Mining Law Regulations, Subchapter K, 30 CRF, Section 816.62, Washington, DC.

Oregon Department of Transportation. 2001. *Rock fall catchment area design guide*. ODOT Research Group Report SPR-3(032), Salem, OR, 77pp. with appendices.

Oriard, L. L. 1971. Blasting effects and their control in open pit mining. *Proceedings of the 2nd International Conference on Stability in Open Pit Mining*, Vancouver, Society of Mining Engineers of AIME, Englewood, CO, pp. 197–222.

Oriard, L. L. 2002. *Explosives Engineering, Construction Vibrations and Geotechnology*. International. Society of Explosives Engineers, Cleveland, OH, 680pp.

Oriard, L. L. and Coulson, J. H. 1980. Blast vibration criteria for mass concrete. In: *Minimizing Detrimental Construction Vibrations*. ASCE, New York, NY, Preprint 80-175, pp. 103–23.

Pahl, P. J. 1981. Estimating the mean length of discontinuity traces. *Int. J. Rock Mech. Min. Sci. Geomech. Abstr.*, 18:221–8.

Palisades Corp. 2012. *@Risk 5.0 – Risk Analysis Using Monte Carlo Simulation*. Palisades Corp., Ithaca, NY.

PanTechnica Corp. 2002. *KbSlope Slope Stability Program for KeyBlock Analysis*. PanTechnica Corporation, Caska, MN, www.pantechnica.com.

Patton, F. D. 1966. Multiple modes of shear failure in rock. *Proceedings of the 1st International Congress on Rock Mechanics*, Lisbon, Vol. 1, pp. 509–13.

Patton, F. D. and Deere, D. U. 1971. Geologic factors controlling slope stability in open pit mines. *Proceedings of the 1st Symposium on Stability in Open Pit Mining*, Vancouver, Canada, Society of Mining Engineers of AIME, Englewood, CO, pp. 23–48.

Paulding, B. W. Jr. 1970. Coefficient of friction of natural rock surfaces. *J. Soil Mech. Found. Div. ASCE*, 96(SM2):385–94.

Peck, R. B. 1967. Stability of natural slopes. *Proc. ASCE*, 93(SM 4):403–17.

Peckover, F. L. 1975. *Treatment of rock falls on railway lines*. American Railway Engineering Association, Bulletin 653, Washington, DC, pp. 471–503.

Peel, M. C., Finlayson, B. L. and McMahon, T. A. 2007. Up-dated world map of the Koppen-Geiger climate classification. *Hydrol. Earth System Sci.*, 11:1633–44.

PEER, P. E. E. R. C. 2015. Gorkha (Nepal) Earthquake: PEER strong motion records. In: *PEER Strong Motion Database Records*. University of California, Berkeley, CA.

Pentz, D. L. 1981. Slope stability analysis techniques incorporating uncertainty in the critical parameters. *3rd International Conference on Stability in Open Pit Mining*, Vancouver, Canada.
Society of Mining Engineers of AIME, Englewood, CO.

Persson, P. A. 1975. Bench drilling – An important first step in the rock fragmentation process. *Atlas
Copco Bench Drilling Symposium*, Stockholm.

Persson, P. A., Holmburg, R. and Lee, J. 1993. *Rock Blasting and Explosive Engineering*. CRC Press,
Boca Raton, FL.

Peterson, J. E., Sullivan, J. T. and Tater, G. A. 1982. The use of computer enhanced satellite imag-
ery for geologic reconnaissance of dam sites. ICOLD, *14th Congress on Large Dams*, Rio de
Janeiro, Q53, R26, Vol. II, pp. 449–71.

Pfeiffer, T. J. and Bowen, T. D. 1989. Computer simulation of rock falls. *Bull. Assoc. Eng. Geol.*,
XXVI(1):135–46.

Pfeiffer, T. J., Higgins, J. D. and Turner, A. K. 1990. Computer aided rock fall hazard analysis.
*Proceedings of the 6th International Congress International Association of Engineering
Geology*, Amsterdam. CRC/Balkema, Leiden, Netherlands, pp. 93–103.

Phillips, F. C. 1971. *The Use of Stereographic Projections in Structural Geology*. Edward Arnold,
London, UK, 90pp.

Pierce, M., Brandshaug, T. and Ward, M. 2001. Slope stability assessment at the Main Cresson Mine.
In: *Slope Stability in Surface Mining*, W. A. Hustralid, M. J. McCarter and D. J. A. Van Zyl,
eds., Society of Mining Engineers of AIME, Englewood, CO, pp. 239–50.

Pierson, L., Davis, S. A. and Van Vickle, R. 1990. *The rock fall hazard rating system, implementation
manual*. Technical Report #FHWA-OR-EG-90-01, Washington, DC.

Poisel, R. and Preh, A. 2008. Modifications of PFC3D for rock mass fall modeling, continuum and
distinct element numerical modeling. In: *Geo-Engineering – 2008*, R. Hart, C. Detournay and
P. Cundall, eds., CRC Press, Boca Raton, FL, Paper: 01-04, 10pp, ISBN 978-0-9767577-1-9.

Post Tensioning Institute (PTI). 2006. *Post-Tensioning Manual*, 6th edition. PTI, Phoenix, AZ, 70pp.

Priest, S. D. and Hudson, J. A. 1976. Discontinuity spacings in rock. *Int. J. Rock Mech. Min. Sci.
Geomech. Abstr.*, 13:135–48.

Priest, S. D. and Hudson, J. A. 1981. Estimation of discontinuity spacing and trace length using scan-
line surveys. *Int. J. Rock Mech. Min. Sci. Geomech. Abstr.*, 18:183–97.

Pritchard, M. A. and Savigny, K. W. 1990. Numerical modelling of toppling. *Can. Geotech. J.*, 27:823–34.

Pritchard, M. A. and Savigny, K. W. 1991. The Heather Hill landslide: An example of a large scale
toppling failure in a natural slope. *Can. Geotech. J.*, 28:410–22.

Protec Engineering. 2002. Rock fall caused by earthquake and construction of GeoRock Wall in
Niijima Island, Japan, http://www.proteng.co.jp.

Pyke, R. 1999. Selection of seismic coefficients for use in pseudo-static slope stability analysis, http://
www.tagasoft.com/opinion/article2.hmtl, 3pp.

Read, R. S., Langenberg, W., Cruden, D. M., Field, M., Stewart, R., Bland, H. 2005. Frank slide a
century later: The Turtle Mountain monitoring project. In: *Landslide Risk Management*, O.
Hungr, R. Fell, R. Couture and E. Eberhardt, eds., Vancouver, Canada, CRC/Balkema, Leiden,
Netherlands, pp. 702–12.

Rengers, N. 1971. *Roughness and friction properties of separation planes in rock. Thesis*, Tech.
Hochschule Fredericiana, Karlsruhe, Inst Bodenmech. Felsmech. Veröff, Vol. 47, 129pp.

Rickemann, D. 2005. Runout prediction methods. In: *Debris-Flow Hazards and Related Phenomena*,
Chapter 13, M. Jakob and O. Hungr, eds., Praxis Springer, Berlin, pp. 305–24.

Ritchie, A. M. 1963. *Evaluation of Rock Fall and Its Control*. Highway Research Record 17, Highway
Research Board, NRC, Washington, DC, pp. 13–28.

Roberds, W. J. 1984. Risk-based decision making in geotechnical engineering: Overview of case studies. *Engineering Foundation Conference on Risk-Based Decision Making in Water Resources*,
Santa Barbara, CA.

Roberds, W. J. 1986. Applications of decision theory to hazardous waste disposal. *ASCE Specialty
Conference GEOTECH IV*, Boston, MA.

Roberds, W. J. 1990. Methods of developing defensible subjective probability assessments.
Transportation Research Board, Annual Meeting, Washington, DC.

Roberds, W. J. 1991. Methodology for optimizing rock slope preventative maintenance programs. *ASCE, Geotechnical Engineering Congress*, Boulder, CO, Geotechnical Special Publication 27, pp. 634–45.

Roberds, W. J., Ho, K. K. S. and Leroi, E. 2002. *Quantitative Risk Assessment of Landslides*. Transportation Research Record 1786, Paper No. 02-3900, 69-75, Transportation Research Board, Washington, DC.

Roberts, D. and Hoek, E. 1971. A study of the stability of a disused limestone quarry face in the Mendip Hills, England. *Proceedings of the 1st International Conference on Stability in Open Pit Mining*, Vancouver, Canada, Society of Mining Engineers of AIME, Englewood, CO, pp. 239–56.

Rockfield Software. 2016. *ELFEN Software*. Rockfield Software Ltd. Technium, Kings Road, Prince of Wales Dock, Swansea, SA1 8PH, UK.

Rocscience. *Dips 7.0 – Stereographic Analysis of Structural Geology*. Rocscience Inc., Toronto, Canada, www.rocscience.com.

Rocscience. *RocFall 5.0 – Analysis of Two-Dimensional Rock Fall Behaviour*. Rocscience Inc., Toronto, Canada, www.rocscience.com.

Rocscience. *RocTopple 1.0 – Analysis of Toppling Slope Stability*. Rocscience Inc., Toronto, Canada, www.rocscience.com.

Rocscience. *RocPlane 3.0 – Two-Dimensional Analysis of Planar Slope Stability*. Rocscience Inc., Toronto, Canada, www.rocscience.com.

Rocscience. *Swedge 7.0 – Probabilistic Analysis of the Geometry and Stability of Surface Wedges*. Rocscience Inc., Toronto, Canada, www.rocscience.com.

Rocscience Inc. *RS2 (Phase2 Ver. 9.0)*, Rocscience Inc., Toronto, Ontario, Canada.

Rocscience Inc. *Slide 7.0 – Two-Dimensional Slope Stability Analysis for Rock and Soil Slopes*. Rocscience Inc., Toronto, Ontario, www.rocscience.com.

Rocscience Inc. *RocData 5.0 – Software for Calculating Hoek-Brown Rock Mass Strength*. Rocscience Inc., Toronto, Ontario, www.rocscience.com.

Rocscience Inc. *RS2 (PHASE)2 9.0 – Two-Dimensional Finite Element Analysis for Slopes and Tunnels*. Rocscience Inc., Toronto, Ontario, www.rocscience.com.

Rodriguez, A. R., Castillo, H. D. and Sowers, G. 1988. *Soil Mechanics in Highway Engineering*. Trans Tech Publications, London, 843pp.

Rohrbaugh, J. 1979. Improving the quality of group judgment: Social judgment analysis and the Delphi technique. *Organ. Behav. Hum. Perform.*, 24:73–92.

Ross-Brown, D. R. 1973. *Slope design in open cast mines*, PhD thesis, University of London, UK, 250pp.

Sagaseta, C., Sánchez, J. M. and Cañizal, J. 2001. A general solution for the required anchor force in rock slopes with toppling failure. *Int. J. Rock Mech. Min. Sci.*, 38:421–35.

Sainsbury, B., Pierce, M. E. and Mas Ivars, D. 2008. Analysis of caving behaviour using a synthetic rock mass-ubiquitous joint rock mass modelling technique. *Proceedings of the 1st Southern Hemisphere International Rock Mechanics Symposium*, Perth, Australia, Vol. 1, pp. 243–53.

Sainsbury, D. P., Sainsbury, B. and Sweeney, E. 2016. Three-dimensional analysis of complex anisotropic slope instability at MMG's Century Mine. In: *Proceedings of Slope Stability 2013, Mining Technology, Section A*, P. M. Dight, ed., Australian Centre for Geomechanics, Perth, pp. 683–96.

Salmon, G. M. and Hartford, N. D. 1995. Risk analysis for dam safety. *Int. Water Power Dam Constr.*, 21:38–9.

Sarma, S. K. 1979. Stability analysis of embankments and slopes. *J. Geotech. Eng. Div. ASCE*, 105(GT12):1511–24.

Savely, J. P. 1987. Probabilistic analysis of intensely fractured rock masses. *6th International Congress on Rock Mechanics*, Montreal, 509–14.

Scheidegger, A. E. 1960. *The Physics of Flow through Porous Media*. Macmillan, New York.

Scholtes, L. and Donze, F. V. 2012. Modelling progressive failure in fractured rock masses using a 3D discrete element method. *Int. J. Rock Mech. Min. Sci.* 52:18–30.

Schuster, R. L. 1992. Keynote paper: Recent advances in slope stabilization. *Session G3, Proceedings of the 6th International Symposium on Landslides*, Auckland, New Zealand, CRC/Balkema, Leiden, Netherlands.

Seed, H. B. 1979. Considerations in the earthquake-resistant design of earth and rockfill dams. *Geotechnique*, 29(3):215–63.

Sepulveda, S. A., Murphy, W., Jibson, R. W. and Petley, D. N. 2005. Seismically induced rock slope failures resulting from topographic amplification of strong ground motions: The case of Pacoima Canyon, California. *Eng. Geol.*, 80(3–4):336–48.

Sharma, J. S., Chu, J. and Zhao, J. 1999. Geological and geotechnical features of Singapore: An overview. *Tunnel. Underground Space Technol.*, 14(4):419–31.

Sharma, S. 1991. *XSTABL, an Integrated Slope Stability Analysis Method for Personal Computers, Version 4.00*. Interactive Software Designs Inc., Moscow, Idaho, USA.

Sharp, J. C. 1970. *Fluid flow through fractured media*, PhD thesis, University of London, UK.

Siskind, D. D., Stachura, V. J. and Raddiffe, K. S. 1976. *Noise and vibrations in residential structures from quarry production blasting*. U.S. Bureau of Mines, Report of Investigations 8168.

Siskind, D. E., Staff, M. S., Kopp, J. W. and Dowding, C. H. 1980. *Structure response and damage produced by ground vibrations from surface blasting*. U.S. Bureau of Mines, Report of Investigations 8507.

Sitar, N., MacLaughlin, M. M. and Doolin, D. M. 2005. Influence of kinematics on landslide mobility and failure mode. *J. Geotech. Geoenviron. Eng.*, 131(6):716–78.

Sjöberg, J. 2000. Failure mechanism for high slopes in hard rock. In: *Slope Stability in Surface Mining*, W. A. Hustralid, M. J. McCarter and D. J. A. Van Zyl, eds., Society of Mining Engineers of AIME, Englewood, CO, pp. 71–80.

Sjöberg, J., Sharp, J. C. and Malorey, D. J. 2001. Slope stability at Aznalcóllar. In: *Slope Stability in Surface Mining*, W. A. Hustralid, M. J. McCarter and D. J. A. Van Zyl, eds., Society of Mining Engineers of AIME, Englewood, CO, pp. 183–202.

Sjöborg, J. 1999. *Analysis of large scale rock slopes*, Doctoral thesis 1999:01, Division of Rock Mechanics, Luleå University of Technology.

Skempton, A. W. 1948. The φ = 0 analysis for stability and its theoretical basis. *Proceedings of the 2nd International Conference on Soil Mechanics and Foundation Engineering*, Rotterdam, Vol. 1, 72pp.

Skempton, A. W. and Hutchinson, J. N. 1948. Stability of natural slopes and embankment foundations. State of the art report. *Proceedings of the 7th International Conference on Soil Mechanics*, Mexico, Vol. 1, pp. 291–340.

Smith, D. D. and Duffy, J. D. 1990. *Field Tests and Evaluation of Rock Fall Restraining Nets*. Division of Transportation Materials and Research, Engineering Geology Branch, California Department of Transportation, Sacramento, CA.

Smithyman, M. 2007. *Distinct-element modelling of time-dependent deformation in two large rockslides*, MSc thesis, University of British Columbia, Canada.

Snow, D. T. 1968. Rock fracture spacings, openings, and porosities. *J. Soil Mech. Found. Div.* American Society of Civil Engineers, 94(1):77–92.

Sonmez, H. and Ulusay, R. 1999. Modifications to the geological strength index (GSI) and their applicability to the stability of slopes. *Int. J. Rock Mech. Min. Sci.*, 36(6):743–60.

Soto, C. 1974. *A comparative study of slope modelling techniques for fractured ground*. MSc thesis. London University, Imperial College.

Spang, K. and Egger, P. 1990. Action of fully-grouted bolts in jointed rock and factors of influence. *Rock Mech. Rock Eng.*, 23(3):201–29.

Spang, R. M. 1987. Protection against rock fall – Stepchild in the design of rock slopes. *Proceedings of the International Conference on Rock Mechanics*, Montreal, Canada, ISRM, Lisbon, Portugal, pp. 551–7.

Spencer, E. 1967. A method of analysis of the stability of embankments assuming parallel inter-slice forces. *Geotechnique*, 17:11–26.

Spencer, E. 1969. Circular and logarithmic spiral slide surfaces. *J. Soil Mech. Found. Div. ASCE*, 95(SM1):227–34.

Srivastava, R. M. and Parker, H. M. 1989. Robust measures of spatial continuity. In: *Geostatistics*, M. Armstrong, ed., Kluwer Academic Publishers, Alphen aan den Rijn, Netherlands, Vol. 1, pp. 295–308.

Stacey, P. F. 1996. Second Workshop on Large Scale Slope Stability. September 13, Las Vegas (no proceedings).

Stagg, M. S., Siskind, D. E., Stevens, M. G. and Dowding, C. H. 1984. *Effects of repeated blasting on a wood frame house*. U.S. Bureau of Mines, Report of Investigations 8896.

Starfield, A. M. and Cundall, P. A. 1988. Towards a methodology for rock slope modelling. *Int. J. Rock Mech. Min. Sci. Geomech. Abstr.*, 25(3):99–106.

Staub, I., Fredriksson, A. and Outters, O. 2002. *Strategy for a Rock Mechanics Site Descriptive Model Development and Testing of the Theoretical Approach*. Swedish Nuclear Fuel and Waste Management Co, Stockholm, Sweden, 219pp.

Stauffer, M. R. 1966. An empirical-statistical study of three-dimensional fabric diagrams as used in structural analysis. *Can. J. Earth Sci.*, 3:473–98.

Stead, D. and Eberhardt, E. 1997. Developments in the analysis of footwall slopes in surface coal mining. *Eng. Geol.*, 46(1):41–61.

Stead, D. and Eberhardt, E. 2013. Understanding the mechanics of large landslides. Invited keynote and paper, Inter. Conf. on Vajont 1963–2013. – Thoughts and analyses after 50 years since the catastrophic landslide, Padua, Italy. *Ital. J. Eng. Geol. Environ. – Book Series*, (6):85–112. doi: 10.4408/IJEGE.2013-06.B-07.

Steffan, O. K. H., Terbrugge, P. J., Wesseloo, J. and Venter, J. 2015. A risk consequence approach to open pit slope design. *Proceedings of the International Symposium on Stability of Rock Slopes in Open Pit Mining and Civil Engineering*, South African Institute of Mining and Metallurgy, Cape Town, South Africa, pp. 81–96.

Strouth, A. and Eberhardt, E. 2009. Integrated back and forward analysis of rock slope stability and rockslide runout at Afternoon Creek, Washington. *Can. Geotech. J.*, 46(10):1116–32.

Sullivan, T. D. 1993. Understanding pit slope movements. In: *Proceedings of the Geotechnical Instrumentation and Monitoring in Open Pit and Underground Mining*, T. Szwedzicki, ed., A. A. Balkema, Rotterdam, pp. 435–45.

TagaSoft 2016. *Three Dimensional Slope Stability Analysis Software, TSlope*. Wellington, New Zealand.

Tang, C., Li, L., Xu, N. and Ma, K. 2015. Microseismic monitoring and numerical simulation on the stability of high-steep rock slopes in hydropower engineering. *J. Rock Mech. Geotech. Eng.*, 7:493–508.

Taylor, D. W. 1937. Stability of earth slopes. *J. Boston Soc. Civil Eng.*, 24:197.

Terzaghi, K. 1943. *Theoretical Soil Mechanics*. John Wiley & Sons, New York.

Terzaghi, K. 1962. Stability of steep slopes on hard unweathered rock. *Geotechnique*, 12(4):1–20.

Terzaghi, K. and Peck R. 1967. *Soil Mechanics in Engineering Practice*. John Wiley & Sons, New York.

Terzaghi, R. 1965. Sources of errors in joint surveys. *Geotechnique*, 15:287–304.

Theis, C. V. 1935. The relation of the lowering of the piezometric surface, and the rate and duration of discharge of a well using ground water storage. *Trans. Am. Geophys. Union*, 16:519–24.

Thomas, M. F. 1966. Some geomorphological implications of deep weathering patterns in cystalline rocks in Nigeria. *Trans. Inst. Br. Geographers*, 40:173–93.

Threadgold, L. and McNichol, D. P. 1984. *Design and Construction of Polymer Grid Boulder Barriers to Protect a Large Public Housing Site for Hong Kong Housing Authority. Polymer Grid Reinforcement*. Thomas Telford, London, UK.

Todd, D. K. 1959. *Ground Water Hydrology*. John Wiley & Sons, New York.

Transportation Research Board (TRB). 1996. *Landslides, investigation and mitigation*. National Research Council, Special Report 247, Washington, DC, 673pp.

Transportation Research Board (TRB). 1999. *Geotechnical related development and implementation of load and resistance factor design (LRFD) methods*. NCHRP Synthesis No. 276, Washington, DC, 69pp.

Transportation Research Board (TRB). 2002. *Evaluation of metal tensioned systems in geotechnical applications*. NCHRP Project No. 24-13, Washington, DC, 102 pages plus figures and appendices.

Transportation Research Board (TRB). 2008a. *Seismic analysis and design of retaining walls, buried structures, slopes, and embankments.* D. G. Anderson, G. R. Martin, I. P. Lam and J. N. J. Wang, authors, NCHRP Report 611, National Cooperative Highway Research Program, Washington, DC.

Transportation Research Board (TRB). 2008b. *Seismic analysis and design of retaining walls, buried structures, slopes, and embankments; recommended specs, commentaries, example problems.* NCHRP Report 12-70, National Cooperative Highway Research Program, Washington, DC.

Travasarou, T., Bray, J. D., Abrahamson, N. A. 2003. Empirical attenuation relationship for Arias intensity. *Earthquake Eng. Struct. Dyn.*, 32(7):1133–55.

Trollope, D. H. 1980. The Vajont slope failure. *Rock Mech.*, 13(2):71–88.

Tse, R. and Cruden, D. M. 1979. Estimating joint roughness coefficients. *Int. J. Rock Mech. Min. Sci. Geomech. Abstr.*, 16:303–7.

Tuckey, Z. 2012. *An integrated field mapping-numerical modelling approach to characterising discontinuity persistence and intact rock bridges in large open pit slopes.* MSc thesis, Simon Fraser University, Burnaby, Canada.

Tuckey, Z., Stead, D. and Eberhardt, E. 2013. An integrated approach for understanding uncertainty of discontinuity persistence and intact rock bridges in large open pit slopes. In: *Proceedings of the Slope Stability 2013*, P. M. Dight, ed., Australian Centre for Geomechanics, Brisbane, pp. 189–204.

Ucar, R. 1986. Determination of shear failure envelope in rock masses. *J. Geotech. Eng. Div. ASCE*, 112(3):303–15.

University of Utah. 1984. Flooding and Landslides in Utah – An Economic Impact Analysis. University of Utah Bureau of Economic and Business Research, Utah Department of Community and Economic Development, and Utah Office of Planning and Budget, Salt Lake City, 123pp.

Ureel, S., Momayez, M. and Oberling, Z. 2013. Rock core orientation for mapping discontinuities and slope stability analysis. *Int. J. Res. Eng. Technol.*, 2(7):1–7.

USEPA, 1993. *Technical Manual: Solid Waste Disposal Facility Criterea.* U. S. Environmental Protection Agency, EPA/530/R-93/017, Washington, DC.

Vamosi, S and Berube, M. 1987. Is Parliament Hill moving? Canadian Department Energy Mines and Resources, *Geos Mag.*, 4:18–22.

Van Velsor, J. E. and Walkinshaw, J. L. 1992. *Accelerated Movement of a Large Coastal Landslide Following the October 17, 1989 Loma Prieta Earthquake in California.* Transportation Research Record 1343, Transportation Research Board, Washington, DC, pp. 63–71.

Vaz, L. F. 1998. A new rock weathering classification system for tropical regions. *8th International Conference for Engineering Geology and the Environment*, Vancouver, Canada, pp. 299–306.

Vivas Becerra, J. 2014. *Groundwater characterization and modelling in natural and open pit rock slopes*, MSc thesis, Simon Fraser University, Burnaby, Canada.

Vivas Becerra, J., Hunt, C., Stead, D., Allen, D. and Elmo, D. 2015. Characterising groundwater in rock slopes using a combined remote sensing-numerical modelling approach. *Proceedings of the International Congress of Rock Mechanics*, ISRM, Montreal, Canada, Paper No. 670.

Vogel, T., Labiouse, V. and Masuya, H. 2009. Rockfall protection as an integral task. *Struct. Eng. Int.*, 19(3):301–12.

Vyazmensky, A., Stead D., Elmo D. and Moss A. 2010. Numerical analysis of block caving-induced instability in large open pit slopes: A finite element/discrete element approach. *Rock Mech. Rock Eng.*, 43(1):21–39.

Wallace, J. M. and Marcum, D. R. 2015. History and mechanisms of rock slope instability along Telegraph Hill, San Francisco, California. *American Rock Mechanics Association, 49th US Rock Mechanics/Geomechanics Symposium*, San Francisco.

Wang, C., Tannant, D. D. and Lilly, P. A. 2003. Numerical analysis of the stability of heavily jointed rock slopes using PFC2D. *Int. J. Rock Mech. Min. Sci.*, 40:415–24.

Watson, J. 2002. ISEE 2002. Electronic blast initiation – A practical users guide. *J. Explosives Eng.*, International Society of Explosives Engineers, Cincinnati, OH, May/June, 6–10.

Wei, Z. Q., Egger, P. and Descoeudres, F. 1995. Permeability predictions for jointed rock masses. *Int. J. Rock Mech. Min. Sci.*, 32(3):251–61.

Whitman, R. V. 1984. Evaluating calculated risk in geotechnical engineering. *J. Geotech. Eng. ASCE*, 110(2):145–88.

Whitman, R. V. and Bailey, W. A. 1967. The use of computers in slope stability analysis. *J. Soil Mech. Found. Eng. ASCE*, 93(SM4):475–98.

Whittall, J. 2015. *Runout exceedance prediction for open pit slope failures*, MSc thesis, University of British Columbia, Canada.

Whyte, R. J. 1973. *A study of progressive hanging wall caving at Chambishi copper mine in Zambia using the base friction model concept*, MSc thesis. London University, Imperial College.

Wiss, J. F. 1981. Construction vibrations: State-of-the-art. *J. Geotech. Eng. Div. ASCE*, 107(2): 167–81.

Wittke, W. W. 1965. Method to analyze the stability of rock slopes with and without additional loading (in German). *Felsmechanick und Ingenieurgeolgie*, 30(Supp. II): 52–79. English translation in Imperial College Rock Mechanics Research Report No. 6, July 1971.

Wolter, A., Gischig, V., Eberhardt, E., Stead, D. and Clague, J. J. 2015. Simulation of progressive rock slope failure due to seismically induced damage. *Proceedings of the Slope Stability 2015, SAIMM*, Cape Town, South Africa, pp. 529–41.

Wright, E. M. 1997. State of the art of rock cut design in eastern Kentucky. *Proceedings of the Highway Geology Symposium*, Knoxville, TN, pp. 167–73.

Wu, S-S. 1984. *Rock Fall Evaluation by Computer Simulation*. Transportation Research Record No. 1031, Transportation Research Board, Washington, DC.

Wu, T. H., Ali, E. M. and Pinnaduwa, H. S. W. 1981. *Stability of slopes in shale and colluvium*. Ohio Department of Transportation and Federal Highway Administration, Prom. No. EES 576. Research Report prepared by Ohio State University, Department of Civil Engineering.

www.terrainsar.com. 2002. Webpage describing features of Synthetic Aperture Radar.

Wyllie, D. C. 1980. Toppling rock slope failures, examples of analysis and stabilization. *Rock Mech.*, 13:89–98.

Wyllie, D. C. 1987. Rock slope inventory system. *Proceedings of the Federal Highway Administration Rock Fall Mitigation Seminar*, FHWA, Region 10, Portland, Oregon.

Wyllie, D. C. 1991. Rock slope stabilization and protection measures. *Association of Engineering Geologists, National Symposium on Highway and Railway Slope Stability*, Chicago, IL.

Wyllie, D. C. 1999. *Foundations on Rock*, 2nd edition. Taylor & Francis, London, UK, 401pp.

Wyllie, D. C. 2014a. Calibration of rock fall modeling parameters. *Int. J. Rock Mech. Min. Sci.*, 67:170–80.

Wyllie, D. C. 2014b. *Rock Fall Engineering*. Taylor & Francis, Boca Raton, FL, 213pp.

Wyllie, D. C., McCammon, N. R. and Brumund, W. F. 1979. *Use of Risk Analysis in Planning Slope Stabilization Programs on Transportation Routes*. Research Record 749, Transportation Research Board, Washington, DC.

Wyllie, D. C. and Munn, F. J. 1979. Use of movement monitoring to minimize production losses due to pit slope failure. *Proceedings of the 1st International Symposium on Stability in Coal Mining*, Vancouver, Canada, Miller Freeman Publications, pp. 75–94.

Wyllie, D. C. and Wood, D. F. 1981. *Preventative Rock Blasting Protects Track*. Railway Track and Structures, Simon-Boardman Publishers, New York, pp. 34–40.

Xanthakos, P. P. 1991. *Ground Anchorages and Anchored Structures*. John Wiley & Sons, New York, 686pp.

Xu, C. and Dowd, P. 2010. A new computer code for discrete fracture network modelling. *Comput. Geosci.*, 36(3):292–301.

Yoshida, H., Ushiro, T., Masuya, H. and Fujii, T. 1991. An evaluation of impulsive design load of rock sheds taking into account slope properties (in Japanese). *J. Struct. Eng.*, 37A(March):1603–16.

Youd, T. L. 1978. Major cause of earthquake damage is ground movement. *Civil Eng. ASCE*, 48(4):47–51.

Youngs, R. R., Chiou, S.-J., Silva, W. J., Humphrey, J. R. 1997. Strong ground motion attenuation relationships for subduction zone earthquakes. *Seismol. Res. Lett.*, 68(1):58–73.

Yu, X. and Vayassde, B. 1991. Joint profiles and their roughness parameters. Technical note, *Int. J. Rock Mech. Min. Sci. Geomech. Abstr.*, 16(4):333–6.

Zanbak, C. 1983. Design charts for rock slope susceptible to toppling. *J. Geotech. Eng. ASCE*, 109(8):1039–62.

Zavodni, Z. M. 2000. Time-dependent movements of open-pit slopes. In: *Proceedings of the Slope Stability in Surface Mining*, W. A. Hustrulid, M. K. McCarter and D. J. A. Van Zyl, eds., Society of Mining, Metallurgy and Exploration, Littleton, CO, pp. 81–7.

Zavodni, Z. M. and Broadbent, C. D. 1980. Slope failure kinematics. *Can. Inst. Min. Bull.*, 73:816.

Zhang, X., Powrie, W., Harness, R. and Wang, S. 1999. Estimation of permeability for the rock mass around the ship locks at the Three Gorges Project, China. *Int. J. Rock Mech. Min. Sci.*, 36:381–97.

Zienkiewicz, O. C., Humpheson, C. and Lewis, R. W. 1975. Associated and non-associated visco-plasticity and plasticity in soil mechanics. *Géotechnique*, 25(4):671–89.

Index